Reactions of Coordinated Ligands

Volume 2

Edited by

Paul S. Braterman

University of North Texas
Denton, Texas

PLENUM PRESS • NEW YORK AND LONDON

Library of Congress Cataloging in Publication Data

(Revised for vol. 2)

Reactions of coordinated ligands.

  Includes bibliographies and indexes.
  1. Coordination compounds. 2. Ligands. 3. Reactivity (Chemistry) I. Braterman,
Paul S.
QD474.R38 1986   541.2′242   85-24443
ISBN-13:978-1-4612-8066-8   e-ISBN-13:978-1-4613-0755-6
DOI: 10.1007/978-1-4613-0755-6

© 1989 Plenum Press, New York
Softcover reprint of the hardcover 1st edition 1989

A Division of Plenum Publishing Corporation
233 Spring Street, New York, N.Y. 10013

Reactions
of
COORDINATED
LIGANDS
Volume 2

PREFACE

This, the second and final volume of Reactions of Coordinated Ligands, describes the chemistry of ligands bound through non-carbon atoms, and of coordinated carbon dioxide. As before, emphasis is on the underlying mechanisms, which provide a unity of understanding for superficially disparate processes.

The wide range of topics covered illustrates well both the versatility and the usefulness of coordination chemistry in the controlled activation of ligands. Looking to the future, carbon dioxide is the feedstock of last resort. The homogeneous reduction of dinitrogen to ammonia now seems unlikely to replace the Haber process, but solution reactions also lead to more complex, varied, and valuable products. Nitrogen monoxide, a "non-innocent" ligand, impinges as pollutant and reagent. Its rich chemistry stems from its linked roles as three-electron donor, and as extremely powerful -acceptor. In the hydrolysis and condensation of complexed amides, esters etc., metals act both as templates and as tunable and polyfunctional Lewis acids. Here the control of hydrophobic and steric interactions begins to model the subtle mechanisms of biological specificity. Finally, phosphorus and sulfur are imporant both as ligand atoms in themselves, and as anchors for other functionalities.

I would like to thank all those who have been involved in the writing and production of this work, and also my colleagues old and new, at Glasgow and the University of North Texas, for their support.

Paul S. Braterman

CONTENTS

REACTIONS OF COORDINATED CARBON DIOXIDE

J.D. Miller

University of Aston, Birmingham

1.   INTRODUCTION

According to current estimates of world fuel reserves there is much more carbon available as carbonate rock than as oil and coal combined. In consequence it is easy to envisage a time when carbon dioxide, from the atmosphere or from such rocks, must replace oil as the major carbon-containing chemical feedstock. The recognition of this eventuality has provided a spur to the study of complexes containing carbon dioxide as a ligand, and into the reactivity of this coordinated ligand when coordinated. A second drive may be detected in the development of this research interest, namely, curiosity about naturally occurring processes.

Photosynthesis[1] is estimated to result in the 'fixation' of 2×10^{14} kg of carbon a year. This is equivalent to more than one per cent of known coal and oil deposits each year, and is achieved under ambient conditions of temperature and pressure according to the overall equation

$$nCO_2 + 2nH_2O \xrightarrow{xh\nu} (CH_2O)_n + nO_2 + nH_2O$$

The liberated oxygen comes from water, which fact is indicated in this equation by showing an extra water molecule on each side. For green plant processes x is approximately 8 and n may well be 6. Taking the absorption of light as the first step in the photosynthetic cycle, we may say that carbon dioxide is reduced at a later stage in the cycle; in a thermal rather than a photochemical process. Energy is first taken into the system and used to produce a good reducing agent, the reduced form of nicotinamide-adenine dinucleotide phosphate (NADPH). A reaction mechanism is then available for the reduction of carbon dioxide to a large organic molecule at an acceptable rate of reaction under ambient conditions. Therefore under laboratory conditions we might be able to parallel and perhaps ultimately improve on this process.

One other important biological process may properly be mentioned within these introductory remarks; the control of the solution/dissolution of carbon dioxide in aqueous media exercised by the enzyme carbonic anhydrase. This process has been intensively studied for almost two decades, and has yielded up much information; see for example References 2 and 3. The enzyme contains a metal ion, Zn^{2+}, at the reaction centre. The replacement of zinc by other metal ions of the same charge usually results in the complete deactivation of the enzyme, or in drastically reduced activity as with Co^{2+}. A recent study[4] of the CO_2-HCO_3^- system in the presence of Cu(II) Bovine Carbonic Anhydrase B using ^{13}C nmr suggests that while HCO_3^- is directly bonded to the metal with a Cu-C distance of 3.2 Å or less, the binding site for carbon dioxide is still within the reaction cavity but has a Cu-C separation of approximately 6Å. The carbon dioxide is therefore not directly attached to the metal, but rather reacts with a coordinated ligand, probably hydroxide. This same pattern of reaction is observed in the reaction of carbon dioxide with hydroxide-containing complexes in non-biological systems, see below.

Many inorganic chemists would guess that at least one transition-metal species will play a part in any process eventually devised for the utilization of carbon dioxide. Such guesses are often self-fulfilling since they direct the attention of the research worker. Certainly in the matter discussed below there is ample evidence of the concentration of attention in this area, with encouraging results. Even so, the original guess might not be correct. It is therefore desirable to draw two morals from our knowledge of carbonic anhydrase. Firstly, carbon dioxide can be converted to another chemical compound in a reaction involving a metal ion without its being directly bonded to that metal. Secondly, where a metal is involved it need not be a transition-metal. Many authors do not treat zinc as a transition-metal since it does not exhibit variable valency. Its role in carbonic anhydrase is the result of a delicate balance of its size, charge and the strength of its bonds with various donor atoms. Changes in the oxidation state of zinc during reaction are extremely unlikely. Thus a reasonable prediction might be that transition metals will be required for reactions involving the reduction of CO_2, but will not always be implicated in reactions which involve no change in the formal oxidation state of the carbon atom. No detailed coverage of biochemical processes is given in this chapter.

2. THE PROPERTIES OF CARBON DIOXIDE

Russian workers have made a large contribution to the current knowledge of complexes containing carbon dioxide and to the related ligand reactivity, not least through their reviews of the subject.[5-7] Within these reviews a large quantity of data is presented relating to CO_2. Only the more immediately relevant data are summarized here, the reader being referred to these reviews, especially References 6 and 7, and the references cited therein for more detail. The most recently published account of the coordination of carbon dioxide is to be found as a section of a wider review by Eisenberg and Hendriksen.[8]

Carbon dioxide is very stable thermodynamically, with

$$\Delta G_f^{\ominus}(298K) = -394.4 \text{ kJ mol}^{-1}$$

Therefore it is unlikely that any new carbon containing compound, even when stabilized by coordination, can be produced under readily accessible reaction conditions unless it also has a large and negative free energy of formation. Thus, the reverse of the water-gas shift reaction,

$$CO_2 + H_2 \longrightarrow H_2O + CO \qquad \Delta G^{\ominus}(298) = +28.5 \text{ kJ mol}^{-1}$$

although producing the very stable H_2O and CO molecules, is unfavorable at room temperature. The use of other standard tabulated data gives free energy changes of +33 and -0.9 kJ mol^{-1} respectively for the production of formic acid and methanol as pure liquids from carbon dioxide and hydrogen gases. In purely thermodynamic terms we can deduce that CO_2 is likely to be capable of conversion to other very stable molecules at or close to ambient conditions, especially if some extra stabilizing factor favors the product. This factor could well be the formation of additional bonds by attachment to a metal ion.

The critical temperature and pressure of carbon dioxide are $31.3°C$ and 75.2 atmospheres. Therefore reactions with the gas at positive pressures are possible, but very high pressures of CO_2 are not available. Carbon dioxide is readily soluble in a wide range of solvents;[9] most of the common organic solvents can give solutions of at least equal volumes of gas per unit of liquid, i.e. approaching 0.1 mol dm^{-3}. Finally, carbon dioxide is cheap and readily available as a solid. Therefore a wide range of experimental conditions are available for the study of its reactions. In 1982 Mason and Ibers reported[10] a new and interesting approach in which they used liquid carbon dioxide as a reagent. In some instances this phase might prove to be very useful, although it is unlikely to be of widespread applicability.

The key factors in a successful strategy for the conversion of carbon dioxide to other species would seem to be three. A suitable energy input will often be required. Usually this energy will need to be in the chemical form, as is widely illustrated in the following pages; but occasionally it may be introduced electrochemically or photochemically. A small number of references are quoted below to the use of these techniques, which seem to hold considerable promises for the future. Methods must be found to provide some extra product stabilization. Lastly and most demandingly, sufficient kinetic and mechanistic knowledge must be acquired to enable one to envisage the complicated reaction pathways which will inevitably be involved in the rearrangement of many atoms.

Carbon dioxide is a linear molecule, OCO, whose bonds are shorter and stronger than the isolated C=O of say acetone, due to the formation of a three-center, rather than two-center, system π-orbitals. The highest energy occupied molecular orbitals are those of the oxygen lone pairs; thus CO_2 is a very poor electron donor and Lewis base. The loss of an electron gives the ion CO_2^+, which retains the linearity of the parent molecule, at least in its ground state. The lowest unoccupied molecular orbitals of CO_2 are the anti-bonding delocalized π-orbitals. The ion CO_2^- is non-linear with the unpaired electron primarily located on C. The angle subtended at carbon is $127\pm8^\circ$ while the infrared spectrum of this radical anion has also been determined.[11-12] The relevance of these data to complexes containing the ligand CO_2 is discussed in the next section.

3.   COMPLEXES CONTAINING CO_2 AS A LIGAND

A range   of M-(CO_2) interactions may be possible, encompassing different bond strengths, donor atoms and even geometries. If there is an extremely weak interaction between a donor and an acceptor, then the partners will be very little changed by their interaction. In the case of carbon dioxide, one might reasonably expect such weak interactions to be demonstrable by infrared spectroscopy since they should result in slight changes in vibrational spectra. When CO_2 is adsorbed on Linde molecular sieve X in various forms, physical adsorption first occurs.[13-14] This moves the characteristic infrared band within the range 2375 to 2350 cm^{-1}; there is some variation as the cation is changed within groups IA and IIA. Weak bands which become stronger with increasing pressure are also observed at 1380 and 1270 cm^{-1}. These bands may be compared with the values for gaseous carbon dioxide of 2349 cm^{-1} (ν_3), 1388 cm^{-1} (ν_1) and 1285 cm^{-1} ($2\nu_2$). The fact that ν_1 has become infrared active is taken to mean that the CO_2 unit remains linear and that adsorption occurs near the O end of the molecule. At temperatures less than or equal to room temperature carbon dioxide also exhibits rapid physical adsorption on MgO.[15] Peaks at 2360 and 2335 cm^{-1} are observed immediately at 20°C. After several hours only the 2360 cm^{-1} peak can still be seen, anything at 2335 cm^{-1} being obscured.

Two slightly different modes of attachment are postulated to account for these two infrared peaks.

When chelating ephedrine ligands are used, the monomeric (but not the trimeric) copper(II) complexes reversibly take up CO_2 in benzene solution. The adduct exhibits medium-strong infrared bands at 2350 and 2380 cm^{-1}, assigned to ν_3 of CO_2. Equilibrium constants of 198 dm^{3}mol^{-1} (at 10°C) and 38 dm^{3}mol^{-1} (at 18°C) were measured. Unfortunately more data are needed for a reliable determination of the enthalpy and entropy changes involved. On the face of it, the temperature variation seems large for what is probably a weak end-on interaction with CO_2.[16] There are two other cases where the presence of a linear CO_2 ligand is claimed. A group of Russian workers report[17] that the gas at one atmosphere can be reversibly taken up by lanthanide complexes with a molar ratio of 2:1. The infrared spectra are interpreted as showing the linearity of the added ligand. For example, addition to Ln[N(SiMe$_3$)$_2$]$_3$.DME (DME = 1,2-dimethoxyethane, Ln = Pr or Nd) results in the appearance of new bands at 2180 and 1520 cm^{-1}. In the second case[18] incompletely characterized compounds, probably containing [(Ru(NH$_3$)$_5$)$_2$CO$_2$]$^{4+}$, are described which absorb at 2330 and 660 cm^{-1}. There is a need for the application of physical techniques less equivocal than infrared spectroscopy to verify these claims.

Usually when an interaction between carbon dioxide and a metal complex is detected, the degree of interaction is much greater than that found for physical adsorption. A comparison with chemisorption is then more appropriate; see for example References 13-15 and 19.

At the time of writing there is little evidence relating to carbon dioxide complexes which incontrovertibly shows the arrangement of the CO_2 unit when it is attached to a metal. Most of the structural suggestions are not based in irrefutable evidence. Aresta, Nobile et al.[20] have published the results of an xray structural study of the compound [Ni(CO$_2$)(PCy$_3$)$_2$]. 0.75C$_7$H$_8$, a compound which can be prepared in various ways as an orange/red solid e.g.

$$Ni(PCy_3)_3 \xrightarrow[C_7H_8]{CO_2} [Ni(CO_2)(PCy_3)_2]0.75C_7H_8$$

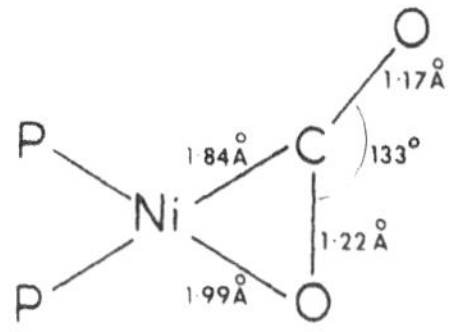

Figure 1.

or from the reaction of $NiBr_2(PCy_3)_2$ and sodium sand under CO_2. The relevant part of the structure is shown in Figure 1. The CO_2 unit is bent with a central angle of $133°$ and the metal interacts both with the carbon atom and with one of the two oxygen atoms. The interacting C-O bond is considerably longer than are the bonds in free CO_2 (1.16Å).

Lappert and coworkers have reported[21] the structure of a compound of niobium in which the ligand is attached in the same way. While the OCO angle is very similar, both the C-O bonds are longer by roughly 0.05Å in the Nb complex than in the Ni complex.

More recently, an xray structural determination has been reported,[22] in which carbon dioxide, behaving as a ligand in a monomeric complex, is attached to a metal only via the central carbon atom. The complex described, $Rh(CO_2)Cl(diars)_2$, has a Rh-C bond length of 2.05Å. The OCO angle is $126°$ while the two C-O bonds differ in length (1.20 and 1.25Å). The longer established examples of this ligand geometry are those where CO_2 is added to cobalt(I) complexes of tetradentate Schiff's base ligands.[23-4] In these instances a Co(II) complex in solution is reduced by an alkali metal before carbon dioxide is introduced, e.g.

$$Co(II)L + M \longrightarrow Co(I)LM \xrightleftharpoons{CO_2} 1:1 \text{ adduct}$$

When L = pr-salen, the condensation product of two molecules of o-hydroxyphenylpropyl ketone and one molecule of ethylenediamine, the deep green solution of the cobalt(I) complex in THF will reversibly take up CO_2 to give deep red crystals of formula $[(CoL)_2K_2(THF)_2(CO_2)_2]_n$. The relevant

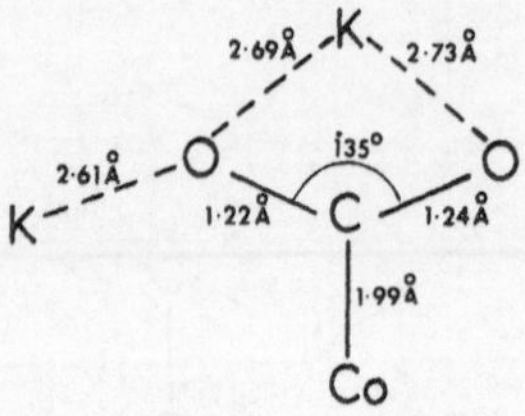

Figure 2.   Part of the structure of $[(CoL)_2K_2(THF)_2(CO_2)_2]_n$

structural details are shown in Figure 2. Carbon dioxide interacts both with cobalt and two different types of six coordinate potassium ion. If the K-O separations are regarded as being too long to be normal bonds, this complex can also be thought of as one in which carbon dioxide is a unidentate ligand bonding through C.

There are also three published descriptions of transition metal-carbonyl compounds in which CO_2 is incorporated as a bent bridging ligand, with each of its three atoms attached to a metal atom. The metal can be Os[25-6], Ru[26] or Re[27]. The C-O bonds are considerably lengthened into the range 1.25 - 1.32Å.

In addition to these few cases where xray crystallographic evidence is cited, a number of other papers containing sound claims for the demonstration of coordinated carbon dioxide have appeared. In these instances the bonding mode of the ligand is less well established, and is best deduced by comparison with the confirmed structures.

The nickel complex reported by Aresta, Nobile et al. is air stable for a few hours. Carbon dioxide is released quantitatively when the complex is treated with triphenylphosphite at room temperature, or when a stream of argon is passed through a toluene solution. Clearly the metal-ligand bond is not a strong bond. When the solid is heated at $83°C$ under argon it decomposes to a mixture of products. This mixture includes $[Ni(CO)_2(PCy_3)_2]$, CO_3^{2-} and toluene oxidation products, but not any phosphine oxide. Presumably the lengthening of one of the C-O bonds renders its cleavage more likely. In the subsequent full paper,[28] Aresta and Nobile present additional information on this compound and other trialkylphosphino products. Carbon dioxide is also released from benzene

Table 1. Assignment of infrared bands (cm^{-1})

Complexes believed to be attached both by C and by O

Compound	ν (C=O)	ν_{asym}(CO)	ν_{sym}(CO)	π	ring deform.	Ref.
[Ni(CO$_2$)(PCy$_3$)$_2$]·0.75C$_7$H$_8$	1740br(vs) 1698(w)	1150(vs)	1094(ms)	845(s)	730(m)	28
[Ni(CO$_2$)(PEt$_3$)$_2$]	1660(vs) 1635(vs)	1203(vs)	1009(s)	828(s)	750(s)	28
[Ni(CO$_2$)(P^{n}Bu$_3$)$_2$]	1660(vs) 1632(vs)	1200(vs)	1008(s)	825(s)	750(s)	28
[RhCl(CO$_2$)(P^{n}Bu$_3$)$_2$]	1668(s) 1630(s)	1165(m)	1120(m)			29

Complexes believed to contain the M-C

$$\begin{matrix} & & O \\ & \diagdown & \\ & C & \\ & \diagup & \\ & & O \end{matrix}$$

grouping

Compound	ν (C=O)	ν_{asym}(CO)	ν_{sym}(CO)	π	ring deform.	Ref.
[(Co[Pr-salen])$_2$K$_2$(THF)$_2$(CO$_2$)$_2$]$_n$	1650(s)	1280(s)	1215(s)	745(s)		23
Na[Co(salen)CO$_2$]	1680	1278	1213			32
Na[Co(salen)(CO$_2$)Py]	1700	1273	1208			32
[RhCl(CO$_2$)(PEt$_2$Ph)$_3$]	1670(vs) 1635sh	1255(m)	963(m)	780(m)		29

solutions of these complexes upon treatment with iodine or sulfuric acid. The infrared absorptions attributable to the CO_2 ligand are listed and assigned in Table 1. The frequency of the band assigned to ν(C=O) is 90 to 100 cm^{-1} higher in toluene solution than in nujol mulls for the ethyl and n-butyl cases. The authors suggest this to be due to an equilibrium involving the complex $[Ni(CO_2)(PR_3)_3]$, even though the compound of proven structure with incorporated toluene is also observed to give a higher frequency band.   Molecular oxygen at room temperature and below atmospheric pressure reacts with the coordinated CO_2 in both solid and solution to yield an Ni(II)-peroxocarbonate.

The same authors have also prepared complexes of type $[RhCl(CO_2)(PR_3)_2]$ and $[RhCl(CO_2)(PR_3)_3]$.   Those complexes with two phosphine ligands per rhodium are the more stable with respect to the loss of CO_2 on evacuation, but still evolve the gas on treatment with acid.   On standing these compounds decompose yielding coordinated carbon monoxide and trialkylphosphine oxide.   The reactions can be represented as

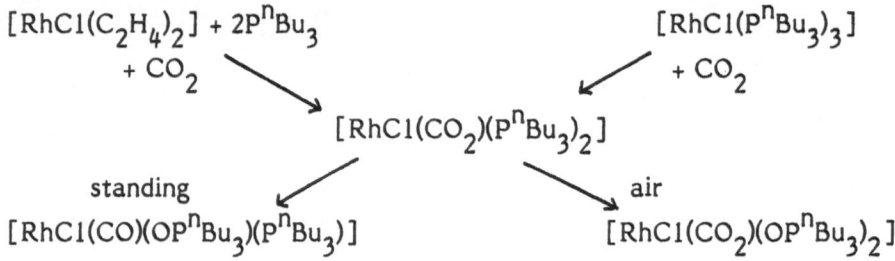

$$[RhCl(C_2H_4)_2] + 2P^nBu_3 + CO_2$$

$$[RhCl(P^nBu_3)_3] + CO_2$$

$$[RhCl(CO_2)(P^nBu_3)_2]$$

standing $\quad$ air

$$[RhCl(CO)(OP^nBu_3)(P^nBu_3)]$$

$$[RhCl(CO_2)(OP^nBu_3)_2]$$

Both in terms of the reactivity pattern and the infrared spectra (Table 1), a side-on presentation of carbon dioxide is indicated.[29]   This same arrangement is suggested[30] for an iron complex, $[Fe(CO_2)(PMe_3)_4]$ formed in non-polar solvents at high temperatures

$$[Fe(PMe_3)_4] + CO_2 \longrightarrow [Fe(CO_2)(PMe_3)_4] + [Fe(CO_3)(CO)(PMe_3)_3]$$

On standing this complex decomposes to $[Fe(CO)_2(PMe_3)_3]$, $[Fe(CO)(PMe_3)_4]$ and Me_3PO.   At present it seems that evidence of the cleavage of one of the C-O bonds of coordinated carbon dioxide may be indicative of bonding between metal and one side of the CO_2 unit. Lappert's work[21] with niobium may also support this deduction as he quotes in his preliminary communication not only the structure of the carbondioxido complex, but also the fact that carbonyl and carbonato complexes are formed.

In the case of one other complex, $[Mo(CO_2)_2(PMe_2Ph)_4]$, it might seem possible to argue from the nature of the decomposition produce to a side-on presentation of CO_2.[31] But here, because of the proximity of two CO_2 ligands, the assumptions involved are very large. Unfortunately, Chatt and his coworkers found that this bis-carbon dioxido complex was unsuitable for an xray structural study. The reactions observed are

$$\underline{cis}\text{-}[Mo(N_2)_2(PMe_2Ph)_4] + CO_2 \longrightarrow [Mo(CO_2)_2(PMe_2Ph)_4]$$

$$[Mo(CO_2)_2(PMe_2Ph)_4] \longrightarrow [(PMe_2Ph)_3(CO)Mo(CO_3)_2Mo(CO)(PMe_2Ph)_4]$$

Aresta and Nobile[29] argue that their compounds of formula $[RhCl(CO_2)(PR_3)_3]$ contain the CO_2 ligand in the more symmetrical unidentate form, e.g.

$$[RhCl(C_2H_4)_2]_2 + \text{excess PEtPh}_2 + CO_2 \underset{\text{vacuum}}{\overset{C_7H_8}{\rightleftarrows}} [RhCl(CO_2)(PEtPh_2)_3]$$

The carbon dioxide is more loosely held than in those compounds with only two phosphine ligands per rhodium, it is fully lost during recrystallization from benzene/pentane under nitrogen. Although this argument is plausible, steric crowding due to the extra ligand might be a sufficient explanation for the difference in the Rh-(CO_2) bond stability. However, since there are significant differences in the infrared bands due to CO_2, the balance of evidence favors the unidentate form of bonding. An examination of the formula of $Rh(CO_2)Cl(diars)_2$, whose structure is known,[22] also suggests that the coordination number on Rh may help to determine the mode of ligand attachment.

An inspection of the infrared data of Table 1 shows that the assignment of a bonding mode on the basis of such spectra is not straightforward. In the absence of crystallographic information, one is reduced to the hints provided by stability and chemical reactivity. Infrared data are probably best regarded at present merely as indicators of the ligand's presence. It is reasonable to assume[33] that the unidentate ligand is present in the complex $[IrCl(CO_2)(Me_2PCH_2CH_2PMe_2)_2]$, but the mode of attachment in the other carbon dioxido complex with only one metal center involved is uncertain. Thus, we can merely list the existence of another 1:1 Ir:CO_2 adduct[34] and various copper complexes of general formulae $[(RCO_2)Cu(CO_2)(PR_3')]$[35] and $[(RCO_2)Cu(CO_2)(PR_3')_2]$.[36] The uncharacterized cobalt complex reported by Misono et al.[37] is probably a formato complex, while the rhodium and iridium complexes claimed[38] to be $[M(OH)(CO_2)(CO)(PPh_3)_2]$ have been reformulated[39] as bicarbonato complexes.

Various dimeric species are known, but no structural information on them is available. $[(Ni(PCy_3)_2)_2CO_2]$ is known,[20,40] as is $[Cu_2(CO_2)(PPh_3)_2(C_6H_4PPh_2)_2]$,[36] a compound containing both triphenylphosphine and its orthometalated derivative as ligands. It seems likely that CO_2 acts as a bridging group in these instances. The following dimeric formulae for rhodium complexes have been reported: $(PPh_3)_3RhCl(CO_2)(PPh_3)_2RhCl$,[41] $Rh_2(CO)_2(CO_2)(PPh_3)_3 \cdot C_6H_6$,[42] $Rh_2H_2(CO_2)(PPh_3)_6 \cdot C_7H_8$,[43] $(PPh_3)_3Rh_2(CO_2)_2(CO)_2 \cdot C_6H_6$[44] and $(PPh_3)_3Rh_2(CO_2)(CO)_2 \cdot C_6H_6$.[44] These materials are reasonably stable, although the CO_2 can be displaced by the action of heat, acid, better ligands or a nitrogen stream. All these materials contain potential bridging ligands other than CO_2 and so some caution is warranted before claiming that bridging CO_2 is present. However, Kolomnikov and Grigoryan[7] believe this to be the case, and in view of the authenticated[25-7] instances of bridging CO_2 they may well be correct.

When organosilver-silver salt coordination compounds are treated with carbon dioxide in ethanol at temperatures below $0^{\circ}C$, a reaction occurs.[45] One of the products is a polymer which contains CO_2 or a derivative.

$$2C_6H_5Ag \cdot nAgNO_3 + CO_2 \longrightarrow 1/x[C_6H_5Ag \cdot CO_2 \cdot Ag]_x + C_6H_6 + 2nAgNO_3$$

Insufficient evidence is available for one to ascertain whether or not discrete CO_2 units are present in this material, although it is unlikely that the CO_2 is attached to the aromatic ring.

Both the reported structures and the number of infrared bands observed suggest that for most, and probably all, complexes containing CO_2 the ligand is bent, and that the carbon atom is involved in the bonding. Given this clue, we now turn to the nature of the metal-ligand interaction. In most complexes steric factors are likely to ensure that any triatomic ligand bonded other than end-on must be non-linear. We may also reason that the C-O bonds in ligating carbon dioxide must remain strong, since the metal-ligand bonds formed are unlikely to be strong enough to compensate for a large amount of C-O bond weakening. Any plausible formal representation of the CO_2 ligand should contain these features. A bent ligand, CO_2^{n-}, whether n = 0, 1 or 2, would satisfy these criteria and might be expected to give a reasonable fit for observed bond angles and infrared spectra. This is certainly the case for the known[12] radical anion CO_2^{2-} which shows a central angle of $127\underline{+}8^{\circ}$ and infrared bands at 1671, 1424(w) and 849(w) cm^{-1}.

Given the present dearth of reliable evidence concerning the nature of the ligand and its bonding, a detailed treatment of the bonding would at present be premature. A simple approach using valence bond diagrams is outlined here, although it should be noted that ab-initio molecular orbital calculations have been made for end-on and side-on CO_2 complexes.[46] The canonical forms of Figure 3a are plausible for the cobalt complex of Fachinetti and Floriani,[23] while one possible representation of the nickel complex of Aresta and Nobile[20] is shown in Figure 3b. It would seem that the unidentate presentation found for the cobalt complex requires the metal to behave as a nucleophile, as is sometimes suggested for cobalt(I) with

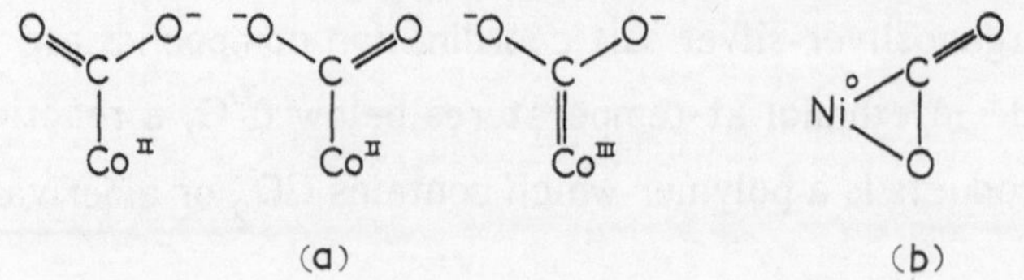

(a)   (b)

Figure 3.   (a) Possible canonical forms for the adduct of CO_2 and Co(I)

(b) One possible representation of the adduct of CO_2 and Ni(0)

planar tetradentate ligands. Such a requirement would clearly limit the range of metal ions capable of forming this type of complex. In the extreme the ligand might be formally represented as CO_2^- or even as CO_2^{2-}, the latter case being comparable with the N-bonded form or the iso-electronic ligand NO_2^-. Since there is a very limited amount of evidence for any nucleophilic character in most metal ions in complexes, we may deduce that the CO_2 ligand will not often be found in this form. The suggestion has been made[29] that rhodium complexes containing CO_2 as a ligand are more stable when the mode of attachment is side-on rather than through the central C atom. If that is a true generalization it could be predominantly due to steric factors. Also the one structure which has been determined contains the unidentate C-bonded ligand.[21] The molecular orbital calculations[46] suggest that back-π-bonding in $(PPh_3)_2Ni(CO_2)$ could staɔilize a side-on presentation, while an end-on mode of attachment would be preferred for the Cu(I) analogue.

In those examples which can be assumed to contain the side-on presentation, proven only for $[Ni(CO_2)(PCy_3)_2]$, it seems that one C-O is significantly weakened and can be broken readily in subsequent reactions. In the reported structure this is the bond attached to the metal. Thus we may suppose that this C-O is a single bond, as implied in Figure 3b; or that synergic bonding occurs with significant population of the ligand's antibonding π-orbital. Neither requires the metal to behave as a nucleophile, in the usual sense of the word, but rather that it can form covalent bonds; in which case, this may be the usual mode of attachment in complexes with CO_2 as a ligand.

Attempts have been made to identify the requirements for carbon dioxide uptake by Ir(I) and Ir(III) complexes[47] and by Fe(II) complexes.[48] The metal center should be easily accessible and, in the case of Fe(II), should carry high electron density. Easy access is obviously needed for a poor ligand. The presence of other free ligands can restrict the formation of CO_2 derived ligands,[49-50] while the steric properties of phosphine ligands are also important.[28,39] The stability of the M-(CO_2) unit in complexes is related to the basicity of the other ligands[20] and the basicity of the metal.[51] Aresta and Nobile[29] suggest that the difference in $v(CO)$ between comparable Rh(I) and Ir(I) adducts, and the greater ease of formation of Ir(I) adducts, is due to the lower iridium d-orbital energy and hence the greater degree of back π-bonding from the metal which is seen as a desirable feature. Lastly we may quote Fachinetti, Floriani and Zanazzi;[23] "In many simple reactions carbon dioxide seems to require for its activation, in addition to a basic center, the assistance of an acidic partner." At present we are unable to identify clearly the bonding properties of complexes containing CO_2 which are required for stability, and can only sketch out likely alternatives. Perhaps, given the poor ligating ability of CO_2, it is impossible to produce a single, simple account of these properties and the demands which they make on the metal and its other ligands.

4. **SUBSEQUENT REACTIONS OF COMPLEXES CONTAINING CO_2**

In the majority of cases where a new chemical product is believed to arise from a reaction of a coordinated CO_2 molecule, no complex containing CO_2 as a discrete unit has been observed. Such reactions are discussed in later sections of this chapter. In this Section attention is restricted to those examples where a CO_2-containing species is observed, and then undergoes a reaction other than the loss of carbon dioxide.

Mention has already been made of the occurrence of C-O bond cleavage in complexes containing side-on bonded CO_2. Thus, when the solid $[Ni(CO_2)(PCy_3)_2].0.75C_7H_8$ is heated at $83^\circ C$ under argon, a mixture of products is obtained.[20] This mixture includes $[Ni(CO)_2(PCy_3)_2]$, CO_3^{2-} and oxidation products of toluene, but no phosphine oxide. Similar cleavages with the transfer of oxygen to another component of the reaction system, seen when solutions are allowed to age, can be summarized as follows:[29-31]

$[RhCl(CO_2)(P^nBu_3)_2] \longrightarrow [RhCl(CO)(OP^nBu_3)(P^nBu_3)]$

$[Fe(CO_2)(PMe_3)_4] \longrightarrow [Fe(CO)_2(PMe_3)_3], [Fe(CO)(PMe_3)_4], Me_3PO$

$2[Mo(CO_2)_2(PMe_2Ph)_4] \longrightarrow [(PMe_2Ph)_3(CO)Mo(CO_3)_2Mo(CO)(PMe_2Ph)_3]$

In the case of the rhodium complex cited above, reaction in the presence of air yeilds $[RhCl(CO_2)(OP^nBu_3)_2]$. This product could either arise as a result of the reoxidation of the carbonyl, or from the formation of a peroxocarbonate intermediate as is seen in the following examples. Treatment of $[Ni(CO_2)(PCy_3)_2]$ with oxygen, at less than one atmosphere pressure, yields a peroxocarbonate complex of Ni(II) which can be isolated and characterized:

$[Ni(CO_2)(PCy_3)_2] + O_2 \longrightarrow [Ni(CO_4)(PCy_3)_2]$

In one of the earliest reports of the activation of carbon dioxide, Iwashita and Hayata[42] describe some results of ^{18}O labeling experiments. They produced coordinated CO_2 by oxidizing a carbonyl ligand with a limited quantity of oxygen.

$$Rh_2(CO)_4(PPh_3)_4 \xrightarrow[C_6H_6]{O_2} Rh_2(CO)_2(CO_2)(PPh_3)_3 \cdot (C_6H_6)_n$$

Probably n=1. When $^{18}O_2$ is used in this process, $Ph_3P^{18}O$ is formed in the residues. Infrared analysis shows an approximately 1:1 ratio for $Ph_3P^{16}O$: $Ph_3P^{18}O$. Also a 1:2:1 ratio is found for $C^{16}O_2$: $C^{16}O^{18}O$: $C^{18}O_2$. Finally,

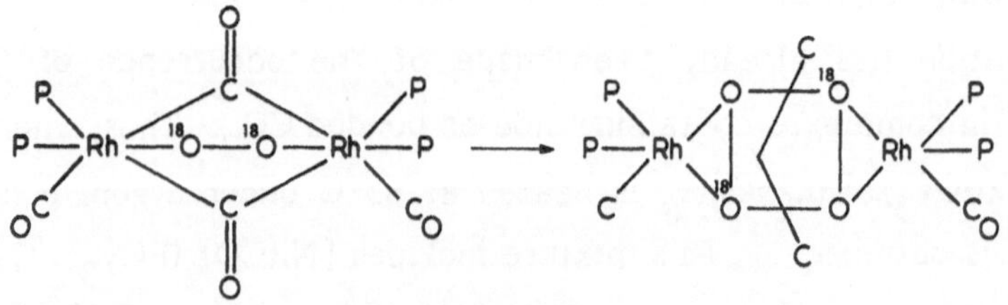

Figure 4.

when the product is recrystallized, oxygen exchange occurs between the coordinated CO_2 and molecular oxygen dissolved in the solvent. In the authors' words, carbon dioxide is in "an activated state, so to say." The isotopic product ratios can be accounted for in terms of the suggested intermediates of Figure 4.

The related dirhodium compound $(PPh_3)_3RhCl(CO_2)(PPh_3)_2RhCl$ yields methanol quantitatively when it is treated with $LiAlH_4$ and methyl acetate when treated with methyl iodide.[7,41] In one case cleavage of C-O has again occurred, while in the other reaction C-C bond formation is seen. One other example of such bond formation occurring with an isolable CO_2 containing complex is known. Herskovitz[33] reports that heating an iridium complex in a closed system at about $120^\circ C$ yields a complex containing a new carboxylate ligand,

$$[Ir(CO_2)L_2]Cl \longrightarrow [Ir(H)(L).O_2C.CH_2.PMe.CH_2CH_2PMe_2]Cl$$

where L represents bisdimethylphosphinoethane.

Maher, Lee and Cooper[52] have made use of isotopic labeling to investigate O transfer in a tungsten(-2) - carbonyl system. The ion $[W(CO)_5CO_2]^{2-}$ is produced as an intermediate when CO_2 reacts with $Li_2[W(CO)_5]$. It readily undergoes reaction with a further molecule of CO_2 to yield $W(CO)_6$ and the carbonate anion. The authors found evidence for the occurrence of three different processes as represented by

$$M(^{13}CO_2)^{2-} + {}^{12}CO_2 \longrightarrow M(^{12}CO_2)^{2-} + {}^{13}CO_2$$
$$M(CO_2)^{2-} + M'(CO) \longrightarrow M(CO) + M'(CO_2)^{2-}$$
$$M(CO_2)^{2-} + M'(CO_2)^{2-} \longrightarrow M(CO) + M'(CO_3)^{4-} \longrightarrow CO_3^{2-} + M'^{2-},$$

where $M = W(CO)_5$. This well executed piece of work fits well into, and helps to support, the general pattern of reactivity which seems to be emerging.

In summary, isolable complexes containing the CO_2 ligand can be converted to carbonyl complexes with an oxygen atom being transferred to

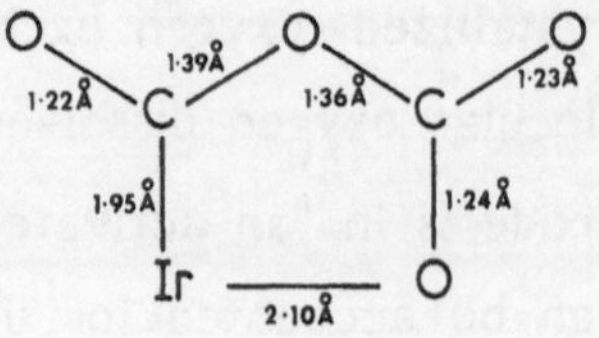

Figure 5.   Part of the structure of $IrCl(C_2O_4)(PMe_3)_3.0.5C_6H_6$

other ligands. The oxygen transfer can even be to another CO_2 ligand converting it to a carbonato complex. New bonds to the carbon atom can also be made involving H, C and O. It is against this background that the more exciting reactions discussed below should be judged.

4.1   Reactions Involving C-O Bond Cleavage

Herskovitz and Guggenberger[34] report the preparation and structure of an iridium complex containing a C_2O_4 chelating ligand, Figure 5, resulting from CO_2 uptake in benzene solution:

$$IrCl(C_8H_{14})(PMe_3)_3 \xrightarrow{\quad CO_2 \quad} IrCl(C_2O_4)(PMe_3)_3.0.5C_6H_6 \text{ etc.}$$

On heating to $150°C$ this solid decomposes giving a mixture which includes a carbonyl and a 1:1 $Ir:CO_2$ adduct. Thus the coupling of two carbon dioxide units can be reversed or can lead to what is presumably a disproportionation reaction. It seems likely that the disproportionation[31] discussed above for a biscarbondioxidomolybdenum complex involves such an intermediate species. Fachinetti, Floriani and coworkers[53] cite another reaction in which the C_2O_4 ligand is probably involved:

$$4Cp_2Ti(CO)_2 + 4CO_2 \longrightarrow [(Cp_2Ti)_2(CO_3)]_2 + 10CO$$

where Cp represents the cyclopentadienyl ligand. When this reaction is carried out using $^{13}CO_2$, it is found that the carbonato ligands, and 20% of the liberated CO, are labeled with ^{13}C. All the carbonate has been formed by the disproportionation of carbon dioxide. An intermediate dimer of the type found by Herskovitz would explain such a result.

A different type of product is obtained from the analogous zirconium compound

$$3Cp_2Zr(CO)_2 + 3CO_2 \longrightarrow [Cp_2ZrO]_3 + 9CO$$

This product is more comparable with the oxo-bridged dimer obtained when carbon dioxide at 10 atmospheres and $90°C$ reacts with a toluene solution of the related titanium chloro complex

$$[(Cp)_2TiCl]_2 + CO_2 \longrightarrow [Cp_2Ti\ Cl]_2O + CO$$

The mechanism for this reaction is suggested to be

$$\tfrac{1}{2}[Cp_2TiCl]_2 + CO_2 \longrightarrow ClCp_2Ti\text{-}O\overset{O}{\underset{O}{\diagdown}}C.$$

$$+ \tfrac{1}{2}[Cp_2TiCl]_2 \longrightarrow ClCp_2Ti\text{-}O\diagdown\underset{\underset{O}{\parallel}}{C}\diagup TiCp_2Cl \longrightarrow etc.$$

Obviously the cleavage of C-O in such a -carbondioxido intermediate could be followed by retention of either O or CO as ligand, depending on the detailed reaction conditions. Carbonyl ligands are retained on the titanium when magnesium[54] or aluminum[55] is present. The following reactions have been reported:

$$Cp_2TiCl_2 + Mg \xrightarrow{\ Ar\ } Cp_2Ti + MgCl_2 \xrightarrow[Mg]{\ CO_2\ } Cp_2Ti(CO)_2$$

and

$$3Cp_2TiCl + Al \xrightarrow{\ Ar\ } 3Cp_2TiCl + AlCl_3$$

$$3Cp_2TiCl_2 + 6CO_2 + 5Al \longrightarrow 3Cp_2Ti(CO)_2 + AlCl_3 + 2Al_2O_3$$

In the aluminum example an 80% yield of the carbonyl was obtained. While the stoichiometry of the reaction remains at $CO_2:Ti = 2:1$, the reaction is autocatalytic with aluminum chloride as catalyst. Initially the rate of reaction is approximately proportional to the aluminum chloride concentration.

The use of zinc rather than aluminum as the reducing agent[55] results in an overall reaction

$$5Cp_2TiCl_2 + 5Zn + 4CO_2 \longrightarrow Cp_2Ti(CO)_2 + 2(Cp_2Ti)_2CO_3 + 5ZnCl_2$$

with the formation of both carbonyl and carbonate ligands. While the addition of zinc chloride has no effect on the reaction, other Lewis acids can alter the yield of carbonyl. The relevant data are reproduced in Table 2. The choice of pathway is related to the strength of the new metal-oxygen bond which is formed as an alternative to carbonate. In the light of these data, Demerseman et al. suggest that the Lewis acid catalyzed route involves the reduction of $Cp_2TiCl(CO_2)(acid)$ by metal. Carbonyl and carbonate are believed to follow the formation of $Cp_2TiCl(CO_2)$ on the zinc surface.

Many reactions are catalyzed by two-metal systems, one metal being a transition metal and the other a typical metal. The conversion of molecular nitrogen and carbon dioxide to the cyanate ion can be achieved[54] using titanium (or vanadium) and magnesium. Thus, in THF, titanium(IV) chloride and magnesium in a 3:1 ratio react in the following manner:

$$TiCl_4 + Mg \xrightarrow[\text{THF}]{N_2} (THF)Cl_2Mg_2TiN + MgCl_2(THF)_2$$

$$(THF)Cl_2Mg_2TiN + CO_2 \xrightarrow{\text{THF}} (THF)_3Cl_2Mg_2OTi(NCO)$$

The use of biscyclopentadienyltitanium dichloride in the place of $TiCl_4$ results in the formation of $Cp_2Ti(NCO)$ and $Cp_2Ti(CO)_2$.

The early transition metals are not unique in their ability to form metal carbonyls from carbon dioxide. Under mild conditions, in the presence of an excess of the silicon compound, the following process has been observed to occur in benzene:[56]

$$RhCl(PPh_3)_3 + HSi(OEt)_3 \longrightarrow RhH[Si(OEt)_3]Cl(PPh_3)_2$$

$$RhH[Si(OEt)_3]Cl(PPh_3)_2 \xrightarrow{\ CO_2\ } Rh(CO)Cl(PPh_3)_2$$

$Ru(CO)Cl_2(PPh_3)_3$ can similarly be derived from $RuCl_2(PPh_3)_3$.

There is a connection between these reactions which produce CO from CO_2, the reverse of the Water Gas Shift Reaction

$$CO_2 + H_2 \longrightarrow H_2O + CO$$

which is unfavorable at room temperature ($\Delta H=+41.2$ kJ mol^{-1}), and the reverse of the reaction between nitric oxide and carbon monoxide

$$2NO + CO \longrightarrow N_2O + CO_2$$

Table 2. The effect of added Lewis acids on the yield of $Cp_2Ti(CO)_2$ from the reaction of CO_2 and Cp_2TiCl_2 in the presence of zinc[55]

Lewis Acid	Mol ratio acid/Ti	% yield	$-\Delta H/kJ\ mol^{-1}$ (a)
$AlCl_3$	0.23	28	
	0.28	36	559
	0.54	63	
	1.00	80	
UCl_4	0.54	52	565
$ZrCl_4$	0.59	52	543
$GaCl_3$	0.30	29	363
$MgCl_2$	1.25	52	602
BCl_3	0.76	21	424
$ZnCl_2$	0 - 3	18	348

(a) ΔH is ΔH_f^o of the oxide related to the Lewis acid per O atom.

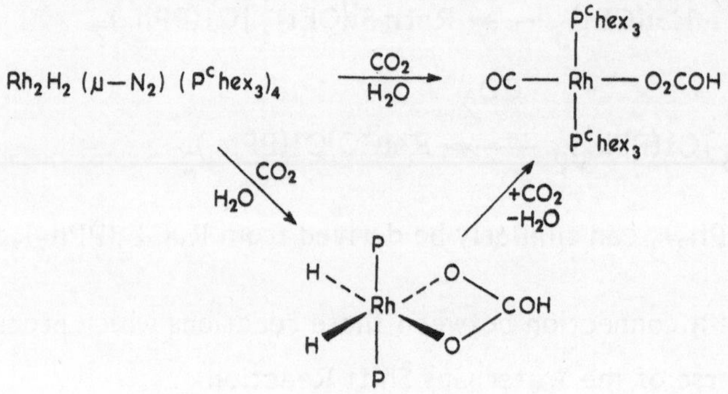

Figure 6.

Coordination of the products of a reaction can alter the thermodynamic parameters sufficiently for a previously unfavorable reaction to become spontaneous. Ibers and his coworkers[39] report several examples of reactions which can be described as coordinated versions of the reversed Water Gas Shift Reaction. For example, in the presence of excess carbon dioxide, the reactions of Figure 6 are found. The reaction probably involves the insertion of CO_2 into an Rh-H bond to give a hydroxycarbonyl ligand followed by the cleavage of the C-OH bond. Such an intermediate has been postulated in the $[Rh(CO)_2X_2]^-$ catalysis of the Shift Reaction,[57] and the reduction of nitric oxide by carbon monoxide.[58-9]

In general, the formation of coordinated or free CO from CO_2 seems to be possible provided that the CO_2 is activated by a metal center, and that a suitably strong bond can be formed by the departing oxygen atom. Failing that, CO_2 can be converted to CO by the provision of energy. Examples of the use of reducing agents to provide that energy are the formation of $[Ru(NH_3)_5CO]^{2+}$ from the reaction of CO_2 with $[Ru(NH_3)_5H_2O]^{2+}$ in the presence of zinc amalgam,[18] and the reduction of $CpFe(CO)_2(CO_2)^-$ to $CpFe(CO)_3^+$ by $HBF_4.OEt_2$.[60] A more interesting way of providing energy is in a photochemical reaction. Several reports describe the production of carbon monoxide by the irradiation of solutions containing carbon dioxide, a sensitizer and certain transition metal complexes. In the case of rhenium(I) or cobalt(I) complexes, Lehn and coworkers[61] deduce that the dissolved CO_2 must coordinate to the metal prior to reduction.

The formation of trialkyl and triaryl carbinols from carbon dioxide also involves C-O bond breaking, but is more appropriately discussed later.

4.2 Reactions Involving C-O Bond Formation: Carbonato- and Bicarbonato- Complexes

Several of the reactions discussed in the previous section under the heading of C-O bond cleavage involved the disproportionation of CO_2, and so also involved C-O bond making. In two more cases, it seems that the extra O atom needed to form carbonate may be derived from carbon dioxide. Both involve closely related complexes. The treatment of $HRh(CO)(PPh_3)_3$ with CO in benzene produces a mixture of two intermediates which can react with CO_2, without the involvement of moisture, to produce a complex containing a bicarbonato complex. The extra O comes from another molecule of CO_2.[62] A slight variation occurs when toluene solutions of the complexes $RhH(PPh_3)_x$, where x=3 or 4, are treated with carbon dioxide at temperatures below 20°C, giving a bridged carbonato product, $(PPh_3)_3Rh(CO_3)Rh(PPh_3)_2$. The rate of reaction is faster when x=3. For x=4, the addition of free triphenylphosphine reduces the rate; but the addition of water or changes in the CO_2 pressure do not affect it.[50] We are tantalizingly short of evidence on which to base a mechanism.

When a better source of oxygen is available, CO_2 readily forms an extra C-O bond, as instanced by solutions of the gas in water. When the formation of this extra bond occurs in the presence of metal complexes the CO_2 might be coordinated or free prior to the bond formation. In only one case can it be stated that the CO_2 had definitely been coordinated before its conversion, and that is in the reaction quoted earlier

$$Ni(CO_2)(PCy_3)_2 + O_2 \longrightarrow Ni(CO_4)(PCy_3)_2$$

Since this reaction does not need a solvent, the release of CO_2 prior to its reaction would seem very unlikely. Most of the known reactions leading to carbonate or bicarbonate seem to involve free carbon dioxide. This has been convincingly demonstrated in the case of copper(II) bovine carbonic anhydrase[4] and in the mechanistic studies on the reactions of CO_2 with hydroxo-complexes discussed below. The chelated carbonato complex of platinum(II) studied by Wilkinson et al.[63-4]

$$Pt(PPh_3)_3 + O_2 + CO_2 \xrightarrow{C_6H_6} (PPh_3)_2Pt(CO_3).C_6H_6 + Ph_3PO$$

can be obtained[65] by the reaction of carbon dioxide with the dioxygen complex $[(PPh_3)_2PtO_2]$. Molecular oxygen is also the source of the additional O atoms in the formation of μ-carbonato-dicopper(II) complexes from copper(I) halides, polydentate N-donor ligands, carbon dioxide and oxygen, according to the equations[66-8]

$$L + CuX \xrightarrow[CH_2Cl_2]{O_2} L_mCu_2Cl_2O \xrightarrow[O_2]{CO_2} LCuX(CO_3)XCuL$$

with various ligands including L = tris(2-pyridyl)amine.

A considerable effort has been made to understand the mechanism of formation of carbonato- and bicarbonato-complexes of Co(II)[66-7,69] and Rh(III)[70-71] by the reaction of carbon dioxide with aquo and hydroxo complexes in aqueous solution. One paper[69] provides a useful review of this topic. All the cobalt studies show that CO_2 can only add to hydroxo ligand, and that the nucleophilicity of this ligand determines the rate of gas uptake. Thus, the following reactions occur in water in the range of pH from 7 to 9, and between 15°C and 25°C; where tren represents β,β',β''-triaminotriethylamine.

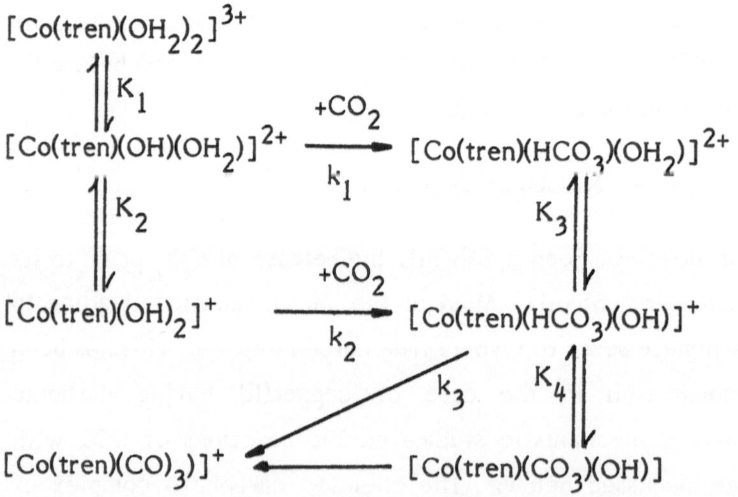

$$[Co(tren)(OH_2)_2]^{3+}$$
$$\Big\updownarrow K_1$$
$$[Co(tren)(OH)(OH_2)]^{2+} \xrightarrow[k_1]{+CO_2} [Co(tren)(HCO_3)(OH_2)]^{2+}$$
$$\Big\updownarrow K_2 \qquad\qquad\qquad\qquad K_3 \Big\updownarrow$$
$$[Co(tren)(OH)_2]^+ \xrightarrow[k_2]{+CO_2} [Co(tren)(HCO_3)(OH)]^+$$
$$\qquad\qquad k_3 \qquad K_4 \Big\updownarrow$$
$$[Co(tren)(CO_3)]^+ \longleftarrow [Co(tren)(CO_3)(OH)]$$

Both in the formation of the range of carbonato products and their reverse decarboxylation reactions the changes in the enthalpy and entropy of activation compensate each other. This common kinetic feature is due to changes in solvation, more specifically in this case to changes in the degree of hydrogen bonding between the solvent and the CO_3 grouping. Data are summarized in Table 3. The decarboxylation reactions proceed by C-O bond cleavage and therefore, applying the Principle of Microscopic Reversibility, we may deduce that the formation involves the reaction of CO_2 with the coordinated ligand.

There are several more examples of the formation of bicarbonato complexes from carbon dioxide in the literature. According to the reformulation[39] of the original claims, Flynn and Vaska[38] were able to produce bicarbonato complexes of rhodium and iridium by treating solid hydroxo complexes with CO_2 gas,

$$trans\text{-}[M(OH)(CO)(PPh_3)_2] + CO_2 \rightleftharpoons [M(OCO_2H)(CO)(PPh_3)_2]$$

but not as originally claimed[72] when ethanol is present. Jewsbury[73] has studied the reaction with water kinetically. His findings are interesting, but too many uncertainties remain for a full mechanism to be proposed. The rate of the forward reaction is dependent on the water content of the solvent, but a second faster pathway becomes feasible in dry dichloromethane. Also, while the rate of reaction is first order in the complex, it is zero order in carbon dioxide. More studies would be worthwhile. Both in dimethylformamide and water,[74] CO_2 is reversibly taken up by a Cu(I) hydroxo complex,

$$Cu(OH)L_n + CO_2 \rightleftharpoons (HOCO_2)CuL_3$$

where $L = PEt_3$ or ^tBuNC. A related reaction has been observed by Yamamoto and coworkers:[75]

Table 3. Kinetic parameters for reactions involving CO_2 in water

Reagent	$k_{25}/$ $dm^3 mol^{-1} s^{-1}$	$\Delta H^{\ddagger}/$ $kJ\ mol^{-1}$	$\Delta S^{\ddagger}/$ $J\ mol^{-1} K^{-1}$	Ref.
H_2O	7×10^{-4}	77	-47	66
OH^-	8.5×10^3	53	8	66
$[Co(NH_3)_5(OH)]^{2+}$	220	64	15	66
$[Co(tren)(OH)(OH_2)]^{2+}$	44	61	-8	67
$[Co(tren)(OH)_2]^+$	170	135	220	67
cis-$[Co(en)_2(OH)(OH_2)]^{2+}$	240	65	14	69
trans-$[Co(cyclam)(OH)(OH_2)]^{2+}$	37	121	193	69
trans-$[Co(cyclam)(OH)_2]^+$	70	118	186	69
cis-$[Co(cyclam)(OH)(OH_2)]^{2+}$	57	62	3	69
cis-$[Co(cyclam)(OH)_2]^+$	196	64	13	69
$[Rh(NH_3)_5(OH)]^{2+}$	490	71	50	71
cis-$[Rh(en)_2(OH)(OH_2)]^{2+}$	69	66	16	70
cis-$[Rh(en)_2(OH)_2]^+$	215	61	5	70
enzymic (a)	4×10^7	25	0	2

(a) the formation of bicarbonate in the presence of bovine carbonic anhydrase.

$$(RO)Cu(PPh_3)_2 + CO_2 \xrightarrow[\text{solvent}]{\text{moist}} [(HOCO_2)Cu(PPh_3)_2]_n$$

$$2[HOCO_2)Cu(PPh_3)_2]_n \underset{}{\overset{\Delta}{\rightleftharpoons}} n(PPh_3)_2CuOCO_2Cu(PPh_3)_2$$

In this case there is no indication of the state of OH^- prior to the formation of bicarbonate. Moisture is also required for the reaction

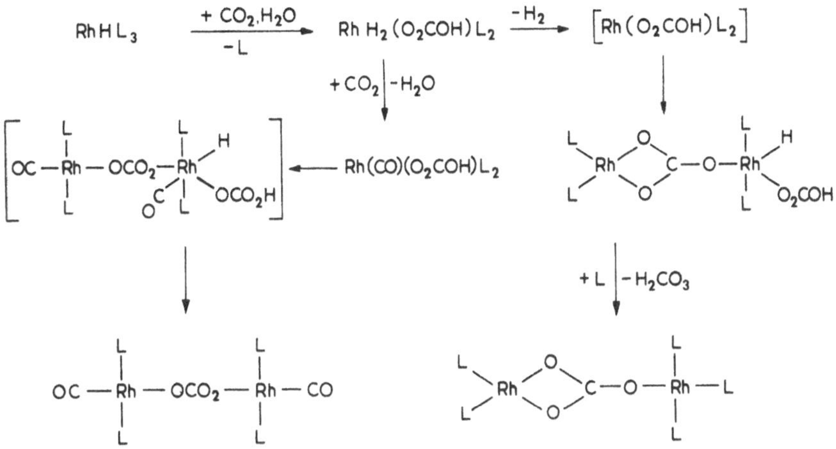

$RhHL_3 \xrightarrow[-L]{+ CO_2 \cdot H_2O} Rh H_2(O_2COH)L_2 \xrightarrow{-H_2} [Rh(O_2COH)L_2]$

$+CO_2 | -H_2O$

$\left[\begin{array}{c} OC-\underset{L}{\overset{L}{Rh}}-OCO_2-\underset{\underset{O}{C}}{\overset{H}{Rh}} \\ OCO_2H \end{array} \right] \longleftarrow Rh(CO)(O_2COH)L_2$

$\underset{L}{\overset{L}{Rh}}\overset{O}{\underset{O}{\diagup}}C-O-\underset{L}{\overset{H}{Rh}}O_2COH$

$+L | -H_2CO_3$

$OC-\underset{L}{\overset{L}{Rh}}-OCO_2-\underset{L}{\overset{L}{Rh}}-CO$

$\underset{L}{\overset{L}{Rh}}\overset{O}{\underset{O}{\diagup}}C-O-\underset{L}{\overset{L}{Rh}}-L$

Figure 7

$\underline{trans}\text{-}[PdMe_2(PEt_3)_2] + CO_2 + H_2O \rightleftharpoons \underline{trans}\text{-}[Pd(HOCO_2)Me(PEt_3)_2] + CH_4$

which gives less than a 1% yield if both the carbon dioxide and the hexane solvent are dry. This product has been reformulated[76] since its first report.[77] The overall reaction is suggested to proceed via a $[Pd(OH)Me(PEt_3)_2]$ intermediate. Water is also present for the first two of the following reactions[39]

$RhH(P^iPr_3)_3 + CO_2 + H_2O \longrightarrow RhH_2(O_2COH)(P^iPr_3)_2 + P^iPr_3$

$RhH_2(O_2COH)(P^iPr_3)_2 + CO_2 \longrightarrow \underline{trans}\text{-}Rh(O_2COH)(CO)(P^iPr_3)_2 + H_2O$

$\underline{trans}\text{-}Rh(O_2COH)(CO)(PCy_3)_2 \underset{\Delta}{\rightleftharpoons} Rh(OH)(CO)(PCy_3)_2 + CO_2$

As was the case[75] for Cu(I), so also with this rhodium system a μ-carbonato-complex can be obtained, presumably via a unidentate bicarbonate. Ibers and his coworkers also advance a plausible mechanism (Figure 7) for the formation of $[Rh_2(O_2CO)(CO)(PPh_3)_5]$.[50]

Both in the solid (xray) and in solution (molecular weight) $RhH_2(O_2COH)(P^iPr_3)_2$ is dimeric due to hydrogen bonding between neighbouring bicarbonate groups. This feature must enhance the possibilities for μ-carbonate formation. Size effects as measured by the cone angles of the phosphine ligands also have an important part to play in controlling this dimer formation.

4.3   The Formation of Other C-O Bonds

The formation of bicarbonate from carbon dioxide both with and without the presence of metal ions is reversible under suitable conditions. It is possible to add an alkoxide rather than a hydroxide group to the gas giving an alkylcarbonate. The aliphatic alkoxides of alkali metals provide such an example.[78]   The alkylcarbonates mirror much of the behavior of bicarbonates. The similarity is well illustrated by the reactions for Cu(I) with L = PEt$_3$ or ^{t}BuNC in organic solvents:[74]

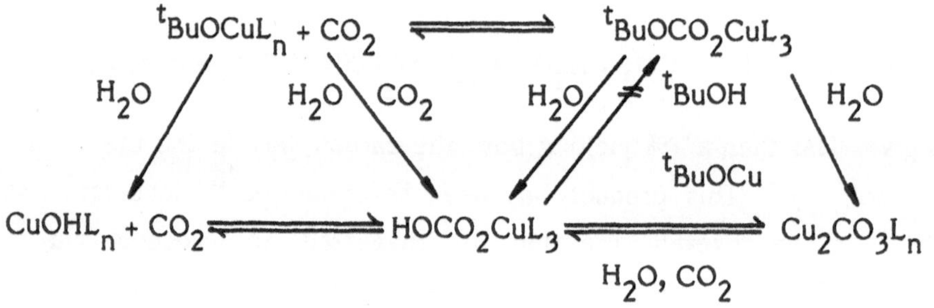

$$^t\text{BuOCuL}_n + CO_2 \rightleftharpoons {}^t\text{BuOCO}_2\text{CuL}_3$$

$$\text{CuOHL}_n + CO_2 \rightleftharpoons \text{HOCO}_2\text{CuL}_3 \rightleftharpoons \text{Cu}_2\text{CO}_3\text{L}_n$$

$$H_2O, CO_2$$

(with H_2O, CO_2, ^{t}BuOH, ^{t}BuOCu labels on the connecting arrows)

Two other papers report the formation of alkylcarbonato-copper(I) complexes from alkoxo complexes. Both reactions involve CO_2 at one atmosphere pressure, but while $\text{ROCu(PPh}_3)_2$ can react in the absence of solvent,[75] the reaction[79]

$$\text{Cu(OMe)}_2 + 2CO_2 \rightleftharpoons \text{Cu(O}_2\text{COMe})_2$$

is carried out using a suspension of the copper compound in pyridine, a potential ligand. The pyridine appears in the product in the case of

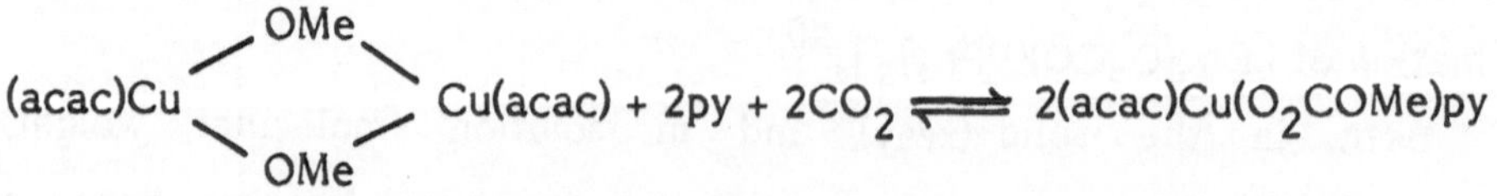

$$(\text{acac})\text{Cu} \underset{\text{OMe}}{\overset{\text{OMe}}{\diagup\diagdown}} \text{Cu(acac)} + 2py + 2CO_2 \rightleftharpoons 2(\text{acac})\text{Cu(O}_2\text{COMe})py$$

Dimeric alkoxides of molybdenum and tungsten are also known to react with CO_2 in this same way,[80-82] according to the general equation

$$M_2(OR)_6 + 2CO_2 \rightleftharpoons M_2(OR)_4(O_2COR)_2$$

The process is completely reversible. In the particular case of $Mo_2(OCH_2CMe_3)_6$ in hydrocarbon solvents, the equilibrium lies well over to the right. In the most recent paper[82] dealing with this topic, a range of molybdenum compounds, R = Me_3Si, Me_3C, Me_2CH and Me_3CCH_2, are reported to undergo this same change both in the solid state and in solution. In the reaction involving the solid, the free energy of activation for the forward reaction is less than or equal to 92 kJ mol^{-1}. The mechanism is believed to proceed by direct attack of CO_2 giving either $Mo-(CO_2)$ or a species in which the C of CO_2 is attacked by the alkoxide O while a bond between Mo and a carbon dioxide O is forming. In solution a more labile pathway is available, i.e.

$$RO-H + CO_2 \rightleftharpoons ROCO_2H$$

$$Mo-OR + ROCO_2H \longrightarrow Mo-O_2COR + ROH$$

Two such alternative mechanisms may often be possible for reactions which cause an increase in the coordination number at carbon. The degree of crowding around the metal and the ease of metal-ligand bond cleavage will then be deciding factors in the choice of route. Other alkylcarbonato complexes have been made. Thus CO_2 readily inserts into Sn-OR bonds.[83-4]

Some reported reactions involving transition metal species include[85,76,86]

trans-$[PtH_2(PCy_3)_2]$ + CO_2 + 2MeOH $\longrightarrow$
trans-$[PtH(O_2COMe)(PCy_3)_2]$.MeOH + H_2

trans-$[PdMe_2(PEt_3)_2]$ + CO_2 + ROH $\longrightarrow$
trans-$[PdMe(O_2COR)(PEt_3)_2]$ + CH_4

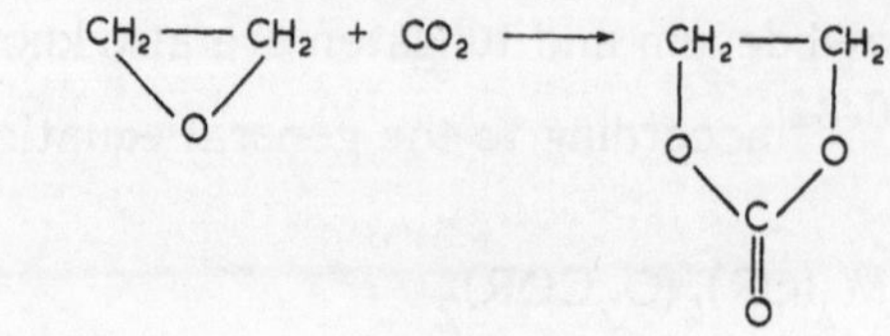

CH₂——CH₂ + CO₂ ⟶ CH₂——CH₂

$$CH_2\text{—}CH_2 + CO_2 \longrightarrow CH_2\text{—}CH_2$$

Figure 8

$[RuH(PMe_2Ph)_5](PF_6)$

CO_2 in ROH

$[Ru(O_2COR)(PMe_2Ph)_4](PF_6)$

from Me_2CO/ROH

recrystallize

$[Ru(O_2CH)(PMe_2Ph)_4](PF_6)$

In the ruthenium example the suggested mechanism involves the initial formation of an Ru-OR bond, although presumably the route from the formato complex could involve nucleophilic attack on carbon. It has also been suggested[39] that the reported bicarbonato complexes of Rh and Ir[72] are really ethylcarbonates, since they were prepared in ethanol.

Cyclic organocarbonates have also been reported as products from carbon dioxide reactions. The earliest report[87] quoted a selectivity in excess of 95% for the reaction of Figure 8. The reaction was carried out in an autoclave in the presence of $Ni(PPh_3)_2$ and benzene. Some other epoxides are said to undergo similar reactions. High temperature and pressure were also used[88] with $CoCl(PPh_3)_3$ as catalyst. This cobalt complex is non-catalytic at 1 atmosphere pressure and a lower temperature,[89] although catalysts for more gentle conditions have been found. Ratzenhofer and Kirsch report that the reaction sequence of Figure 9 can be achieved at 20°C and 1 atmosphere using a metal halide plus a Lewis base as catalyst. The best catalysts found were the halides of Al, Mo or Fe with triphenylphosphine. The halides of Ti, W, Co and Ni may also be used. The most successful experiments gave yields, after isolation, up to 76% and turnover up to 67 molecules/MX_n. Under argon, the strongly exothermic reaction of the epoxide reagent and $AlCl_3$ forming $Al(OCH_2CHClCH_3)_3$ was found to occur. Treatment of this first product with carbon dioxide results in the appearence of new strong infrared bands

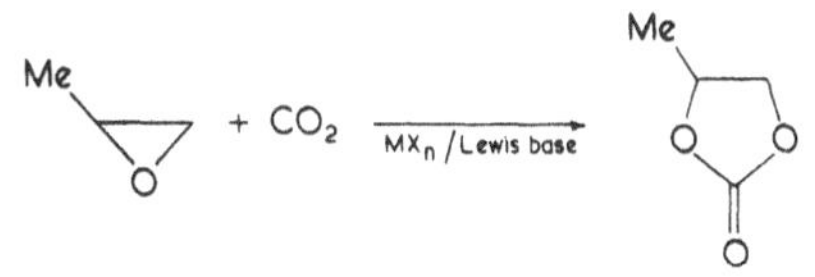

Figure 9

at 1610 and 1430 cm^{-1}, probably due to an $ROCO_2Al$ unit. This material can be converted to the final dioxolanone product on treatment with a Lewis base such as PPh_3, pyridine or NEt_3. Preliminary experiments with chiral bases did not yield chiral products.

The bicarbonato-copper(I) complexes derived from CO_2 are capable of taking part in trans-carboxylation;[74] thus $HOCO_2Cu(^tBuNC)_3$ in dimethylformamide catalyzes the formation of dioxolanones with 67% yields at 80°C, and 82% at 130°C.

Various reports have been published of the formation of alkyl formates,[51,90-92] as in the reaction at 125°C and 250 psi of each gas in the presence of $HRu_3(CO)_{11}^-$

$$CO_2 + H_2 + CH_3OH \longrightarrow HCO_2CH_3 + H_2O$$

Obviously one could propose several alternative mechanisms to account for this process. The possibility of making a $C-OCH_3$ bond and then cleaving one of the other bonds might be worthy of consideration.

4.4 Formato Complexes

Carbon dioxide can be converted into formate without the intermediacy of complex formation. For example, CO_2 in solution can be reduced electrochemically. Normally it is formed as a minor product, but in the presence of some transition metal complexes as catalysts formate is preferentially produced.[93] We are likely to see more reports of electrocatalyzed reductions of CO_2, with and without illumination, in the near future; but up to now most formate production has occurred during more conventional coordination chemistry. In this reporter's opinion, one of the strangest features of the chemistry of CO_2 is the formation of formate

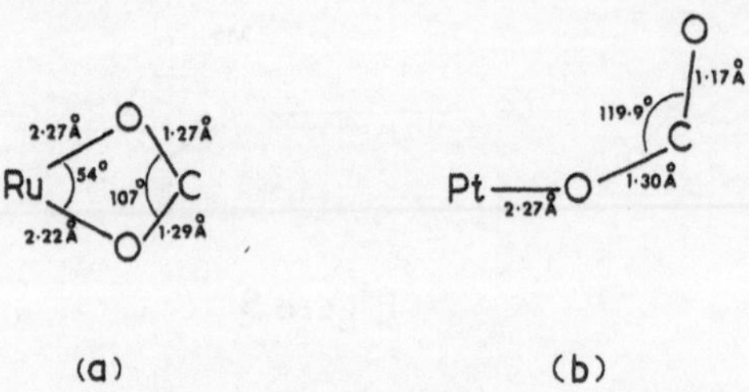

(a)   (b)

Figure 10.   Parts of the structures of (a) $RuH(O_2CH)(PPh_3)_3$ and (b) trans-$[PtH(O_2CH)(PCy_3)_2].2C_6H_6$

ligands and their ready reversion back to carbon dioxide. Two xray structural examinations of formato complexes derived from CO_2 have been reported. The reaction

$$RuH_2(PPh_3)_4 + CO_2 \longrightarrow RuH(O_2CH)(PPh_3)_3 + PPh_3$$

has been reported by two separate research groups.[43,49,94] The reaction, which is readily reversible, occurs in an aromatic solvent at room temperature. The structure of the solid product shows the metal to be 6-coordinate, with both oxygen atoms of the CO_2 derived ligand attached to Ru as shown in Figure 10a. The xray data do not show the presence of hydrogen, but the ligand is presumed to be formate.

The other known structure is that of the product obtained when a solution of trans-$[PtH_2(PCy_3)_2]$ in benzene at $8°C$ takes up CO_2 to give trans-$[PtH(O_2CH)(PCy_3)_2].2C_6H_6$ (see Figure 10b). In this case the structure appears to contain the grouping PtOC(O)H.

While reasonable empirical formulations seem to require the CO_2 derivative to be formate, more solid evidence is needed; especially as both the products mentioned above lose CO_2 easily, even when an inert gas is bubbled through their solutions. Usually one thinks of formate as a kinetically stable group. Such evidence has been obtained from two independent physical techniques, infrared and nuclear magnetic resonance spectroscopy. The existence of infrared bands in the range 1300–1650 cm^{-1}, assigned to OCO stretching, is poor evidence for formato complexes since

Table 4. Infrared frequencies/cm^{-1} of some formate species

Assignment	$Rb(O_2CH)$	$(HCO_2)RuH(PPh_3)_3$	$(HCO_2)Co(PPh_3)_3$	$(HCO_2)Pt(H)(PCy_3)_2$	$(HCO_2)_2Fe(PEt_2Ph)_2$
C-H stretch	2801(w)	2895(w) 2805(w)	2910(m) 2820(m)	2795(w) 2700(w)	
asym OCO stretch	1592(s)	1553(s)	1565(s)	1620(s)	1590(s)
asym OCO deform.	1381	1365(m) 1347(m)	1420(m) 1370(m) 1355(m)		
sym OCO stretch	1370(s)	1310(s)	1340(m) 1300(s)	1310(s)	1370(s)
sym OCO deform.	774(s)	795(s)	800(s) 690(vs)		
Reference	49	49	94 96	85	97

most possible ligands derived from CO_2 will show bands in this same region. However, in the ruthenium case new bands appear at 2910 and 2820 cm^{-1} [94] or at 2895 and 2805 cm^{-1},[49] which are shifted to 2150 cm^{-1} when the deuteriated ruthenium reagent is used. These are credibly assigned to ν(C-H), supporting the idea of formate. Table 4 shows some infrared data on this topic.

The ^{1}H nmr spectrum of the related ruthenium complex $[Ru(O_2CH)(PMe_2Ph)_4]^+$ shows[86] a broad hydride signal at 17.4 τ which is taken to demonstrate the equilibrium

$$L_4Ru(O_2CH) \rightleftharpoons L_4Ru(H)(CO_2)$$

The nmr spectrum of the platinum complex mentioned above, measured in toluene-d_8 is more convincing.[85] Even using a CO_2 saturated solution the dominant nmr spectrum is that due to the reagent $[PtH_2(PCy_3)_2]$. A much weaker spectrum of the product can also be identified. This shows a doublet due to O_2CH at 0.81τ , and a high field sextet due to the hydride. Coupling constants involving the formate proton are J_{Pt-H} = 50 Hz and J^4_{H-H} = 4.5 Hz. Paonessa and Trogler[95] have studied the equilibrium for the complex containing triethylphosphine ligands. In toluene they determined the equilibrium constant to be approximately 2 atm^{-1} at 25°C. They also found that changing the solvent could result in different final products. In polar solvents uncoordinated formate ions were found.

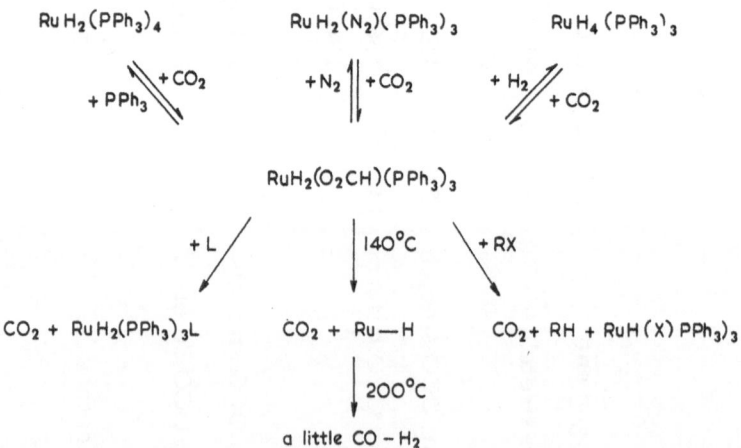

Figure 11.   Reactions of a ruthenium-formato complex[43,49,94]

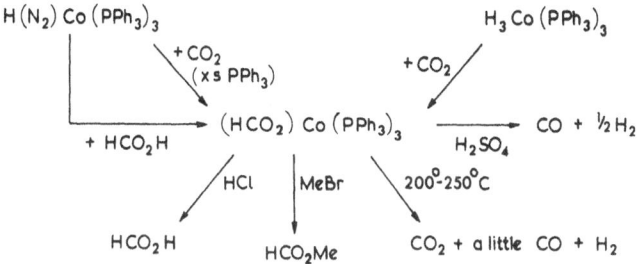

Figure 12.   Reactions of a cobalt-formato complex[96]

Chemical evidence can also be found to support the formato representation. The known reactions of the various formato complexes are displayed in Figures 11 to 13. In addition to the complexes mentioned in text and in figures, formate also seems to be produced from CO_2 in the case of another Fe complex,[30] Rh,[39] and Mo and W.[98]

The reported reactions of formato-complexes show a ready reversion to CO_2, the formation of CO on treatment with heat or acid, and sometimes the liberation of formic acid and formates. Several instances occur where the postulation of a formato-intermediate is reasonable without its having been detected. This is especially so for the catalytic formation of alkyl formates from carbon dioxide.[51,90-91] The general reaction can be written as

$$ROH + CO_2 + H_2 \longrightarrow HCO_2R + H_2O$$

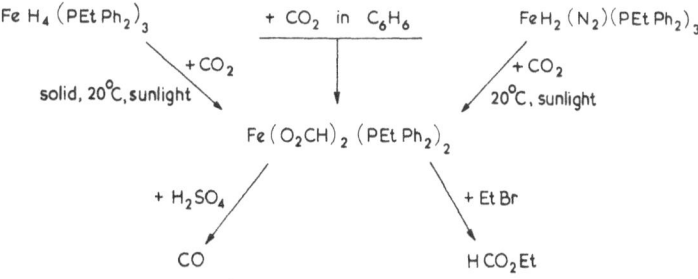

Figure 13.   Reactions of an iron-formato complex[97]

Moderate pressures of gas are used, with temperatures above $100°C$. Various catalyst systems have been studied. In the earliest report[90] a metal complex and BF_3 were used. The complexes $RuCl_2(PPh_3)_4$, $RuH_2(PPh_3)_4$, $RuHCl(PPh_3)_4$, $RuCl_2(PPh_3)_3$ and $IrH_3(PPh_3)_3$ showed catalytic activity. The suggested mechanistic scheme is

$$L_nMX + H_2 \xrightarrow{} L_nMH \xrightarrow{CO_2} L_nM(O_2CH) \xrightarrow{MeOH}$$

$$HCO_2Me + L_nMOH \xrightarrow{H_2} L_nMH + H_2O$$

Mechanistic studies using ^{18}O would be interesting here. Subsequently[91] complexes plus trialkylamines were found to catalyze the reaction. In this case the alkyl group in the ester originates with the alcohol. Low oxidation state complexes of Fe, Co, Ni, Ru, Rh, Ir and Pt are active but the most effective catalysts found are $Pd(dppe)_2$, $RuH_2(PPh_3)_4$ and $RhCl(PPh_3)_3$. A more recent report quotes $[Ni(PPh_3)_2][HFe_3(CO)_{11}]$ and $[HNEt_3]$ $[HFe_3(CO)_{11}]$ as good catalysts. The reactions are definitely catalytic with turnover numbers well in excess of unity per metal ion.

Under similar conditions of temperature and pressure, but not with benzene present as solvent, the reaction

$$R_2NH + CO_2 + H_2 \longrightarrow HCONR_2 + H_2O$$

is catalyzed by a similar range of transition metal complexes.[99] Again the active complexes listed are strongly reminiscent of the complexes discussed already in this chapter. Using trans-$[Ir(PPh_3)_2(CO)Cl]$ for mechanistic studies, Haynes et al. found that in the early stages of reaction the

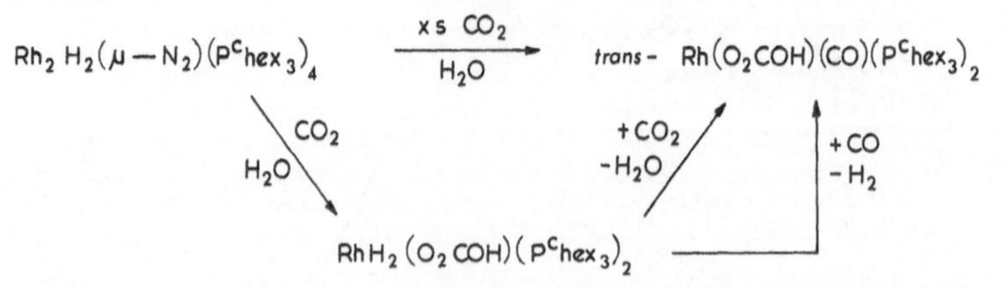

Figure 14

deuterium content of product dimethylformamide paralleled that of the amine and not that of the molecular hydrogen. With $RhCl(PPh_3)_3$ as catalyst no $Rh(PPh_3)_2(CO)Cl$ was detected. Under the experimental conditions this carbonyl does not release CO; therefore reduction of CO_2 to CO is unlikely. Hence a reasonable mechanism is

$$L_nMH + CO_2 \longrightarrow L_nMO_2CH \xrightarrow[\text{+ } H_2, \text{ } -H_2O]{HNR_2} L_nMOH + HCONR_2$$

identical to that for alkyl formate production, except for the identity of the nucleophile which attacks the formate ligand. This involvement of a formato complex has also been postulated[100] in the formation of an N,N-dialkylcarbamato complex, but see below.

Lastly, carbonyl formato and carbonyl bicarbonato complexes of rhodium can be obtained from appropriate reactions with carbon dioxide, as shown in Figure 14.[39] The mechanism probably proceeds by CO_2 insertion into an Rh-H bond to give an $Rh-(CO_2H)$ grouping. If the CO_2H ligand is to give a carbonyl readily, then its formulation as hydroxycarbonyl rather than as formate is appealing. The hydroxycarbonyl ligand has already been suggested in other systems, e.g. Reference 58. Thus it would be unwise to neglect this isomer of formate in other circumstances. Perhaps the two forms are capable of ready interconversion in some cases. Certainly the ease with which formato complexes lose CO_2 points to the kinetic lability of the C-H bond.

4.5   Formation of N,N-Dialkylcarbamates

Amongst the most commonly reported products of CO_2 and metal complexes are N,N-dialkylcarbamato complexes, $M-O_2CNR_2$. Such preparations have been reported for Ti,[101-2] V,[101] Cu,[75,103] Zr,[101-2] Nb,[101-2,104] Mo,[101] Ru,[100] Pd,[105] Ta,[101-2] W[63,101,106] and U,[107] and Si and Sn.[108-112] In the light of the generally accepted mechanism, there is no obvious reason why this list of metals should not be substantially extended; indeed ionic amides undergo this same reaction.[113-16] The conversion of a dialkylamine and carbon dioxide to the corresponding dialkyl carbamate is generally regarded as occurring away from any metal ion. The first clue to this came from the observation that in hydrocarbon solvents the reaction

$$W(NMe_2)_6 + CO_2 \longrightarrow W(NMe_2)_3(O_2CNMe_2)_3$$

occurs quantitatively and rapidly at room temperature, even though the tungsten reagent is extremely congested sterically. Intermediate complexes with 5 and 4 amine ligands were observed as transients by nmr spectroscopy. Most of the carbamato products are known to be labile with respect to CO_2 exchange. Thus when the product of the reaction in benzene solution[104]

$$Nb(NMe_2)_5 + 5CO_2 \longrightarrow NbL_5 \qquad\qquad (L = O_2CNMe_2)$$

is studied by nmr in toluene-d_8, significant concentrations of $Nb(NMe_2)L_4$ are observed unless a positive pressure of CO_2 is used.

The reaction in solution between ^{13}C-labeled $W(NMe_2)_3L_3$ and CO_2 is approximately first order in the tungsten complex and zeroth order in carbon dioxide, while the equivalent reactions involving ZrL_4 and NbL_5 are approximately first order in CO_2. Thus for CO_2 exchange we can rule out a scheme such as

$$MI_n \underset{+CO_2}{\overset{-CO_2}{\rightleftharpoons}} M(NMe_2)L_{n-1} \underset{-^{13}CO_2}{\overset{+^{13}CO_2}{\rightleftharpoons}} M^{13}LL_{n-1}$$

The favored mechanism both for CO_2 exchange and CO_2 insertion involves trace amounts of free amine, according to

$$Me_2NH + CO_2 \rightleftharpoons Me_2NCO_2H$$

$$Me_2NH + Me_2NCO_2H \rightleftharpoons Me_2NH_2^+ + Me_2NCO_2^-$$

$$M^{13}L + LH \rightleftharpoons ML + {}^{13}LH$$

$$M^{13}L + L^- \rightleftharpoons ML + {}^{13}L^-$$

Methyl esters of the carbamates can be obtained from a copper complex:[103]

$$Cu(O^tBu) + R^1R^2NH + CO_2 \longrightarrow R^1R^2NCO_2Cu$$

$$\Big\downarrow + L$$

$$R^1R^2NCO_2Me \xleftarrow{\quad +MeI \quad} R^1R^2NCO_2CuL_n$$

where $L = {}^tBuNC$ or P^nBu_3. Ethyl carbamate can also be made:[75]

$$MeCu(PPh_3)_2 \cdot 0.5(Et_2O) + NH_3 + CO_2 \longrightarrow [uncharacterized]$$

$$\Big\downarrow + EtBr$$

$$NH_2CO_2Et \ (20\%)$$

A mono-N-phenyl carbamato complex has also been reported:[75]

$$[(HOCO_2)Cu(PPh_3)_2]_n \xrightarrow{\quad C_6H_5NH_2 \quad} (C_6H_5NHCO_2)Cu(PPh_3)_2$$

and

$$(C_6H_5NH)Cu(PPh_3)_2 + CO_2 \longrightarrow (C_6H_5NHCO_2)Cu(PPh_3)_2$$

There seems to be no evidence of any carbamate formation on metal ions.

Bellon and coworkers have reported[117] what they claim to be the only authentic example of CO_2 insertion into an M-N bond, when carbon dioxide at one atmosphere reacts with $Pt(PhNO)(PPh_3)_2$ in dichloromethane. The reaction produces a five membered chelate ring, which unfortunately cleaves at N-C rather than at N-O.

4.6   Carbon–Carbon Bond Formation

Insertion of CO_2 into metal-carbon bonds has been widely reported for a range of metal ions and organic ligands. Little definitive mechanistic information is available, although it is reasonable to assume that a normal insertion mechanism operates in most cases. For convenience, the material in this section is arranged according to the organic reagents and products of reaction, rather than by metal or type of complex involved.

The most widely known reactions of carbon dioxide in this area involve organo-lithium and Grignard reagents. These reactions have been reviewed by Vol'pin and Kolomnikov.[6] In the case of the Grignard reagents the usual reaction can be written as

$$RMgX + CO_2 \longrightarrow RCO_2MgX \xrightarrow{H_2O} RCO_2H$$

In these reactions, the Grignard reagent reacts with cold CO_2, often as the solid. As with most Grignard reactions, this process is usually accompanied by side reactions. Alkanes and ketones are quoted as side products. When the temperature of the reaction is raised, significant yields (37-47%) of carbinols can be obtained,[118] and the overall reaction after hydrolysis can be described as

$$3PhMgBr + CO_2 + 2H_2O \longrightarrow Ph_3COH + 3MgBrOH$$

In recent years examples of this same type of product formation using different metals have been reported. A 90% yield of triethylcarbinol is reported[119] for the sequence

$$AlEt_3 + CO_2 \longrightarrow Et_2AlOCOEt$$

$$Et_2AlOCOEt + 2AlEt_3 \longrightarrow Et_2AlOAlEt_2 + Et_3COAlEt_2$$

$$Et_3COAlEt_2 + H_2O \longrightarrow Et_3COH$$

while at 13 atmospheres pressure, two mols of CO_2 react with one mol of tetrabenzyltitanium(IV) or zirconium(IV) in toluene to give, after hydrolysis, essentially equimolar amounts of phenylacetic acid and tribenzylcarbinol:[119]

$$M(CH_2Ph)_4 \xrightarrow[\text{ii) } H_2O]{\text{i) } CO_2} PhCH_2CO_2H + (PhCH_2)_3COH$$

Carbinol formation would seem more likely to occur in systems with a high local concentration of alkyl or aryl groups near the reaction center.

Therefore it is not surprising that the cited examples involve 3 and 4 M-C bonds per reagent metal ion, rather than the one bond per Grignard reagent, which leads less commonly to carbinol production.

More usually the product of reaction with organometallics is either the carboxylic acid or a carboxylato complex. In addition to the organolithium and Grignard reagents already mentioned, carboxylic acids can be obtained from other typical metal-alkyl bonds, e.g. M = Be,Ca,Sr,Ba,Zn,Al,Tl(I). The relevant references may be found in the cited review.[6] When a solution of an alkylcopper(I) complex in ether or tetrahydrofuran is treated with carbon dioxide at low temperatures, insertion into the Cu-C bond occurs:[36,120]

$$MeCu(PPh_3)_2 \cdot 0.5Et_2O + CO_2 \longrightarrow (MeCO_2)Cu(PPh_3)_2$$

A further molecule of gas is taken up upon prolonged reaction at room temperature:

$$(MeCO_2)Cu(PPh_3)_2 + CO_2 \longrightarrow (MeCO_2)Cu(CO_2)(PPh_3)_2$$

This additional molecule of CO_2 is readily lost on recrystallization, heating etc., as one would expect, but the acetato ligand does not decompose. The most likely intermediate is obviously one in which both methyl and carbondioxido ligands are attached to the same copper ion, but attempts to isolate this intermediate failed. Similar reactions are also reported for other alkyl groups, Et, ^nPr and ^iBu,[36] and with other phosphine ligands, e.g. $(EtCu)_2(dppe)_3$[36] and $MeCu(PCy_3)$.[35] In this last example the carbon dioxide is specifically stated to be dry, thus ruling out bicarbonato intermediates (cf. Ref. 74); again, an extra CO_2 molecule can become attached to give $(MeCO_2)Cu(CO_2)(PCy_3)$. Carbon dioxide has also been inserted into a M-CH_3 bond when $W(CO)_5CH_3$ in THF is treated with CO_2 at high pressure.[121] The formation of carboxylato complexes has also been reported for Rh(I)[122]

$$Rh(Ph)(PPh_3)_3 + CO_2 \longrightarrow Rh(O_2CPh)(PPh_3)_3$$

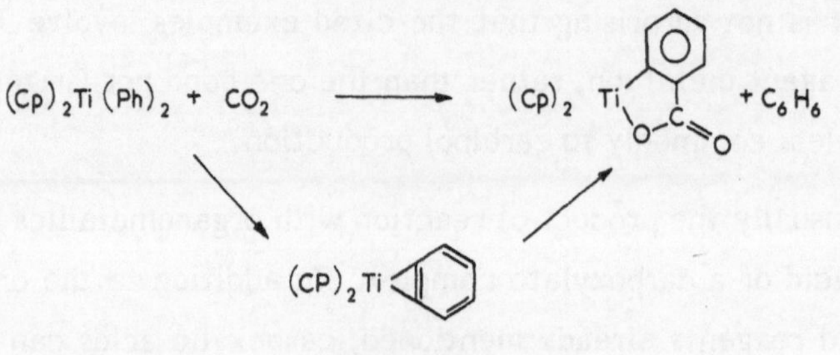

Figure 15

using benzene as solvent and 20 atmospheres pressure at room temperature. No reaction occurs with the Rh(III) complex $RhCl_2(Et)(PPh_3)_3$, probably due to the higher coordination number and greater kinetic inertness of this compound. A related reaction does occur when CO_2 reacts with a slightly different rhodium(I) complex in benzene/ethanol. Again coordinated benzoate can be made, but this time its phenyl group originated as a solvent molecule.[123] There is one report[124] of the conversion of a phenyl to an ortho-metalated benzoato ligand. The reaction occurs in xylene at 80-90°C and at one atmosphere pressure as shown in Figure 15. Kolomnikov and his coworkers speculate that the reaction may proceed through the intermediate shown.

Metal acetylides, M = Cu or Ag, also undergo conversion to carboxylate,[125-6] but now the process reaches equilibrium in THF under ambient conditions

$$PhC{\equiv}CCu(P^nBu_3)_3 + CO_2 \rightleftharpoons PhC{\equiv}CCO_2Cu(P^nBu_3)_3$$

The equilibrium constant varies with the phosphine used, carboxylation being promoted by ligands with high σ-donating strength. This reaction is not restricted to transition metals; both rubidium and caesium phenylacetylides react with carbon dioxide in a similar manner.[127]

The copper acetylides mentioned above appear to react in a straightforward manner. Several other unsaturated hydrocarbon ligands can also be carboxylated, but product analysis often shows more than one organic product. The insertion of CO_2 into a methallyl-nickel compound

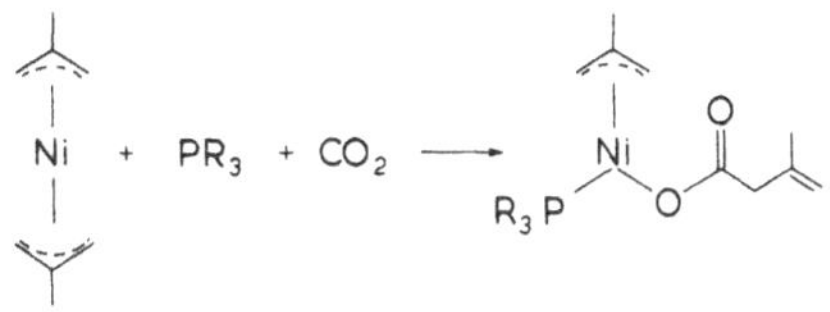

Figure 16

appears to follow a simple scheme (Figure 16), although other allyl ligands can undergo more complicated reactions,[128] as indeed can this same ligand on palladium.[129] In acetonitrile as solvent, treatment with CO_2 at 50 atmospheres and 80°C for four hours allows the processes of Figure 17 to occur. Santi and Marchi suggest that the reaction proceeds through the intermediate shown in Figure 18. This suggestion is supported by the observation that no insertion occurs with trans-[PdI(Me)(PPh$_3$)$_2$]; or for the complexes [(η^3-C$_4$H$_7$)PdCl(PR$_3$)$_n$]Cl where R = Et, Cy or Ph and n = 0, 1 or 2; or when Cl is replaced by PF$_6$. Insertion does however occur with [PdCl(Me)(dppe)]. This example clearly demonstrates that the presence of further metal-carbon bonds may allow the carboxylato-product to react further. The ester formation observed here arises from an intermolecular

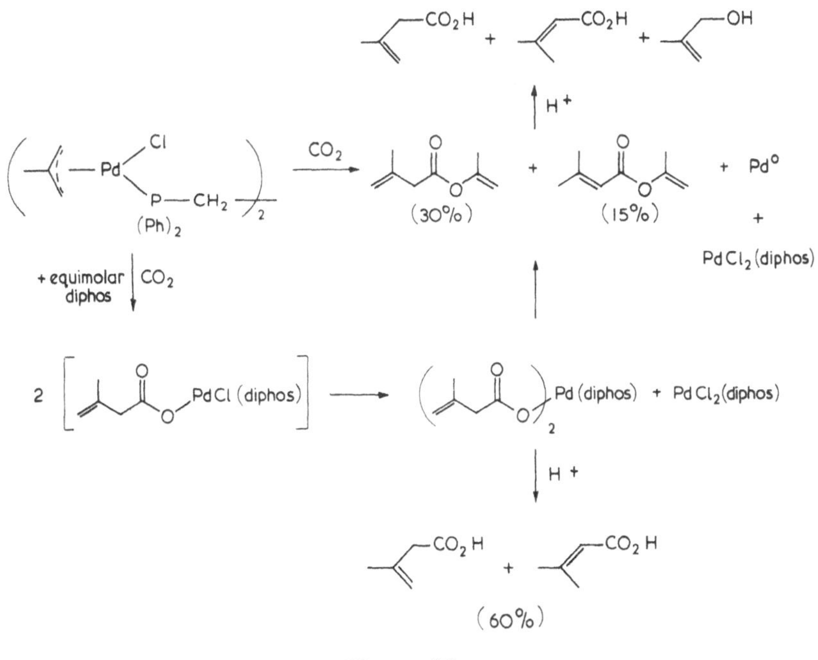

Figure 17

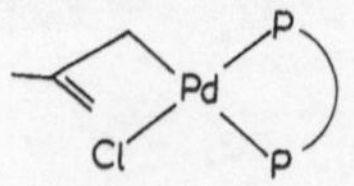

Figure 18

reaction, but there are known examples of intramolecular processes. The most obvious are those which yield lactones.[130-33] Thus, using triphenylphosphine and a palladium complex as catalyst the process of Figure 19 occurs[133] in benzene at 126°C and a pressure in excess of 40 atmospheres. Similarly the palladium complex/phosphine catalyst systems result in lactone production when butadiene and carbon dioxide at high pressure react together. With Pd(dppe)$_2$ as catalyst[132] the major reaction is butadiene dimerization and only small yields of lactone are obtained. In contrast, Pd(0) complexes with non-chelating phosphine ligands enable both lactone and open chain octadienyl esters to be obtained.[131] Thus inter- and intra-molecular esterification reactions can be in competition with each other.

At elevated temperature and pressure, phenoxide ions add on CO_2 in the Kolbe-Schmitt carboxylation, giving predominantly the corresponding ortho-hydroxy carboxylate

$$PhO^- + CO_2 \longrightarrow o\text{-}HOC_6H_4CO_2^-$$

The ortho-product probably predominates because of the involvement of the counter ion, typically Na^+, in the transition state. Hirao and Kito[134] show that an initial complex is formed, but are unable to distinguish between the various possibilities for its structure. An analogous reaction involving

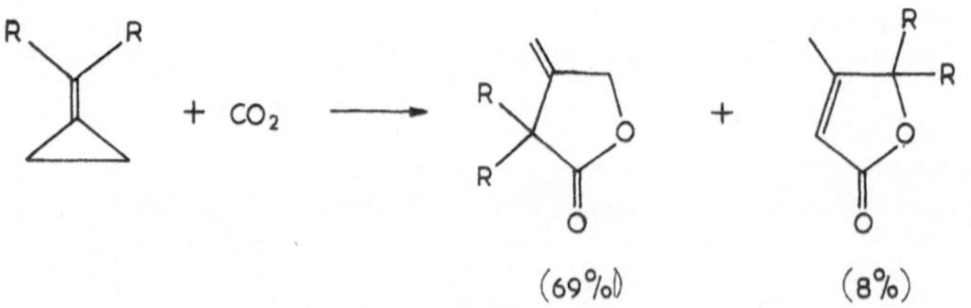

Figure 19

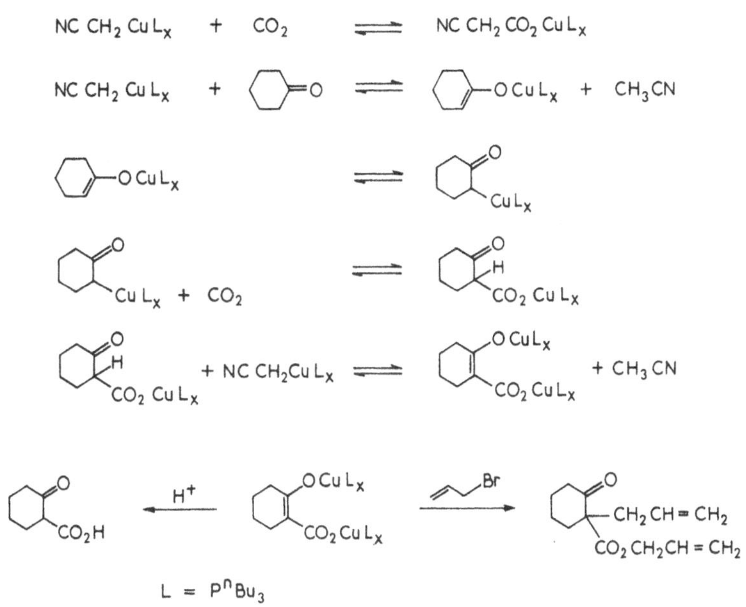

$$NC\ CH_2\ Cu\ L_x\ +\ CO_2 \rightleftharpoons NC\ CH_2\ CO_2\ Cu\ L_x$$

$$NC\ CH_2\ Cu\ L_x\ +\ [C_6H_{10}=O] \rightleftharpoons [C_6H_9]-O\ Cu\ L_x\ +\ CH_3CN$$

Figure 20 (reaction scheme)

L = P^{n}Bu$_3$

Figure 20

cyclohexanone, carbon dioxide and copper(I) has been reported, although it is described as _trans_-carboxylation. The reaction involving a cyanoacetate-copper(I) complex is quoted here.[135] The same process occurs when the cyanoacetato ligand is replaced by hydroxide.[74] In dimethylformamide the scheme of Figure 20 is suggested. This scheme commences with a complex obtained by the replacement of the active hydrogen of acetonitrile by a metal ion, which then reversibly inserts CO_2. This sequence has also been reported[136] for Fe(II) in a series of reactions in which a 2-naphthyl ligand is firstly replaced during reaction in neat acetonitrile to give $[HFe(CH_2CN)(dmpe)_2]$. Subsequent treatment with CO_2 under ambient conditions yields a cyanoacetato-complex from which the free acid can be liberated by heat.

English and Herskovitz[137] have studied the carboxylation reactions of acetonitrile, acrylonitrile, phenylacetylene and other activated hydrocarbons, brought about by rhodium and iridium complexes of type $[M(QR_3)_4]^+$, where Q = P or As. If all their systems follow a common mechanism, then it seems that the processes do not involve CO_2 insertion into a M-C bond. Thus while the reactions

$$[Ir(PMe_3)_4]Cl + CH_3CH \longrightarrow [IrH(CH_2CN)(PMe_3)_4]Cl$$

and

$$[Ir(PMe_3)_4]Cl + CH_3CN + CO_2 \longrightarrow [IrH(O_2CCH_2CN)(PMe_3)_4]Cl$$

occur readily, the first listed product is not an intermediate in the formation of the second. Similarly, although $[Rh(dmpe)_2]Cl$ does not react with CH_3CN or CH_3NO_2, nitromethane is reversibly carboxylated in the presence of this complex. Nmr studies on carboxylation systems containing $[Ir(depe)_2]^+$ show that all four P atoms equivalent and coupled with a single hydridic proton. The authors prefer an explanation in terms of a _trans-_ complex, and write for their reaction scheme

$$ML_4^+ + A\text{-}H \rightleftharpoons MH(A)L_4^+ ------ K_1$$
$$ML_4^+ + A\text{-}H + CO_2 \rightleftharpoons MH(O_2CA)L_4^+ ------ K_2$$

these equilibria being competitive rather than sequential.

In contrast, the reported carboxylation of trimethylphosphine[30] probably involves insertion into M-C. In solution the following equilibrium occurs:

$$Fe(0)(PMe_3)_4 \rightleftharpoons Fe(II)H(CH_2PMe_2)(PMe_3)_3$$

The position of equilibrium is both solvent- and temperature-dependent. In non-polar solvents, such as pentane, at high temperatures carbon dioxide is taken up to yield $Fe(CO_2)(PMe_3)_4$ and $(PMe_3)_3(CO)Fe(CO_3)$. However, the use of polar solvents such as tetrahydrofuran and lower temperatures gives mainly $[FeH(O_2CCH_2PMe_2)(PMe_3)_3]$.

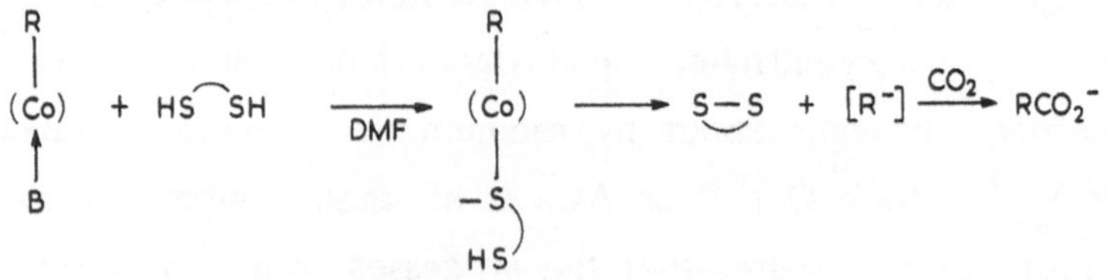

Figure 21.

Finally, low yields for CO_2 fixation[138-9] and for isotopically labeled aminoacid production[140] have been obtained from mixtures containing CO_2 and alkyl cobaloximes in reactions which might mimic microbial carbon dioxide fixation. The scheme proposed, involving dithioerythritol, is shown in Figure 21.

REFERENCES

1. R.M. Devlin and A.V. Barker, "Photosynthesis", Van Nostrand Reinhold, New York (1971).

2. R.H. Prince and P.R. Wooley, Angew. Chem. Int. Ed. Engl. 11:408 (1972).

3. G.L. Eichhorn, "Inorganic Biochemistry", Elsevier, Amsterdam (1973).

4. I. Bertini, E. Borghi and C. Luchinat, J. Am. Chem. Soc. 101:7069 (1979).

5. M.E. Vol'pin and I.S. Kolomnikov, XIV Int. Conf. Coord. Chem., Toronto (1972).

6. M.E. Vol'pin and I.S. Kolomnikov, Organomet. Reactions 5:313 (1975).

7. I.S. Kolomnikov and M. Kh. Grigoryan, Russ. Chem. Rev. 47:334 (1978).

8. R. Eisenberg and D.E. Hendriksen, Adv. Catal. 28:119 (1979).

9. G. Just, Z. Phys. Chem. 37:342 (1901).

10. M.G. Mason and J.A. Ibers, J. Am. Chem. Soc. 104:5153 (1982).

11. J.R. Morton, Chem. Rev. 64:453 (1964).

12. K.O. Hartman and I.C. Isatune, J. Chem. Phys. 44:1913 (1966).

13. L. Bertsch and H.W. Habgood, J. Phys. Chem. 67:1621 (1963).

14. J.W. Ward and H.W. Habgood, J. Phys. Chem. 70:1178 (1966).

15. S.J. Gregg and J.D. Ramsay, J. Chem. Soc. (A) 2784 (1970).

16. J. Vlčková and J. Bartoň, J.C.S. Chem. Commun. 306 (1973).

17. M.N. Bochkarev, E.A. Fedorova, Yu. F. Radkov, S. Ya. Khorshev, G.S. Kalinina and G.A. Razuvaev, J. Organomet. Chem. 258:C29 (1983).

18. S. Ashuri and J.S. Miller, in press.

19. L.H. Little, "Infrared Spectra of Adsorbed Species", Academic Press, London (1966).

20. M. Aresta, C.F. Nobile, V.G. Albano, E. Forni and M. Manassero, J.C.S. Chem. Commun. 636 (1975).

21. G.S. Bristow, P.B. Hitchcock and M.F. Lappert, J.C.S. Chem. Commun. 1145 (1981).

22. J.C. Calabrese, T. Herskovitz and J.B. Binney, J. Am. Chem. Soc. 105:5914 (1983).

23. G. Fachinetti, C. Floriani and P.F. Zanazzi, J. Am. Chem. Soc. 100:7405 (1978).

24. S. Gambarotti, F. Arena, C. Floriani and P.F. Zanazzi, J. Am. Chem. Soc. 104:5082 (1982).

25. G.R. Eady, J.J. Guy, B.F.G. Johnson, J. Lewis, M.C. Malatesta and G.M. Sheldrick, J.C.S. Chem. Commun. 602 (1976).

26. G.R. John, B.F.G. Johnson, J. Lewis and K.C. Wong, J. Organomet. Chem. 169:C23 (1979).

27. W. Beck, K. Raab, U. Nagel and M. Steimann, Angew. Chem. Int. Ed. Engl. 21:526 (1982).

28. M. Aresta and C.F. Nobile, J.C.S. Dalton Trans. 708 (1977).

29. M. Aresta and C.F. Nobile, Inorg. Chim. Acta 24:L49 (1977).

30. H.H. Karsch, Chem. Ber. 110:2213 (1977).

31. J. Chatt, M. Kubota, G.J. Leigh, F.C. March, R. Mason and D.J. Yarrow, J.C.S. Chem. Commun. 1033 (1974).

32. C. Floriani and G. Fachinetti, J.C.S. Chem. Commun. 615 (1974).

33. T. Herskovitz, J. Am. Chem. Soc. 99:2391 (1977).

34. T. Herskovitz and L.J. Guggenberger, J. Am. Chem. Soc. 98:1615 (1976).

35. T. Ikariya and A. Yamamoto, J. Organomet. Chem. 72:145 (1974).

36. A. Miyashita and A. Yamamoto, J. Organomet. Chem. 113:187 (1976).

37. A. Misono, Y. Uchida, M. Hidai and T. Kuse, Chem. Commun. 981 (1968).

38. B.R. Flynn and L. Vaska, J.C.S. Chem. Commun. 703 (1974).

39. T. Yoshida, D.L. Thorn, T. Okano, J.A. Ibers and S. Otsuka, J. Am. Chem. Soc. 101:4212 (1979).

40. P.W. Jolly, K. Jonas, C. Kruger and Y.H. Tsay, J. Organomet. Chem. 33:109 (1971).

41. M.E. Vol'pin, I.S. Kolomnikov and T.S. Lobeeva, Izv. Akad. Nauk SSR, Ser. Khim. 2084 (1969).

42. Y. Iwashita and A. Hayata, J. Am. Chem. Soc. 91:2525 (1969).

43. S. Komiya and A. Yamamoto, J. Organomet. Chem. 46:C58 (1972).

44. I.S. Kolomnikov, T.S. Belopotapova, T.V. Lysyak and M.E. Vol'pin, J. Organomet. Chem. 67:C25 (1974).

45. C.D.M. Beverwijk and G.J.M. Van der Kerk, J. Organomet. Chem. 49:C59 (1973).

46. S. Sakaki, K. Kitaura and K. Morokuma, Inorg. Chem. 21:760 (1982).

47. V.D. Bianco, S. Doronzo and N. Gallo, Inorg. Nucl. Chem. Lett. 15:187 (1979).

48. V.D. Bianco, S. Doronzo and N. Gallo, Inorg. Nucl. Chem. Lett. 16:97 (1980).

49. S. Komiya and A. Yamamoto, Bull. Chem. Soc. Jpn 49:784 (1976).

50. S. Krogsrud, S. Komiya, T. Ito, J.A. Ibers and A. Yamamoto, Inorg. Chem. 15:2798 (1976).

51. G.O. Evans and C.J. Newell, Inorg. Chim. Acta 31:L387 (1978).

52. J.M. Maher, G.R. Lee and N.J. Cooper, J. Am. Chem. Soc. 104:6797 (1982).

53. G. Fachinetti, C. Floriani, A. Chiesi-Villa and C. Guastini, J. Am. Chem. Soc. 101:1767 (1979).

54. P. Sabota, B. Jezowska-Trzebiatowska and Z. Janas, J. Organomet. Chem. 118:253 (1976).

55. B. Demerseman, G. Bouguet and M. Bigorgne, J. Organomet. Chem. 145:41 (1978).

56. P. Svoboda, T.S. Belopotapova and J. Hetflejs, J. Organomet. Chem. 65:C37 (1974).

57. E.C. Baker, D.E. Hendriksen and R. Eisenberg, J. Am. Chem. Soc. 102:1020 (1980).

58. D.E. Hendriksen and R. Eisenberg, J. Am. Chem. Soc. 98:4662 (1976).

59. C.D. Meyer and R. Eisenberg, J. Am. Chem. Soc. 98:1364 (1976).

60. T. Bodnar, E. Coman, K. Menard and A. Cutler, Inorg. Chem. 21:1275 (1982).

61. J. Haweckar, J-M. Lehn and R. Ziessel, J.C.S. Chem. Commun. 536 (1983).

62. S.F. Hossain, K.M. Nicholas, C.L. Teas and R.E. Davis, J.C.S. Chem. Commun. 268 (1981).

63. C.J. Nyman, C.E. Wymore and G. Wilkinson, Chem. Commun. 407 (1967).

64. C.J. Nyman, C.E. Wymore and G. Wilkinson, J. Chem. Soc. (A) 561 (1968).

65. P.J. Hayward, D.M. Blake, G. Wilkinson and C.J. Nyman, J. Am. Chem. Soc. 92:5873 (1970).

66. E. Chaffee, T.P. Dasgupta and G.M. Harris, J. Am. Chem. Soc. 95:4169 (1973).

67. T.P. Dasgupta and G.M. Harris, J. Am. Chem. Soc. 97:1733 (1975).

68. M.R. Churchill, G. Davies, M.A. El-Sayed, M.F. El-Shazly, J.P. Hutchinson and M.W. Rupich, Inorg. Chem. 19:201 (1980).

69. T.P. Dasgupta and G.M. Harris, J. Am. Chem. Soc. 99:2490 (1977).

70. D.A. Palmer, R. van Eldik, H. Kelm and G.M. Harris, Inorg. Chem. 19:1009 (1980).

71. D.A. Palmer and G.M. Harris, Inorg. Chem. 13:965 (1974).

72. B.R. Flynn and L. Vaska, J. Am. Chem. Soc. 95:5081 (1973).

73. R.A. Jewsbury, Inorg. Chim. Acta 49:141 (1981).

74. T. Tsuda, Y. Chujo and T. Saegusa, J. Am. Chem. Soc. 102:431 (1980).

75. T. Yamamoto, M. Kubota and A. Yamamoto, Bull. Chem. Soc. Jpn 53:680 (1980).

76. R.J. Crutchley, J. Powell, R. Faggiani and C.J.L. Lock, Inorg. Chim. Acta 24:L15 (1977).

77. T. Ito, H. Tsuchiya and A. Yamamoto, Chem. Lett. 851 (1976).

78. V.I. Kirov, Trudy Leningr. Tekstil. Inst. 6:99 (1955).

79. T. Tsuda and T. Saegusa, Inorg. Chem. 11:2561 (1972).

80. M.H. Chisholm and M. Extine, J. Am. Chem. Soc. 97:5625 (1975).

81. M.H. Chisholm, W.W. Reichert, F.A. Cotton and C.A. Murillo, J. Am. Chem. Soc. 99:1652 (1977).

82. M.H. Chisholm, F.A. Cotton, M.W. Extine and W.W. Reichert, J. Am. Chem. Soc. 100:1727 (1978).

83. A.J. Bloodworth, A.G. Davies and S.C. Vasishtha, J. Chem. Soc. (C) 1309 (1967).

84. A.G. Davies and P.G. Harrison, J. Chem. Soc. (C) 1313 (1967).

85. A. Immirzi and A. Musco, Inorg. Chim. Acta 22:L35 (1977).

86. T.V. Ashworth and E. Singleton, J.C.S. Chem. Commun. 204 (1976).

87. R.J. De Pasquale, J.C.S. Chem. Commun. 157 (1973).

88. A. Koinuma, H. Kato and H. Hirai, Chem. Lett. 517 (1977).

89. M. Ratzenhofer and H. Kirsch, Angew. Chem. Int. Ed. Engl. 19:317 (1980).

90. I.S. Kolomnikov, T.S. Lobeeva and M.E. Vol'pin, Izv. Akad. Nauk SSR, Ser. Khim. 2329 (1972).

91. Y. Inoue, Y. Sasaki and H. Hashimoto, J.C.S. Chem. Commun. 718 (1975).

92. D.J. Darensbourg, C. Ovalles and M. Pala, J. Am. Chem. Soc. 105:5937 (1983).

93. M. Tezuka, T. Yajima, Y. Matsumoto, Y. Uchida and M. Hidai, J. Am. Chem. Soc. 104:6834 (1982).

94. I.S. Kolomnikov, A.I. Gusev, G.G. Aleksandrov, T.S. Lobeeva, Yu.T. Struchkov and M.E. Vol'pin, J. Organomet.Chem. 59:349 (1973).

95. R.S. Paonessa and W.C. Trogler, J. Am. Chem. Soc. 104:3529 (1982).

96. L.S. Pu, A. Yamamoto and S. Ikeda, J. Am. Chem. Soc. 90:3896 (1968).

97. V.D. Bianco, S. Doronzo and M. Rossi, J. Organomet. Chem. 35:337 (1972).

98. D.J. Darensbourg, A. Rokicka and M.Y. Darensbourg, J. Am. Chem. Soc. 103:3223 (1981).

99. P. Haynes, L.H. Slaugh and J.F. Kohnle, Tet. Lett. 365 (1970).

100. T.V. Ashworth, M. Nolte and E. Singleton, J. Organomet. Chem. 121:C57 (1976).

101. M.H. Chisholm and M. Extine, J. Am. Chem. Soc. 96:6214 (1974).

102. M.H. Chisholm and M. Extine, J. Am. Chem. Soc. 99:782 (1977).

103. T. Tsuda, H. Washita, K. Watanabe, M. Miwa and T. Saegusa, J.C.S. Chem. Commun. 815 (1978).

104. M.H. Chisholm and M. Extine, J. Am. Chem. Soc. 97:1623 (1975).

105. F. Ozawa, T. Ito and A. Yamamoto, Chem. Lett. 735 (1979).

106. M.H. Chisholm, M. Extine, F.A. Cotton and B.R. Stults, J. Am. Chem. Soc. 98:4683 (1976).

107. F. Calderazzo, G. dell'Amico, R. Netti and M. Pasquali, Inorg. Chem. 17:471 (1978).

108. H. Breederveld, Rec. Trav. Chim. Pays-Bas 81:276 (1962).

109. K. Jones and M.F. Lappert, Proc. Chem. Soc. 358 (1962).

110. T.A. George, K. Jones and M.F. Lappert, J. Chem. Soc. 2157 (1965).

111. K. Jones and M.F. Lappert, J. Organomet. Chem. 3:295 (1965).

112. R.H. Cragg and M.F. Lappert, J. Chem. Soc. (A) 82 (1966).

113. J.L. Frahn and J.A. Mills, Aust. J. Chem. 17:256 (1964).

114. C. Faurholt, J. Chim. Phys. 22:1 (1925).

115. M.B. Jensen and C. Faurholt, Acta Chem. Scand. 14:2240 (1960).

116. M.B. Jensen and C. Faurholt, Acta Chem. Scand. 18:377 (1964).

117. P.L. Bellon, S. Cenini, F. Demartin, M. Pizzotti and F. Porta, J.C.S. Chem. Commun. 265 (1982).

118. H. Gilman and N.B. St. Lohn, Rec. Trav. Chim. Pays-Bas 49:1172 (1930).

119. K. Ziegler, F. Krupp, K. Weijer and W. Larbig, Liebigs Ann. Chem. 629:251 (1960).

120. A. Miyashita and A. Yamamoto, J. Organomet. Chem. 49:C57 (1973).

121. D.J. Darensbourg and A. Rokicka, J. Am. Chem. Soc. 104:349 (1982).

122. I.S. Kolomnikov, A.O. Gusev, T.S. Belopotapova, M. Kh. Grigoryan, T.V. Lyryak, Yu. T. Struchkov and M.E. Vol'pin, J. Organomet. Chem. 69:C10 (1974).

123. P. Albano, M. Aresta and M. Manassero, Inorg. Chem. 19:1069 (1980).

124. I.S. Kolomnikov, T.S. Lobeeva, V.V. Gorbachevskaya, G.G. Aleksandrov, Yu. T. Struchkov and M.E. Vol'pin, Chem. Commun. 972 (1971).

125. T. Tsuda, K. Ueda and T. Saegusa, J.C.S. Chem. Commun. 380 (1974).

126. T. Tsuda, Y. Chujo and T. Saegusa, J.C.S. Chem. Commun. 963 (1975).

127. H. Gilman and R. Young, J. Org. Chem. 1:315 (1936).

128. P.W. Jolly, S. Stobbe, G. Wilke, R. Goddard, C. Krüger, J.C. Sekutowski and Y.H. Tsay, Angew. Chem. Int. Ed. Engl. 17:124 (1978).

129. R. Santi and M. Marchi, J. Organomet. Chem. 182:117 (1979).

130. Y. Inoue, Y. Itoh and H. Hashimoto, Chem. Lett. 855 (1977).

131. A. Musco, C. Perego and V. Tartiari, Inorg. Chim. Acta. 28:L147 (1978).

132. Y. Inoue, Y. Sasaki and H. Hashimoto, Bull. Chem. Soc. Jpn 51:2375 (1978).

133. Y. Inoue, T. Hibi, M. Satake and H. Hashimoto, J.C.S. Chem. Commun. 982 (1979).

134. I. Hirao and T. Kito, Bull. Chem. Soc. Jpn 46:3470 (1973).

135. T. Tsuda, Y. Chujo and T. Saegusa, J. Am. Chem. Soc. 100:630 (1978).

136. S.D. Ittel, C.A. Tolman, A.D. English and J.P. Jesson, J. Am. Chem. Soc. 100:7577 (1978).

137. A.D. English and T. Herskovitz, J. Am. Chem. Soc. 99:1648 (1977).

138. G.N. Schrauzer and J.W. Sibert, J. Am. Chem. Soc. 92:3509 (1970).

139. G.N. Schrauzer, J.S. Seck, R.J. Holland, T.M. Beckman, E.M. Rubin and J.W. Sibert, Bioinorg. Chem. 2:93 (1972).

140. G.L. Blackmer and C.W. Tsai, J. Organomet. Chem. 155:C17 (1978).

REACTIONS OF COORDINATED DINITROGEN AND RELATED SPECIES

M. Hidai and Y. Mizobe

Department of Synthetic Chemistry

University of Tokyo

1. INTRODUCTION

Since the discovery of the novel dinitrogen complex in $[Ru(NH_3)_5(N_2)]X_2$ (X = Br, I, BF_4 or PF_6) by Allen and Senoff in 1965,[1] extensive studies have been carried out on the preparation of dinitrogen complexes and their reactivities and nitrogen coordinating ability is now known in almost all transition metals. Detailed knowledge of the bonding of dinitrogen to transition metals and the reactivity of coordinated dinitrogen may contribute not only to our understanding of the function of the active site of the natural enzyme, nitrogenase, but also to the design of effective catalysts for the conversion of molecular nitrogen into such compounds as nitrogen hydrides and organo-nitrogen compounds under mild conditions. In this review we will describe the recent advances in the studies of dinitrogen and related complexes with major emphasis on the reactions of coordinated dinitrogen in well characterized dinitrogen complexes. The reader is also referred to a number of reviews for chemical and biological studies on nitrogen fixation.[2-7]

2. METHODS FOR PREPARING DINITROGEN COMPLEXES

The first example of a dinitrogen complex, $[Ru(NH_3)_5(N_2)]X_2$, (X = Br, I, BF_4 or PF_6) was isolated from the reaction of $RuCl_3$ with hydrazine in aqueous solution in the presence of the appropriate anions, where the dinitrogen has its origin in hydrazine.[1] In 1967 three groups independently succeeded in preparing the complex $[CoH(N_2)(PPh_3)_3]$ directly from gaseous dinitrogen.[8-10] Since then over a hundred different dinitrogen complexes have been prepared and for those which show interesting reactivities improved methods of preparation have been developed. These methods may be classified roughly in three types, of which the most common (type A below) is to treat high valence complexes with reducing agents in the presence of ligands such as phosphines under a nitrogen atmosphere. The second type of procedure is to exchange easily dissociating ligands for dinitrogen. Direct binding of dinitrogen to coordinatively unsaturated complexes also gives dinitrogen complexes, and we include both in type B. The third method, where the dinitrogen ligand is originally derived from other nitrogen compounds such as nitric oxide, hydrazine and azide is summarized as type C. Typical examples are given in eqs. 1-11.

Type A

$$[Mo(acac)_3] \xrightarrow[N_2]{dppe, AlEt_3} trans\text{-}[Mo(N_2)_2(dppe)_2]^{11\text{-}12} \tag{1}$$

$$[WCl_4(PMe_2Ph)_3] \xrightarrow[N_2]{PMe_2Ph, Mg} cis\text{-}[W(N_2)_2(PMe_2Ph)_4]^{13} \tag{2}$$

$$[Co(acac)_3] \xrightarrow[N_2]{PPh_3, Al(^iBu)_3} [CoH(N_2)(PPh_3)_3]^{8\text{-}9} \tag{3}$$

$$[Cp^*_2ZrCl_2] \xrightarrow[N_2]{Na\text{-}Hg} [\{Cp^*_2Zr(N_2)\}_2(N_2)]^{14} \tag{4}$$

Type B

$$[CoH_3(PPh_3)_3] \xrightarrow{N_2} [CoH(N_2)(PPh_3)_3]^{10} \tag{5}$$

$$[(Cp*_2Ti)_2] \xrightarrow{N_2} [(Cp*_2Ti)_2(N_2)]^{15} \tag{6}$$

$$[Mo(CO)(dppe)_2] \xrightarrow{N_2} \underline{trans}\text{-}Mo(CO)(N_2)(dppe)_2]^{16\text{-}17} \tag{7}$$

Type C

$$[Ru(NH_3)_6]^{3+} \xrightarrow[\text{Ar}]{\text{NO,NaOH}} \xrightarrow[\text{Ar}]{\text{HCl}} [Ru(NH_3)_5(N_2)]Cl_2{}^{18} \tag{8}$$

$$[CpMn(CO)_2(N_2H_4)] \xrightarrow{H_2O_2,\ Cu^{2+}} [CpMn(CO)_2(N_2)]^{19} \tag{9}$$

$$[IrCl(CO)(PPh_3)_2] \xrightarrow{ArCON_3} [IrCl(N_2)(PPh_3)_2]^{20} \tag{10}$$

$$[CpMn(CO)_2(N_2R)][BF_4] \xrightarrow{I^-} [CpMn(CO)_2(N_2)]^{21} \tag{11}$$

3.   BONDING MODES AND CHARGE DISTRIBUTION OF LIGATING DINITROGEN

A number of physico-chemical investigations and theoretical studies on dinitrogen complexes of transition metals have made a great contribution to clarifying the bonding of coordinated dinitrogen to metals. Two types of bonding mode are known for dinitrogen, end-on and side-on, which are further divided into terminal and bridging. Typical examples of dinitrogen complexes are given in Table 1 and Fig. 1. Dinitrogen usually binds terminally end-on to transition metals in the same way as carbon monoxide, nitric oxide and isocyanides. Theoretical calculations for some dinitrogen complexes of transition metals indicated that end-on bonding is more advantageous than side-on.[22-4] The end-on bonding of dinitrogen consists of σ-donation of the lone pair of dinitrogen to the unoccupied orbitals of the metal and π-back-donation from the occupied d orbitals of the metal to the unoccupied π* orbitals of dinitrogen. Side-on bonding, on the other hand, is

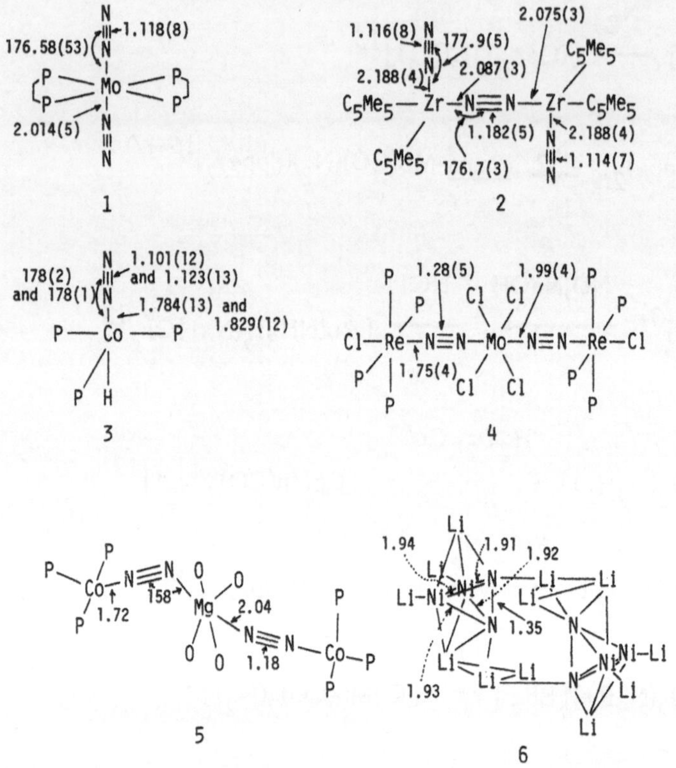

Figure 1.   Structures of some representative dinitrogen complexes.

1: trans-[Mo(N$_2$)$_2$(dppe)$_2$][131-2]

2: [{Cp*$_2$Zr(N$_2$)}$_2$(N$_2$)][133]

3: [CoH(N$_2$)(PPh$_3$)$_3$][134]

4: [MoCl$_4${(N$_2$)ReCl(PMe$_2$Ph)$_4$}$_2$][46]

5: [{Co(N$_2$)(PMe$_3$)$_3$}$_2$Mg(THF)$_4$][92]

6: [(C$_6$H$_5$Li)$_6$Ni$_2$(N$_2$)(Et$_2$O)$_2$]$_2$[32]

composed of electron donation from the bonding σ orbital (and π orbital) of the dinitrogen ligand to the unoccupied orbitals of the metal and π-back-donation from the occupied orbitals of the metal to the antibonding π* orbital of the dinitrogen ligand. While this latter type of bonding is very familiar with acetylenes and olefins, only a few examples of side-on bonded dinitrogen have been reported so far.

Table 1. NN Stretching Frequencies and Bond Distances of some Dinitrogen Complexes

Complex	ν_{NN} (cm^{-1})a	NN distance (Å)
end-on terminal		
trans-$[Mo(N_2)_2(dppe)_2]$[11-12]	1970	1.118(8)[131-2]
$[\{ZrCp*_2(N_2)\}_2(N_2)]$[14]	2040^N, 2003^N	1.114(7), 1.116(8)[133]
$[CoH(N_2)(PPh_3)_3]$[8-10]	2088(toluene)	1.101(12), 1.123(13)[134]
$[RhCl(N_2)(P^iPr_3)_2]$[135]	2100	0.958(5)
$[RhH(N_2)(PPh^tBu_2)_2]$[136]	2155	1.074
end-on bridging		
$[(NH_3)_5RuN_2Ru(NH_3)_5][(BF_4)_4]$	2100^R [137]	1.124(15)[138]
$[\{Co(PMe_3)_3(N_2)\}_2Mg(THF)_4]$[92]	2068(hexane)	1.18
$[Co(PPh_3)_3(N_2)Li(THF)_3]$[91]	1890^N	1.153(26)
$[MoCl_4\{Re(PMe_2Ph)_4Cl(N_2)\}_2]$[46]	1800^N	1.28(5)
$K[(PMe_3)_3Co(N_2)]$[139]	1795^N, 1758^N	1.16-1.18
$[\{ZrCp*_2(N_2)\}_2(N_2)]$[14]	1556^N	1.182(5)[133]
side-on-bridging		
$[(C_6H_5Li)_6Ni_2(N_2)(Et_2O)_2]_2$[32]		1.35
$[C_6H_5(NaOEt_2)_2\{(C_6H_5)_2Ni\}_2(N_2)$- $NaLi_6(OEt)_4OEt_2]_2$[140]		1.359(18)

cf. N≡N: 2331 cm^{-1} ^{R}, 1.098; RN=NR: 1.25±0.02; R_2N-NH_2 1.451±0.005 Å.[145]

a. KBr disks unless stated otherwise, N: nujol, R: Raman spectrum.

Dinitrogen is activated by coordination to metals to an extent which can be estimated from the N-N bond length observed by the xray analysis, or, more conveniently, by the N-N stretching frequency observed in the IR or Raman spectrum. Table 1 shows these data for pertinent dinitrogen complexes. The N-N bond weakening generally follows the order side-on bridging > end-on bridging > end-on terminal, and in the case of side-on bridging the N-N bond is remarkably lengthened to suggest a bond order of about 1.3.

It is generally accepted that in end-on bound dinitrogen complexes the N-N bond weakening is caused mainly by the back-donation of d electrons from the metal to the anti-bonding π^* orbitals of dinitrogen.[24-5] The σ donation from the lone pair orbital of dinitrogen to the vacant orbitals of the central metal may be largely responsible for the stability of dinitrogen complexes. Thus, for species $[IrCl(PR_3)_2(N_2)]$, the complexes with the more basic phosphines show the lower $\nu(N_2)$, but are less stable with respect to N_2 loss.[26] However, for species $[OsCl_2(PR_3)_3(N_2)]$, stability with respect to N_2 loss seems to correlate directly with the Os-N_2 π bond strength.[27-8] Extended Hückel and _ab initio_ SCF calculations have been performed for some dinitrogen complexes. For the model complexes _cis_-$[Mo-(N_2)_2(PH_3)_4]$ and _trans_-$[Mo(N_2)_2(PH_3)_4]$, Dubois and Hoffmann[29] have calculated the charge distribution as Mo(+0.15) - N(-0.11) - N(-0.88) and Mo(-0.59) -N(+0.05) - N(-0.67), respectively. This clearly indicates that the dinitrogen ligand is overall negatively charged and that the terminal nitrogen atom may be susceptible to electrophilic attack. This is totally consistent with the isolation of hydrazido(2-) complexes from the reaction of $[M(N_2)_2(L)_4]$ (M = Mo or W; L = phosphines) with protic acids, and acyl- or aroyldiazenido complexes with acyl or aroyl chlorides, as described later. Recent _ab initio_ SCF molecular orbital calculations on complexes such as _trans_-$[Mo(N_2)_2(PH_3)_4]$ and $[Ru(NH_3)_5(N_2)]^{2+}$ showed[30] that the dinitrogen ligand in the molybdenum complex has a negative charge whereas in the ruthenium complex the dinitrogen ligand is rather positively charged and the polarization is described as Ru(δ+)-N(δ-)-N(δ+). This result might explain the fact that the former complex is protonated at the dinitrogen ligand but the latter is not. The relative polarization of ligating dinitrogen seems to change from one type of complex to another.

In the side-on type complexes it is presumed that the N-N bond is considerably weakened because of electron donation from the bonding π orbital (and σ orbital) of the dinitrogen ligand to the unoccupied molecular orbital of the metal as well as electron acceptance from the occupied d orbitals of the metal by the unoccupied antibonding π^* orbital. This inference completely agrees with the fact that the side-on bound dinitrogen complexes of Zr and Ni, $[Zr(\eta^5-C_5H_4R')_2(N_2)R]$ (R = $(Me_3Si)_2CH$, R' = H or Me)[31] and $[\{(LiC_6H_5)_3(Et_2O)_{1-1.5}Ni\}_2(N_2)]$,[32] give hydrazine on treatment with halogen acids and ammonia on hydrolysis, respectively.

The relative extent of charge withdrawal, and the charge distribution on coordinated dinitrogen, may be directly measured by the xray photoelectron spectrum (XPE) of the nitrogen atom 1s electron. Almost all N(1s) spectra measured appear in the 398.3 - 400.4 eV region which corresponds to a slight negative charge.[6] The N(1s) binding energies of end-on bound dinitrogen complexes are in general resolved as a doublet. The higher energy peak is assigned to the inner nitrogen atom, while the lower energy one is due to the terminal one, since the terminal nitrogen atom shows the more basic character towards Lewis acids and is in some cases attacked by protons to give hydrazido(2-) complexes containing the $MNNH_2$ group. On the other hand, in the bridging dinitrogen complex $[MoCl_4(OMe)(N_2)ReCl(PMe_2Ph)_4]$, where the oxidation states of the two metals are quite different, the nitrogen atoms could not be distinguished in the XPE spectrum.[33] It is of great interest to note that among the measured dinitrogen complexes only the manganese complex $[Cp(Mn(CO)_2)(N_2)]$ shows relatively high N(1s) binding energies, at 403.0 and 401.8 eV.[34] This is consistent with the fact that the coordinated dinitrogen undergoes nucleophilic attack by carbanions such as Me^- or Ph^- to give complexes $[CpMn(CO)_2(RNN^-Li^+)]$.[35-7] The electron-poor character of ligating dinitrogen in this complex is also deduced from the fairly high primary oxidation potential (Es) calculated for the 16-electron species $[CpMn(CO)_2]$ to which dinitrogen binds (see below).[38]

Chemical shifts of ^{15}N nuclei in the nmr spectra are expected to give useful information about the electronic state of the coordinated dinitrogen, but ^{15}N-nmr studies of dinitrogen complexes are still at an early stage and data have so far been collected only for three kinds of complexes, trans-$[M(N_2)_2(dppe)_2]$ (M = Mo or W),[39] cis-$[M(N_2)_2(PMe_2Ph)_4]$ (M = Mo or W),[39] and $[\{Cp*_2M(N_2)\}_2(N_2)]$ (M = Ti[15,40-41] or Zr[42]). In the spectra of the molybdenum and tungsten complexes the coordinated dinitrogen exhibits two distinct bands, lower by 10 - 40 ppm than in molecular $^{14}N^{15}N$, in which the metal-bound nitrogen probably resonates upfield of the terminal nitrogen. On the other hand, the bridging dinitrogen in the titanium and zirconium complexes shows resonances downfield of the terminal dinitrogen. The chemical shifts of both the dinitrogen ligands lie at remarkably lower field compared with the molybdenum and tungsten complexes. Interpretation of these results is a task for the future.

A number of attempts to compare the relative σ-donor and π-acceptor properties of dinitrogen with those of such analogous ligands as carbon monoxide, organic cyanides, and isocyanides, have been made by using XPE, infrared, and Mössbauer spectroscopy. On the basis of these investigations it is generally believed that coordinated dinitrogen is overall negatively charged to an extent about equal to carbon monoxide, although dinitrogen is weaker than carbon monoxide in both its σ-donor and π-acceptor functions. Dinitrogen is the weakest σ-donor among these ligands whereas the π-acceptor capacity lies between carbon monoxide and organic cyanides and isocyanides. Leigh, Pickett and coworkers[38] have quantitatively estimated the electron-withdrawing or -donating properties of a wide variety of ligands by calculating a scale of ligand constants P_L from the reversible one-electron oxidation potentials of octahedral complexes $[Cr(CO)_5L]$ according to the equation

$$P_L = E^{ox}_{1/2}[Cr(CO)_5L] - E^{ox}_{1/2}[Cr(CO)_6]$$

Thus weak σ-donors with strong π-acceptor properties such as NO^+, CO, and N_2 have P_L constants of +1.40, 0.00, and -0.07 V, respectively, whereas strong σ-donors with weak π-acceptor capacities such as PhNC and PhCN have P_L constants of -0.38 and 0.40 V, respectively. The $E_{1/2}^{ox}$ of various closed-shell octahedral complexes [MsL] is expressed as

$$E_{1/2}^{ox}[MsL] = Es + \beta P_L$$

where Es is a measure of the electron richness of the site Ms and β a measure of its polarizability. They concluded that dinitrogen will give thermally stable complexes at sites with Es values ranging from +1.3 to -1.3 V, and that $Ms-N_2$ binding interactions may be the strongest at sites with low Es and high β values.

Coordinated dinitrogen, carbon monoxide, and organic isocyanides show essentially similar types of chemical reactivity, in the sense that the metal bound atom of the ligands is susceptible to nucleophilic attack whereas the atom β to the metal is attacked by electrophiles. Thus as described later,

$$
\begin{array}{ccc}
M \cdots N \cdots N & M \cdots C \cdots O & M \cdots C \cdots N - R \\
\uparrow \quad \uparrow & \uparrow \quad \uparrow & \uparrow \quad \uparrow \\
Nu \quad E & Nu \quad E & Nu \quad E
\end{array}
$$

the terminal nitrogen of coordinated dinitrogen is attacked by electrophiles such as H^+ and Lewis acids while the nitrogen adjacent to the metal undergoes nucleophilic attack by carbanions. Similar reactions are well known in the chemistry of transition metal carbonyl and isocyanide complexes. However, both the relative extent of charge withdrawal by coordinated dinitrogen, and the charge distribution within the ligand, seem to change from one type of complex to another, and further investigations are needed for quantitative understanding.

4.   REACTIONS OF LIGATING DINITROGEN

The previous Section has described the bonding modes and charge distribution of ligating dinitrogen, indicating that ligating dinitrogen in some complexes undergoes electrophilic attack and is, in some cases, susceptible to nucleophilic attack. In this Section the reactions of ligating dinitrogen are described in detail, including substitution reactions with other ligands.

4.1   Replacement of Ligating Dinitrogen by Other Ligands

Since the bonding of dinitrogen to metals is relatively weak,

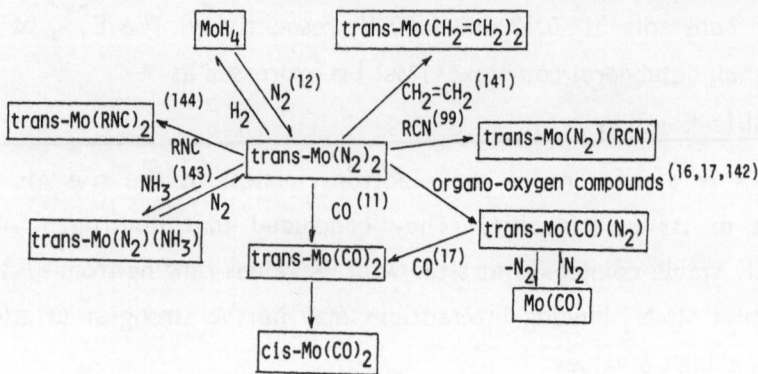

Figure 2.   Replacement of ligating dinitrogen in <u>trans</u>-[Mo(N₂)₂(dppe)₂]
             (dppe ligands omitted for clarity).

coordinated dinitrogen is easily replaced by other ligands such as carbon
monoxide, organocyanides, isocyanides, and ammonia. Typical examples of
substitution reactions using <u>trans</u>-[Mo(N₂)₂(dppe)₂] are summarized in
Figure 2, where several reactions are reversible.

4.2   Reactions of Ligating Dinitrogen with Lewis Acids

As described in the previous Section, the terminal nitrogen in end-on
bound dinitrogen has in general a weakly basic character. This is
demonstrated by formation of adducts with Lewis acids of both non-
transition metals and transition metals. The rhenium complex [ReCl(N₂)
(PMe₂Ph)₄] reacts with a variety of Lewis acids to give dinitrogen-bridged
dinuclear and trinuclear complexes. Thus, the dinuclear complex
[MoCl₄(OMe)(N₂)ReCl(PMe₂Ph)₄] is isolated from the reaction of the above
complex with [MoCl₄(THF)₂] in dichloromethane in the presence of
methanol.[43-4] The xray structure shows a considerable legthening of the N-
N bond (to 1.21 Å) and a shortening of the Re-N and Mo-N bonds in the linear
Re-N-N-Mo grouping. This is explained by donation of π-bonding electrons
from dinitrogen to the molybdenum in addition to its conventional σ-donor/
π-acceptor ability.[45]   The trinuclear complex [MoCl₄{(N₂)ReCl
(PMe₂Ph)₄}₂] is also prepared by a similar reaction;[46] its xray structure is
given in Fig. 1.

The relative σ-donor strengths of the terminal nitrogen in some end-on
dinitrogen complexes towards trimethylaluminum have been estimated by

measuring the equilibrium constant, K, of a series of equilibria (eq. 12).[47] The K values qualitatively increase with decrease of $\nu(N_2)$, indicating that the relative extent of back donation of electrons from the metal into the $\pi*$ orbital of dinitrogen is reflected in the basicity of the terminal nitrogen.

$$L_nMN_2 + Me_3Al.OEt_2 \overset{K}{\rightleftharpoons} [L_nMN_2AlMe_3] + Et_2O \qquad (12)$$

4.3 Formation of Ammonia and Hydrazine from Ligating Dinitrogen

In 1964 Vol'pin and Shur found that the combination of some transition metal chlorides and organometallic reducing agents, especially the $[Cp_2TiCl_2]$/RMgBr system, can fix molecular nitrogen under mild conditions to give ammonia after hydrolysis of the product.[48] Since then the protonation of ligating dinitrogen in well-defined complexes has been earnestly investigated. However, only a limited number of dinitrogen complexes form N-H bonds to give nitrogen hydrides, whereas in most dinitrogen complexes the metals are oxidized by protons with evolution of dinitrogen. In this Section formation of ammonia and hydrazine from ligating dinitrogen is described, together with mechanisms supported by the isolation of several kinds of intermediate.

4.3.1 Titanium and zirconium

Many attempts have been made to isolate intermediates from the $[Cp_2TiCl_2]$/RMgBr system, which has a remarkable ability to fix dinitrogen. However, a unifying mechanism has still not been presented, since the isolated intermediates are very labile and their characterization is difficult.

Shilov and his coworkers[49] reported the formation of an unstable binuclear complex, tentatively formulated as $[(Cp_2Ti^iPr)_2N_2]$, from the reaction of $[Cp_2TiCl]$ and ^iPrMgX under dinitrogen at low temperatures (-100 to -80°C). This complex, after treatment with an excess of ^iPrMgCl at -60°C and hydrolysis, produces hydrazine in almost quantitative yield. If the reduction is carried out at temperatures > 0°C, ammonia is formed. In a subsequent study, another intermediate, $[(Cp_2Ti)_2N_2MgX]$, was also reportedly isolated from the system.[50] The same group also isolated the paramagnetic complex $[(Cp_2Ti)_2N_2]$ from the system $[Cp_2TiCl]/MeMgI$ under dinitrogen.[51] From the remarkably low $\nu(N_2)$ at 1280 cm^{-1}, they postulated the diazenido structure Ti-N=N-Ti. This complex gives hydrazine on treatment with methanolic HCl and ammonia with etherial HCl. In related work, Brintzinger and his coworkers[52] reported that $[Cp_2Ti]$ and its permethylcyclopentadienyl analogue reversibly absorb molecular nitrogen to give $[(Cp_2Ti)_2N_2]$ and $[(Cp*_2Ti)_2N_2]$, respectively. Xray structural analysis was later performed on the latter complex and the linear bonding of the Ti-N-N-Ti group established.[15] From this complex another titanium dinitrogen complex $[(Cp*_2TiN_2)_2N_2]$ is formed at lower temperatures. These complexes react with HCl in toluene at -80°C to afford hydrazine and a little ammonia.[53]

The complexes $[(Cp_2TiR)_2N_2]$ can be isolated as stable solids when R is C_6H_5, o-, m-, p-$CH_3C_6H_4$, C_6F_5, or $C_6H_5CH_2$. They give ammonia on thermolysis followed by treatment of the residue with etherial HCl (0.21 mol/Ti atom), or on reaction with sodium naphthalene, ^nBuLi or ^iPrMgCl, and working up with HCl in ether (0.2 - 0.1 mol/Ti atom).[54] It should be noted that the Cp_2Ti structure is almost completely destroyed if excess sodium naphthalene or ^nBuLi is used.

In contrast to the above titanium complexes, the zirconium dinitrogen complex $[(Cp*_2ZrN_2)_2N_2]$ is fairly stable and the structure has been well established by xrays.[55] The terminal dinitrogen in this complex can be readily displaced by $^{15}N_2$ atmosphere. The complex reacts with HCl in toluene to give hydrazine (0.86 mol/dimer) with a little ammonia. On the basis of the ^{15}N distribution in the hydrazine produced from the reaction of $[\{Cp*_2Zr(^{15}N_2)\}_2(^{14}N_2)]$ with HCl the mechanism of (eq. 13) has been proposed.[53,56] Yields of hydrazine and ammonia depend considerably on the solvents and acids used.

$$
\begin{array}{c}
\overset{15\text{N}_2}{} \\
\underset{\overset{|}{{}^{15}\text{N}_2}}{\text{Zr}-\text{N}\equiv\text{N}}-\underset{\overset{|}{{}^{15}\text{N}_2}}{\text{Zr}} \longrightarrow \underset{\overset{|}{{}^{15}\text{N}_2\text{H}}}{\text{Zr}-\text{N}\equiv\text{N}}-\underset{\overset{|}{\text{Cl}}}{\text{Zr}} \longrightarrow \text{Zr}\overset{\diagup \text{N}_2\text{H}}{\underset{\diagdown {}^{15}\text{N}_2\text{H}}{}} + \text{ZrCl}_2
\end{array}
$$

$$\longrightarrow \text{N}_2 + \text{N}_2\text{H}_4 \ (1/2\,{}^{15}\text{N}_2 + 1/2\text{N}_2 + 1/2\,{}^{15}\text{N}_2\text{H}_4 + 1/2\text{N}_2\text{H}_4) \qquad (13)$$

A zirconium complex with side-on coordinated dinitrogen, $[\text{Cp}_2\text{Zr}(\text{N}_2)\{\text{CH}(\text{SiMe}_3)_2\}]$, is also known to afford hydrazine on treatment with aqueous HBr or HCl but in lower yield (25%), and with a trace amount of ammonia.[31]

4.3.2 Molybdenum and tungsten

It is well known that molybdenum plays an important role as a key metal in nitrogenase. The molybdenum atoms probably provide the site or sites where dinitrogen is reduced to ammonia. Thus, the reactions of molybdenum dinitrogen complexes and the tungsten analogues have extensively been studied for their possible relevance to the nitrogenase reaction.

In 1969 the molybdenum dinitrogen complexes $[\text{Mo}(\text{N}_2)(\text{PPh}_3)_2\text{PhMe}]$ and trans-$[\text{Mo}(\text{N}_2)_2(\text{dppe})_2]$ were prepared for the first time in our laboratory by the reduction of $[\text{Mo}(\text{acac})_3]$ with AlR_3 in the presence of phosphines under a nitrogen atmosphere.[11-12,57] Since then many dinitrogen complexes of molybdenum and tungsten containing diphosphines, a triphosphine, and monophosphines have been isolated by choosing the appropriate combination of molybdenum and tungsten precursors and reducing agents.[13,58-62]

By selecting the combination of acids and solvents, the dinitrogen ligands in these complexes may be protonated to ammonia and hydrazine. Chatt and coworkers found that the monophosphine complexes cis-$[\text{M}(\text{N}_2)_2(\text{PMe}_2\text{Ph})_4]$ and trans-$[\text{M}(\text{N}_2)_2(\text{PMePh}_2)_4]$ (M = Mo or W) gave, on treatment with H_2SO_4 in methanol at room temperature, followed by base distillation for M = Mo, high yields of ammonia together with a little hydrazine for M = W and a trace for M = Mo.[63-4] The yield of ammonia was essentially 2 mol/M atom for M = W, but only ca. 0.66 mol/M atom for M = Mo. The remaining nitrogen atoms were essentially evolved as molecular nitrogen. The yields of ammonia and hydrazine and their ratio depended remarkably upon the solvent and on the acid used (Table 2). Treatment of the complex cis-$[W(N_2)_2(PMe_2Ph)_4]$ with H_2SO_4 in THF or N-methylpyrrolidone instead of methanol slightly increases the yield of hydrazine. A more dramatic change was observed when the combination of HCl gas and 1,2-dimethoxyethane (DME) was used in the protonation reactions with the complexes cis-$[M(N_2)_2(PMe_2Ph)_4]$ (M = Mo or W), where hydrazine was formed in relatively high yields in addition to ammonia.[65-8] When HBr or HI was used instead of HCl in the above reactions, the yields of hydrazine were decreased with an increase in the yields of ammonia.

$$H_2SO_4/MeOH$$

$[M(N_2)_2(L)_4]$

$$\longrightarrow NH_3 \text{ (M = Mo or W; L = PMe}_2\text{Ph or PMePh}_2)$$

$$\text{(14)}$$

$$HCl/DME$$

$$\longrightarrow N_2H_4 + NH_3 \text{(M = Mo or W; L = PMe}_2\text{Ph)}$$

The reaction of the diphosphine dinitrogen complexes, trans-$[M(N_2)_2(dppe)_2]$ (M = Mo or W), with H_2SO_4 in MeOH at room temperature stops at the stage of hydrazido(2-) complexes $[M(HSO_4)(dppe)_2(NNH_2)]$ $[HSO_4]$ and further protonation does not proceed.[63] However, it is possible to obtain ammonia and hydrazine to some extent if the combination of acid and solvent is well chosen. Thus, treatment of $[Mo(N_2)_2(dppe)_2]$ with aqueous HBr in N-methylpyrrolidone or propylene carbonate at 25°C, followed by evaporation of the solvent and the successive distillation of the reside under Kjeldahl conditions, gives ammonia in the yield of 0.286 or 0.190 mol/Mo atom, respectively.[69]

The molybdenum triphosphine complex $[Mo(N_2)_2(triphos)(PPh_3)]$ (triphos = $PhP(CH_2CH_2PPh_2)_2$) gives, on treatment with HBr gas in THF, ammonia (0.7 mol/Mo atom) and the complex $[MoBr_3(triphos)]$, which is the precursor for the preparation of the starting dinitrogen complex, is almost quantitatively recovered from the reaction mixture.[70]

$$2[Mo(N_2)_2(triphos)(PPh_3)] + 8HBr$$
$$\longrightarrow 2NH_4Br + 2[MoBr_3(triphos)] + 3N_2 + 2PPh_3 \qquad (15)$$

A variety of complexes corresponding to several intermediate stages towards ammonia and hydrazine have been isolated by use of appropriate acids, phosphines, metals and solvents.

The complexes trans-$[M(N_2)_2(dppe)_2]$ (M = Mo or W) react with an excess of halogen acids HX to give hydrazido(2-) complexes $[MX(NNH_2)(dppe)_2]X$.[71] This type of reaction is also shown by cis-$[M(N_2)_2(PMe_2Ph)_4]$ (M = Mo or W) with an excess of aqueous HX, affording a series of hydrazido(2-) complexes $[MX_2(NNH_2)(PMe_2Ph)_3]$.[13] From these, a variety of other hydrazido(2-) derivatives can be prepared by exchanging the labile anions of $[MX(NNH_2)(dppe)_2]X$ with non-coordinating anions such as BPh_4^- and ClO_4^-, or by displacing some ligands in $[MX_2(NNH_2)(PMe_2Ph)_3]$ by other ligands. The former reaction gives salts $[MX(NNH_2)(dppe)_2]X'$ (M = W: X = Cl or Br, X' = BPh_4, ClO_4 or PF_6; M = Mo: X = Br, X' = BF_4),[13] while the latter produces three new types of hydrazido(2-) complexes, $[MX(NNH_2)(PMe_2Ph)_3L^1]X$ (M = Mo or W; X = Cl, Br or I), $[M(NNH_2)L^2(PMe_2Ph)_3]X$ (M = Mo or W; X = Cl, Br or I), and $[WBr(NNH_2)L^3(PMe_2Ph)_2]$, where neutral ligands such as phosphines and pyridines, quinoline-8-olate ion, and dithiocarbamate ion act as L^1, L^2, and L^3, respectively.[72]

$$\text{trans-}[M(N_2)_2(dppe)_2] + 2HX \longrightarrow [MX(NNH_2)(dppe)_2]X + N_2 \qquad (16)$$
(M = Mo: X = Br; M = W: X = Cl or Br)

$$\text{trans-}[M(N_2)_2(dppe)_2] + 2HBF_4 \xrightarrow{\text{THF}} [MF(NNH_2)(dppe)_2][BF_4]$$
$$+ BF_3 \cdot THF + N_2 \qquad (17)$$
(M = Mo or W)

$$\text{cis-}[M(N_2)_2(PMe_2Ph)_4] + 3HX \xrightarrow{\text{MeOH}} [MX_2(NNH_2)(PMe_2Ph)_3]$$
$$+ [PHMe_2Ph]X + N_2 \qquad (18)$$
(M = Mo or W; X = Cl, Br or I)

Table 2. Yields of Ammonia and Hydrazine from Treatment of Molybdenum and Tungsten Complexes with Acids

Complex	Acid	Solvent	$NH_3{}^a$	$N_2H_4{}^a$	Ref.
cis-$[W(N_2)_2(PMe_2Ph)_4]$	H_2SO_4	MeOH	1.86	0.02	64
	H_2SO_4	THF	0.89	0.20	64
	HCl	DME	0.22	0.63	67
	HBr	DME	0.34	0.16	67
	HI	DME	0.92	0.26	67
cis-$[Mo(N_2)_2(PMe_2Ph)_4]$	H_2SO_4	MeOH	0.64	<0.005	64
	HCl	DME	0.31	0.32	67
trans-$[W(N_2)_2(PMePh_2)_4]$	H_2SO_4	MeOH	1.80	0.07	64
trans-$[Mo(N_2)_2(PMePh_2)_4]$	H_2SO_4	MeOH	0.66	tr.	64
$[WCl_2(NNH_2)(PMe_2Ph)_3]$	H_2SO_4	MeOH	1.26	0.12	75
$[WBr_2(NNH_2)(PMe_2Ph)_3]$	H_2SO_4	MeOH	1.58	0.05	75
$[WI_2(NNH_2)(PMe_2Ph)_3]$	H_2SO_4	MeOH	1.88	0.04	75
$[MoCl_2(NNH_2)(PMe_2Ph)_3]^b$	H_2SO_4	MeOH	0.61	0.0	75
$[WCl_2(NNH_2)(PMe_2Ph)_3]$	HCl	DME	0.47	0.50	67
$[WBr_2(NNH_2)(PMe_2Ph)_3]$	HCl	DME	0.65	0.50	67
$[MoCl_2(NNH_2)(PMe_2Ph)_3]$	HCl	DME	0.24	0.52	67
$[WHClBr(NNH_2)(PMe_2Ph)_3]Br$	HCl	DME	0.50	0.55	67
$[WHClBr(NNH_2)(PMe_2Ph)_3][BPh_4]$	HCl	DME	0.32	0.53	67

a. mol/M atom.

b. liberation of 0.44 mol N_2 gas/Mo atom.

Xray structural analyses have been performed for several hydrazido(2-) complexes. These complexes include $[WCl(NNH_2)(dppe)_2][BPh_4]$,[73] $[MoF(NNH_2)(dppe)_2][BF_4]$,[74] $[M(quin)(NNH_2)(PMe_2Ph)_3]I$ (M = Mo or W; quin = quinoline-8-olate),[75,146] $[WBr(NNH_2)(4-Me-py)(PMe_2Ph)_3]Br$,[72] and $[WCl_3(NNH_2)(PMe_2Ph)_2]$,[76,147] the structural parameters of which are summarized in Table 3. In all of these complexes the M-N-N linkages are essentially linear and the N-N bond orders are between 1 and 2. In some structures the N-hydrogen atoms have been located and the hydrazido(2-) ligands lie essentially in one plane, indicating strong conjugation along the M-N-N chain. Thus the structure of hydrazido(2-) ligand can be described as the combination of two resonance structures:

$$[M \equiv N - N \diagdown^H_H] \qquad [\bar{M} \doteq N = \overset{+}{N} \diagup^H_H]$$

The hydrazido(2-) complexes $[MX_2(NNH_2)(PMe_2Ph)_3]$ (M = Mo or W; X = Cl, Br or I), isolated from the reaction of monophosphine dinitrogen complexes with protic acids, give ammonia and hydrazine in essentially similar yields to those obtained from the parent dinitrogen complexes under the same conditions.[66-7,148] Even the complex $[WBr(NNH_2)(dppe)_2]Br$ decomposes in the presence of HBr in CH_2Cl_2 at 80°C to give 0.40 mol ammonia and 0.44 mol hydrazine/W atom.[77] Typical data are shown in Table 2 together with those of parent dinitrogen complexes. These clearly show that the hydrazido(2-) ligand is an intermediate for the reduction of ligating dinitrogen towards ammonia and hydrazine.

A single protonation at the terminal nitrogen atom would give the diazenido complex with the NNH ligand, but no complex of this type has yet been prepared directly from the bis(dinitrogen) complexes, probably because the second protonation is too fast for it to be isolated. In fact, when the protonation of cis-$[W(^{15}N_2)_2(PMe_2Ph)_4]$ with H_2SO_4 in THF has been followed by ^{15}N-NMR spectroscopy, no intermediate between the bis(dinitrogen) complex and the hydrazido(2-) complex has been detected (see below).[78] Two types of diazenido complexes have so far been prepared. One of them is prepared by the careful deprotonation of hydrazido(2-) complexes, $[MX(NNH_2)(dppe)_2]X$, by weak base under vacuum or argon and has the formula $[MX(NNH)(dppe)_2]$ (M = W: X = F, Cl or Br; M = Mo: X = F or Br).[79] These complexes show a strong band assigned to $\nu(N_2)$ in the 1820 - 1940 cm^{-1} region in their IR spectra, which is somewhat higher than that

Table 3. Xray Structural Parameters and NH Stretching Frequencies of Hydrazido(2−) Complexes

Complex	M-N (Å)	N-N (Å)	M-N-N (°)	(NH) (cm⁻¹)	Ref.
[WCl(NNH₂)(dppe)₂][BPh₄]	1.73(1)	1.37(2)	171(1)	3340, 3240	13
[MoF(NNH₂)(dppe)₂][BF₄]	1.762(12)	1.333(24)	176.4(13)	3350, 3250	74
[W(quin)(NNH₂)(PMe₂Ph)₃]I	1.753(10)	1.360(17)	174.7(9)	3145, 3045	75,146
[Mo(quin)(NNH₂)(PMe₂Ph)₃]I	1.743(4)	1.347(7)	172.3(5)		75,146
[WBr(NNH₂)(4-Me-py)(PMe₂Ph)₃]Br	1.75	1.34	177	3090, 2692	72
[WCl₃(NNH₂)(PMe₂Ph)₂]₂	1.752(10)	1.300(17)	178.7(9)		76,147
[WBr(NNHMe)(dppe)₂]Br	1.768(14)	1.32(2)	174(1)	2980, 2820	109,98
[MoI(NNHC₈H₁₇)(dppe)₂]I	1.801(11)	1.259(14)	174(1)	3275	105

expected for the structure described as M-N=N-H with a bent NNH grouping. Furthermore, the complexes (X = Cl or Br) do not exhibit any bands assigned to ν(NH) although a band assigned to ν(NH) has been observed for the complex M = W; X = F. Recently, $[WX(H)(N_2)(dppe)_2]$ (X = Cl or Br) has been proposed as a possible alternative formulation.[38] It is relevant that addition of acids to the complexes trans-$[M(CNR)_2(dppe)_2]$ under carefully controlled conditions gives the hydrido complexes $[MH(CNR)_2(dppe)_2]X$ (M = Mo or W; R = Me, ^tBu, Ph or p-MeC$_6$H$_4$; X = BF$_4$, FSO$_3$, HSO$_4$, ClSO$_3$, Cl or Br).[80]

The other diazenido complex has been isolated as its adduct with the Lewis acid BPh$_3$.[67] The complex $[WClBr\{N_2H(BPh_3)\}(PMe_2Ph)_3]$ is obtained as by-product in low yield from the anion exchange reaction of the hydrido–hydrazido(2-) complex $[WClBr(NNH_2)(PMe_2Ph)_3]Br$ with NaBPh$_4$ (see below). This complex has been well characterized by xray analysis. A hydrogen atom and a boron atom are attached to the terminal nitrogen atom and the atoms W, N, N, B and H lie on the same plane. The W-N-N linkage is essentially linear and the N-N bond order is ca. 1.6.

Recently intermediates of another type which give hydrazine and ammonia by protonation have been isolated. These complexes were at first regarded as hydrazido(1-) complexes with NHNH$_2$ ligands[59] or as hydrazinium complexes with NNH$_3^+$ ligands,[66,68] but have now been characterized as hydrido-hydrazido(2-) complexes containing the MH(NNH$_2$) moiety by a study of ^{1}H- and ^{15}N-nmr spectroscopy and xray analysis.[76,78,147] The dinitrogen complex trans-$[W(N_2)_2(PMePh_2)_4]$ reacts with 10 molar equivalents of HCl gas in dichloromethane to give the hydrido-hydrazido(2-) complex $[WCl_3H(NNH_2)(PMePh_2)_2]$.[59,147] On the other hand, treatment of the hydrazido(2-) . complexes $[WX_2(NNH_2)(PMe_2Ph)_3]$ (X = Cl or Br) with one molar equivalent of anhydrous halogen acid in THF or 1,2-dimethoxyethane produces the complexes $[WClBr(NNH_2)(PMe_2Ph)_3]Br$,[66] $[WBr_3H(NNH_2)(PMe_2Ph)_3]$, and $[WCl_3H(NNH_2)(PMe_2Ph)_3]$.[76,147] This last complex loses PMe$_2$Ph with more HCl to give $[WCl_3H(NNH_2)(PMe_2Ph)_2]$. An xray structural determination of $[WClBr(NNH_2)(PMe_2Ph)_3]Br$[66] shows the presence of the NNH$_2$ ligand and both inter- and intramolecular hydrogen bonding between the N-hydrogen

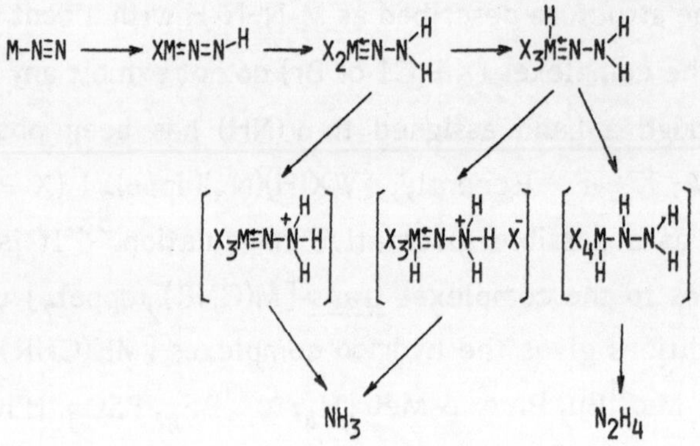

Figure 3.   Proposed protonation mechanism for cis-$[M(N_2)_2(PMe_2Ph)_4]$.[67]

atoms   and   the   halide   ions.   This   material   is   converted   to
$[WClBr(NNH_2)(NNH_2)(PMe_2Ph)_3][BPh_4]$ by anion exchange with $Na[BPh_4]$.
These hydrido-hydrazido(2-) compounds are further protonated to produce
nitrogen hydrides. The compounds $[WClBr(NNH_2)(NNH_2)(PMe_2Ph)_3]X$ (X =
Br or BPh_4) provide hydrazine and ammonia on treatment with HCl gas in
DME in almost similar yields to those of the complexes $[W(N_2)_2(PMe_2Ph)_4]$
and $[WX_2(NNH_2)(PMe_2Ph)_3]$ (X = Cl or Br).[67] This clearly indicates that the
hydrido-hydrazido(2-) complex is the intermediate next to the hydrazido(2-)
complex. Furthermore, it is noteworthy that the hydrido-hydrazido(2-)
complexes give hydrazine in relatively high yields even when treated with
H_2SO_4 in methanol, although ammonia is almost exclusively formed in a
similar reaction with the hydrazido(2-) complexes $[WX_2(NNH_2)(PMe_2Ph)_3]$
(X = Cl or Br). This suggests that the $MH(NNH_2)$ stage is probably on the
route to hydrazine.

To get·further information on the pathways, a ^{15}N-nmr study of the
protonation of cis-$[W(^{15}N_2)_2(PMe_2Ph)_4]$ with H_2SO_4 in THF has been
carried out. As the reaction proceeds, the solution at first exhibits a
spectrum characteristic of the $^{15}N - ^{15}NH_2$ ligand (^{15}N at -78.2 ppm and
^{15}N at -243.0 ppm relative to nitromethane) which is replaced by spectra
showing four further distinct stages without any change in the NNH_2 group.
These species are probably the above hydrido-hydrazido(2-) intermediate and
hydrazido(2-) intermediates in which the monophosphines are displaced by

the HSO_4 anion or by solvent. No further intermediates have been detected until the final product $^{15}NH_4^+$ appears.[78,81,148]

On the basis of these results, possible pathways for the reduction of ligating dinitrogen of the bis(dinitrogen) complexes towards ammonia and hydrazine are proposed as shown in Fig. 3. The first step is probably the protonation of ligating dinitrogen at the terminal nitrogen to give the NNH ligand. However, this mechanism is not yet clear and two pathways may be possible. One is a mechanism in which a proton attacks at the terminal nitrogen atom to produce $[M(NNH)(N_2)(P)_4]^+$. This causes electron withdrawal from the metal, resulting in the liberation of the second dinitrogen ligand as nitrogen gas and the coordination of the acid anion. It is highly relevant that the compounds $[M(N_2)(NC^nPr)(dppe)_2]$ (M = Mo or W) containing the more electron-rich propyl cyanide instead of dinitrogen are protonated at the dinitrogen ligand without loss of the trans-^{n}PrCN ligand to give $[M(N_2H_2)(NC^nPr(dppe)_2]$ $[HSO_4]_2$.[82] The other mechanism involves attack of the acid anion or the undissociated acid on the central metal. For example, displacement of one of the dinitrogen ligands by the anion X will give $[MX(N_2)(P)_4]^-$, in which the remaining dinitrogen ligand receives more electronic charge with the aid of the more electron-donating acid anion and becomes susceptible to electrophilic attack by protons. Anionic dinitrogen complexes of the type trans-$[M(N_2)X(dppe)_2]^-$ (M = Mo or W; X = SCN, CN or N_3) have been prepared by the reaction of $[M(N_2)_2(dppe)_2]$ with $[NBu_4]X$.[82] However, the reaction of the anionic dinitrogen complex with M = Mo, X = SCN with H_2SO_4 in THF leads unexpectedly to loss of dinitrogen, with oxidation of the metal as the major pathway of attack by the protic acids.

The second protonation also occurs on the terminal nitrogen to give the NNH_2 ligand. The course of the third protonation is probably determined by the nature of the acid anion coordinated to the metal. Ammonia may be formed if the terminal nitrogen is further attacked by a proton. Thus the NNH_2 ligand is converted to ammonia and a nitrido ligand which readily hydrolyzes to give another ammonia molecule. Good yields of ammonia obtained by use of H_2SO_4 probably depend on the effectiveness of the anion as a π-electron donor to the metal which makes the terminal nitrogen in the NNH_2 ligand more electron-rich and susceptible to attack by protons. This is supported by the fact that the complex $[W(NNH_2)(PMe_2Ph)_3(HSO_4)_2]$, which can be isolated by the careful treatment of cis-$[M(N_2)_2(PMe_2Ph)_4]$ with H_2SO_4 in THF, is easily protonated even by methanol to give ammonia in 95% yield.[78,148] When treated with HCl in DME, the third protonation is probably directed to the metal to give the hydrido-hydrazido(2-) complex which finally leads to the predominant formation of hydrazine. However, the mechanism of the protonation reaction seems to be more complex, since the ratio of ammonia to hydrazine is sharply affected by the sort of solvent (see above) and treatment of closely related complexes $[M(8\text{-quin})(NNH_2)(PMe_2Ph)_3]I$ with H_2SO_4 in methanol produces completely different distributions of nitrogen hydrides depending upon the metals (0.39 mol N_2H_4 and 0.0 NH_3/M atom for M = W, 0.55 mol NH_3 and 0.0 N_2H_4/metal atom for M = Mo).[75]

In these reactions, the metals are eventually oxidized to higher valent states after providing electrons to the coordinated dinitrogen and converting the ligand into nitrogen hydrides. Six electrons are required for the conversion of dinitrogen into $2NH_3$. On treatment of cis-$[W(N_2)_2(PMe_2Ph)_4]$ with H_2SO_4 in THF, the final tungsten product appears to be a W(VI) oxide species. Thus, the six electrons required for the formation of ammonia come from W(0) $\longrightarrow$ W(VI) + 6e⁻. In contrast to the tungsten complex, $[Mo(N_2)_2(triphos)(PPh_3)]$ gives quantitatively, on treatment with HBr in THF, the Mo(III) compound $[MoBr_3(triphos)]$ as the final product together with ca. 0.7 mol of ammonia per metal atom (see above).[70] This strongly suggests that the six electrons stem from $2Mo(0) \longrightarrow 2Mo(III) + 6e⁻$. The reaction seems to proceed in two steps. The first step is the relatively rapid loss of 1 mol of dinitrogen accompanied by formation of the NNH_2 ligand. The second, slower, step involves further loss of 1/2 mol of dinitrogen and formation of ammonia. This latter step may formally be described as a valence disproportionation involving two $MoNNH_2$ units, but the details remain unsolved. On the other hand, the W(IV) compound $[WCl_4(PMe_2Ph)_2]$ is recovered as a final product from the reaction mixture of $[W(N_2)_2(PMe_2Ph)_4]$ and an excess of HCl in DME,[83] which gives hydrazine in moderate yield.   This suggests that the four

electrons required for the formation of hydrazine come from

$W(0) \longrightarrow W(IV) + 4e^-$.

Not only protic acids but also alcohols or alkaline alcohols can convert

the coordinated dinitrogen into nitrogen hydrides.   The compound cis-

$[W(N_2)_2(PMe_2Ph)_4]$ gives ammonia in high yield (1.6 mol/W atom) on

treatment with methanol alone, either at reflux or under irradiation.[84,64]

The yield of ammonia sharply decreases from methanol to ethanol.

However, alcohols coupled with KOH produce ammonia in moderate yield at

$50^\circ C$.[85] It is of great interest that treatment of the dinitrogen complex with

alcohols in the presence of ketones such as acetone yields hydrazine as the

main nitrogen hydride (compare eq. 36).[149]

$$\text{cis-}[W(N_2)_2(PMe_2Ph)_4] \quad \begin{array}{l} \xrightarrow[\text{alcohol/KOH at } 50^\circ C]{CH_3OH \text{ at reflux or}} NH_3 \\[2em] \xrightarrow[\text{at } 50^\circ C]{\text{alcohol/ketone/KOH}} N_2H_4 \end{array} \tag{19}$$

The transition metal and non-transition metal hydrides can also react

with cis-$[M(N_2)_2(PMe_2Ph)_4]$ (M = Mo or W) to give ammonia after base

distillation of the product. Transition metal carbonyl hydrides such as

$[HCo(CO)_4]$, $[H_2Fe(CO)_4]$ and $[HFeCo_3(CO)_{12}]$ having acidic character

produce ammonia in low yield (0.2 - 0.3 mol/M atom under H_2).[85-6]

Surprisingly, hydridic metal hydrides such as $[Cp_2ZrHCl]$,

$NaAlH_2(OCH_2CH_2OCH_3)_2$ and $BH_3 \cdot THF$ also react at $50^\circ C$ with the

dinitrogen complexes to yield ammonia in moderate yield.[85]

$[NaAlH_2(OCH_2CH_2OCH_3)_2]$ in particular gives a high yield of ammonia (1.8

mol/W atom), although the ruthenium hydride $[RuHCl(PPh_3)_3]$ produces no

ammonia under a similar condition. Elucidation of the mechanisms of these

reactions may lead to clues for catalytic conversion of dinitrogen into

nitrogen hydrides under mild conditions.

cis-$[M(N_2)_2(PMe_2Ph)_4]$

$\quad$ (1) metal carbonyl hydrides,

$\qquad$ Cp_2ZrHCl or $NaAlH_2(OCH_2CH_2OCH_3)_2$

$$\xrightarrow{\hspace{6cm}} \quad NH_3 \qquad (20)$$

$\quad$ (2) base distillation

The reactivities of molybdenum and tungsten dinitrogen complexes have been described here in some detail, because they have been most extensively examined and the reaction mechanisms of the conversion of coordinated dinitrogen into nitrogen hydrides can be discussed relatively precisely with reference to several isolable intermediates. It is very interesting to note that these complexes, although simple compared to nitrogenase, react with protic acids, alcohols/KOH and metal hydrides to give ammonia and hydrazine and that the reactions are remarkably sensitive to subtle changes in electronic circumstances around the central metals.

4.3.3 Iron, cobalt and nickel

A very unstable compound isolated at -50°C from the reaction of $[(Ph_3P)_2FeCl_3]$ and ^iPrMgCl under dinitrogen has been formulated as $[(Ph_3P)_2H(^iPr)FeN_2Fe(^iPr)(PPh_3)_2]$, with (N_2) in the IR spectrum at 1761 cm^{-1}. The compound gives hydrazine on treatment with anhydrous HCl in approximately 10% yield per complexes nitrogen molecule. The following mechanism involving a diazene intermediate has been proposed:[87]

$$
\begin{array}{cc}
^iPr \quad ^iPr & \qquad\qquad H \quad H \\
| \qquad | & \qquad\qquad | \qquad | \\
(Ph_3P)_2Fe-N{\equiv}N-Fe(PPh_3)_2 \xrightarrow{H^+} L_nFe-N{=}N-FeL_n \\
| & \qquad\qquad\qquad + \\
H &
\end{array}
$$

$$\xrightarrow{\hspace{1cm}} \overset{+}{L_n}FeNHNHFeL_n \xrightarrow{H^+} N_2H_4 \qquad\qquad (21)$$

Sellman and his coworkers have prepared a series of N_2, N_2H_2, N_2H_4 and NH_3 complexes of Cr, Mo, W, Mn and Re by the reduction or oxidation of hydrazine complexes.[88] In the complexes $[\{CpM(CO)_2\}_2N_2H_2]$ (M = Mn or

Re) and [{M(CO)$_5$}$_2$N$_2$H$_2$] (M = Cr, Mo or W), diazene, very unstable in the free state, is stabilized by its coordination to the metal. Almost all of these complexes were unfortunately unsuitable for xray structural analysis but the complex trans-[{Cr(CO)$_5$}$_2$N$_2$H$_2$·2THF] has been analyzed at -30°C. In this complex, the diazene coordinates to the metals as a bridging ligand in the trans form, and the N-N distance (1.25 Å) and HNN angle (122.3°) well correspond to the values expected for sp^2 hybridized N atoms.[89] This diazene complex disproportionates in the presence of a catalytic amount of strong base (e.g. NaOMe) to give [{Cr(CO)$_5$}$_2$N$_2$H$_4$] and [{Cr(CO)$_5$}$_2$N$_2$].

The reduction of FeCl$_3$ with Mg in THF under dinitrogen gives a dimeric compound formulated as [{FeMgCl$_3$(THF)$_{1.5}$}$_2$N$_2$], which reacts with aqueous HCl to give hydrazine.[90] On the other hand, a heteronuclear complex [{Fe(Et)(N$_2$)(PPh$_3$)$_2$}$_2$Mg(THF)$_4$], obtained from the reaction of [Fe(acac)$_3$] with MgEt$_2$ in the presence of PPh$_3$ under dinitrogen gives both hydrazine and ammonia on hydrolysis with H$_2$SO$_4$.[91]

A novel heteronuclear complex formulated as [{Co(PMe$_3$)$_3$(N$_2$)}$_2$ Mg(THF)$_4$] has been prepared by reduction of [CoCl$_2$(PMe$_3$)$_2$] with Mg in THF under dinitrogen.[92] The xray crystal structure is shown in Fig. 1. A similar complex [{Co(PPh$_3$)$_3$(N$_2$)}$_2$Mg(THF)$_4$] is produced by the reaction of [CoH(N$_2$)(PPh$_3$)$_3$] with MgEt$_2$. When treated with H$_2$SO$_4$, the latter compound gives 0.26 mol N$_2$H$_4$ and 0.06 mol NH$_3$ per Co atom, although the coordinated dinitrogen of [CoH(N$_2$)(PPh$_3$)$_3$] is released as dinitrogen gas under similar conditions.[93] The analogous complexes [Co(PPh$_3$)$_3$(N$_2$) M(THF)$_3$] (M = Li or Na) also produce hydrazine and ammonia on hydrolysis with H$_2$SO$_4$.[91,151] It seems likely that the coordinated dinitrogen of these cobalt(-1) complexes receives more electronic charge from the metal than that of [CoH(N$_2$)(PPh$_3$)$_3$], and hence undergoes electrophilic attack by protons. This is reflected in the low ν(N$_2$) at 1840 cm^{-1} (Nujol) for [{Co(PPh$_3$)$_3$(N$_2$)}$_2$Mg(THF)$_4$] compared with 2090 cm^{-1} (Nujol) for [CoH(N$_2$)(PPh$_3$)$_3$]. Alkali cation-dinitrogen interactions may play an important role in some cases.

The N-N bond distance of the side-on dinitrogen nickel complex [{(LiPh)$_3$(Et$_2$O)$_{1-1.5}$Ni]$_2$N$_2$]$_2$ is remarkably lengthened, implying a bond

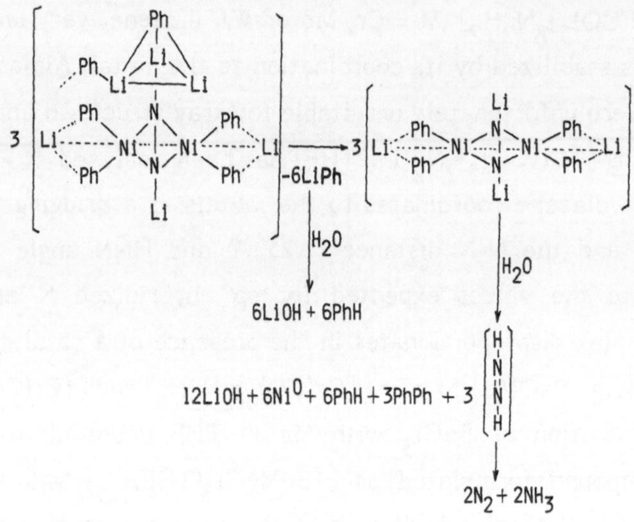

Figure 4.   Conversion of side-on bonded dinitrogen into ammonia in the $[(C_6H_5Li)_3Ni]_2N_2$ unit.[94]

order of ca. 1.3, by the interaction of the dinitrogen ligand with both Ni and Li metals.[32] This compound is easily hydrolysed even by water in THF at room temperature to give ammonia (30 - 38% of complexed dinitrogen).[94] No hydrazine or dihydrogen is formed in this reaction. The mechanism of hydrolysis proposed is shown in Fig. 4. The $[\{(LiPh)_3Ni\}_2N_2]$ unit is converted into $[\{(PhLi)_2Ni\}_2N_2]$ after loss of two LiPh molecules from the Li_3Ph_2 group. A conceivable exchange of Li atoms in this unit with H^+, with concomitant transfer of two electrons from the side-on bonded $[NiPh_2Li]$ units to nitrogen, would lead to biphenyl and diazene, which disproportionates to ammonia and nitrogen gas.

4.4   Nitrogen-Carbon Bond Formation Reactions

In 1966 Vol'pin and Shur discovered the production of aniline (0.03 mol/Ti atom) together with ammonia (0.17 mol/Ti atom) after hydrolysis of the products given by the reaction of $[Cp_2TiCl_2]$ with excess PhLi in ether under dinitrogen. The highest capacity for amine formation is displayed by the system $[Cp_2TiPh_2]/Li/N_2$ in THF (0.10 - 0.15 mol aniline and 1.1 mol NH_3 per Ti atom).[95] Following this discovery, several systems have been found to give organic nitrogen compounds, such as amines and organocyanides, from fixed dinitrogen.[96] Although the intimate details of the mechanisms remain ambiguous, it seems likely that the formation of a nitrogen-carbon bond from ligating dinitrogen is involved in some of these reactions. The discovery of catalysts producing organo-nitrogen compounds from dinitrogen and a relatively cheap organic feedstock is of industrial importance and definitely one of the final goals in the chemistry of nitrogen fixation. The catalytic cycles in these reactions will involve essentially the formation of nitrogen-carbon bonds from ligating dinitrogen as important elementary steps. This type of reaction will be reviewed in this Section.

4.4.1 Organo–diazenido and organo–hydrazido(2-) complexes

The first formation of nitrogen-carbon bonds from ligating dinitrogen in well characterized complexes was achieved by the reaction of trans-$[M(N_2)_2(dppe)_2]$ (M = Mo or W) with acyl or aroyl chlorides in the presence of HCl.[97-8] The complexes obtained are converted into acyl- or aroyl-diazenido complexes by weak base such as triethylamine and potassium carbonate, as shown in eq. 22. The molybdenum complex of this type $[MoCl(NNCOPh)(dppe)_2]$, is also obtained from the reaction of trans-$[Mo(N_2)(RCN)(dppe)_2]$ (RCN = alkyl- or benzonitrile) with benzoyl chloride.[99]

$$[M(N_2)_2(dppe)_2] + RCOCl + HCl \longrightarrow [MCl(N_2HCOR)(dppe)_2]Cl$$

$$\xrightarrow{\text{base}} [MCl(NNCOR)(dppe)_2] \ (M = Mo \text{ or } W) \qquad (22)$$

$$[Mo(N_2)(RCN)(dppe)_2] + PhCOCl \longrightarrow [MoCl(NNCOPh)(dppe)_2] \qquad (23)$$

Chelation of acyl- or aroyldiazenido ligands via the carbonyl oxygen was initially suggested, because of the relatively low $\nu(C=O)$ at 1550 - 1575 cm^{-1} in the IR spectra.[97] However, xray structural analysis of the complex $[MoCl(NNCOPh)(dppe)_2]$ shows that it has a linear Mo-N-N linkage with a nitrogen-carbon bond and that the carbonyl oxygen does not interact in any way with the metal.[99-100] The bond orders of N-N, N-C and C-O are estimated as ca. 1.85, 1.3 and 1.85, respectively. The structure is hence described by the combination of the two resonance forms below. The shift of ν(C=O) to low frequency is understood by the contribution of the right-hand structure with a single C-O bond. On the other hand, the acyl- or aroyldiazene structure NH=NCOR, rather than the acyl- or aroylhydrazido(2-) structure NNHCOR, has been proposed for the complexes $[MCl_2(N_2HCOR)(dppe)_2]$ (M = Mo or W) since they show the ν(NH) bands in the IR spectra at 2780 - 2500 cm^{-1} and the NH resonance at 12.0 - 14.1 ppm in the nmr spectra, and their molar conductivities are low (2 - 30 cm^2 mol^{-1}).[98] However, these results may be explained by postulating a hydrazido(2-) structure in which a strong hydrogen bonding interaction exists between the hydrazido(2-) proton and the anion of the outer coordination sphere.

The complexes $[ReCl(N_2)(PMe_2Ph)_4]$ and $[ReCl(N_2)(py)(PMe_2Ph)_3]$, preferably the latter because the pyridine is more labile, react with an excess of RCOCl (R = Me or Ph) to yield the diazenido complexes cis-mer-$[ReCl_2(N_2COR)(PMe_2Ph)_3]$. However, acylated products could not be obtained from the osmium complex $[OsCl_2(N_2)(PMe_2Ph)_3]$.[98] In these reactions it is presumed that loss of a neutral ligand, such as a dinitrogen in the case of $[M(N_2)_2(dppe)_2]$ (M = Mo or W) and a PMe$_2$Ph or pyridine in the rhenium complexes, is rate-determining and that the key intermediates are $[M(N_2)(dppe)_2]$ and $[ReCl(N_2)(PMe_2Ph)_3]$, respectively. To these coordinatively unsaturated species acid halides easily coordinate through the chlorine.[101] In the rhenium complex $[ReCl(N_2)(PMe_2Ph)_4]$ there is considerable crowding of the four phosphine ligands and the loss of a phosphine is assisted by this steric overcrowding. However, since there is no corresponding strain in the osmium complex and thus ligand loss does not occur under these conditions, the acylation or aroylation reaction does not occur.

The hydrazido(2-) complex $[WCl_2(N_2H_2)(dppe)_2]$, derived from trans-$[W(N_2)_2(dppe)_2]$, is readily acetylated by acetyl chloride or acetic anhydride in boiling THF solution to give a mono-acetylated complex $[WCl_2(N_2HCOMe)(dppe)_2].$[13]   This reaction may be initiated by nucleophilic attack of the terminal nitrogen atom on the electron deficient carbonyl carbon atom, followed by elimination of HCl. In contrast, when succinyl chloride is used in the reaction with $[WF(NNH_2)(dppe)_2][BF_4]$, the diacylated complex $[WF(\overline{NNCOCH_2CH_2CO})(dppe)_2]$ $[BF_4]$ is smoothly produced at room temperature.[102] Preliminary xray analysis shows that the W-N-N linkage is essentially linear and that all the nitrogen, carbon, and oxygen atoms in the diacylhydrazido(2-) ligand lie nearly on the same plane, indicating sp^2 character for the nitrogen atom bearing the carbonyl groups.

$$[WF(NNH_2(dppe)_2][BF_4] + ClCOCH_2CH_2COCl$$
$$\longrightarrow [WF(\overline{NNCOCH_2CH_2CO})(dppe)_2][BF_4] + 2HCl \qquad (24)$$

This complex exhibits a strong band, assigned to $\nu(C=O)$, at 1720 cm^{-1} which is higher by 25 cm^{-1} than that of succinimide. Thus the diacylhydrazido (2-) ligand is described by the two resonance structures:

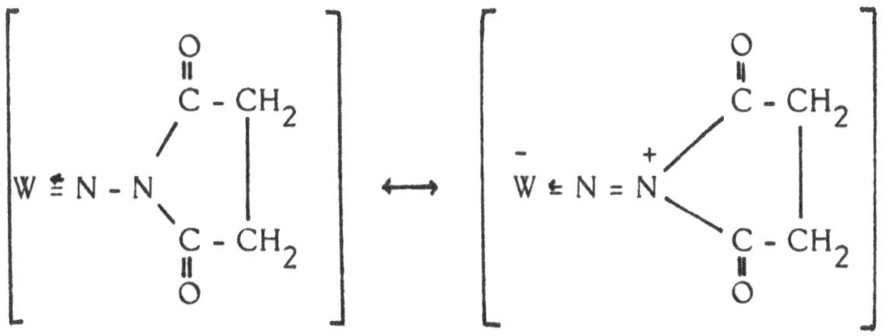

The reactions of the complexes trans-$[M(N_2)_2(dppe)_2]$ (M = Mo or W) with alkyl halides under tungsten filament irradiation produce a series of organodiazenido complexes which are converted into alkylhydrazido(2-) complexes by protonation with protic acids (eq. 25). Xray structural analysis of the alkyldiazenido complex $[MoI(NNC_6H_{11})(dppe)_2]$ has shown that the alkyldiazenido ligand is bonded to the metal in singly bent fashion with Mo-N-N and N-N-C bond angles of 176° and 142°, respectively.[104] ^{15}N-nmr shifts also provide an unambiguous technique for distinguishing the singly-bent or doubly-bent geometry of diazenido ligands ($-N_\alpha N_\beta R$) in solution. Thus the diazenido complexes trans-$[MX(^{15}N_2R')(dppe)_2]$ (M = Mo or W; X = Cl or Br; R' = Et or MeCO), with singly-bent geometry, show resonances in the ^{15}N-nmr spectra at -28 to -36 ppm for N_α and at -123 to -165 ppm for N_β, whereas the N_α resonance for the doubly-bend diazenido ligands appears at very low field, for example at 327.1 ppm for $[RhCl_2(^{15}NNC_6H_4NO_2-4)(PPh_3)_2]$ and 298.4 ppm for $[RhCl_2(^{15}NNC_6H_5)(PPh_3)_2]$, relative to CD_3NO_2.[108]

$$[M(N_2)_2(dppe)_2] + RX \xrightarrow{h\nu} [MX(NNR)(dppe)_2]$$

$$\xrightarrow{HX} [MX(NNHR)(dppe)_2]X \qquad (25)$$

(M = Mo or W; R = Me, Et, ^{n}Pr, ^{i}Pr,[98,150] ^{n}Bu,[103,150] ^{t}Bu,[98] C_6H_{11}[104,150] C_8H_{17},[105,150] $Me_3CCH_2CH_2$[103] or $EtOCOCH_2$;[106-7] X = Cl, Br or I).

Since the organodiazenido ligands derived from the monoalkylation of coordinated dinitrogen have the singly-bent geometry, the protonation is expected to occur on N_β to give alkylhydrazido(2-) ligands. The structures of alkylhydrazido(2-) complexes $[WBr(NNHMe)(dppe)_2]Br$[109] and $[MoI(NNHC_8H_{17})(dppe)_2]I$[105] have been established by xray diffraction. The structural parameters are shown in Table 3.

The mechanism of Fig. 5 for the monoalkylation of ligating dinitrogen by alkyl halides is well established.[101] The reaction is initiated by the rate-determining dissociation of a dinitrogen ligand from trans-$[M(N_2)_2(dppe)_2]$ (M = Mo or W). This ligand loss is accelerated by irradiation, which is essential for M = W. Addition of RX to the coordinatively unsaturated metal through the halogen gives the unstable intermediate $[M(N_2)(RX)(dppe)_2]$. Homolysis of the carbon halogen bond and subsequent attack of the organic

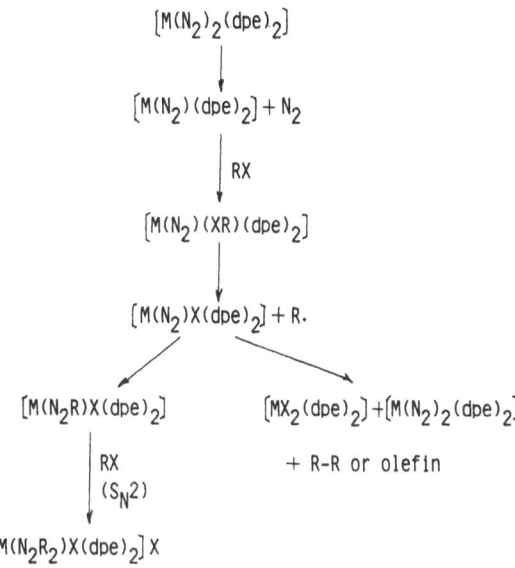

Figure 5. Mechanism of alkylation of ligating dinitrogen in _trans_-$[M(N_2)_2(dppe)_2]$ (M = Mo or W) under irradiation.[101]

free radical so generated on the terminal nitrogen atom of ligand dinitrogen produces the alkyldiazendio complexes $[MX(NNR)(dppe)_2]$. This free radical process is strongly supported by the fact that the reaction of _trans_-$[M(N_2)_2(dppe)]$ and 6-bromo-1-hexene gives the cyclopentylmethyldiazenido complex $[MBr(NNCH_2\overline{CH(CH_2)_3CH_2})(dppe)_2]$,[110,103] since the 6-hexenyl radical is known to cyclize rapidly to form the cyclopentylmethyl radical. The reactions of $[M(N_2)_2(dppe)_2]$ with acid chlorides described above seem to follow an analogous mechanism,[98] since the reaction of ^tBuCOCl with $[M(N_2)_2(dppe)_2]$ (M = Mo or W) in benzene yields a t-butyldiazenido complex and, though the radical ^tBuCO could not be trapped, it is known to break down very rapidly to give $^tBu.$ and CO.

In the scheme shown in Fig. 5, the generation of the alkyl free radical proceeds via an inner-sphere redox process. In contrast, the alkylation of the species $[Mo(N_2)(SCN)(dppe)_2]^-$, which has a more negative oxidation potential, shows a rate with ^nBuI which is proportional to the concentrations of both the complex and the halide. Furthermore, the product is

[Mo(SCN)(NN^{n}Bu)(dppe)$_2$] which retains the thiocyanate ligand. These results suggest a mechanism involving an outer sphere electron transfer reaction (eq. 26).[82]

$$[Mo(N_2)(SCN)(dppe)_2]^- + RX \longrightarrow [Mo(N_2)(SCN)(dppe)_2] + R\cdot + X^- \qquad (26)$$

In some cases, the terminal nitrogen atom of coordinated dinitrogen is dialkylated by alkyl halides. Thus the reaction of trans-[W(N$_2$)$_2$(dppe)$_2$] and trans-[Mo(N$_2$)$_2$(Et$_2$PCH$_2$CH$_2$PEt$_2$)$_2$] with alkyl bromides RBr produces the dialkylated products [WBr(NNR$_2$)(dppe)$_2$]Br and [MoBr(NNR$_2$)(Et$_2$PCH$_2$-CH$_2$PEt$_2$)$_2$]Br (R = Me or Et), respectively.[111] Kinetic studies on these reactions have suggested that the second alkylation of coordinated dinitrogen is achieved by a typical S$_N$2 reaction of the alkyldiazenido ligand, formed by the scheme of Fig. 5, with alkyl halides. Generally, electron-poorness in the system, which increases the lability of coordinated dinitrogen, encourages the first alkylation. However, this inhibits the second alkylation since the nucleophilicity of an alkyldiazenido ligand will be decreased.

When α,ω -dibromides are used as alkyl halides, products depend strongly on the value of n in Br(CH$_2$)$_n$Br(eq. 27).[112] These reactions do not generally give a single clean product, although the dialkylhydrazido(2-) complexes of tungsten with rings N(CH$_2$)$_n$ (n = 4 or 5) are isolated with yields reaching 80%. For n = 3, complexes containing the group N$_2$(CH$_2$)$_3$Br are formed whereas for n = 6 - 12 two series of complexes containing either {N$_2$(CH$_2$)$_n$N$_2$} or N$_2$(CH$_2$)$_n$Br are isolated.

Aryldiazenido complexes,[113] which are generally accessible from existing aryldiazene or diazonium precursors, do not seem to be obtainable by the arylation reaction of coordinated dinitrogen with aryl halides. Thus the reaction of trans-[W(N$_2$)$_2$(dppe)$_2$] with bromobenzene produces only [WBr$_2$(dppe)$_2$].[101] The reactions of the dinitrogen-derived hydrazido(2-) compounds [WX(NNH$_2$)(dppe)$_2$]Y [X = F, Br, CF$_3$COO; Y = BF$_4$, Br, or (CF$_3$COO)$_2$H] with 2,4-dinitrofluorobenzene in the presence of aqueous

$$[M(N_2)_2(dppe)_2] + Br(CH_2)_nBr$$

$n = 2$

$$\longrightarrow [MBr_2(dppe)_2] + CH_2 = CH_2$$

$n = 3$

$$\longrightarrow [MBr\{N_2(CH_2)_3Br\}(dppe)_2]$$

$n = 4, 5$

$$\longrightarrow [MBr\{\overline{NN(CH_2)_{n-1}CH_2}\}(dppe)_2]Br$$

$n = 6\text{-}12$

$$\longrightarrow [MBr\{N_2(CH_2)_nBr\}(dppe)_2] +$$
$$[\{N_2(CH_2)_nN_2\}\{MBr(dppe)_2\}_2] \qquad (27)$$

K_2CO_3 give 2,4-dinitrophenyldiazenido complexes $[WX[NNC_6H_3(NO_2)_2]$ $(dppe)_2]$ in moderate yield.[114] This reaction is obviously initiated by the formation of the diazenido ligand from the deprotonation of the hydrazido(2-) ligand, since the corresponding diazenido complexes $[WX(N_2H)(dppe)_2]$ (X = F or Br) react with 2,4-dinitrofluorobenzene in the absence of base. The 2,4-dinitrofluorobenzene then undergoes nucleophilic substitution by the diazenido ligand. In this dinitrophenyldiazenido ligand, the π-electrons are probably highly delocalized, since the complexes show intense purple color and lower values of $\nu_{sym}(NO_2)$ in their IR spectra than those commonly observed for aromatic nitro groups. The 2,4-dinitrophenyldiazenido ligand shows relatively low basicity compared with the alkyldiazenido ligand. Thus the arylhydrazido complexes obtained by treatment of the diazenido complexes with acids react with mild bases, such as water or methanol, to revert to the original diazenido complexes.

When the hydrazido(2-) compounds $[MBr(NNH_2)(dppe)_2]Br$ (M = Mo or W) are treated with phenylisocyanate in refluxing THF, organohydrazido(2-) compounds of the semicarbazido type $[MBr(NNHCONHPh)(dppe)_2]Br$ are obtained.[102] These semicarbazido type complexes are easily converted into new organodiazenido complexes by deprotonation with base. Thus the compounds $[MBr(NNHCONHPh)(dppe)_2]Br$ (M = Mo or W) react with triethylamine to give $[MBr(NNCONHPh)(dppe)_2]$. These organodiazenido complexes are also prepared from the nucleophilic addition of the diazenido complexes $[MBr(NNH)(dppe)_2]$ (M = Mo or W) to phenylisocyanate. These reactions are summarized in eq. 28. The organodiazenido complexes show ν (C=O) in their IR spectra lower by ca. 90 cm^{-1} than those of the corresponding organohydrazido complexes. This is explained by the increased ability of the metal atom in the diazenido complexes to donate electrons to the NNCONHPh ligand. Similar delocalization was also found in acyl- or aroyldiazenido and 2,4-dinitrophenyldiazenido complexes (above).

$$
\begin{array}{ccc}
 & \xrightarrow{\text{PhNCO}} & \\
[MBr(NNH_2)(dppe)_2]Br & \longrightarrow & [MBr(NNHCONHPh)(dppe)_2]Br \\
\downarrow \text{NEt}_3 & & \downarrow \text{NEt}_3 \qquad\qquad (28) \\
 & \xrightarrow{\text{PhNCO}} & \\
[MBr(NNH)(dppe)_2] & \longrightarrow & [MBr(NNCONHPh)(dppe)_2]
\end{array}
$$

4.4.2 Diazoalkane complexes

The alkylation of coordinated dinitrogen by alkyl halides under irradiation is carried out in inert solvents such as benzene, giving the alkyldiazenido complexes (see above). However, when heterocyclic compounds such as tetrahydrofuran and N-methylpyrrolidine are used as solvents, organo-nitrogen complexes derived from the solvents are obtained in the reaction with the lower alkyl halides, especially methyl bromide.[115-7] The mechanism of this reaction has been well established, especially in the case of tungsten. The methyl radical generated as shown in Fig. 5 abstracts an α-hydrogen from heterocyclic solvents, such as THF. The solvent radical then reacts with $[MBr(N_2)(dppe)_2]$ to give a solvent-derived diazenido complex, which undergoes reversible ring-opening with protic acids to afford diazoalkanol (diazoalkane) complexes (eq. 29). Similar diazoalkanol complexes can also be obtained by use of tetrahydrophyran, 1,4-dioxane and 2-methyltetrahydrofuran. These diazoalkanol complexes are converted by LiAlH$_4$ into   the   diazenido   complexes,   for   example

$[WBr\{NN(CH_2)_4OH\}(dppe)_2]$ in the case of THF, and subsequent protonation of these diazenido complexes with HBr affords alkylhydrazido(2-) complexes such as $[WBr\{NNH(CH_2)_4OH\}(dppe)_2]Br$. The reaction of $[W(N_2)_2(dppe)_2]$ with CH_3Br in tetrahydrothiophene or N-methylpyrrolidine similarly forms diazenido derivatives which react with HBF_4 to give $[WBr\{NNH\overline{CHS(CH_2)_2}CH_2\}(dppe)_2]$ $[BF_4]$ and $[WF\{NNH\overline{CHNMe(CH_2)_2}\overline{CH_2}\}(dppe)_2]Br$, respectively. In these cases ring opening of the cyclic diazenido ligands does not occur.

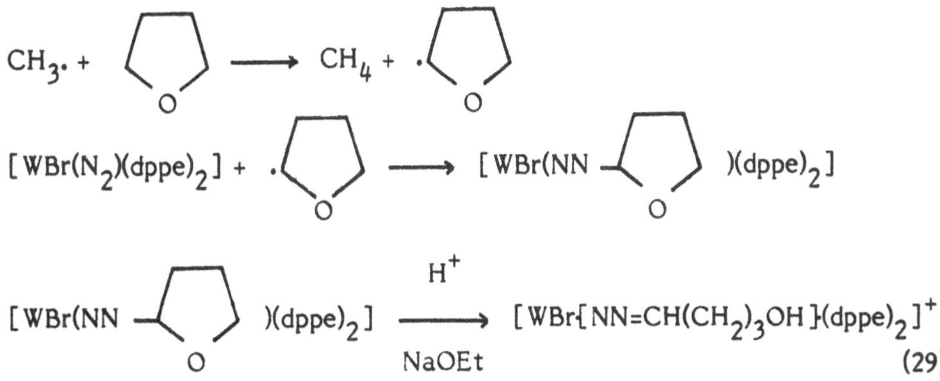

$$\tag{29}$$

The complex trans-$[W(N_2)_2(dppe)_2]$ reacts with gem-dibromides $RR'CBr_2$ (R = R' = H or Me; R = H, R' = Me or Ph) in benzene under tungsten-filament irradiation to yield the diazoalkane compounds $[WBr(N_2CRR')(dppe)_2]Br$.[118-9] The reactions of the molybdenum analogue with gem-dibromides are more complex and diazoalkane complexes are, at best, minor products. The compounds trans, trans-$[(dppe)_2BrMo(NNHCH_2-NHN)MoBr(dppe)_2]Br_2$ and trans-$[MoBr(N_2HCH_2CH_2Br)(dppe)_2]Br$ have reportedly been isolated from the reaction of trans-$[Mo(N_2)_2(dppe)_2]$ with CH_2Br_2 in benzene under no special irradiation conditions.

A more generally useful method to prepare diazoalkane complexes has been discovered, which involves the condensation reactions of the dinitrogen-derived hydrazido(2-) complexes with a series of aldehydes and ketones including diketones and unsaturated aldehydes.[120-21] This method of forming a nitrogen–carbon bond can be applied even to the monophosphine complexes of both molybdenum[122-3] and tungsten,[124] although these complexes evolve two dinitrogens in the reactions with alkyl halides under irradiation. Formation of diazoalkane complexes proceeds readily at room temperature when aldehydes are used as the carbonyl compounds but hardly proceeds in the case of ketones. The presence of a trace amount of acid, however, remarkably accelerates the rate of ketone condensation.

$$\underline{trans}\text{-}[M(N_2)_2(dppe)_2] + HBF_4 \longrightarrow [MF(NNH_2)(dppe)_2][BF_4]$$

$$[MF(NNH_2)(dppe)_2][BF_4] + RR'C=O \longrightarrow [MF(NN=CRR')(dppe)_2][BF_4]$$

M = Mo: R = H, R' = H, Me, Et, Ph, $PhCH_2$, MeCH=CH or PhCH=CH;
R = Me, R' = Me or $MeCOCH_2$; R, R' = $(CH_2)_5$
M = W: R = H, R' = Et or Ph; R = Me, R' = $MeCOCH_2$ (30)

$$\underline{cis}\text{-}[M(N_2)_2(PMe_2Ph)_4] + HX \longrightarrow [MX_2(NNH_2)(PMe_2Ph)_3]$$

$$[MX_2(NNH_2)(PMe_2Ph)_3] + RR'C=O \longrightarrow [MX_2(NN=CRR')(PMe_2Ph)_3]$$

M = Mo: X = Cl: R = Ph, R' = H; R = Me, R' = Me or Ph; R, R' = $(CH_2)_5$
X = Br: R = Me, R' = Ph; R, R' = $(CH_2)_5$
M = W: X = Cl: R = R' = Me; X = Br: R = Me, R' = Me, Et, Ph, $MeCOCH_2$
or $MeCOCH_2CH_2$; R = R' = CD_3; R, R' = $(CH_2)_5$; X = I: R = R' = Me   (31)

The carbon atom of the C=N bond in $[WBr(NNCH_2)(dppe)_2]Br$ is susceptible to attack by nucleophiles such as hydride, carbonions, or alkoxides to give diazenido complexes $[WBr(NNCH_2R)(dppe)_2]$ (R = H, Me, Ph, OMe or O^tBu).[118-9] A similar reaction occurs in the case of $[WBr(NNCHMe)(dppe)_2]Br$. The benzyldiazenido complex obtained here is not accessible from the alkylation of coordinated dinitrogen with benzyl bromide since the benzyl radical formed in the initial stage of this reaction is not sufficiently reactive towards the coordinated dinitrogen and eventually dimerizes (Fig. 5). These reactivities of diazoalkane complexes towards nucleophiles are understood by describing the diazoalkane ligand as a combination of the resonance structures (ii) and (iii). However, in the case of tungsten aryldiazomethane derivatives $[WF(NN=CHC_6H_4X-p)(dppe)_2]$ $[BF_4]$, the contribution of the resonance structure (i) rather than (iii) to the structure (ii) probably predominates since more electron-withdrawing substituents X of the para position in the aryl group lower the C=N stretching frequency.[125] This electronic effect of the substituent transmitted to the nitrogen atoms in the aryldiazomethane ligand and the central metal is reflected in the linear correlation of the Hammett σ constants with the oxidation potentials and the reduction potentials. Cyclic voltammetric studies have shown that aryldiazomethane derivatives $[MF(NN=CHC_6H_4X-p)(dppe)_2][BF_4]$ (M = Mo or W) undergo consecutive one- and two-electron oxidations and reductions. The esr spectra of the species generated by controlled potential electrolysis indicate that primary oxidation occurs on the metal atom (M = Mo) and reduction on the two nitrogen atoms in the diazoalkane ligands (M = Mo or W).[125]

$$[\overset{+}{M} \leq N = N \diagdown \overset{-}{C} - R] \longleftrightarrow [M \leq N - N \approx C - R] \longleftrightarrow [\overset{-}{M} \leq N = N \diagdown \overset{+}{C} - R]$$
$$\qquad\quad | \qquad\qquad\qquad\qquad | \qquad\qquad\qquad\qquad |$$
$$\qquad\quad R' \qquad\qquad\qquad\qquad R' \qquad\qquad\qquad\qquad R'$$

$$(i) \qquad\qquad\qquad\qquad (ii) \qquad\qquad\qquad\qquad (iii)$$

Recently the new dichlorodiazomethane complex $[WBr(NN=CCl_2)-(dppe)_2]^+$ has been isolated in low yield as its PF_6^- salt from the reaction between $[WBr(NNH_2)(dppe)_2]Br$ and $Ph_2I^+Cl^-$ in the two phase system $CHCl_3$ - aq. 5% w/v K_2CO_3.[126] Replacement of $CHCl_3$ by $CBrCl_3$ increases the yield of the complex to ca. 60%. The reaction sequence of eq. 32 has been proposed.

$$[WBr(NNH_2)(dppe)_2]^+ \xrightarrow{K_2CO_3} [WBr(NNH)(dppe)_2]$$

$$[WBr(NNH)(dppe)_2] + Ph_2I^+ \longrightarrow [WBr(N_2)(dppe)_2] + Ph_2I\cdot + H^+$$

$$Ph_2I\cdot \longrightarrow PhI + Ph\cdot$$

$$Ph\cdot + CHCl_3(CBrCl_3) \longrightarrow PhH(PhBr) + CCl_3\cdot$$

$$CCl_3\cdot + [WBr(N_2)(dppe)_2] \longrightarrow [WBr(NNCCl_3)(dppe)_2]$$

$$\longrightarrow [WBr(NN=CCl_2)(dppe)_2]^+ \qquad\qquad (32)$$

This dichlorodiazomethane complex undergoes rapid reaction with a variety of nucleophiles to yield a series of novel organo-nitrogen ligands:

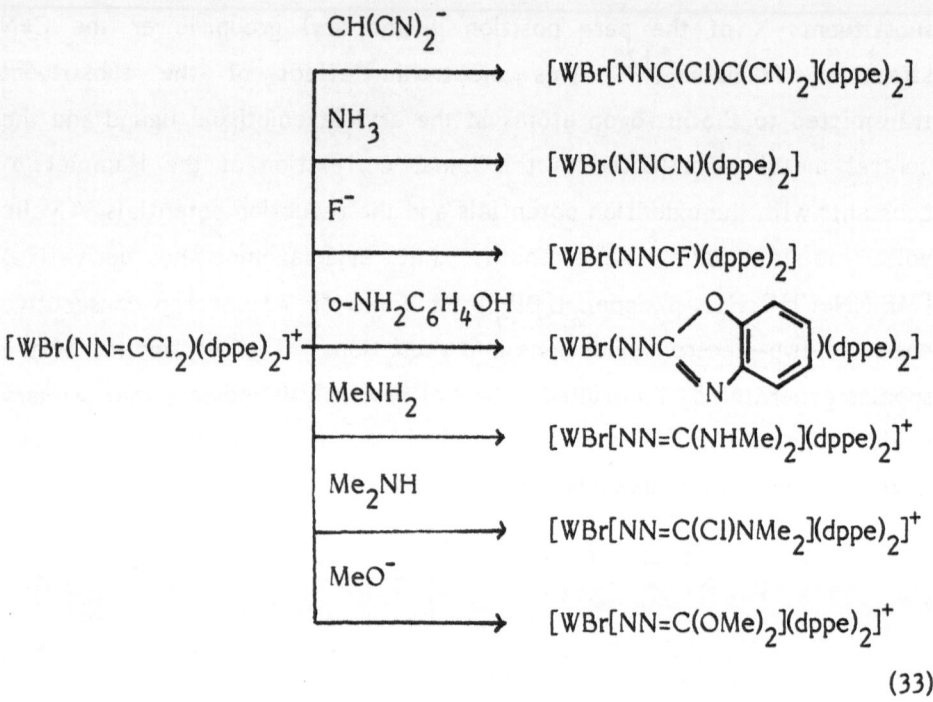

$$(33)$$

The xray structure of a vinyldiazenido complex, which is derived from the reaction of the dichlorodiazomethane complex with dicyanomethanide ion, is shown in Fig. 6. All three C-C bonds in the vinyl group are of essentially the same length and the C-N and N-N linkage are of bond order between 1 and 2. This implies extensive delocalization of electron density within the planar metal-ligand π-system.

The structures of four diazoalkane compounds have been determined by xray analysis, namely $[WBr[NNCH(CH_2)_3OH](dppe)_2]$ $[PF_6]$,[115,127] $[WF[NNCMe(CH_2COMe)](dppe)_2]$ $[BF_4]$,[121] $[WBr(NNCMe_2)(dppe)_2]Br$,[127-8] and $[WBr(NN=CCl_2)(dppe)_2]$ $[PF_6]$.[126] The structures of the diazoalkane ligands are essentially similar. The endo-methyl or chlorine attached to the C=N bond in the latter three complexes lies in a sterically crowded sandwich position among the two phenyl groups of the diphosphine ligand. Selected bond lengths and angles are shown in Table 4.

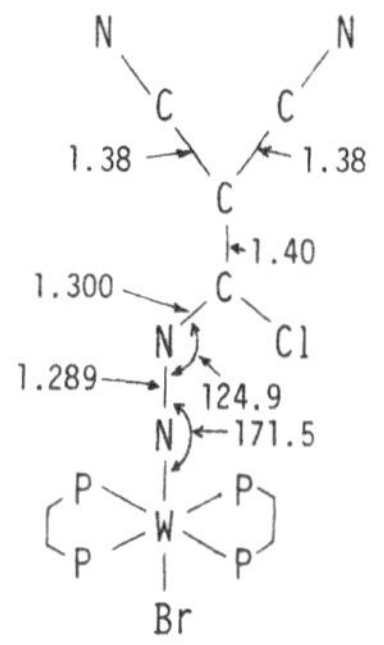

Figure 6. Xray structure of the vinyldiazenido complex [WBr[NNC(Cl)C(CN)$_2$](dppe)$_2$].[126]

Table 4. Selected Bond Lengths and Angles of Diazoalkane Complexes

Complex	N-N	C-N	M-N-N	N-N-C	Ref.
	(Å)		(°)		
[WF[NN=CMe(CH$_2$COMe)](dppe)$_2$] [BF$_4$]	1.317	1.300	173.8	125.2	121
[WBr[NN=CH(CH$_2$)$_3$OH](dppe)$_2$]Br	1.32	1.30	172.6	116.2	127
[WBr(NN=CMe$_2$)(dppe)$_2$]Br	1.355	1.290	171.3	123.9	127
[WBr(NN=CCl$_2$)(dppe)$_2$][PF$_6$]	1.35	1.28	169	122	126

4.4.3 Displacement of organo–nitrogen ligands from the metals

Some dialkylhydrazido(2-), alkyldiazenido and diazoalkane complexes of molybdenum and tungsten are now known to give organo-nitrogen compounds after reaction with acids or bases (Table 5). The dialkylhydrazido(2-) complexes [MBr(NNMe$_2$)(dppe)$_2$]Br and [MBr(NN(CH$_2$)$_3$CH$_2$)(dppe)$_2$]Br are reduced by LiAlH$_4$ at 80°C in ether to yield dimethylamine and pyrrolidine, respectively, after treatment of the product with methanol and HBr.[129] Direct base distillation of these complexes from 40% aqueous KOH also gives the corresponding amines but the yields are not so reproducible since the system is heterogeneous. On the

Table 5. Yields of Nitrogen Compounds from Organo–Nitrogen Ligands

Complex	Reagents/solvents	Products (yield[a])	Ref.
[WBr(NNMe$_2$)(dppe)$_2$]Br	H$_2$SO$_4$/propylene carbonate	MeNH$_2$ (0.08), Me$_2$NH (0.19)	129
[WBr{NN(CH$_2$)$_3$CH$_2$}(dppe)$_2$]Br	LiAlH$_4$/ether	pyrrolidine (0.80)	129
[MoBr{NN(CH$_2$)$_3$CH$_2$}(dppe)$_2$Br	LiAlH$_4$/ether	pyrrolidine (0.87)	129
[WBr(NNMe$_2$)(dppe)$_2$]Br	LiAlH$_4$/ether	Me$_2$NH (0.95)	129
[WBr(NNMe$_2$)(dppe)$_2$]Br	b	Me$_2$NH (0.60)	129
[WBr{NN(CH$_2$)$_3$CH$_2$}(dppe)$_2$]Br	b	pyrrolidine (0.27–0.38)	129
[MoBr{NN(CH$_2$)$_3$CH$_2$}(dppe)$_2$]Br	b	pyrrolidine (0.27–0.42)	129
[MoBr(NNC$_4$H$_9$)(dppe)$_2$]	NaBH$_4$/MeOH-benzene (100°C)	butylamines (0.32), NH$_3$ (0.25)	103
[MoBr(NNC$_4$H$_9$)(dppe)$_2$]	NaOMe/MeOH-benzene (100°C)	total nitrogen base (0.30)	103
[WBr$_2$(NN=CMe$_2$)(PMe$_2$Ph)$_3$]	LiAlH$_4$/ether	^{i}PrNH$_2$ (0.93–0.95), NH$_3$ (0.55–0.60)	124
[WBr$_2${NN=C(CH$_2$)$_5$}(PMe$_2$Ph)$_3$]	LiAlH$_4$/ether	C$_6$H$_{11}$NH$_2$ (0.88–0.91), NH$_3$ (0.65)	124
[WBr$_2$(NN=CMe$_2$)(PMe$_2$Ph)$_3$]	HBr/CH$_2$Cl$_2$	N$_2$H$_4$ (0.57–0.64), Me$_2$C=NN=CMe$_2$ (0.25–0.30)	124
[MoCl$_2$(NN=CMe$_2$)(PMe$_2$Ph)$_3$]	HCl/CH$_2$Cl$_2$	N$_2$H$_4$ (0.32), NH$_3$ (0.22)	65,122

a. mol/M atom.   b. direct base distillation from 40% aq. KOH.

other hand, the reaction of $[WBr(NNMe_2)(dppe)_2]Br$ with H_2SO_4 yields very small amounts of Me_2NH and $MeNH_2$. The yields of amines in these reactions are at best 0.95 mol/M atom. No ammonia was detected from these reactions and hence only one-half of the theoretical amount of nitrogen present per molecule was reduced.[129]

Recently the hydrogenolysis of the alkyldiazenido complex $[MoBr(NNC_4H_9)(dppe)_2]$ has been examined in detail.[103] Although no hydrogenolysis occurs with dihydrogen in the presence of palladium/charcoal, sodium borohydride in a benzene-methanol solution is very effective at reducing the diazenido ligand at 100°C into ammonia and amines (butylamine and N-methylbutylamine) in ca. 60% yield (Table 5). This hydrogenolysis is initiated by the rapid reaction of sodium borohydride with methanol to produce $B(OCH_3)_4^-$ and dihydrogen. The fact that the reaction temperature is an important factor in controlling the extent to which the reduction occurs, coupled with the necessity for an induction period (1.25 h at 90°C), may point to the existence of a slow rate-controlling step in the reduction, which probably corresponds to the replacement of bromide ion by methoxide ion (eq. 34). In fact, the replacement of the bromide ion of $[MoBr(NNC_4H_9)(dppe)_2]$ by azide or thiocyanate ion, which proceeds very slowly at room temperature, is facilitated by raising the temperature to give the corresponding azido- or thiocyanate-alkyldiazenido complexes $[MoX(NNC_4H_9)(dppe)_2]$ ($X = N_3$ or NCS). Replacement of the bromide ion by the much harder oxygen atom of a methoxide ion will cause the metal to feed more electron density into the diazenido ligand, and hence the diazenido ligand may be easily reduced by methanol. When sodium methoxide is used in place of sodium borohydride, formation of nitrogen base is also observed but in significantly lower yield.

$$[MoBr(NNC_4H_9)(dppe)_2] + B(OCH_3)_4^- \longrightarrow$$

$$Br^- + B(OCH_3)_3 + [Mo(OCH_3)(NNC_4H_9)(dppe)_2] \qquad (34)$$

Diazoalkane complexes with the monophosphine PMe_2Ph release organo-nitrogen compounds on treatment with hydrogen halides or $LiAlH_4$ at room temperature. The reaction of $[WBr_2(NN=CMe_2)(PMe_2Ph)_3]$ or $[WBr_2[NN=C(CH_2)_5](PMe_2Ph)_3]$ with $LiAlH_4$, after several color changes, produces $^iPrNH_2$ or $C_6H_{11}NH_2$ in over 90% yield together with about 60% ammonia. In contrast, when HBr gas is introduced to a CH_2Cl_2 solution of $[WBr_2(NN=CMe_2)(PMe_2Ph)_3]$, hydrazine and acetone azine are formed. In the latter reaction, the tungsten-containing complex is recovered from the reaction mixture as $[WBr_4(PMe_2Ph)_2]$.[124] Some tentatively formulated intermeidates have been isolated but the reaction scheme is not yet clear. In the case of molybdenum diazoalkane complexes similar treatment with HCl gas gives a more complicated result; the product consists of hydrazine, ammonia and a little azine[65,122] (Table 5).

On the other hand, no organo-nitrogen compounds have been detected from the reaction mixture of diazoalkane complexes containing dppe with acids or bases. As already mentioned, the reaction of $[WBr(NN=CH_2)(dppe)_2]Br$ with $LiAlH_4$ or PhLi gives diazenido complexes $[WBr(NNCH_3)(dppe)_2)]$ or $[WBr(NNCH_2Ph)(dppe)_2]$, and further reaction hardly proceeds.[119] The reaction with alkoxide ions is more complicated. When $[WBr(NN=CH_2)(dppe)_2]Br$ is treated with NaOMe, the methoxymethyldiazenido complex $[WBr(NNCH_2OMe)(dppe)_2]$ can be isolated. However, the reaction with NaO^tBu in place of NaOMe affords not only the analogous product $[WBr(NNCH_2O^tBu)(dppe)_2]$ but also $[W(N_2)_2(dppe)_2]$. Similar treatment of $[WBr(^{15}N_2CHMe)(dppe)_2]Br$ under $^{14}N_2$ gives $[W(^{15}N_2)(^{14}N_2)(dppe)_2]$, indicating that one dinitrogen in the recovered bis(dinitrogen) complex comes from atmospheric dinitrogen and the other is derived from C-N bond cleavage in the alkoxymethyldiazenido ligand.

4.4.4 Diazene complexes

Almost all the nitrogen-carbon bond formation reactions described so far are concerned with molybdenum and tungsten dinitrogen complexes. The other dinitrogen complex known to form a nitrogen-carbon bond in a novel way is the manganese complex $[CpMn(CO)_2(N_2)]$. Though no reactions occur with hydrides such as $LiAlH_4$, $Li[BEt_3H]$, LiH, NaH and KH, an organolithium, phenyllithium for instance, reacts with this complex in THF at -30°C to give $[CpMn(CO)_2(NC_6H_5=NH)]$ after quenching with H_2SO_4.[35,37] The ^{1}H-nmr spectrum of this complex shows a band at 14.50 ppm assigned to the proton bound to the sp^2-hybridized nitrogen atom. The first step of this reaction is obviously nucleophilic attack of the phenyl carbanion at the nitrogen atom bound to metal, resulting in the formation of $[CpMn(CO)_2(NC_6H_5=N^-Li^+)]$. It is interesting that the coordinated dinitrogen is attacked by the carbanion in preference to the carbonyl ligand. The second step is protonation at the negatively charged terminal nitrogen atom to afford the phenyldiazene complex $[CpMn(CO)_2(NC_6H_5=NH)]$. The binuclear diazene complex $[\{CpMn(CO)_2\}_2(NC_6H_5=NH)]$ can be prepared by the reaction of this diazene complex with $[CpMn(CO)_2(THF)]$. These phenyldiazene complexes are very unstable and decompose in the solid state above -15°C to give the dinitrogen complexes $[CpMn(CO)_2(N_2)]$ and $[\{CpMn(CO)_2\}_2(N_2)]$. When methyllithium is used instead of phenyllithium, the analogous reaction proceeds to give $[CpMn(CO)_2(NMe=N^-Li^+)]$. Protonation of the latter complex by HCl or H_2SO_4 solution at -30°C, however, results in the formation of $[CpMn(CO)_2(N_2)]$ and CH_4. On the other hand, treatment of the complex with $Me_3O^+BF_4^-$ gives the dimethyldiazene complex $[CpMn(CO)_2(NMe=NMe)]$.[36-7] The complex exhibits two resonances in the ^{1}H-nmr spectrum at 4.00 and 4.18 ppm, assigned to two non-equivalent methyl protons, and $\nu(N=N)$ at 1518 cm^{-1} in the IR spectrum. This indicates coordination of dimethyldiazene to the metal through the lone pair of one nitrogen atom. The dimethyldiazene complex reacts with dinitrogen (ca. 100 atm) at 20°C to give the parent dinitrogen complex $[CpMn(CO)_2(N_2)]$ and free dimethyldiazene. On the basis of these reactions an elegant catalytic cycle (Fig. 7) can, in principle, be written, in which ligating dinitrogen is reduced to dimethyldiazene by consecutive nucleophilic and electrophilic attack, while the site of dinitrogen coordination stays intact. However, in practice, the catalyst $[CpMn(CO)_2]$ decomposes through a side-reaction after a short period.

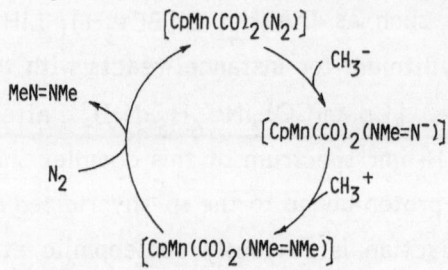

Figure 7.   Conversion of ligating dinitrogen into diazene.[36-7]

Another   electron-poor   complex,   <u>trans</u>-[ReCl(CO)$_2$(N$_2$)(PPh$_3$)$_2$],
reacts with MeLi/H$^+$ to give a hydroxy-carbene dinitrogen complex <u>trans-</u>
[ReCl{C(OH)Me}(CO)(N$_2$)(PPh$_3$)$_2$] whereas under similar conditions <u>trans-</u>
[ReCl(CO)$_3$(PPh$_3$)$_2$] is attacked at the metal by MeLi to give <u>trans-</u>
[ReMe(CO)$_3$(PPh$_3$)$_2$].[130] Further investigations are needed to elucidate the
factors determining the actual site of RLi attack (N$_2$, CO or the metal).

5.   MORE RECENT RESULTS

Tantalum   and   niobium   complexes   containing   diimido   bridging
dinitrogen ligands have been reported.[152-5]   An xray crystallographic
study[152] of the complex [{Ta(=CHCMe$_3$)(CH$_2$CMe$_3$)(PMe$_3$)$_2$}$_2$(μ-N$_2$)]
prepared according to eq. 35 indicated that the linear Ta-(μ-N$_2$)-Ta
fragment is best described as a Ta=NN=Ta system (Ta-N = 1.840(4) Å and N-
N = 1.298(12) Å) and the N-N bond is significantly activated compared to
such other end-on bridging dinitrogen complexes of the early transition
metals as [{Cp*$_2$Ti}$_2$(μ-N$_2$)] or [{Cp*$_2$Zr(N$_2$)}$_2$(μ -N$_2$)]. Diimido bridging
dinitrogen complexes of tantalum and niobium are also obtained from
metathesis-like reactions between alkylidene complexes and a diimine
reagent PhCH=NN=CHPh.   The complex [{TaCl$_3$(P(CH$_2$Ph)$_3$)(THF)}$_2$(μ-
N$_2$)] prepared by the latter method has also been shown by xrays to contain
a Ta=NN=Ta bridge.[153] Interestingly, these tantalum μ-N$_2$ complexes such
as [{Ta(=CHCMe$_3$)(PMe$_3$)$_2$Cl}$_2$(μ-N$_2$)] and [{TaCl$_3$(PEt$_3$)$_2$}$_2$(μ-N$_2$)]
react not only with HCl to give hydrazine in high yield, but also with
acetone to give dimethylketazine in moderate yield.[154-5]

$$\text{[Ta(=CHCMe}_3\text{)(PMe}_3\text{)}_2\text{Cl}_3\text{]} \xrightarrow{\text{2Na/Hg, PMe}_3\text{, N}_2} \text{[\{Ta(=CHCMe}_3\text{)(PMe}_3\text{)}_2\text{Cl\}}_2\text{(}\mu\text{-N}_2\text{)]}$$

$$\xrightarrow{\text{LiCH}_2\text{CMe}_3} \text{[\{Ta(=CHCMe}_3\text{)(CH}_2\text{CMe}_3\text{)(PMe}_3\text{)}_2\text{\}}_2\text{(}\mu\text{-N}_2\text{)]} \qquad (35)$$

Recently, another synthetic route to μ-N$_2$ complexes of niobium and tantalum has been developed. The reaction of NbCl$_5$ with (Me$_3$Si)$_2$NN(SiMe$_3$)$_2$ in the presence of THF gives [\{NbCl$_3$(THF)$_2$\}$_2$(μ-N$_2$)] in high yield. The μ-N$_2$ complex or its tantalum analogue reacts with Me$_3$SiS$_2$CNEt$_2$ to afford the dinitrogen complexes with exclusively sulfur co-ligands [\{M(S$_2$CNEt$_2$)$_3$\}$_2$(μ-N$_2$)] (M = Nb or Ta). The dinitrogen ligand in these compounds is quantitatively converted into hydrazine on acidolysis.[156]

The complex [\{W(N$_2$)$_2$(PEt$_2$Ph)$_3$\}$_2$(μ-N$_2$)], which reacts with HCl to give hydrazine, has been prepared and structurally characterized by xray analysis. The coordination mode of the dinitrogen ligands to the metal is essentially the same as in [\{Cp*$_2$M(N$_2$)\}$_2$(μ-N$_2$)] (M = Ti or Zr).[157]

A novel titanium complex containing a triply-coordinated dinitrogen ligand of type I has been prepared from the reaction of [[μ-(η^1:η^5-C$_5$H$_4$)(η^5-C$_5$H$_5$)$_3$Ti$_2$] with dinitrogen. The xray crystallographic analysis showed that the complex consists of a μ-N$_2$ complex (μ-N$_2$)[η^5:η^5-C$_{10}$H$_8$)(η^5-C$_5$H$_5$)$_2$Ti$_2$][(η^1:η^5-C$_5$H$_4$)(η^5-C$_5$H$_5$)$_3$Ti$_2$] and a titanocene unit with two molecules of diglyme. The dinitrogen ligand in the complex is essentially σ-bonded to the one formally divalent titanium atom in \{μ-(η^1:η^5-C$_5$H$_4$)\}(η^5-C$_5$H$_5$)$_3$Ti$_2$ and is also coordinated in a (σ+π) fashion to the two titanium atoms in (η^5:η^5-C$_{10}$H$_8$)(η^5-C$_5$H$_5$)$_2$Ti$_2$. This multiple metal bonding to dinitrogen results in considerable lengthening of the N-N distance (1.301 Å). Aqueous hydrolysis of the complex in ether solvent gives ammonia in high yield.[158] On the other hand, type (II) multiple-bonding of dinitrogen has been demonstrated in bimetallic diazenido complexes

$[\{(PhMe_2P)_3(Py)ClW(\mu\text{-}N_2)AlCl_2\}_2]C_6H_6$[159] and $[\{(Me_3P)_3Co(\mu\text{-}N_2)AlMe_2\}_2]$,[160] where the dinitrogen ligand is triply end-on bridged by the one W or Co atom on one side and the two Al atoms on the other.

M ===== N ===== N ===== M   M ===== N ===== N⟨ M / M

(I)   (II)

More recently, nitrogen-15 chemical shifts and coupling constants have been reported for terminal dinitrogen complexes of Mo, W, Re, Fe, Ru, Os, and Rh.[161] Spectrophotometric titration of a dilute solution of cis-$[M(N_2)_2(PMe_2Ph)_4]$ (M = Mo or W) with HX (X = Cl, Br, or I) in methanol has shown that a protic solvent plays a unique role in this reaction and a common product $[M(NNH_2)(OCH_3)_2(PMe_2Ph)_3]$ is formed.[162] Alkoxy-hydrazido(2-) derivatives $[W(OR)(NNH_2)(dppe)_2][Co(CO)_4]$ (R = CH_3, C_2H_5, etc.) have been prepared by reaction of trans-$[W(N_2)_2(dppe)_2]$ with $HCo(CO)_4$.[163] Solvent effects on the protonation of ligating dinitrogen may be caused by such kinds of reaction. Recent ^{31}P nmr studies on the reaction of eq. 15 showed that a mixture of two hydrazido(2-) complexes was at first formed with rapid evolution of dinitrogen (ca. 1.0 mole) and resonances assigned to one of them rapidly disappeared, with appearance of free PPh_3, whereas resonances due to the other disappeared more slowly. Dissociation of PPh_3 from the former hydrazido(2-) complex seems to be a key step for formation of ammonia.[164] Interestingly, hydrazine was detected upon early quenching of the reaction with water, although it is not a product in the reaction.[165] Elucidation of the structure of the hydrazine-forming intermediate is eagerly awaited since hydrazine was detected upon quenching the nitrogenase reaction in vitro with acid or base.[166]

Detailed studies on the reaction of cis-$[W(N_2)_2(PMe_2Ph)_4]$ with a mixture of alcohol (ROH) and ketone ($R^1R^2C=O$) have recently revealed that ketazine ($R^1R^2C=NN=CR^1R^2$) is formed as the major product, which is converted into hydrazine by hydrolysis (see eq. 19).[167] It has been postulated that the reaction proceeds via the diazoalkane complex $[W(OR)_2(N_2CR^1R^2)(PMe_2Ph)_3]$, formed by the condensation reaction of $[W(OR)_2(NNH_2)(PMe_2Ph)_3]$ with ketone. The diazoalkane ligand is then liberated by alcohol as $H_2NN=CR^1R^2$ which reacts further with ketone to give ketazine as the final product.

$$\underline{cis}\text{-}[W(N_2)_2(PMe_2Ph)_4] \xrightarrow[\text{(KOH)}]{ROH/R^1R^2C=O} R^1R^2C=NN=CR^1R^2 \qquad (36)$$

(R = Me or Et; $R^1 = R^2$ = Me, Et, or Ph)

Reaction of $\underline{trans}$-$[M(N_2)_2(dppe)_2]$ (M = Mo or W) with trifluoroacetic anhydride gives trifluoroacetyldiazenido complexes $[M(dppe)_2(N_2COCF_3)]$ $(O_2CCF_3)]$ in high yield.[168] Treatment of $[WBr(dppe)_2(NNH_2)]Br$ with $[Ph_2I]$ $[Br]$ in the two phase system CH_2Cl_2 - 5% aqueous K_2CO_3 yields a formyldiazenido complex $[WBr(dppe)_2(N_2CHO)]$. The reaction using $CHFBr_2$ as the organic phase produces a dinuclear complex $[\{WBr(dppe)_2\}_2 \, [\mu\text{-} CH(N_2)_2\}]^+$ which contains a formazanido(3-) ligand $[N=N=CH=N=N]^{3-}$, bridging two tungsten atoms in a fully conjugated 7-atom chain.[169]

Electrochemical reduction of $\underline{trans}$-$[MoBr\{\overline{N_2CH_2(CH_2)_3CH_2}\}$ $(dppe)_2]^+$, which is obtained from $\underline{trans}$-$[Mo(N_2)_2(dppe)_2]$ and $Br(CH_2)_5Br$, at a Pt electrode in THF-0.2 M $[NBu_4]$ $[BF_4]$ under dinitrogen yields N-aminopiperidine and $\underline{trans}$-$[Mo(N_2)_2(dppe)_2]$ via a four-electron reduction pathway. Thus a cycle for the fixation of dinitrogen as an organohydrazine is plausible.[170] By contrast, two-electron electrochemical reduction of the organo-hydrazido(2-) complex gives $[Mo\{\overline{NNCH_2(CH_2)_3CH_2}\}(dppe)_2]$, which reacts rapidly with anhydrous HBr to give $[MoBr(NH)(dppe)_2]Br$ and piperidine.[171] On the other hand, one-electron reduction of $\underline{trans}$-$[WF(N_2CH_2)(dppe)_2]$ $[BF_4]$ in a MeCN electrolyte at a mercury pool cathode leads to carbon-carbon coupling of the diazomethane ligand giving $[(dppe)_2FW(N_2CH_2CH_2N_2)WF(dppe)_2]$ in moderate yield.[172]

Vinyldiazenido complexes $[WBr(N_2CR^3=CR^1R^2)(dppe)_2]$ have been prepared by reaction of $[WBr(NNH_2)(dppe)_2]Br$ with cyanoalkenes $R^1R^2C=CR^3Cl$ ($R^1 = R^2$ = CN, R^3 = H; $R^1 = R^2$ = CN, R^3 = Cl; R^1 = CN, R^2 = COOEt, R^3 = H) in the presence of triethylamine.[173] Electrochemical reduction of $[WF(dppe)_2[NNH(CH=C(CN)_2)]][BF_4]$, which is derived from $[WF(NNH_2)(dppe)_2][BF_4]$ by a similar reaction, gives 5-amino-4-cyanopyrazole in up to 50% yield.[174] Trimethylsilylation of coordinated dinitrogen has been achieved by reaction of tungsten dinitrogen complexes such as cis-$[W(N_2)_2(PMe_2Ph)_4]$ with Me_3SiI. The xray structures of two silylated compounds, trans-$[WI(N_2SiMe_3)(PMe_2Ph)_4]$ and mer-$[WI_2(NNHSiMe_3)(PMe_2Ph)_3]$, showed that they contain the trimethylsilyl-diazenido and trimethylsilylhydrazido(2-) ligands, respectively.[175]

The linear geometry of type III as a four-electron donor is very common in hydrazido(2-) complexes. Recently, the bent geometry of type IV as a formally two-electron donor has been demonstrated by xray analysis in the organohydrazido(2-) complexes $[Cp_2WH(p-NNHC_6H_4F)][PF_6]\cdot Me_2CO$[176] and $[CpRe(CO)_2\{NNMe(p-C_6H_4OMe)\}]$.[177] The former complex and its analogues rearrange in solution to give arylhydrazido(1-) complexes $[Cp_2W(H_2NNAr)]X$ (Ar = Ph, p-C_6H_4F, etc.; X = BF_4 or PF_6) in which the hydrazido(1-) ligand is bound side-on to the metal through both nitrogen atoms (type V).[178] On the other hand, the latter rhenium complex can be protonated by HBF_4 to yield the unidentate hydrazido(1-) complex $[CpRe(CO)_2\{NHNMe(p-C_6H_4OMe)\}][BF_4]$ having the structure of type VI.[177] Simple electron counting does not always provide an explanation for the geometry of a hydrazido(2-) ligand. Thus, the bis-hydrazido(2-) complex $[Mo(NNMePh)_2(S_2CNMe_2)_2]$ show linear geometry in both the hydrazido(2-) ligands although the complex formally acquires a 20-electron count.[179] Protonation of the complex with one mole of HCl gives $[Mo(NNMePh)(NHNMePh)(S_2CNMe_2)_2]^+$, containing a side-on hydrazido(1-) ligand of type V. The latter complex reacts further with an excess of acid to give $[MoCl_2(NNMePh)(S_2CNMe_2)_2]$ and $PhMeNNH_2$.[180] Side-on dihapto-bonding of diazenido(1-) to titanium has been demonstrated by xray crystallography in the complex [CpTiCl$_2$(NNPh)].[181]   The chemistry of these organo-diazenido(1-), hydrazido(2-), and hydrazido(1-) complexes might shed some light on the mechanism for reduction of ligating dinitrogen to nitrogen hydrides.

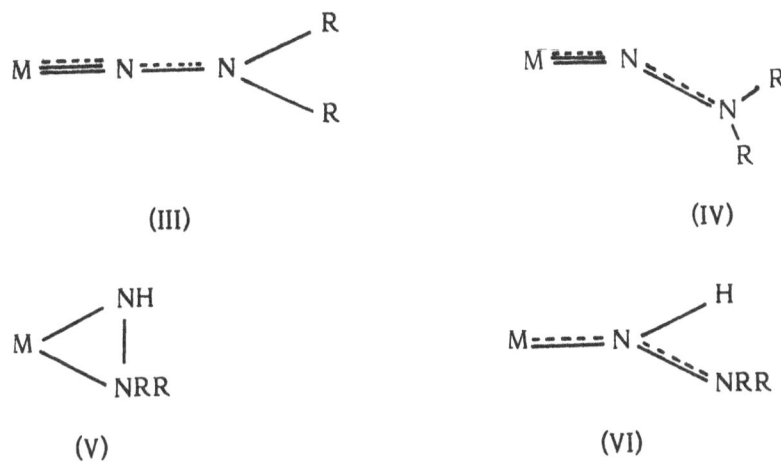

A series of mono-dinitrogen complexes [Mo(N$_2$)(triphos)(L$_2$)] (L$_2$ = 2PMe$_2$Ph, dppe, etc.) have been prepared by careful treatment of [MoCl$_3$(triphos)] with L$_2$ under a limited volume of N$_2$. Treatment of these complexes with HBr in CH$_2$Cl$_2$ gives ammonia and hydrazine in moderate yields.[182]

Reduction of WCl$_6$ or [WCl$_4$(PR$_2$Ph)$_2$] with Mg in THF in the presence of PR$_2$Ph (R = Et or ^{n}Pr) gives a mixture of three types of dinitrogen complexes which can be crystallized independently from the reaction solution.[183]   The first of these has the well known structure trans-[W(N$_2$)$_2$(PR$_2$Ph)$_4$], while the second has an interesting dinuclear structure with bridging dinitrogen, [{W(N$_2$)$_2$(PR$_2$Ph)$_3$}$_2$(μ-N$_2$)], already described in this Section for R = Et, and the third is a complex with arene ligands, [W(η^6-C$_6$H$_5$PR$_2$)(N$_2$)(PR$_2$Ph)$_2$]. As expected by analogy with the related PMe$_2$Ph and PPh$_2$Me species, the first (R = Et) gives ammonia (1.8 mol/W atom) on treatment with H$_2$SO$_4$/MeOH. On the other hand, the second series of compounds (R = Et or ^{n}Pr) give hydrazine (0.2 mol/W atom) and no ammonia and the third (R = ^{n}Pr) loses N$_2$ quantitatively when reacted with HCl gas in THF.

Reduction of $[MoCl_3(THF)_3]$ by Mg in the presence of PPh^nPr_2 results in the formation of trans-$[Mo(N_2)_2(PPh^nPr_2)_4]$, which in the absence of an excess of PPh^nPr_2 is unstable in solution under N_2, being converted into the novel mononuclear tris-dinitrogen complex mer-$[Mo(N_2)_3(PPh^nPr_2)_3]$.[184]

Electrolytic reduction is a most attractive way to design catalytic nitrogen fixing systems, by coupling protonation to electronation. Recently electroreduction of the hydrazido(2-) complex cation trans-$[W(NNH_2)(OTs)(dppe)_2]^{2+}$ (Ts = p-$MeC_6H_4SO_2$) was carried out in THF with 0.2 M $[Bu_4N][BF_4]$ as supporting electrolyte, and free ammonia (0.22 - 0.24 mol/W atom) and hydrazine (0.01 - 0.02 mol/W atom) were detected in the reaction mixture concurrent with the regeneration of trans-$[W(N_2)_2(dppe)_2]$ (0.85 - 0.95 mol/W atom).[185] The total stoichiometry of this reaction can be represented as in (37). Since the regenerated dinitrogen complex is protonated:

$$8[W(NNH_2)(OTs)(dppe)_2]^+ + 16e^- + 9N_2$$

$$\longrightarrow 2NH_3 + 8[W(N_2)_2(dppe)_2] + 8TsO^- + 5(H_2) \qquad (37)$$

to reproduce the hydrazido(2-) complex in situ by stoichiometric addition of TsOH, successive electrolysis-protonation cycles can be performed on a single catholyte. After three cycles, the yield of ammonia reaches 0.73 mol/W atom and that of hydrazine, 0.01 mol/W atom, indicating the cyclic conversion of dinitrogen to ammonia. Electrosynthesis of trans-$[M(N_2)_2(dppe)_2]$ was also achieved by the controlled-potential reductions of trans-$[MX_2(dppe)_2]$ (M = Mo: X = Cl, Br, I, SPh or S^nBu; M = W: X = Cl) under N_2.[186]

An unsubstituted hydrazido(1-) complex has been reported as the product of the reaction of $[Cp*WMe_4][PF_6]$ with 2 equivalents of hydrazine.[187] This complex, with proposed formula $[Cp*WMe_4(NHNH_2)]$ decomposes quantitatively in polar solvents such as CH_2Cl_2 into the W(IV) dinitrogen complex $[Cp*WMe_3]_2(\mu-N_2)$ on treatment with $[Cp*WMe_4][PF_6]$.

Though cis-$[W(N_2)_2(PMe_2Ph)_4]$ reacts with alcohol/ketone systems at 50°C to give ketazines according to eq. 36, treatment of this complex with acetylacetone under the same conditions affords a novel alkenyldiazenido complex (eq. 38):[188]

$$\text{cis-}[W(N_2)_2(PMe_2Ph)_4] \quad \xrightarrow{\text{acacH, MeOH 50°C}} \quad \text{mer-}[W(acac)(NNCMe=CHCOMe)(PMe_2Ph)_3] \quad (38)$$

Analogous dppe complexes trans-$[MF(NNCMe=CHCOMe)(dppe)_2]$ (M = W or Mo) are produced by the reaction of the diazoalkane complexes trans-$[MF(NN=CMeCH_2COMe)(dppe)_2]$ $[BF_4]$ with NaOMe. When these diazenido complexes are treated with 1 equivalent of HCl gas, the alkenyl-hydrazido(2-) complex mer-$[W(acac)(NNHCMe=CHCOMe)(PMe_2Ph)_3]$ and the diazoalkane complex trans-$[WF(NN=CMeCH_2COMe)(dppe)_2]Cl$ are formed as crystals. However, the ^{1}H nmr spectrum of the former complex shows the existence of an equilibrium between the alkenylhydrazido(2-) structure and two diazoalkane structures WNN=CMeCH$_2$COMe and WNN=CMe=CHC(OH)Me, although the latter exists only in the keto-diazoalkane structure in solution.

An optically active dinitrogen complex with asymmetric phosphine ligands, trans-$[Mo(N_2)_2(Chiraphos)_2]$ (Chiraphos = (-)-(2S,3S)-bis(diphenylphosphino)butane) has been prepared.[189] This complex reacts with racemic 2-bromooctane according to the mechanism shown in Fig. 5 to afford a high yield of the diazenido complex $[MoBr(NNCMeC_6H_{13})(Chiraphos)_2]$, in which one diastereomer is in excess by about 10%. However, hydrogenolysis of the mixture of diastereomeric alkyldiazenido complexes by sodium borohydride in benzene/methanol at 100°C results in the formation of 2-aminooctane that is totally racemized.

Extensive ^{15}N nmr studies have been reported on a wide range of dinitrogen complexes and their derivatives. The nmr spectra of terminal dinitrogen complexes of Mo, W, Re, Fe, Ru, Os and Rh show that the nitrogen shielding increases with increasing atomic number across the transition metal series, or down a particular group, the increase being steeper for N_α than for N_β. For complexes of metals later in the series N_α is more shielded than N_β but a crossover of resonance positions appears at Mo and at or near W.[190] The replacement of one N_2 ligand in trans-$[M(^{15}N_2)_2(dppe)_2]$ (M = Mo or W) by RCN leads to deshielding of both nitrogens in the remaining $^{15}N_2$ ligand. Among a series of trans-$[M(^{15}N_2)_2(dppe)_2]$ (R = alkyl or aryl) the shielding of N_α (but not of N_β) increases with increasing the σ-donor properties of the trans nitrile ligand.[191] As for bridging dinitrogen ligands, three types of complexes have been studied. In the case of Mo, W, Re and Os complexes with a $^{15}N_2$ bridge to a main group acceptor $AlMe_3$, the shielding of N_β increases upon attachment of $AlMe_3$ in the order of Os > Re > W but the increase is negligible for a Mo complex. On the other hand, in binuclear adducts of trans-$[ReCl(^{15}N_2)(PMe_2Ph)_4]$ with $TiCl_4$, $TaCl_5$ and $NbCl_5$ both N_α and N_β are deshielded on adduct formation. The Ti, Zr, Ta and W complexes with symmetrical N_2 bridges are characterized by low nitrogen shielding and the shifts follow the normal periodicity, despite the differences in the coordination sphere.[192] A ^{15}N-nmr study of hydrazido(2-) complexes has also been carried out, and shows that N_β is more highly shielded than N_α and that the NNH_2 group is nearly planar rather than pyramidal.[193]

Some theoretical calculations on protonation of ligating dinitrogen have been carried out,[194-6] and over the past years several further reviews on dinitrogen complexes of transition metals have appeared.[199-202]

6.   CONCLUSION

In the past fifteen years extensive studies on dinitrogen complexes have remarkably increased our information about coordination modes and reactivities of ligating dinitrogen.   It has already been shown that dinitrogen, which was considered as one of the most inert molecules, can be converted into ammonia, hydrazine and some kinds of organonitrogen compounds under ambient conditions.   Although these reactions are at present far from being practical catalytic processes, further extensive studies will open a new chemistry of dinitrogen and finally lead to the development of new industrial processes of nitrogen fixation in the near future.

REFERENCES

1. A.D. Allen and C.V. Senoff, Chem. Commun. 621 (1965).

2. "Recent Developments in Nitrogen Fixation", W.E. Newton, J.R. Postgate and C. Rodriguez-Barrueco, Eds., Academic Press, New York (1977).

3. "A Treatise on Dinitrogen Fixation", R.W.F. Hardy, F. Bottomley and R.C. Burns, Eds., John Wiley and Sons, New York (1979).

4. "Molybdenum Chemistry of Biological Significance", W.E. Newton and S. Otsuka, Eds., Plenum Press, New York (1980).

5. E.I. Stiefel, Prog. Inorg. Chem. 22:1 (1977).

6. J. Chatt, J.R. Dilworth and R.L. Richards, Chem. Rev. 78:589 (1978).

7. "Nitrogen Fixation", W.P.P. Stewart and J.P. Gallen, Eds., Academic Press, New York (1980).

8. A. Yamamoto, S. Kitazume, L.S. Pu and S. Ikeda, Chem. Commun. 79 (1967).

9. A. Misono, Y. Uchida and T. Saito, Bull. Chem. Soc. Jpn 40:700 (1967).

10. A. Sacco and M. Rossi, Chem. Commun. 316 (1967).

11. M. Hidai, K. Tominari, Y. Uchida and A. Misono, Chem. Commun. 1392 (1969).

12. M. Hidai, K. Tominari and Y. Uchida, J. Am. Chem. Soc. 94:110 (1972).

13. J. Chatt, G.A. Heath and R.L. Richards, J.C.S. Dalton Trans. 2074 (1974).

14. J.M. Manriquez and J.E. Bercaw, J. Am. Chem. Soc. 96:6229 (1974).

15. R.D. Sanner, D.M. Duggan, T.C. McKenzie, R.E. Marsh and J.E. Bercaw, J. Am. Chem. Soc. 98:8358 (1976).

16. T. Tatsumi, H. Tominaga, M. Hidai and Y. Uchida, J. Organomet. Chem. 114:C27 (1976).

17. M. Sato, T. Tatsumi, T. Kodama, M. Hidai, T. Uchida and Y. Uchida, J. Am. Chem. Soc. 100:4447 (1978).

18. S. Pell and J.N. Armor, J. Am. Chem. Soc. 94:686 (1972).

19. D. Sellmann, Angew. Chem. Int. Ed. Engl. 10:919 (1971).

20. J.P. Collman, K. Kubota, F.D. Vastine, J.Y. Sun and J.W. Kang, J. Am. Chem. Soc. 90:5430 (1968).

21. C. Barrientos-Penna and D. Sutton, J.C.S. Chem. Commun. 111 (1980).

22. R. Hoffmann, M. M-L. Chen and D.L. Thorn, Inorg. Chem. 16:503 (1977).

23. H. Veillard, Nouveau J. Chim. 2:215 (1978).

24. T. Yamabe, K. Hori, T. Minato and K. Fukui, Inorg. Chem. 19:2154 (1980).

25. F. Bottomley, in Ref. 3, p. 109, and references therein.

26. J. Chatt, D.P. Merville and R.L. Richards, J. Chem. Soc. (A) 2841 (1969).

27. D.J. Darensbourg and C.I. Hyde, Inorg. Chem. 10:431 (1971).

28. P.K. Maples, F. Basolo and R.G. Pearson, Inorg. Chem. 10:765 (1971).

29. D.L. Dubois and R. Hoffmann, Nouveau J. Chim. 1:479 (1977).

30. J.N. Murrell, A. Al-Derzi, G.J. Leigh and M.F. Guest, J.C.S. Dalton Trans. 1452 (1980).

31. J. Jeffery, M.F. Lappert and P.I. Riley, J. Organomet. Chem. 181:25 (1979).

32. C. Kruger and Y-H. Tsay, Angew. Chem. Int. Ed. Engl. 12:998 (1973).

33. J. Chatt, C.M. Elson, N.E. Hooper and G.J. Leigh, J.C.S. Dalton Trans. 2392 (1975).

34. H. Binder and D. Sellmann, Angew. Chem. Int. Ed. Engl. 12:1017 (1973).

35. D. Sellmann and W. Weiss, Angew. Chem. Int. Ed. Engl. 16:880 (1977).

36. D. Sellmann and W. Weiss, Angew. Chem. Int. Ed. Engl. 17:269 (1978).

37. D. Sellmann and W. Weiss, J. Organomet. Chem. 160:183 (1978).

38. J. Chatt, C.T. Kan, G.J. Leigh, C.J. Pickett and D.R. Stanley, J.C.S. Dalton Trans. 2032 (1980).

39. J. Chatt, M.E. Fakley, R.L. Richards, J. Mason and I.A. Stenhouse, J. Chem. Res. (S) 44 (1979).

40. J.E. Bercaw, E. Rosenberg and J.D. Roberts, J. Am. Chem. Soc. 96:612 (1974).

41. J.E. Bercaw, J. Am. Chem. Soc. 96:5078 (1974).

42. J.M. Manriquez, D.R. McAlister, E. Rosenberg, A.M. Schiller, K.L. Williamson, S.I. Chan and J.E. Bercaw, J. Am. Chem. Soc. 100:3078 (1978).

43. M. Mercer, R.H. Crabtree and R.L. Richards, J.C.S. Chem. Commun. 808 (1973).

44. M. Mercer, J.C.S. Dalton Trans. 1637 (1974).

45. J. Chatt, R.C. Fay and R.L. Richards, J. Chem. Soc. (A) 702 (1971).

46. P.D. Cradwick, J. Chatt, R.H. Crabtree and R.L. Richards, J.C.S. Chem. Commun. 351 (1975)

47. J. Chatt, R.H. Crabtree, E.A. Jeffery and R.L. Richards, J.C.S. Dalton Trans. 1167 (1973).

48. M.E. Vol'pin and V.B. Shur, Dokl. Akad. Nauk. SSSR 156:1102 (1964).

49. A.E. Shilov, A.K. Shilova, E.F. Kvashina and T.A. Vorontsova, Chem. Commun. 1590 (1971).

50. Yu. G. Borodko, I.N. Ivleva, L.M. Kachapina, E.F. Kvashina, A.K. Shilova and A.E. Shilov, J.C.S. Chem. Commun. 169 (1973).

51. Yu. G. Borodko, I.N. Ivleva, L.M. Kachapina, S.I. Salienko, A.K. Shilova and A.E. Shilov, J.C.S. Chem. Commun. 1178 (1972)

52. J.E. Bercaw, R.H. Marvich, L.G. Bell and H.H. Brintzinger, J. Am. Chem. Soc. 94:1219 (1972).

53. J.E. Bercaw, in Ref. 2., p. 129.

54. F.W. van der Weij and J.H. Teuben, J. Organomet. Chem. 105:203 (1976).

55. R.D. Sanner, J.M. Manriquez, R.E. Marsh and J.E. Bercaw, J. Am. Chem. Soc. 98:8351 (1976).

56. J.M. Manriquez, R.D. Sanner, R.E. Marsh and J.E. Bercaw, J. Am. Chem. Soc. 98:3042 (1976).

57. M. Hidai, K. Tominari, Y. Uchida and A. Misono, Chem. Commun. 814 (1969).

58. B. Bell, J. Chatt and G.J. Leigh, J.C.S. Dalton Trans. 2492 (1972).

59. J. Chatt, A.J. Pearman and R.L. Richards, J.C.S. Dalton Trans. 2139 (1977).

60. M.W. Anker, J. Chatt, G.J. Leigh and A.G. Wedd, J.C.S. Dalton Trans. 2369 (1975).

61. T.A. George and C.D. Seibold, Inorg. Chem. 12:2544 (1973).

62. T.A. George and M.E. Noble, Inorg. Chem. 17:1678 (1978).

63. J. Chatt, A.J. Pearman and R.L. Richards, Nature (Lond.) 253:39 (1975).

64. J. Chatt, A.J. Pearman and R.L. Richards, J.C.S. Dalton Trans. 1852 (1977).

65. M. Hidai, Y. Mizobe, T. Takahashi and Y. Uchida, Chem. Lett. 1187 (1978).

66. T. Takahashi, Y. Mizobe, M. Sato, Y. Uchida and M. Hidai, J. Am. Chem. Soc. 101:3405 (1979).

67. T. Takahashi, Y. Mizobe, M. Sato, Y. Uchida and M. Hidai, J. Am. Chem. Soc. 102:7461 (1980).

68. M. Hidai, Y. Mizobe, T. Takahashi, M. Sato and Y. Uchida, in Ref. 4, p. 255 (1980).

69. C.R. Brulet and E.E. van Tamelen, J. Am. Chem. Soc. 97:911 (1975).

70. J.A. Baumann and T.A. George, J. Am. Chem. Soc. 102:6153 (1980).

71. J. Chatt, G.A. Heath and R.L. Richards, J.C.S. Chem. Commun. 1010 (1972).

72. J. Chatt, A.J. Pearman and R.L. Richards, J.C.S. Dalton Trans. 1766 (1978).

73. G.A. Heath, R. Mason and K.M. Thomas, J. Am. Chem. Soc. 96:259 (1974).

74. M. Hidai, T. Kodama, M. Sato, M. Harakawa and Y. Uchida, Inorg. Chem. 15:2694 (1976).

75. J. Chatt, M.E. Fakley, R.L. Richards, I.R. Hanson and D.L. Hughes, J. Organomet. Chem. 170:C6 (1979).

76. J. Chatt, M.E. Fakley, P.B. Hitchcock, R.L. Richards, N.T. Luong-Thi and D.L. Hughes, J. Organomet. Chem. 172:C55 (1979).

77. J. Chatt, R.A. Head and G.J. Leigh, J.C.S. Dalton Trans. 1129 (1980).

78. J. Chatt, in Ref. 4, p. 241.

79. J. Chatt, A.J. Pearman and R.L. Richards, J.C.S. Dalton Trans. 1520 (1976).

80. J. Chatt, A.J.L. Pombeiro and R.L. Richards, J.C.S. Dalton Trans. 1585 (1979).

81. J. Chatt, M.E. Fakley, R.L. Richards, J. Mason and I.A. Stenhouse, J. Chem. Res. (S) 322 (1979).

82. J. Chatt, G.J. Leigh, H. Neukomm, C.J. Pickett and D.R. Stanley, J.C.S Dalton Trans. 121 (1980).

83. M. Hidai, T. Takahashi and Y. Uchida, unpublished results.

84. J. Chatt, A.J. Pearman and R.L. Richards, Nature (Lond.) 259:204 (1976).

85. M. Hidai, T. Takahashi, I. Yokotake and Y. Uchida, Chem. Lett. 645 (1980).

86. H. Nishihara, T. Mori, T. Saito and Y. Sasaki, Chem. Lett. 667 (1980).

87. Yu. G. Borodko, M.O. Boitman, L.M. Kachapina, A.E. Shilov and L. Yu. Ukhin, Chem. Commun. 1185 (1971).

88. D. Sellmann, in Ref. 2, p. 53, and references therein.

89. G. Huttner, J. Organomet. Chem. 91:47 (1975).

90. B. Jezowska-Trzebiatowska and P. Sobota, J. Organomet. Chem. 46:339 (1972).

91. T. Itoh, Y. Miura, A. Yamamoto, Y. Sei, K. Miki, N. Tanaka and N. Kasai, Abstr. 26th Symposium Organomet. Chem., Jpn 67 (1979).

92. R. Hammer, H-F. Klein, U. Schubert, A. Frank and G. Huttner, Angew. Chem. Int. Ed. Engl. 15:612 (1976).

93. Y. Miura and A. Yamamoto, Chem. Lett. 937 (1978).

94. K. Jonas and C. Kruger, Angew. Chem. Int. Ed. Engl. 19:520 (1980).

95. M.E. Vol'pin, J. Organomet. Chem. 200:319 (1980) and references therein.

96. E.E. van Tamelen and H. Rudler, J. Am. Chem. Soc. 92:5253 (1970).

97. J. Chatt, G.A. Heath and G.J. Leigh, J.C.S. Chem. Commun. 444 (1972).

98. J. Chatt, A.A. Diamantis, G.A. Heath, N.E. Hooper and G.J. Leigh, J.C.S. Dalton Trans. 688 (1977).

99. T. Tatsumi, M. Hidai and Y. Uchida, Inorg. Chem. 14:2530 (1975).

100. M. Sato, T. Kodama, M. Hidai and Y. Uchida, J. Organomet. Chem. 152:239 (1978).

101. J. Chatt, R.A. Head, G.J. Leigh and C.J. Pickett, J.C.S Dalton Trans. 1638 (1978).

102. K. Iwanami, Y. Mizobe, T. Takahashi, T. Kodama, Y. Uchida and M. Hidai, Bull. Chem. Soc. Jpn 54:1773 (1981).

103. G.E. Bossard, D.C. Busby, M. Chang, T.A. George and S.D.A. Iske, Jr., J. Am. Chem. Soc. 102:1001 (1980).

104. V.W. Day, T.A. George and S.D.A. Iske, J. Am. Chem. Soc. 97:4127 (1975).

105. V.W. Day, T.A. George, S.D.A. Iske and S.D. Wagner, J. Organomet. Chem. 112:C55 (1976).

106. D.C. Busby and T.A. George, J. Organomet. Chem. 118:C16 (1976).

107. D.C. Busby and T.A. George, Inorg. Chem. 18:3164 (1979).

108. J.R. Dilworth, C-T. Kan, R.L. Richards, J. Mason and I.A. Stenhouse, J. Organomet. Chem. 201:C24 (1980).

109. F.C. March, R. Mason and K.M. Thomas, J. Organomet. Chem. 96:C43 (1975).

110. J. Chatt, R.A. Head, G.J. Leigh and C.J. Pickett, J.C.S. Chem. Commun. 299 (1977).

111. J. Chatt, W. Hussain, G.J. Leigh, H. Neukomm, C.J. Pickett and D.A. Rankin, J.C.S. Chem. Commun. 1024 (1980).

112. J. Chatt, W. Hussain, G.J. Leigh and F.P. Terreros, J.C.S. Dalton Trans. 1408 (1980).

113. D. Sutton, Chem. Soc. Rev. 4:433 (1975).

114. H.M. Colquhoun, J. Chem. Res. (S) 325 (1979).

115. P.C. Bevan, J. Chatt, R.A. Head, P.B. Hitchcock and G.J. Leigh, J.C.S. Chem. Commun. 509 (1976).

116. P.C. Bevan, J. Chatt, A.A. Diamantis, R.A. Head, G.A. Heath and G.J. Leigh, J.C.S. Dalton Trans. 1711 (1977).

117. J. Chatt, R.A. Head and G.J. Leigh, J.C.S. Dalton Trans. 1129 (1980).

118. R. Ben-Shoshan, J. Chatt, W. Hussain and G.J. Leigh, J. Organomet. Chem. 112:C9 (1976).

119. R. Ben-Shoshan, J. Chatt, C.J. Leigh and W. Hussain, J.C.S. Dalton Trans. 771 (1980).

120. M. Hidai, Y. Mizobe and Y. Uchida, J. Am. Chem. Soc. 98:7824 (1976).

121. M. Hidai, Y. Mizobe, M. Sato, T. Kodama and Y. Uchida, J. Am. Chem. Soc. 100:5740 (1978).

122. M. Hidai, Y. Mizobe, T. Takahashi, M. Sato and Y. Uchida, in 3rd Int. Conf. Chem. Uses Molybdenum, H.F. Barry and P.C.H. Mitchell, Eds., p. 287 (1979).

123. Y. Mizobe, Y. Uchida and M. Hidai, Bull. Chem. Soc. Jpn 53:1781 (1980).

124. P.C. Bevan, J. Chatt, M. Hidai and G.J. Leigh, J. Organomet. Chem. 160:165 (1978).

125. Y. Mizobe, R. Ono, Y. Uchida, M. Hidai, M. Tezuka, S. Moue and A. Tsuchiya, J. Organomet. Chem. 204:377 (1981).

126. H.M. Colquhoun and T.J. King, J.C.S. Chem. Commun. 879 (1980).

127. R.A. Head and P.B. Hitchcock, J.C.S. Dalton Trans. 1150 (1980).

128. J. Chatt, R.A. Head, P.B. Hitchcock, W. Hussain and G.J. Leigh, J. Organomet. Chem. 133:C1 (1977).

129. P.C. Bevan, J. Chatt, G.J. Leigh and E.G. Leelamani, J. Organomet. Chem. 139:C59 (1977).

130. J. Chatt, G.J. Leigh, C.J. Pickett and D.R. Stanley, J. Organomet. Chem. 184:C64 (1980).

131. T. Uchida, Y. Uchida, M. Hidai and T. Kodama, Bull. Chem. Soc. Jpn 43:2883 (1971).

132. T. Uchida, Y. Uchida, M. Hidai and T. Kodama, Acta Cryst. B31:1197 (1975).

133. R.D. Sanner, J.M. Manriquez, R.E. Marsh and J.E. Bercaw, J. Am. Chem. Soc. 98:8351 (1976).

134. B.R. Davies, N.C. Payne and J.A. Ibers, J. Am. Chem. Soc. 91:1240 (1969).

135. D.L. Thorn, T.H. Tulip and J.A. Ibers, J.C.S. Dalton Trans. 2022 (1979).

136. P.R. Hoffman, T. Yoshida, T. Okano, S. Otsuka and J.A. Ibers, Inorg. Chem. 15:2462 (1976).

137. J. Chatt, A.B. Nikolsky, R.L. Richards and J.R. Sanders, J.C.S. Chem. Commun. 154 (1969).

138. I.M. Treitel, M.T. Flood, R.E. Marsh and H.B. Gray, J. Am. Chem. Soc. 91:6512 (1969).

139. R. Hammer, H-F. Klein, P. Friedrich and G. Huttner, Angew. Chem. Int. Ed. Engl. 16:485 (1977).

140. K. Jonas, D.J. Brauer, C. Kruger, P.J. Roberts and Y-H. Tsay, J. Am. Chem. Soc. 98:74 (1976).

141. J.W. Byrne, H.V. Blaser and J.A. Osborn, J. Am. Chem. Soc. 97:3871 (1975).

142. T. Tatsumi, H. Tominaga, M. Hidai and Y. Uchida, Chem. Lett. 37 (1977).

143. T. Tatsumi, H. Tominaga, M. Hidai and Y. Uchida, J. Organomet. Chem. 198:63 (1980).

144. J. Chatt, C.M. Elson, A.J.C. Pombeiro, R.L. Richards and G.H.D. Royston, J.C.S. Dalton Trans. 165 (1978).

145. "The Chemist's Companion", A.J. Gordon and R.A. Ford, Eds., John Wiley and Sons, New York, p. 107 (1972).

146. I.R. Hanson and D.L. Hughes, J.C.S. Dalton Trans. 390 (1981).

147. J. Chatt, M.E. Fakley, P.B. Hitchcock, R.L. Richards and N.T. Luong-Thi, J.C.S. Dalton Trans. 345 (1982).

148. S.N. Anderson, M.E. Fakley and R.L. Richards, J.C.S. Dalton Trans. 1973 (1981).

149. M. Hidai, I. Yokotake, T. Takahashi and Y. Uchida, Chem. Lett. 453 (1982).

150. D.C. Busby, T.A. George, S.D.A. Iske, Jr. and S.D. Wagner, Inorg. Chem. 20:22 (1981).

151. A. Yamamoto, Y. Miura, T. Ito, H-L. Chen, K. Iri, F. Ozawa, K. Miki, T. Sei, N. Tanaka and N. Kasai, Organometallics 2:1429 (1983).

152. M.R. Churchill and H.J. Wassermann, Inorg. Chem. 20:2899 (1981).

153. M.R. Churchill and H.J. Wassermann, Inorg. Chem. 21:218 (1982).

154. S.M. Rocklage and R.R. Schrock, J. Am. Chem. Soc. 104:3077 (1982).

155. S.M. Rocklage, H.W. Turner, J.D. Fellmann and R.R. Schrock, Organometallics 1:703 (1982).

156. J.R. Dilworth, S.J. Harrison, R.A. Henderson and D.R.M. Walton, J.C.S. Chem. Commun. 176 (1984).

157. S.N. Anderson, R.L. Richards and D.L. Hughes, J.C.S. Chem. Commun. 1291 (1982).

158. G.P. Pez, P. Apgar and R.K. Crissey, J. Am. Chem. Soc. 104:482 (1982).

159. T. Takahashi, T. Kodama, A. Watakabe, Y. Uchida and M. Hidai, J. Am. Chem. Soc. 105:1680 (1983).

160. H.F. Klein, K. Ellrich and K. Ackermann, J.C.S. Chem. Commun. 888 (1983).

161. S. Donovan-Mtunzi and R.L. Richards, J.C.S. Dalton Trans. 469 (1984).

162. R.A. Henderson, J. Organomet. Chem. 208:C51 (1981).

163. H. Nishihara, T. Mori, Y. Tsurita, K. Nakano, T. Sato and Y. Sasaki, J. Am. Chem. Soc. 104:4367 (1982).

164. G.E. Bossard, T.A. George, D.B. Howell, L.M. Koczon and R.K. Lester, Inorg. Chem. 22:1968 (1983).

165. T.A. George and L.M. Koczon, J. Am. Chem. Soc. 105:6334 (1983).

166. R.N.F. Thornley, R.R. Eady and D.J. Lowe, Nature (Lond.) 272:557 (1978).

167. A. Watakabe, T. Takahashi, Dou-Man Jin, I. Yokotake, Y. Uchida and M. Hidai, J. Organomet. Chem. 254:75 (1983).

168. H.M. Colquhoun, Transition Met. Chem. 6:57 (1981).

169. H.M. Colquhoun, J.C.S. Chem. Commun. 85 (1981).

170. C.J. Pickett and G.J. Leigh, J.C.S. Chem. Commun. 1033 (1981).

171. W. Hussain, G.J. Leigh and C.J. Pickett, J.C.S. Chem. Commun. 747 (1982).

172. C.J. Pickett, J.E. Tolhurst, A. Copenhaver, T.A. George and R.K. Lester, J.C.S. Chem. Commun. 1071 (1982).

173. H.M. Colquhoun, A.E. Crease, S.A. Taylor and D.J. Williams, J.C.S. Chem. Commun. 736 (1982).

174. H.M. Colquhoun, A.E. Crease and S.A. Taylor, J.C.S. Chem. Commun. 1158 (1983).

175. M. Hidai, K. Komori, T. Kodama, Dou-Man Jin, T. Takahashi, S. Sugiura, Y. Uchida and Y. Mizobe, J. Organomet. Chem. 272:155 (1984).

176. F.W.B. Einstein, T. Jones, A.J.L. Hanlan and D. Sutton, Inorg. Chem. 21:2585 (1982).

177. C.F. Barrientos-Penna, F.W.B. Einstein, T. Jones and D. Sutton, Inorg. Chem. 21:2578 (1982).

178. J.A. Carrol and D. Sutton, Inorg. Chem. 19:3137 (1980).

179. J. Chatt, B.A.L. Crichton, J.R. Dilworth, P. Dahlstrom, R. Gutkoska and J. Zubieta, J.C.S. Dalton Trans. 2383 (1982).

180. J. Chatt, J.R. Dilworth, P.L. Dahlstrom and J. Zubieta, J.C.S. Chem. Commun. 786 (1980).

181. J.R. Dilworth, I.A. Latham, G.J. Leigh, G. Huttner and I. Jibril, J.C.S. Chem. Commun. 1368 (1983).

182. T.A. George and R.C. Tisdale, J. Am. Chem. Soc. 107:5157 (1985).

183. S.N. Anderson, R.L. Richards and D.L. Hughes, J.C.S. Dalton Trans. 245 (1986).

184. S.N. Anderson, D.L. Hughes and R.L. Richards, J.C.S. Chem. Commun. 958 (1986).

185. C.J. Pickett and J. Talavmin, Nature 317:652 (1985).

186. T.I. Al-Salih and C.J. Pickett, J.C.S. Dalton Trans. 1255 (1985).

187. R.C. Murray and R.R. Schrock, J. Am. Chem. Soc. 107:4557 (1985).

188. M. Hidai, S. Aramaki, K. Yoshida, T. Kodama, T. Takahashi, Y. Uchida and Y. Mizobe, J. Am. Chem. Soc. 108:1562 (1986).

189. G.E. Bossard, T.A. George, R.K. Lester, R.C. Tisdale and R.L. Turcotte, Inorg. Chem. 24:1129 (1985).

190. S. Donovan-Mtunzi, R.L. Richards and J. Mason, J.C.S. Dalton Trans. 469 (1984).

191. S. Donovan-Mtunzi, R.L. Richards and J. Mason, J.C.S. Dalton Trans. 2729 (1984).

192. S. Donovan-Mtunzi, R.L. Richards and J. Mason, J.C.S. Dalton Trans. 2429 (1984).

193. S. Donovan-Mtunzi, R.L. Richards and J. Mason, J.C.S. Dalton Trans. 1329 (1984).

194. T. Yamabe, K. Hori and K. Fukui, Inorg. Chem. 21:2046 (1982).

195. T. Yamabe, K. Hori and K. Fukui, Inorg. Chem. 21:2816 (1982).

196. J.R. Dilworth, A. Garcia-Rodriguez, G.J. Leigh and J.N. Murrell, J.C.S. Dalton Trans. 455 (1983).

197. "New Trends in the Chemistry of Nitrogen Fixation", J. Chatt, G.L.M. da Camara Pina and R.L. Richards, Eds., Academic Press, London (1980).

198. "Comprehensive Organometallic Chemistry", J.R. Dilworth and R.L. Richards, in G. Wilkinson, F.G.A. Stone and E.W. Abel, Eds., Vol.8, Pergamon, Oxford, p. 1073 (1982).

199. J. Chatt and R.L. Richards, J. Organomet. Chem. 239:65 (1982).

200. R.A. Henderson, G.J. Leigh and C.J. Pickett, Adv. Inorg. Chem. Radiochem. 27:197 (1983).

201. H.M. Colquhoun, Acc. Chem. Res. 17:23 (1984).

202. M. Hidai, in "Molybdenum Enzymes", T.G. Spiro Ed., John Wiley and Sons, New York, p. 285 (1985).

REACTIONS OF NITROSYLS

Frank Bottomley

Department of Chemistry, University of New Brunswick

Fredericton, Canada

1.   INTRODUCTION

Nitrosyls are complexes containing the nitrogen monoxide (NO) ligand. The oldest known nitrosyl is $[Fe(CN)_5(NO)]^{2-}$, first prepared by Playfair in 1849.[1] Reactions of the coordinated NO ligand in this complex were first reported by Boedeker in 1861.[2] Therefore, in both discovery and reactivity nitrosyls predate most other ligands including CO. These early beginnings were not followed by rapid growth; until the middle of the 1960's progress in transition metal nitrosyl chemistry was mainly confined to three fronts. Firstly, the preparation of nitrosyls which were isoelectronic to known carbonyls was pursued. An example is [Co (CO)$_3$(NO)], isoelectronic with $[Ni(CO)_4]$ and first prepared by Mond in 1922.[3] Hieber and his coworkers prepared many compounds of this type. Apart from the occasional inexplicable presence of nitro (NO$_2$) complexes in the reaction products, or the rare but equally mysterious loss of NO on treatment of the nitrosyl with a ligand,[4] no reactions of the NO ligand were observed. Secondly, the preparation of nitrosyls of ruthenium, which date back at least to 1860,[5] was continued. The names of Charonnat and Gleu are prominent in this work, which received a boost after 1945 from the discovery that ruthenium constituted a significant fraction of uranium fission products. Treatment of these products with nitric acid invariably yielded nitrosyls of ruthenium. These were remarkable only for the wide variety of chemical and physical changes through which the {RuNO} group could survive completely unchanged.[6] Thirdly, the reactivity of $[Fe(CN)_5(NO)]^{2-}$ towards basic reagents was investigated, principally by Cambi and Scagliarini and their coworkers.[7] Other $[M(CN)_5(NO)]^{n-}$ complexes were prepared and their solution chemistry, particularly electrochemistry, investigated.[8] However, in terms of reactions at the NO ligand, $[Fe(CN)_5(NO)]^{2-}$ was the exception to apparent boring inertness.

There were several reasons for the change in the status of the nitrosyl ligand in the mid 1960's. One was that prior to then NO had been regarded as at best a somewhat undesirable nuisance, whereas the closely related CO took part in a variety of industrially important processes catalyzed by metals. There was no incentive to investigate nitrosyls until it was realised that high temperature combustion processes using air invariably produced NO, and this was environmentally a very undesirable nuisance indeed. Unlike CO, which could be "removed" by simple oxidation to CO_2, NO was not easily disposed of,[9] and a rather sudden interest in metal nitrosyl chemistry resulted. Further impetus has been provided by the realisation that the "undesirable" NO actually represents a source of fixed nitrogen, and may be converted to useful products such as caprolactam.

A second reason for an increase in interest in nitrosyls was the development of better models for the bonding between NO and a metal. The radical nature of NO was clearly much modified, and usually completely lost, on complexation. As early as 1934 Sidgwick had suggested that the NO ligand could be regarded as NO^+ or NO^-,[10] thus raising implicitly the possibility of a variable MNO angle and a chemistry dependent on this angle. However, until the determination of the IrNO angle of 124° in $[IrCl(Ph_3P)_2(CO)(NO)]^+$ by Ibers and Hodgson in 1968,[11] no well-documented case of a bent {MNO} group was known. This discovery sparked attempts to explain the "non-innocent" nature of the NO ligand, and to predict when a nitrosyl would have a linear MNO group and when a bent. It is inappropriate to go into these models here and the reader is referred to three papers which discuss present views in detail.[12-14] Reference to these models will be made as often as is necessary to understand the reactivity of nitrosyls. Also in the mid 60's nitrosyls became of considerable interest to many who would scarcely have known of their existence if it had not been discovered that $[Fe(CN)_5(NO)]^{2-}$ was an extremely effective anti-hypertensive drug. It is now widely used as such, particularly in surgery requiring lowered blood pressure. Finally nitrosyls benefited from the general interest in nitrogen containing ligands, of which the most spectacular example is N_2 (see Chapter 2).

The above introduction explains the organisation of the present chapter. The chemistry of $[Fe(CN)_5(NO)]^{2-}$ and its $[M(CN)_5(NO)]^{n-}$ relatives will be treated separately, though the principles common to all nitrosyls will be stressed. Reactions of other nitrosyl complexes will be considered in sections on the type of reaction.

In considering reactions of nitrosyls one is always confronted by the problem of the non-innocent nature of the ligand referred to above. Is the reaction:

$$[Co(pdma)_2(NO)]^{2+} + SCN^- \longrightarrow [Co(SCN)(pdma)_2(NO)]^+ \qquad (1)$$
$$[pdma = \underline{o}\text{-phenylenebis (dimethylarsine)}]$$

in which the CoNO angle changes from $178°$ to $132°$[17] a reaction of coordinated NO? For this chapter it has been arbitrarily decided that a reaction of a nitrosyl involves conversion of {MNO} into $\{MN(A_n)O(B_m)\}$, where A and B represent some atom or group of atoms and n and/or m are not zero. Loss of the NO ligand is also discussed. Reaction (1) and fluxional processes of metal nitrosyls will therefore not be explicitly discussed. Wherever possible the {MNO} angle will be indicated and the problem of assigning electrons to M or NO will be circumvented by using the {MNO}^{n} convention of Enemark and Feltham.[12]

To conform to the limitations of this book, this chapter will be confined to reactions of nitrosyls, or to reactions of NO with complexes which lead to $MN(A_n)O(B_m)$ products <u>in solution</u>. The burgeoning literature on reactions of NO on metal surfaces, particularly those reactions involved in the removal of automobile exhaust pollution, will not be discussed. A review devoted to this topic exists.[9]

Previous general reviews of nitrosyls have been written,[18-21] as have reviews of their preparation,[22-23] their catalytic reactions with CO,[24] their use in organic synthesis and pollution control[464] and their electrophilic behavior.[8-25] Specialized reviews devoted to $[Fe(CN)_5(NO)]^{2-}$[7] and to ruthenium nitrosyls[6] have also appeared. Two reviews devoted to the reactivity of nitrosyls have also been written previously.[26-27]

2.   THE NITROPRUSSIDE ION, $[Fe(CN)_5(NO)]^{2-}$, AND RELATED PENTACYANONITROSYLS

2.1   General Remarks

As described in the introduction, $[Fe(CN)_5(NO)]^{2-}$ was first prepared in the middle of the 19th century. Over the intervening years it has been joined by the 18 electron analogues $[Ru(CN)_5(NO)]^{2-}$, $[Os(CN)_5(NO)]^{2-}$, $[Mn(CN)_5(NO)]^{3-}$, $[Cr(CN)_5(NO)]^{4-}$ and $[Mo(CN)_5(NO)]^{4-}$; as well as the 17 electron $[Cr(CN)_5(NO)]^{3-}$ and $[Mo(CN)_5(NO)]^{3-}$, and $[V(CN)_5(NO)]^{3-}$, a 16 electron complex. The formally 20 electron $[Co(CN)_5(NO)]^{3-}$ has also been isolated, as have $[Fe(CN)_4(NO)]^{2-}$ and $[V(CN)_6(NO)]^{4-}$. Other cyanonitrosyls such as $[Fe(CN)_5(NO)]^{3-}$ have a transitory existence in solution. The reactivities of these $[M(CN)_5(NO)]^{n-}$ complexes are markedly different. The NO ligand in $[Fe(CN)_5(NO)]^{2-}$ is readily attacked by nucleophiles, and this complex also has an extensive reduction chemistry.[7] It appears that $[Ru(CN)_5(NO)]^{2-}$ and $[Os(CN)_5(NO)]^{2-}$ have a chemistry similar to that of $[Fe(CN)_5(NO)]^{2-}$, but the low solubility of these complexes has restricted exploration of the reactions until recently.[508] None of the other $[M(CN)_5(NO)]^{n-}$ complexes reacts with nucleophiles at the NO ligand, and reduction of these complexes is also more difficult. There is some slight evidence for attack by the electrophiles H^+ and O_2 at some $[M(CN)_5(NO)]^{n-}$ complexes, and it is claimed that a linear {FeNOM} bridge is present in $M[Fe(CN)_5(NO)]$ (M=Mn, Cu).[46] However $[Fe(CN)_5(NO)]^{2-}$ behaves predominantly as an electrophile and the present section is therefore divided into three parts, devoted to (i) nucleophilic attack at $[Fe(CN)_5(NO)]^{2-}$; (ii) reduction of $[Fe(CN)_5(NO)]^{2-}$; and (iii) reactions of other $[M(CN)_5(NO)]^{n-}$ complexes.

2.2   Reactions of $[Fe(CN)_5(NO)]^{2-}$ with Nucleophiles

Reactions of $[Fe(CN)_5(NO)]^{2-}$, which has an {MNO} angle of 175.7,[28] with nucleophiles may be summarized by the general equation:

$$[Fe(CN)_5(NO)]^{2-} + :A \rightarrow [Fe(CN)_5(N(=O)A)]^{2-} \rightarrow products \qquad (2)$$

In very few reactions has the complex obtained by initial attack of :A been isolated. The simplest and best studied reaction is with OH^-:[29]

$$[Fe(CN)_5(NO)]^{2-} + OH^- \rightleftharpoons [Fe(CN)_5(N(O)OH)]^{3-} \qquad (3)$$

$$[Fe(CN)_5(N(O)OH)]^{3-} + OH^- \rightleftharpoons [Fe(CN)_5(NO_2)]^{4-} + H_2O \qquad (4)$$

The overall equilibrium constant for the total reaction:

$$[Fe(CN)_5(NO)]^{2-} + 2OH^- \rightleftharpoons [Fe(CN)_5(NO_2)]^{4-} + H_2O \qquad (5)$$

depends on ionic strength but is close to 1.0×10^6 M^{-2},[7,29-30] with $k_{f(3)}$ = 0.2 m^{-1} $1 s^{-1}$[30] (at an ionic strength of 0.35 M) or 0.55 m^{-1} $1 s^{-1}$[29] (at an ionic strength of 1M) and $k_{f(4)} > 10^6$ m^{-1} $1 s^{-1}$.[30] A more recent polarographic investigation gives $K_{(5)}$ as 3.17×10^4; $k_{f(3)}$ as 0.216 m^{-1} 1

s^{-1}; K_3 as 3.17 m^{-1} l; $k_{f(4)}$ is estimated as 10^{10} m^{-1} l s^{-1} and $k_{r(4)}$ as 1 x 10^6 s^{-1}, all at an ionic strength of 0.5 M.[75] The final product, $[Fe(CN)_5(NO_2)]^{4-}$ undergoes hydrolysis:

$$[Fe(CN)_5(NO_2)]^{4-} + H_2O \rightarrow [Fe(CN)_5(H_2O)]^{3-} + NO_2^- \tag{6}$$

with $k_{f(6)}$ = 1.4 x 10^{-4} m^{-1} l s^{-1} at 25°.[29] The reaction sequence (5) - (6) is important for the reactions of $[Fe(CN)_5(NO)]^{2-}$ with other nucleophiles, since most nucleophiles are bases in water, the solvent of necessity for $[Fe(CN)_5(NO)]^{2-}$. It has been shown that the blue color formed on reaction of $[Fe(CN)_5(NO)]^{2-}$ with NCS$^-$ is due to $[Fe(CN)_5(NCS)]^{4-}$ produced by substitution of NCS$^-$ at $[Fe(CN)_5(H_2O)]^{3-}$, which is in turn produced by the reaction sequence (5) - (6).[31] It has however been suggested that the reaction of NCS$^-$ is not with $[Fe(CN)_5(H_2O)]^{3-}$, but with the $[Fe(CN)_5(NO_2H)]^{3-}$ intermediate produced by reaction (3).[75] A similar situation occurs with $[Os(CN)_5(NCS)]^{4-}$.[32] Formation of $[Fe(CN)_5(NH_3)]^{3-}$ from $[Fe(CN)_5(NO)]^{2-}$ and aqueous NH$_3$ proceeds via (5) - (6) and the substitution reaction

$$[Fe(CN)_5(H_2O)]^{3-} + NH_3 \rightleftharpoons [Fe(CN)_5(NH_3)]^{3-} + H_2O \tag{7}$$

However at high NH$_3$ concentration N$_2$ is evolved, suggesting direct attack of NH$_3$ at $[Fe(CN)_5(NO)]^{2-}$:[33]

$$[Fe(CN)_5(NO)]^{2-} + 2NH_3 \rightarrow [Fe(CN)_5(NH_3)]^{3-} + N_2 + H_3O^+ \tag{8}$$

Further evidence for direct attack of amines is the production of N-nitrosamines when secondary amines such as diethylamine attack $[Fe(CN)_5(NO)]^{2-}$[34] and the production of $[Fe(CN)_5(N(O)NPh)]^{4-}$ from [Fe(CN)$_5$(NO)]$^{2-}$ and NaNHPh in liquid NH$_3$:[35]

$$[Fe(CN)_5(NO)]^{2-} + 2NHPh^- \rightarrow [Fe(CN)_5(N(O)NPh)]^{4-} + PhNH_2 \tag{9}$$

On the other hand, deprotonated secondary amines NR$_2^-$ (R = Ph or Si(CH$_3$)$_3$) do not react in liquid ammonia to give nitrosamines; only

$[Fe(CN)_5(NH_3)]^{3-}$ is obtained.[35] Neither do tertiary amines react in aqueous solution.[33] Some amines, e.g. benzylamine, give alcohols and olefins derived from the initially formed N-nitrosamine by elimination of water and subsequent liberation of N_2 from the intermediate diazonium complex:[473-4]

$$[Fe(CN)_5(NO)]^{2-} + RNH_2 \rightleftharpoons [Fe(CN)_5(N(O)NH_2R)]^{2-} \tag{10}$$

$$[Fe(CN)_5(N(O)NH_2R)]^{2-} \rightarrow [Fe(CN)_5(N_2R)]^{2-} + H_2O \tag{11}$$

$$[Fe(CN)_5(N_2R)]^{2-} + OH^- \rightarrow [(Fe(CN)_5(N_2)]^{3-} + ROH \tag{12}$$

$$[Fe(CN)_5(N_2R)]^{2-} + OH^- \rightarrow [Fe(CN)_5(N_2)]^{3-} + (R-H) + H_2O \tag{13}$$

$$[Fe(CN)_5(N_2)]^{3-} + H_2O \rightarrow [Fe(CN)_5(H_2O)]^{3-} + N_2 \tag{14}$$

Direct attack by N_3^- [36] and NH_2OH[36-37] at the NO ligand has been proved by kinetic and labeling studies:

$$[Fe(CN)_5(NO)]^{2-} + N_3^- + H_2O \rightarrow [Fe(CN)_5(H_2O)]^{3-} + N_2 + N_2O \tag{15}$$

$$[Fe(CN)_5(NO)]^{2-} + NH_2OH + H_2O \rightarrow [Fe(CN)_5(H_2O)]^{3-} + N_2O + H_3O^+ \tag{16}$$

The labeling studies show that the N and O atoms from NH_2OH become the central N and the O of N_2O.[36] These arise from decomposition of the intermediate produced by nucleophilic attack of N_2O according to the scheme:

$$[(NC)_5Fe\overset{\overset{\displaystyle O}{\|}}{-N}-N\,H_2\,O\,H]^{2-} \longrightarrow N_2O + H_3O^+ + [Fe(CN)_5]^{3-} \tag{17}$$

Note that it has been suggested that N_3^- attacks not at $[Fe(CN)_5(NO)]^{2-}$ directly, but at the $[Fe(CN)_5(NO_2H)]^{3-}$ intermediate produced by reaction (3).[75]

Attack by SH^- at $[Fe(CN)_5(NO)]^{2-}$ superficially follows that of OH^-:[38]

$$[Fe(CN)_5(NO)]^{2-} + SH^- \rightleftharpoons [Fe(CN)_5(N(O)SH)]^{3-} \tag{18}$$

$$[Fe(CN)_5(N(O)SH)]^{3-} + OH^-(SH^-) \rightarrow [Fe(CN)_5(N(O)S)]^{4-} +$$
$$H_2O\ (H_2S) \tag{19}$$

However reaction (19) is irreversible and has a rate constant of 1.3×10^2 s^{-1}, which would give the $[Fe(CN)_5(N(O)SH)]^{3-}$ intermediate a half-life of 55 s.[38] Such rates are not compatible with simple proton transfer in aqueous solution. The kinetic studies suggest a rearrangement from

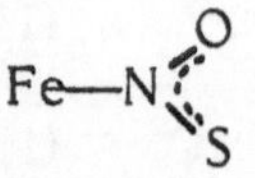

$$Fe-N\overset{O}{\underset{S}{\lessdot}} \quad \text{to} \quad Fe-S-N-O \quad \text{as being the explanation for the}$$

slow rate.[7] An infrared study of the final product was interpreted on the basis of a dimeric $[(Fe(CN)_5)_2(N_2S_2O_2)]^{8-}$ complex,[39] various structures for the $(N_2S_2O_2)$ bridge having been proposed.[39,26] Dimer formation would be expected to have been observed in the kinetic study however. A crystallographic investigation of this product is urgently needed before more can be said.

The initial rapid reaction of SR^- ($R = CH_2CH_2OH$, $CH_2CO_2^-$, Ph, CH_2Ph, or RSH = cysteine) with $[Fe(CN)_5(NO)]^{2-}$ gives $[Fe(CN)_5(N(O)SR)]^{3-}$. These adducts decompose moderately rapidly to give RSSR and $[Fe(CN)_5(NO)]^{3-}$,[40] the latter complex reacting further to give $[Fe(CN)_4(NO)]^{2-}$ (see Section 2.3 below). The kinetics of the reaction with cysteine have been determined.[44, 513]

Reactions of ketones with $[Fe(CN)_5(NO)]^{2-}$ generally proceed according to equations (20) and (21) which are for acetone:[41]

$$[Fe(CN)_5(NO)]^{2-} + CH_3COCH_3 + OH^- \rightleftharpoons [Fe(CN)_5(C_3H_5NO_2)]^{3-} + H_2O \tag{20}$$

$$[Fe(CN)_5(C_3H_5NO_2)]^{3-} + H_2O \rightarrow [Fe(CN)_5(H_2O)]^{3-} + CH_3C(O)CH=NOH \tag{21}$$

The structure of the $C_3H_5NO_2$ ligand in the red intermediate is unknown; since a proton is readily lost from acetone in basic solution, and given the final oxime product, it is logical to assume initial formation of $[(NC)_5Fe(N(O)CH_2C(O)CH_3)]^{3-}$ (see also attack of ketones at other nitrosyls, Section 4), followed by proton transfer and/or proton loss to give $[(NC)_5Fe(N(OH)=CHC(O)CH_3)]^{3-}$ and $[(NC)_5Fe(N(-O^-)=CHC(O)CH_3)]^{4-}$.[41-42] The structures of the various intermediates in these reactions with ketones are not known however, and even though attack on NO by the ketonic C atom is most reasonable, initial attack by oxygen in the enolate $CH_3C(-O^-)=CH_2$ cannot be completely eliminated from consideration.

The reaction of SO_3^{2-} with $[Fe(CN)_5(NO)]^{2-}$ (the Boedeker reaction) was the first reaction of a coordinated nitrosyl, discovered in 1861:[2]

$$[Fe(CN)_5(NO)]^{2-} + SO_3^{2-} \rightleftharpoons [Fe(CN)_5N(O)(SO_3)]^{4-} \tag{22}$$

The equilibrium (22) is well established in solution,[43] but the structure of the adduct was long in dispute. It has recently been shown that analogous reactions to (22) occur with other electrophilic nitrosyls, and an X-ray structure determination of $[RuCl(bpy)_2(N(O)SO_3)]$ establishes conclusively that the adducts contain the $\{MN(O)SO_3\}$ unit with a long N - S bond.[463]

In addition to the above reactions, which have been investigated in some detail, reactions of a large number of other bases with $[Fe(CN)_5(NO)]^{2-}$ have been described.[7] One of the most interesting is the formation of a red color on reaction of cysteine with $[Fe(CN)_5(NO)]^{2-}$ (see above). The structure of the red product is unknown. The effect of $[Fe(CN)_5(NO)]^{2-}$ as an anti-hypertensive agent in humans[15-16] must, given the known chemistry of the ion, be related to a reaction at the coordinated nitrosyl ligand.[47-48] The clinical use of $[Fe(CN)_5(NO)]^{2-}$, which has proved remarkably non-toxic,[15-16,49] does lead to an increasee in NCS$^-$ in patients,[15] implying attack of a sulfur-containing amino acid at the nitrosyl. It is to be expected that detailed studies of such reactions will appear in the biochemical and medical literature in the near future.[48]

The fact that nucleophilic attack occurs at the nitrogen atom of the nitrosyl ligand in $[Fe(CN)_5(NO)]^{2-}$ shows that the lowest unoccupied molecular orbital (LUMO) in the complex is largely localized on this atom, and molecular orbital calculations agree with this conclusion. The LUMO is a $\pi*\{MNO\}$ orbital having contributions from p_x, $p_y(N)$ (67%); d_{xz}, $d_{yz}(Fe)$ (7.5%); and p_x, $p_y(O)$ (5.5%) with the remaining 20% being π from the CN^- ligands. The LUMO is much lower in energy than other unoccupied orbitals.[50-51]

2.3   Reduction of $[Fe(CN)_5(NO)]^{2-}$

The one-electron reduction of $[Fe(CN)_5(NO)]^{2-}$ has been investigated many times over many years, with a surprisingly large variety of products being reported. The earlier work will not be reviewed here and only the presently accepted results (which appear to be definitive, at least to this author) will be presented.

The initial one-electron reduction of $[Fe(CN)_5(NO)]^{2-}$ is reversible,[52-53] and the added electron is largely localized in the $\pi*$ $\{MNO\}$ orbital:[54-55]

$$[Fe(CN)_5(NO)]^{2-} + e \rightleftharpoons [Fe(CN)_5(NO)]^{3-} \tag{23}$$

This conclusion is supported by esr evidence[54-55] and the reactions of the radical $[Fe(CN)_5(NO)]^{3-}$ with organic radicals (the two radical species being generated simultaneously by radiolysis) to give $[Fe(CN)_5(N(O)R)]^{3-}$ (R = $CH_2C(CH_3)_2X$, X = OH, NH_3^+, CO_2^-; R = $CH_2C(CH_3)XY$, X = NH_3^+ Y = CO_2^-, X = OH Y = CO_2^-, and R = $CH_2(CH_3)NC(O)CH_3$).[56] These products would be expected if R^- nucleophilically attacked $[Fe(CN)_5(NO)]^{2-}$ (see Section 2.2 above).

Unless trapped by another radical the reduced $[Fe(CN)_5(NO)]^{3-}$ is unstable with respect to loss of a CN^- ligand:[52-53]

$$[Fe(CN)_5(NO)]^{3-} \rightleftharpoons [Fe(CN)_4(NO)]^{2-} + CN^- \tag{24}$$

For reaction (24) K = 6.8×10^{-5} l mol^{-1} and $k_f = 2.8 \times 10^2$ s^{-1}. The question of which ligand,   a cis or a trans CN^-,   is lost is not settled.[50,52-53,57] The paramagnetic blue nitrosyl $[Fe(CN)_4(NO)]^{2-}$ is unusual in having a square pyramidal structure with a linear {MNO} group[57] instead of the bent apical {MNO} expected for such an {MNO}[7] complex.[57-59] The odd electron was originally believed to be mainly localized in a $d_{z^2}(Fe)$ orbital.[59] More recent molecular orbital calculations on this {MNO}[7] but formally 17 electron complex suggest: (a) that the odd electron may lie in an orbital having considerable π(CN) character[50] and (b) that the potential energy curve for bending of the {FeNO} group is very flat,[60] so that the FeNO angle may be determined by crystal packing forces not considered in simple theoretical models of nitrosyls.

Further reduction of $[Fe(CN)_4(NO)]^{2-}$ takes place, but the product formed on one-electron reduction rapidly decomposes, possibly by multi-electron transfer, to unknown products.[53] Reduction of $[Fe(CN)_5(NO)]^{2-}$ by alkali metals in liquid ammonia gives not only $[Fe(CN)_5(NO)]^{3-}$ [57] and $[Fe(CN)_4(NO)]^{2-}$ [57,61] but also $[Fe(CN)_4(NO)]^{3-}$ and $[Fe(CN)_3(NO)]^{4-}$ [62] Rapid electrochemical multi-electron reduction of $[Fe(CN)_5(NO)]^{2-}$ appears to result finally in production of $[Fe(CN)_5(NH_2OH)]^{3-}$ via a number of ill-characterized intermediates.[40,63-65]

2.4   Reactions of Other Pentacyanonitrosyls

All $[M(CN)_5(NO)]^{n-}$ complexes so far investigated have linear {MNO} groups.[66] Nucleophilic attack of OH^- at $[Ru(CN)_5(NO)]^{2-}$ [67] or $[Os(CN)_5(NO)]^{2-}$ [68] gives the corresponding $[M(CN)_5(NO_2)]^{4-}$ complex, analogous to $[Fe(CN)_5(NO)]^{2-}$. The reactions of $[Ru(CN)_5(NO)]^{2-}$ with SH^-, SO_3^{2-}, N_2H_4 and NH_3 have been briefly investigated and seem similar to those of $[Fe(CN)_5(NO)]^{2-}$.[508] Despite purposeful attempts[35] nucleophilic attack at other $[M(CN)_5(NO)]^{n-}$ complexes has not been observed. The existence of both $[V(CN)_5(NO)]^{3-}$ [69] and $[V(CN)_6(NO)]^{4-}$ [70] seems to indicate that nucleophilic attack at $[V(CN)_5(NO)]^{3-}$ may be possible, but would take place at the vanadium atom rather than the nitrosyl ligand. Reactions with electrophiles have been observed in a very few cases. The nitrosyl $[Cr(CN)_5(NO)]^{4-}$, isoelectronic to $[Fe(CN)_5(NO)]^{2-}$ but with a very low ν(NO) of 1515 cm^{-1} (though even this low value is still compatible with a linear {CrNO} group), is only stable in alkaline solution. In acid it is protonated giving a complex assumed to be $[Cr(CN)_5(NOH)]^{3-}$;[8] $[Cr(CN)_5(NO)]^{4-}$ is very readily oxidized to $[Cr(CN)_5(NO)]^{3-}$.[8,71-72] The analogous $[Mo(CN)_5(NO)]^{4-}$ has only been observed in solution, and is also rapidly oxidized, e.g. by O_2, to the isolable $[Mo(CN)_5(NO)]^{3-}$.[73]

Electrochemical reduction of $[Cr(CN)_5(NO)]^{3-}$ ultimately gives Cr^{2+}, $5CN^-$ and NH_3, though initial reversible one-electron reduction has been demonstrated.[72] This reaction is paralleled exactly by $[Mn(CN)_5(NO)]^{3-}$.[8] Note that exhaustive reduction of $[Fe(CN)_5(NO)]^{2-}$ gives $[Fe(CN)_5(NH_2OH)]^{3-}$;[8] the reason for the difference between Cr or Mn and Fe is not clear. Reduction of $[Cr(CN)_5(NO)]^{3-}$ by potassium in liquid ammonia gives the NO-bridged dimer $[Cr_2(CN)_6(NO)_2]^{8-}$.[74]

3.   REDUCTION OF NITROSYLS

The standard reduction potential for the one-electron reduction of NO:

$$NO + H^+ + e \rightleftharpoons NOH \tag{25}$$

has been estimated as -0.3V.[76] This highly negative value indicates that only powerful reducing agents will directly reduce NO. Complexes containing a linear {MNO} group, where the NO is formally bonded as NO^+ and the π^*-{MNO} orbital lies at low energy, are expected to be more easily reduced since E° for the reaction

$$HNO_2 + H^+ + e \rightleftharpoons NO + H_2O \tag{26}$$

is +1.00 V.[76] Reversible electrochemical one-electron reduction of several nitrosyls which are known or can be confidently expected to contain linear $\{MNO\}$ groups has been reported:

$$[Ru(NH_3)_5(NO)]^{3+} + e \rightleftharpoons [Ru(NH_3)_5(NO)]^{2+} \qquad (27)^{77\text{-}78}$$

$$[RuX(bpy)_2(NO)]^{n+} + e \rightleftharpoons [RuX(bpy)_2(NO)]^{(n-1)+} \qquad (28)^{79\text{-}80}$$

$$[Fe(S_2C_2R_2)_2(NO)]^{-} + e \rightleftharpoons [Fe(S_2C_2R_2)_2(NO)]^{2-} \qquad (29)^{81}$$

$$[Co(S_2C_2(CN)_2)_2(NO)]^{2-} + e \rightleftharpoons [Co(S_2C_2(CN)_2)_2(NO)]^{3-} \qquad (30)^{81}$$

$$[Tc(NH_3)_4(H_2O)(NO)]^{3+} + e \rightleftharpoons [Tc(NH_3)_4(H_2O)(NO)]^{2+} \qquad (31)^{143}$$

$$[MY(dppe)_2(NO)]^{m+} + e \rightleftharpoons [MY(dppe)_2(NO)]^{(m-1)+} \qquad (32)^{85\text{-}6}$$

where $X = N_3$, Cl or NO_2 for n = 2; X = NH_3 or pyridine or CH_3CN for n = 3; R = CN, CF_3 or $R_2 = C_6R'_4$ [R' = Cl, Ph or $R'_4 = H_3(CH_3)$]; M = Cr or Mo and Y = CH_3CN for m = 2 or Cl for m = 1. The reduced products of equations (28) - (32) have been isolated. For $[RuCl(bpy)_2(NO)]^{+}$ (an $\{MNO\}^6$ complex) it is clear from the esr, ir, and $E°$ parameters that reduction occurs at the NO ligand,[80] as would be expected if the LUMO was mainly located on the N atom of the $\{RuNO\}$ group. In the case of the $\{MNO\}^5$ complex $[Tc(NH_3)_4(H_2O)(NO)]^{3+}$ the site of reduction is unknown, but since nucleophilic attack at this complex does not occur,[143] the LUMO, if it is localized in the TcNO group, must be of high energy. For $[Fe(S_2C_2R_2)_2(NO)]^{2-}$ the added electron also appears to have entered a $\pi*\{FeNO\}$ orbital, but this orbital includes a considerable contribution of sulfur from the diothiolate ligands.[81] In addition to the esr spectrum,[54] evidence that the added electron in $[Ru(NH_3)_5(NO)]^{2+}$ is largely localized in the $\pi*\{RuNO\}$ orbital is given by the reaction of this complex with simultaneously generated organic radicals to give $[Ru(NH_3)_5(N(O)R)]^{2+}$ [56,82] (see also reactions of $[Fe(CN)_5(NO)]^{3-}$, Section 2.3).

As with $[Fe(CN)_5(NO)]^{3-}$, unless trapped by another radical $[Ru(NH_3)_5(NO)]^{2+}$ is unstable with respect to loss of an NH_3 ligand:[77]

$$[Ru(NH_3)_5(NO)]^{2+} + H_2O \rightarrow [Ru(NH_3)_4(H_2O)(NO)]^{2+} + NH_3 \qquad (33)$$

The $\{MNO\}^7$ complex of formula $[Ru(NH_3)_4(H_2O)(NO)]^{2+}$ (equation (33))
has not been isolated. It is an intermediate in the redox catalyzed
substitution of ammonia in $[Ru(NH_3)_5(NO)]^{3+}$:[77]

$$[Ru(NH_3)_5(NO)]^{3+} + e \rightleftharpoons [Ru(NH_3)_5(NO)]^{2+} \tag{34}$$

$$[Ru(NH_3)_5(NO)]^{2+} + H_2O \rightleftharpoons [Ru(NH_3)_4(H_2O)(NO)]^{2+} + NH_3 \tag{35}$$

$$[Ru(NH_3)_4(H_2O)(NO)]^{2+} + [Ru(NH_3)_5(NO)]^{3+} \rightleftharpoons [Ru(NH_3)_5(NO)]^{2+}$$
$$+ [Ru(NH_3)_4(H_2O)(NO)]^{3+} \tag{36}$$

$$[Ru(NH_3)_4(H_2O)(NO)]^{3+} + H_2O \rightleftharpoons [Ru(OH)(NH_3)_4(NO)]^{2+} + H_3O^+ \tag{37}$$

Reduction of $[Ru(NH_3)_5(NO)]^{3+}$ by Cr^{2+} in the presence of Cl^-
rapidly gives $[Ru(NH_3)_6]^{2+}$ via a six-electron reduction.[87-88] The
reduction is proposed to occur via a nitrene intermediate which can be
scavenged by Cl^-, ultimately giving $[Ru(NH_3)_6]^{2+}$ in the aqueous medium.
In the absence of Cl^-, the nitrene remains attached to chromium, and the
reaction is slower and leads to $[(NH_3)_5Ru(NH)Cr(H_2O)]^{5+}$ as the isolable
product.[88] $[Ru(H_2O)_5(NO)]^{3+}$ gives an analogous nitrene (imido) bridged
complex on reduction by Cr(II),[88] indicating that the bridging nitrene does
indeed originate with the nitrosyl ligand. Bridging nitride has also been
obtained by Sn(II) reduction of chloronitrosylruthenium complexes, giving
diamagnetic $[Ru_2Cl_8(H_2O)_2(N)]^{3-}$.[89-90]

The complete reduction of NO to NH_3 in $[Ru(trpy)(bpy)(NO)]^{3+}$
(trpy = 2,2',2"-terpyridine) has been shown by cyclic voltammetry to occur
via the following steps:[471]

$$[Ru(trpy)(bpy)(NO)]^{3+} \underset{-e}{\overset{+e}{\rightleftharpoons}} [Ru(trpy)(bpy)(NO)]^{2+} \tag{38}$$

$$[Ru(trpy)(bpy)(NO)]^{2+} \underset{-e}{\overset{+e}{\rightleftharpoons}} [Ru(trpy)(bpy)(NO)]^{+} \tag{39}$$

$$[Ru(trpy)(bpy)(NO)]^+ \underset{-H^+}{\overset{+H^+}{\rightleftharpoons}} [Ru(trpy)(bpy)(NHO)]^{2+} \quad (40)$$

$$[Ru(trpy)(bpy)(NHO)]^{2+} \underset{+H_2O, -H^+, -e}{\overset{+H^+, +e, -H_2O}{\rightleftharpoons}} [Ru(trpy)(bpy)(N)]^{2+} \quad (41)$$

$$[Ru(trpy)(bpy)(N)]^{2+} \underset{-e, -H^+}{\overset{+e, +H^+}{\rightleftharpoons}} [Ru(trpy)(bpy)(NH)]^{2+} \quad (42)$$

$$[Ru(trpy)(bpy)(NH)]^{2+} \underset{-e, -2H^+}{\overset{+e, +2H^+}{\rightleftharpoons}} [Ru(trpy)(bpy)(NH_3)]^{3+} \quad (43)$$

$$[Ru(trpy)(bpy)(NH_3)]^{3+} \underset{-e}{\overset{+e}{\rightleftharpoons}} [Ru(trpy)(bpy)(NH_3)]^{2+} \quad (44)$$

The overall reduction also occurs for the osmium analogue, and in the case of $[Ru(NO_2)(bpy)_2(NO)]^{2+}$ a 12-electron reduction giving $[Ru(bpy)_2(NH_3)_2]^{2+}$ is found.[471-472]

The reduction of $[CrL_5(NO)]^{2+}$ (L = NH_3 or H_2O) by $[Cr(H_2O)_6]^{2+}$ produces uncoordinated NH_2OH, but no NH_3.[91] It was suggested that linear nitrosyls such as $[Ru(NH_3)_5(NO)]^{3+}$ [92] were reduced to NH_3 whereas bent nitrosyls, such as $[CrL_5(NO)]^{2+}$ [$\nu(NO)$ 1730 cm^{-1}] might be, were reduced to NH_2OH.[88] However $[Fe(CN)_5(NO)]^{2-}$, which is linear,[28] produces $[Fe(CN)_5(NH_2OH)]^{2+}$ on reduction,[40,63-65] and addition of Cr(II) to $[Co(NH_3)_5(NO)]^{2+}$ or $[Co(en)_2(H_2O)(NO)]^{2+}$ (en = ethylendiamine), which do have bent CoNO groups[93-94] results not in reduction but in transfer of the NO group intact to Cr(II) (see Section 8.4).[87,95] In addition the reduction of $[Cr(NH_3)_5(NO)]^{2+}$ by $[Cr(edta)(H_2O)]^{2-}$ gives at least 70% $[Cr(NH_3)_6]^{3+}$.[104] It is clear that the point of attachment of protons, and the stage in the reduction when the protons become attached, is crucial to the product obtained, as are the reduction potentials of the metals involved.

The identification of various (NH_xOH_y) intermediates in the reduction of coordinated nitrosyl has usually been attempted by indirect methods such as esr spectroscopy.[40] The reverse of nitrosyl reduction, namely hydroxylamine oxidation, is a potential route to such intermediates. Nitrosyls were prepared using hydroxylamine as the source of NO many years ago;[96-97] this method has recently been reactivated and many nitrosyls prepared from a high-valent metal oxide, a suitable ligand, and hydroxylamine.[465,475] The reactions always produce NH_3 as a by-product of the formal disproportionation:

$$5NH_2OH \rightarrow 3NH_3 + 2NO + 3H_2O \tag{45}$$

Recently it has proved possible to use NH_2OH to prepare nitrosyls without the co-production of NH_3, and a number of very interesting complexes containing (NH_xOH_y) ligands have been isolated and crystallographically characterized.[98-99, 468, 506] Perhaps the most remarkable is the seven-coordinate complex[98]

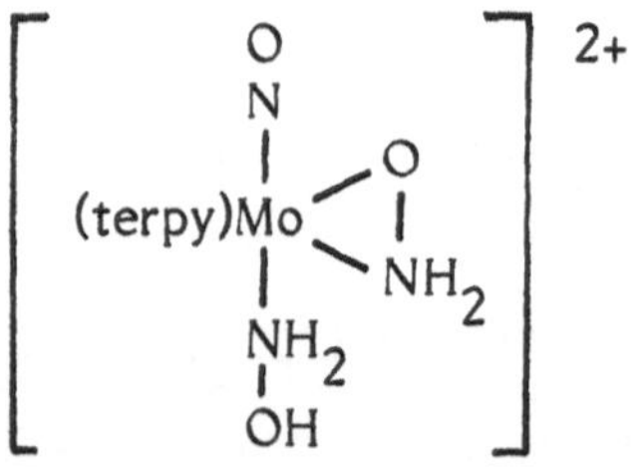

$$
\left[
\begin{array}{c}
O \\
\| \\
N \\
| \\
(terpy)Mo \diagdown{\begin{array}{c} O \\ | \\ NH_2 \end{array}} \\
| \\
NH_2 \\
| \\
OH
\end{array}
\right]^{2+}
$$

where terpy is the tridentate 2,2',6',2''-terpyridine ligand and the $\{MoNO\}^4$ group is linear. This is obtained from the reaction between MoO_4^{2-}, terpy and NH_2OH in acid solution. The N-coordinated NH_2OH ligand is labile, and the derived aquo complex reacts with CN^- in aqueous solution to give the deprotonated NHO^{2-} complex[99]

The analogous vanadium complexes have also been prepared,[100-102] and it has been shown that reversible intramolecular conversion of "side-on" η^2-coordinated NH_2O^- to coordinated NO occurs in these complexes.[103] The facts that the (HNO^-) and (NH_2O^-) ligands in these complexes have the hydrogen attached to N and not O, and that these are η^2-ligands, imply that the ability to expand the coordination sphere around the metal, or to lose other ligands to accommodate an $\eta^1 \rightarrow \eta^2$ shift on reduction and protonation is important in nitrosyl chemistry.

Intermediates of the (NH_xOH_y) type have been isolated in other systems, but not in such wealth and always by simple protonation rather than the reductive protonation described here. They are therefore considered under the heading of attack by electrophiles at nitrosyls (Section 5 below).

One-electron electrochemical reduction of $[Co(CO)_3(NO)]$ is reversible at high scan rates,[105-106] but the anion $[Co(CO)_3(NO)]^-$ undergoes a rapid irreversible chemical transformation.[105-7] It has been suggested that this process is loss of the NOH^- ion,[107] but this conclusion has been questioned.[105] Reduction of the five-coordinate $[CoXL_3(NO)]^+$ nitrosyls (X = I for L = $P(OEt)_3$ or $P(OMe)_3$ and X = Cl for L = $P(OEt)_3$) is reversible, but a second electron is readily added, with slow loss of X^- and formation of $[CoL_3(NO)]$.[108] The nitrosyl ligand in a variety of $[(\eta^3\text{-allyl)}Fe(CO)_2(NO)]$ complexes also remains intact on reduction, at the expense of loss of the other three-electron donor ligand as allene.[109] Reduction of $[Co(Ph_3P)_2(NO)_2]^+$ with BH_4^- gives $[Co(Ph_3P)_3(NO)]$ and NH_3.[110,122] Electrochemical reduction of dinitrosyls (beyond addition of a single electron which then resides mainly on $NO^{[111-112]}$) has lead to very active and versatile catalysts for the dimerization of dienes.[113-121] The most active catalyst, derived from dinitrosyls of iron, is believed to be $[Fe(NO)_2]$ stabilized by weakly coordinated solvent ligands.

Several reactions of nitrosyls with powerful nucleophilic reducing agents lead to reduction of the nitrosyl ligand. The initial step in the reduction may actually involve nucleophilic attack at the coordinated nitrosyl though the evidence for this is tenuous (nucleophilic attack is discussed in Section 4).

The nitrosyl bridged dimer $[Cp_2Cr_2(\mu_2 - NO)_2(NO)_2]$ **(Ia)**

Cp   NO
  \   A   /
   \   /   \   /
    Cr ═══════ Cr
   /   \   /   \
  /   B   \
ON   Cp

I (a - g)

(a A = B = NO; b A = NH_2, B = NO; c A = B = NH_2; d A = NH_2, B = OH; e A = $N(Et)BEt_2$, B = NO; f A = $N(OH)Bu^t$, B = NO; g A = N:CHR (R = H or n-C_3H_7), B = NO) can be reduced by a wide variety of reagents. Even simple refluxing in toluene for 24 h gives **Ib** in 15% yield.[123] **Ib** is obtained in higher yield when BH_4^- is used as reductant;[124-126] $AlH_2(OCH_2CH_2OCH_3)_2^-$ gives initially **Ib** and subsequently **Ic** or **Id**;[127] $LiEt_3BH$ gives a mixture of **Ib, Ic, Id, Ie** and $[CpCr(C_2H_5)(NO)_2]$.[128] With $LiBu^t$ (followed by hydrolysis) a mixture of **Ib** and **If** is formed, but with $LiCH_3$ or $LiBu^n$ an α-elimination occurs and **Ig** is obtained.[129] It was suggested that all of these reactions take place by initial nucleophilic attack of H^- or R^- at a μ_2 - NO group.[127,129] This implies that the bridging nitrosyl is more electrophilic than the terminal nitrosyls which are linear.[130] It seems more likely that there is initial electrophilic attack of Li^+ at the oxygen atom of the bridging nitrosyl followed by nucleophilic attack at the nitrogen atom (though both terminal and bridging NO groups are equally basic toward Cp_3Sm[138]). It has also been suggested that nucleophilic attack occurs at the terminal nitrosyls, and that the products rearrange to the thermodynamically stable bridged complexes.[127] In this connection it is worth noting that reduction of $[CpCrCl(NO)_2]$ (which contains only terminal nitrosyls) with BH_4^- gives $[CpCr(H)(NO)_2]$, and with AlR_3, $[CpCr(R)(NO)_2]$ is formed. No attack at the nitrosyl ligand occurs.[131-132,135]

Reactions of $LiBu^t$ with $[CpCo(NO)]_2$ (which is isoelectronic to $[CpCr(NO)_2]_2$) apparently gives a mixture of $[Cp_2Co_2(\mu_2\text{-}NHBu^t)$ $(\mu_2$ - $NO)]$, $[Cp_3Co_3(\mu_3\text{-}NO)_2]$ and $[Cp_3Co_3(\mu_3\text{-}NBu^t)_2]$;[136] since $[Cp_3Co_3(\mu_3\text{-}NO)_2]$ can be obtained thermally from $[CpCo(NO)]_2$,[137] the mixture of products may represent a disproportionation of first-formed $[Cp_2Co_2(\mu_2\text{-}NHBu^t)(\mu_2\text{-}NO)]$. Reaction of LiR (R = Ph, Bu^t) with $[CpNi(NO)]$ gave $[Cp_3Ni_3(\mu_3\text{-}NBu^t)]$, and again nucleophilic attack of R^- at the nitrosyl group was postulated as a likely first step.[139] Reduction of $[CpNi(NO)]$,[140] $[CpCo(NO)]_2$[141] or $[CpFe(NO)]_2$[140] by $LiAlH_4/AlCl_3$ gives the appropriate Cp_2M complex, along with $[Cp_4Ni_4H_3]$ and $[Cp_4Co_4H_4]$ in the cases of the nickel and cobalt complexes. The fate of the nitrosyl ligand is unknown. On the other hand $[Cp_3Mn_2(NO)_3]$, which has the structure

is reduced by $LiAlH_4/AlCl_3$ to $[Cp_3Mn_3(NO)_4]$[142] and with LiPh the complex suffers only replacement of $\eta^1\text{-}C_5H_5$ by $\eta^1\text{-}C_6H_5$.[148] The nitrosyl ligands in $[Cp_3Mn_2(NO)_3]$ are obviously much less reactive than those of the Cr, Co or Ni complexes discussed above.

It is worth noting that reduction of mixed carbonyl-nitrosyl complexes involves the carbonyl ligand, as in the reduction of $[CpRe(CO)_2(NO)]^+$ with $K^+HB(OCH(CH_3)_2)_3^-$ which   gives $[CpRe(CO)(CHO)(NO)]$,[149] or even an aromatic ligand e.g. reduction of

TABLE I

Reactions of Nitrosyls with Nucleophiles

Complex	ν(NO) (cm^{-1})	Nucleophile	Products	Ref.
[IrCl$_5$(NO)]$^-$	2006	OH$^-$ NH$_3$ N$_3^-$ NH$_2$OH	[IrCl$_5$(NO$_2$)]$^{3-}$ [IrCl$_5$(NH$_3$)]$^-$ + N$_2$ [IrCl$_5$(H$_2$O)]$^{2-}$ + N$_2$ + N$_2$O [IeCl$_5$(H$_2$O)$^-$ + N$_2$O	151
[IrBr$_5$(NO)]$^+$	1953	As for [IrCl$_5$(NO)]$^-$	Br analogues of [IrCl$_5$(NO)]$^-$ products	151
[IrCl$_3$(Ph$_3$P)$_2$(NO)]$^+$	1945	CH$_3$OH C$_2$H$_5$OH	[IrCl$_3$(Ph$_3$P)$_2$(N(O)OR] R = CH$_3$, C$_2$H$_5$	152
cis- [RuX(HpVi)$_2$(NO)]$^{2p-4}$	1885– 1950	$\overline{\text{H}_2\text{C. CO. NR. CO. NR. CO.}}$ R = H, CH$_3$		153–160
		CH$_3$COCH$_2$COCH$_3$	[RuX(HpV)$_2$(oxime)]$^{2p-5}$	156
		CH$_3$CHOCH$_2$COOC$_2$H$_5$	[(Oxime is ligand	156
		Compounds containing acidic methylene groups	containing (N(OH)=C<)]	156
		OH$^-$	[RuX(HpV(NO$_2$)]$^{2p-6}$	154,158

HpV = $\overline{\text{Ō - O - N:C.CO.NR.CO.NR.CO}}$,
X = Cl, Br, NO, OH

where R = H for 2p = 0–4
R = CH$_3$ for p =2

[RuCl(pdma)₂(NO)]²⁺	1886	OH⁻	[RuCl(pdma)₂(NO₂)]	183
		N₃⁻	[RuCl(pdma)₂(N₃)] + N₂ + N₂O	184
		N₂H₄	[RuCl(pdma)₂(N₃)]	183, 185
		PhNHNH₂	[RuCl(pdma)₂(N(O)NNHPh)]	183
cis-[RuCl(acac)₂(NO)] (acac = MeCOCHCOCMe)	1884	MeCOCH₂COMe	[(acac)₂Ru.O.C.Me:(COMe).N(O)]	182
[OsCl(pdma)₂(NO)]²⁺	1861	OH⁻	[OsCl(pdma)₂(NO₂)]	186
		N₂H₄	[OsCl(pdma)₂(N₃)]	186
		NH₂OH	[Os(OH)(pdma)₂(N(O)NNHPh)]	186
		PhNHNH₂	[OsCl(pdma)₂(N(O)NNHPh)]	186
[RuX(py)₄(NO)]²⁺ (X=Cl, Br, OH)	1868 (OH⁻) to 1910 (Cl⁻)	OH⁻	[RuX(py)₄(NO₂)]	462
		N₃⁻	[RuCl(N₃)(py)₄] + [RuCl(H₂O)(py)₄]⁺ +[RuCl(py)₄(N₂)]⁺	462
		SO₃²⁻	[RuCl(py)₄(N(O)SO₃)]	463
[Os(trpy)(bpy)(NO)]³⁺	1904	OH⁻	[Os(trpy)(bpy)(NO₂)]⁺	515
[OsCl(bpy)₂(NO)]²⁺	1888	OH⁻	[OsCl(bpy)₂(NO₂)]	515
[Mo(acac)₂(NO₂)]	1763, 1645 acac		[Mo(acac)₂(hia)(NO)]	503
[Mo(CN)₄(NO₂)]²⁻	1765, 1620 CN⁻/OH⁻		[Mo(CN)₅(NO)]⁴⁻ + NO₂⁻	503

TABLE I (continued)

TABLE I (continued)

Complex	(NO) (cm⁻¹)	Nucleophile	Products	Ref.
$[Ru(NH_3)_5(NO)]^{3+}$	1925	OH^-	$[Ru(NH_3)_5(NO_2)]^+$	161, 164
		NH_2OH	$[Ru(NH_3)_5(N_2O)]^{2+}$	162–3
		$[RuNH_2(NH_3)_4(NO)]^{2+}$	$[Ru\,(NH_3)_5(N_2)]^{2+}\,+$	165–6
			$[RuOH(NH_3)_4(NO)]^{2+}$	167
		$R'COCH_2R$	$[Ru(NH_3)_5(N=CR)]^{2+}$	
		$CH_3NH_2,\ C_2H_5NH_2$	$[Ru(NH_3)_5(N_2)]^{2+}\,+\,ROH\,+\,RNH_3^+$	168
$[RuCl(bpy)_2(NO)]^{2+}$	1927	OH^-	$[RuCl(bpy)_2(NO_2)]$	169, 170
		N_3^-	$[RuCl(bpy)_2(H_2O)]^+\,+\,N_2\,+\,N_2O$	171, 179
		$ArNH_2$	$[RuCl(bpy)_2(N_2Ar)]^{2+}$	174–5
		$ArNR_2$	$[RuCl(bpy)_2(N(O)ArNR_2)]^+$	176–7
		RO^-	$[RuCl(bpy)_2(N(O)OR)]^+$	178
		$CH_3COCH_2COCH_3$	$[Ru(H_2O)(bpy)_2(N(O)C(C(O)CH_3)_2)]^+$	181
		SO_3^{2-}	$[RuCl(bpy)_2(N(O)SO_3)]$	463
		$PhC(O)CH_3$	$[\overline{Ru(bpy)_2(N(O)CHC(O)Ph)}]^+$	477
		$RCH=NNHR'$	$[Ru(bpy)_2(\overline{N(O)C(R)NN(R')})]^+$	476
$[Ru(OH)(NO_2)_4(NO)]^{2-}$	1902	OH^-	$[Ru(OH)(NO_2)_5]^{4-}$	180
		$H_2\overline{C.CO.NMe.CO.NMe.CO}$	$[Ru(DMV)_3OH]^-$	157

$DMV = N(O^-){:}C.\overline{CO.NMe.CO.NMe.CO}$

$[(C_6H_n(CH_3)_{6-n})Cr(CO)_2(NO)]^+$ gives $[C_6H_{(n+1)}(CH_3)_{6-n})Cr(CO)_2(NO)]$ (n = 0 to 2).[150] No reactions of the nitrosyl ligand are observed.

The nitrosyl ligand appears to be more readily reduced when bound to a cluster than when bound to a single metal. Reaction of $NOBF_4$ with $[Ru_4(H)_3(CO)_{12}]^-$ gives $[Ru_4(H)_3(CO)_{11}(\mu_4\text{-N})]$ and $[Ru_4(H)$ $(CO)_{12}(N)]$.[489-490] Analogous products are found with $[Os_4(H_3)_3$ $(CO)_{12}]^-$.[490] It has been shown that $[Ru_3(H)(CO)_{10}(NO)]$ can be hydrogenated to give $[Ru_3(H)(CO)_{10}(NH_2)]$, $[Ru_4(H)(CO)_{12}(N)]$, $[Ru_3(H)_2$ $(CO)_9(NH)]$ and $[Ru_4(H)_4(CO)_{12}]$.[491] Reaction of $[Fe(CO)_3(NO)]^-$ with $Fe_3(CO)_{12}$ gives $[Fe_4(CO)_{12}(N)]^-$; the oxygen of NO is removed by CO as CO_2.[492] Other interesting reactions of cluster bound nitrosyl are discussed under the heading of protonation in Section 5.

4.   REACTIONS OF NITROSYLS WITH NUCLEOPHILES

In Table I are listed the nitrosyl complexes (other than $[M(CN)_5(NO)]^{2-}$) which react with nucleophiles. Also listed are the nitrosyl ligand stretching frequencies $\nu(NO)$ in their ir spectra, the nucleophiles with which they react, and the products of the reactions. Only those reactions which have clearly defined products are listed in Table I. It is seen that all complexes have $\nu(NO) > 1860$ cm^{-1} (the $[M(CN)_5(NO)]^{2-}$ complexes which also react with nucleophiles have $\nu(NO)$ 1938 cm^{-1} for M = Fe, 1927 for M = Ru and 1905 for M = Os). The structurally characterized complexes ($[IrX_5(NO)]^-$,[187] $[Ru(NH_3)_5(NO)]^{3+}$,[92] and $[Ru(OH)(NO_2)_4(NO)]^{2-}$[188] as well as $[Fe(CN)_5(NO)]^{2-}$[28]) all have linear {MNO} groups. Hence the suggestion has been made that the complexes in Table I formally contain coordinated NO^+ and that the high $\nu(NO)$ indicates a low electron density at the {MNO} nitrogen atom, since the greater the donation of metal electrons into π^* orbitals on NO^+ the lower will be the $\nu(NO)$ frequency. There is therefore a relationship between $\nu(NO)$ and the electrophilic behavior of coordinated nitrosyl.[25,180] This approach is clearly very over-simplified.

There is now much X-ray photoelectron spectroscopic and electrochemical evidence which shows that the nitrogen atom in metal nitrosyls carries at best no charge and more usually a negative charge, even when ν(NO) is high.[189-196] In the simplified approach, electrophilic behavior is treated as being solely a function of the π-bonding; σ-bonding and the synergic effects between σ- and π-bonding are ignored. In addition, electrophilic behavior is treated as a function of electron density alone. Since attack of a nucleophile is ipso facto bimolecular, the energy and orbital constitution of the acceptor orbital on the {MNO} group is clearly very important. Despite the glaring deficiences, the simplified approach of suggesting that those nitrosyls with high ν(NO) frequencies will behave as electrophiles is likely to be in use for a long time. It is the easiest practical approach, it is successful, and the reactions actually undergone by the nitrosyls listed in Table I do in fact bear a close resemblance to those undergone by NO^+ or HNO_2.

A more intellectually satisfying, if less practically useful, analysis of the electrophilic behavior of nitrosyls using the molecular orbital model has been made recently.[50] It is calculated that the LUMO (which accepts the electrons from the nucleophile) in $[ML_5(NO)]$ complexes is composed of $d\pi(M)$, $p\pi(N)$, $p\pi(O)$ and $p\pi(cis-L)$ orbitals. The contribution of $p\pi(N)$ to the LUMO increases at the expense of $d\pi(M)$ and the energy of the LUMO decreases as: (1) M moves from left to right across the transition elements; and (2) the electronegativity of the L ligand increases. The composition of the LUMO is essentially unaffected by the number of d electrons on M or the charge on the complex. It is therefore concluded that nucleophilic attack at nitrosyls of the early transition metals will require more energy and will take place at M, whereas attack at nitrosyls of Group VIII will be energetically more favorable and will take place at the nitrosyl ligand. These conclusions are in agreement with the experimental results found to date.

Some of the reactions in Table I require a more detailed discussion because they themselves or their products are unusually interesting.

The attack of NH_2OH at $[Ru(NH_3)_5(NO)]^{3+}$ gives $[Ru(NH_3)_5$ $(N_2O)]^{2+}$, the only known complex containing N_2O as a ligand.[162-163] A mechanism has been proposed similar to that of the reactions between NO^+ and NH_2OH,[197] and that between NH_2OH and $[Fe(CN)_5(NO)]^{2-}$ discussed in Section 2.2. It leads to N-bonded N_2O in which the oxygen atom is derived from NH_2OH.[163,198]

Attack of N_2H_4 at $[MCl(pdma)_2(NO)]^{2+}$ (M = Ru, Os) gives rise to $[MCl(pdma)_2(N_3)]$;[183,186] this parallels the reaction of N_2H_4 with NO^+ to give N_3^-.[199] It is believed that an initial adduct of the type $[MCl(pdma)_2(N(O)NH_2NH_2)]^{2+}$ is formed, and in fact use of $PhNHNH_2$ as the nucelophile gives the isolable "intermediate" $[MCl(pdma)_2$ $(N(O)N_2HPh)]$. It is suggested that the $N(O)N_2HPh^-$ ligand has the structure : $N(O^-)=N-NHPh$, though the precise location of the proton is uncertain.[183,186]

The formation of C - N bonds by nucleophilic attack of carbon bases yields nitroso compounds. The simplest example of such a reaction is that between $[RuCl(bpy)_2(NO)]^{2+}$ and activated aromatic systems:[176-177]

$$[RuCl(bpy)_2(NO)]^{2+} + H-C_6H_4NR(CH_3) \longrightarrow$$

$$[RuCl(bpy)_2(N(O)-C_6H_4NR(CH_3))]^+ + H^+ \qquad (46)$$

$$R = H, CH_3$$

a reaction which is again paralleled by the behavior of NO^+ itself.[200] It is possible that initial nucleophilic attack is by the amine group (primary amines such as $p-CH_3C_6H_4NH_2$ give diazonium complexes by amine attack and H_2O elimination)[174] producing an N-nitrosamine which then undergoes a Fischer-Hepp rearrangement.[177] Phenoxide ion also forms a C - N bond:[177]

$$[RuCl(bpy)_2(NO)]^{2+} + PhO^- \longrightarrow [RuCl(bpy)_2(N(O)C_6H_4OH)]^+ \qquad (47)$$

but again initial attack by the oxygen atoms seems probable, since aliphatic oxides react with $[RuCl(bpy)_2(NO)]^{2+}$ [178] and $[IrCl_3(Ph_3P)_2(NO)]^+$ [152] to give O-nitroso compounds, $MN(O)OR$. Allylidenearylhydrazones, $RCH=NNHR'$, form chelates with $[RuCl(bpy)_2(NO)]^{2+}$: [476]

$$[RuCl(bpy)_2(NO)]^{2+} + RCH = NNHR' \rightarrow [(bpy)_2\overline{Ru(N(O)=CR-N=NR')}]^+$$
$$+ 2H^+ + Cl^- \qquad (48)$$

Reactions of ketones or aldehydes with $[Ru(NH_3)_5(NO)]^{3+}$ give nitrile complexes:

$$[Ru(NH_3)_5(NO)]^{3+} + R'COCH_2R \rightarrow [Ru(NH_3)_5(N \equiv CR)]^{2+} + R'CO_2^- + 2H^+$$
$$(49)$$

though in the case of the reaction between $PhCOCH_3$ and $[RuCl(bpy)_2(NO)]^{2+}$, a low yield of the deprotonated chelate complex $[\overline{Ru(bpy)_2(N(O)-CH = CPh-O)}]^+$ is obtained. [477]

The proposed mechanism is similar to that for production of the isoelectronic $[Ru(NH_3)_5(N_2O)]^{2+}$: [167]

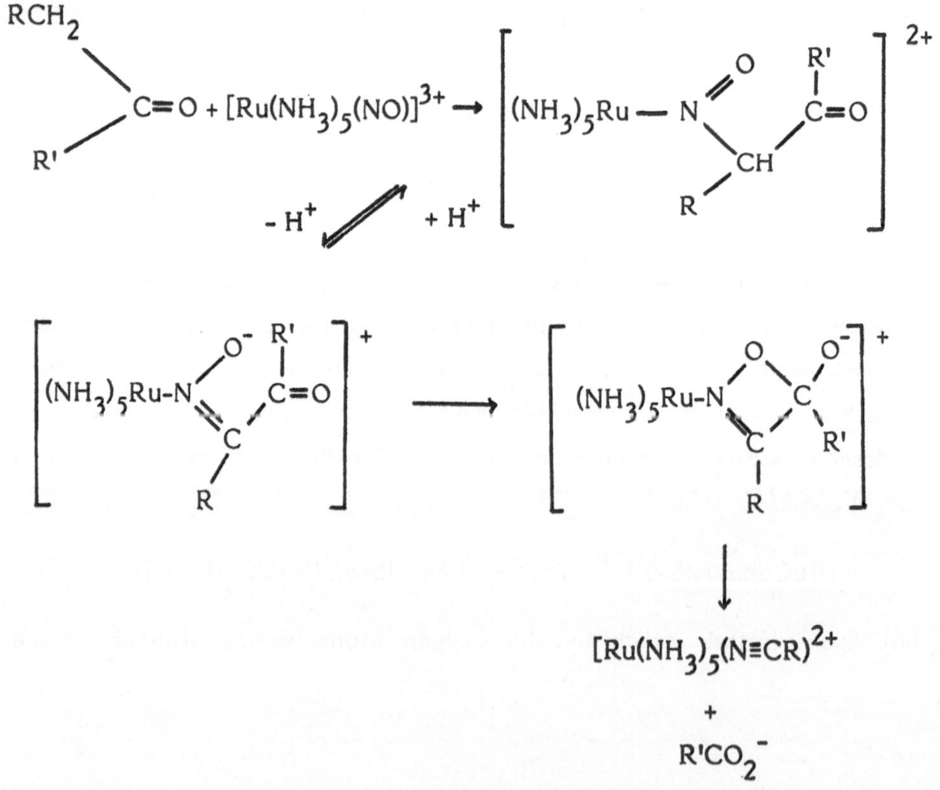

Reaction of barbituric acid with $[Ru(OH)(NO_2)_4(NO)]^{2-}$ gives a variety of complexes of the violurato ligand.[153] The essential reaction occurring is

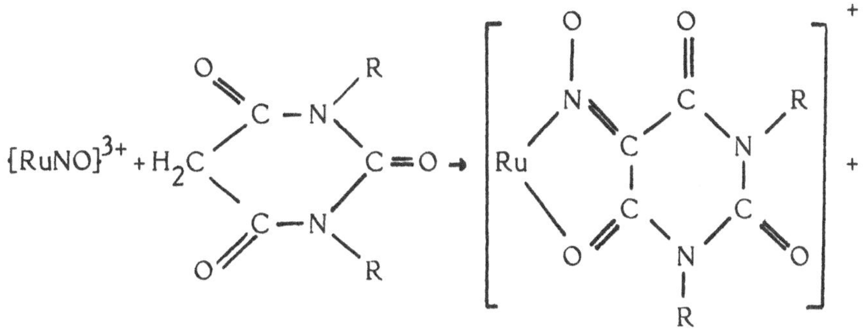

$$R = H, CH_3$$ (51)

R = H, CH$_3$

The bidentate $\underline{N}$, $\underline{O}$ coordinated violurato ligand (abbreviated here as η^2-V when showing $\underline{N},\underline{O}$-coordination) can also act as a unidentate $\underline{O}$-coordinated ligand (abbreviated as η^1-V):

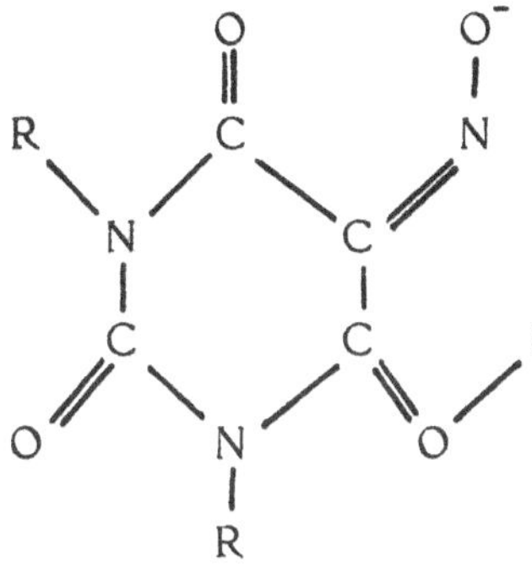

From the reaction between barbituric acid and $[Ru(OH)(NO_2)_4(NO)]^{2-}$ the simple $[Ru(\eta^2\text{-}V)_3]^-$ is obtained, but other complexes such as $[Ru(\eta^2\text{-}V)_2$ $(\eta^1\text{-}V)(OH)]^{2-}$, cis-$[Ru(\eta^2\text{-}V)_2(\eta^1\text{-}V)(NO)]$ and cis-$[RuX(\eta^2\text{-}V)_2(NO)]$ (X = Cl, Br) have been obtained also.[154,155] Both nitrosyls react with nulceophiles,[154] and the formal reaction:[155]

$$[Ru(\eta^2\text{-}V)_2(\eta^1\text{-}V)(NO)] \rightarrow [Ru(\eta^2\text{-}V)_3]^- + NO^+$$ (52)

presumably occurs via prior conversion of $\{RuNO\}^{3+}$ into $\{RuNO_2\}^+$ with subsequent substitution of the NO_2^- ligand:

$$[Ru(\eta^2\text{-}V)_2(\eta^1\text{-}V)(NO)] + 2OH^- \rightleftharpoons [Ru(\eta^2\text{-}V)_2(\eta^1\text{-}V)(NO_2)]^{2-} + H_2O \qquad (53)$$

$$[Ru(\eta^2\text{-}V)_2(\eta^1\text{-}V)(NO_2)]^{2-} \rightarrow [Ru(\eta^2\text{-}V)_3]^- + NO_2^- \qquad (54)$$

A very interesting kinetic study of the relative electrophilicity of the NO ligand in cis-[RuCl$(\eta^2$-V)$_2$(NO)] (R = H) and its deprotonated derivatives has been made.[156] In the reaction of this complex with barbituric acid four distinct steps are evident:

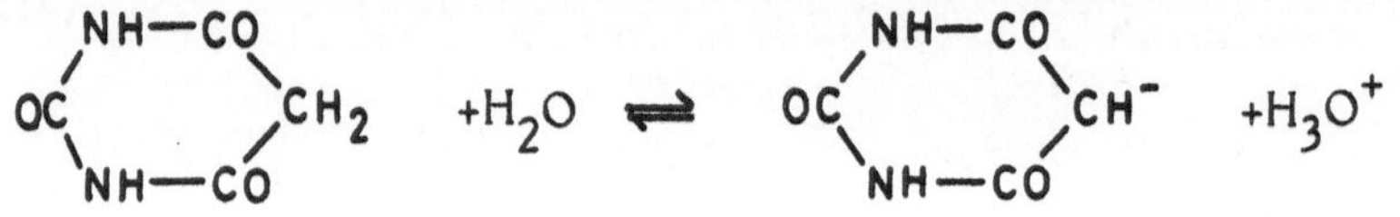

$$\qquad (55)$$

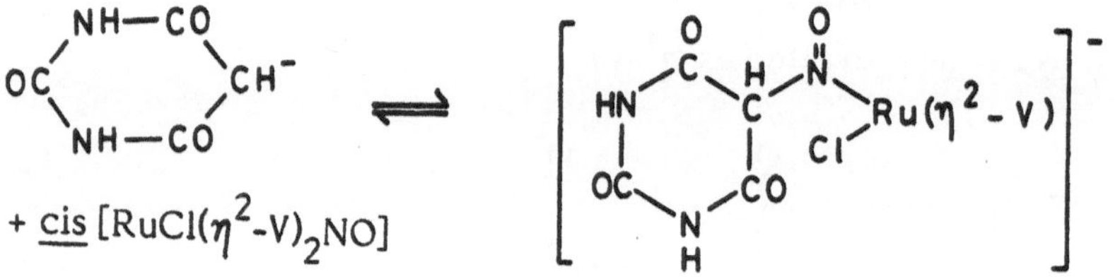

$$\qquad (56)$$

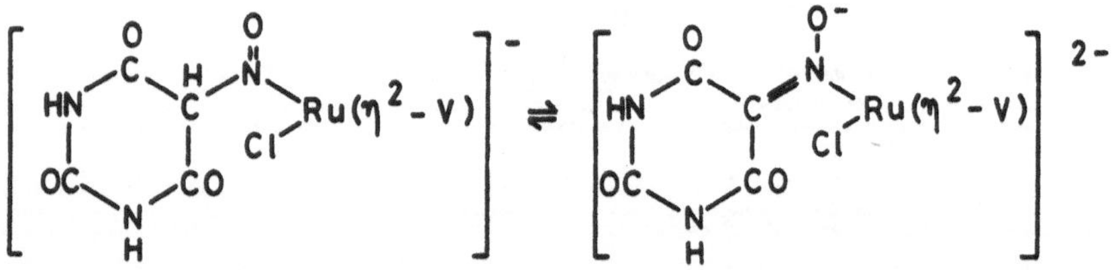

$$\qquad (57)$$

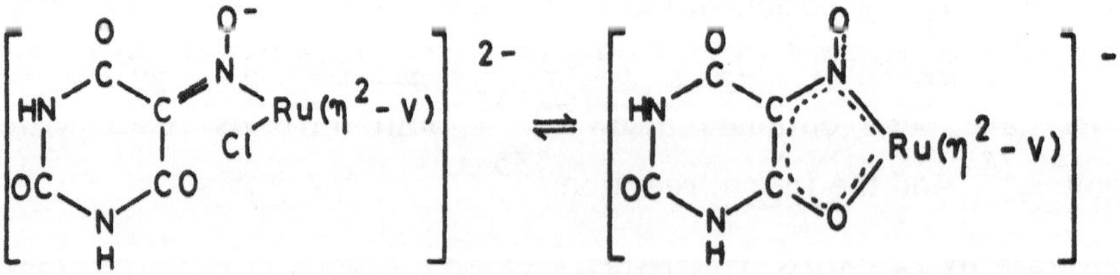

$$\qquad (58)$$

The (η^2-V) ligands can be successively deprotonated at the nitrogen atoms to yield the starting complexes [RuCl(η^2-V - H)(η^2-V)(NO)]$^-$, [RuCl(η^2 -V - H)$_2$(NO)]$^{2-}$, [RuCl(η^2-V - 2H)(η^2-V - H)]$^{3-}$ and [RuCl(η^2-V - 2H)$_2$(NO)]$^{4-}$, where η^2-V - H and η^2-V - 2H represent the monoacid and diacid bases respectively. As deprotonation increases, the electron density at the (η^2-V) ligand increases, and this increased electron density is transferred through Ru to the nitrosyl ligand (ν(NO) decreases). The rate constant for reaction (56), the actual attack of the nucleophile, decreases as deprotonation increases. The dependency of the rate on the acidity of the methylene group of barbituric acid can be observed by changing the R group on this nucleophile, and by changing the solvent.[156] This study is the most thorough so far on the effect of gradual and controllable changes in the electron density of the nitrosyl on nucleophilic attack.

Reaction of (CH$_3$COCHCOCH$_3$)$^-$ (acac) with "nitrosyl ruthenium" in alkaline solution gives [Ru(acac)$_2$(hia)], where hia is the bidentate ligand:[182]

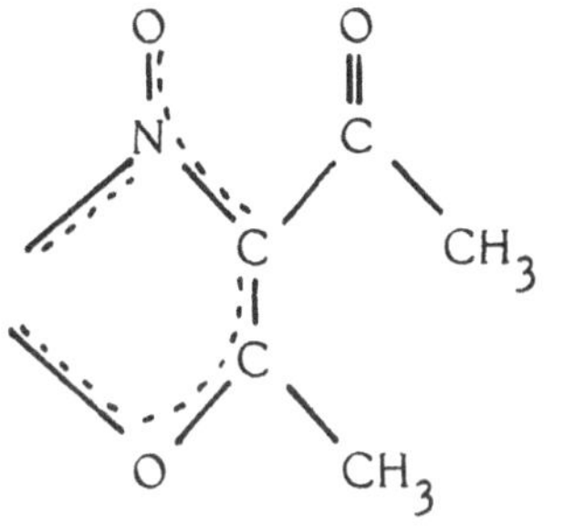

Reaction of [RuCl(bpy)$_2$(NO)]$^{2+}$ with acac gives the same ligand, but this time N-unidentate.[181] There is apparently a similar reaction with [Mo(acac)$_2$(NO)$_2$], which has very low ν(NO) frequencies (1765 and 1620 cm^{-1}) to give [Mo(acac)$_2$(hia)(NO)].[503]

The reaction between [RuCl(bpy)$_2$(NO)]$^{2+}$ and N$_3^-$ is presumed to lead to the unstable intermediate [RuCl(bpy)$_2$(N(O)N$_3$)]$^+$ which decomposes to [RuCl(bpy)$_2$(H$_2$O)]$^+$, N$_2$ and N$_2$O. The ready formation of a labile aquo complex has been exploited for the synthesis of

TABLE II

Reactions of Nitrosyls with Hydroxide Ion:

$$\{MNO\}^{n+} + 2OH^- \rightleftharpoons \{MNO_2\}^{(n-2)+} + H_2O$$

Complex	K (l^2 mol^{-2})	ν(NO) (cm^{-1})	Ref.
NO$^+$	2.3×10^{31}	2220	7, 144
[IrCl$_5$(NO)]$^-$	$>6 \times 10^{29}$	2006	151
[IrBr$_5$(NO)]$^-$	6.8×10^{27}	1953	151
[RuOH(η^2-V)$_2$(NO)]	1.0×10^{13} a	1930	154
[Ru(η^2-V)$_2$(η^1-V)(NO)]	1.0×10^{13} a	1945	154
[Os(trpy)(bpy)(NO)]$^{3+}$	7.0×10^{10}	1904	515
[Ru(trpy)(bpy)(NO)]$^{3+}$	2.1×10^{23}	1952	516
[RuCl(bpy)$_2$(NO)]$^{2+}$	1.6×10^9	1927	170
[Fe(CN)$_5$(NO)]$^{2-}$	1.5×10^6	1938	7, 180

a K is in l mol^{-1} because the nitrosyl is insoluble.

[RuCl(bpy)$_2$L]$^{n+}$ complexes.[173,179] The complex [RuCl(bpy)$_2$(N(O)SO$_3^-$)] produced by reaction of [RuCl(bpy)$_2$(NO)]$^{2+}$ with SO$_3^{2-}$ contains the otherwise unknown {N(O)SO$_3^-$} ligand with a very long (1.82 Å) N-S bond.[463]

The most common reaction of coordinated nitrosyl with a nucleophile is that with OH$^-$ to form coordinated NO$_2^-$. In Table II are listed the complexes for which the equilibrium constants for this reaction are known, together with the ν(NO) frequencies. It is seen that the equilibrium constants cover a huge range and are therefore a good measure of the electrophilicity of the nitrosyl, even though they include the thermodynamic stability of the nitro product as well as of the nitrosyl. The charge on the complex is not an important factor to the K_{equil} values, which follow quite closely the much more restricted range of $\gamma(NO)$ frequencies.

The reaction of $[Ru(NH_3)_5(NO)]^{3+}$ with OH^- is not as simple as those in Table II. Although $[Ru(NH_3)_5(NO_2)]^+$ is formed, the deprotonated species $[RuNH_2(NH_3)_4(NO)]^{2+}$ also appears. The amido ligand acts as a nucleophile towards $[Ru(NH_3)_5(NO)]^{3+}$, ultimately forming $[Ru(NH_3)_5(N_2)]^{2+}$ via the intramolecular reaction of one coordinated ligand with another:[165-166]

$$[Ru(NH_3)_5(NO)]^{3+} + H_2O \rightleftharpoons [Ru(NH_2)(NH_3)_4(NO)]^{2+} + H_3O^+ \quad (59)$$

$$[Ru(NH_2)(NH_3)_4(NO)]^{2+} + [Ru(NH_3)_5(NO)]^{3+} \rightleftharpoons$$
$$[(NH_3)_5Ru[N(O)NH_2]Ru(NH_3)_4(NO)]^{5+} \quad (60)$$

$$[(NH_3)_5Ru[N(O)NH_2]Ru(NH_3)_4(NO)]^{5+} + H_2O \rightarrow [Ru(NH_3)_5(N_2)]^{2+} +$$
$$\underline{cis}\text{-}[Ru(OH)(NH_3)_4(NO)]^{2+} + H_3O^+ \quad (61)$$

Even $[Ru(NH_3)_6]^{3+}$, which is known to be acidic[201] and whose conjugate base $[Ru(NH_3)_5(NH_2)]^{2+}$ acts as a nucleophile towards carbonyl groups,[202] can act as a source of $[Ru(NH_3)_5(N_2)]^{2+}$ in reaction with $[Ru(NH_3)_5(NO)]^{3+}$ via an analogous reaction scheme to (59) - (61).[168] An unknown reducing agent must also be involved. On the other hand $C_2H_5NH_2$ and CH_3NH_2 react directly with $[Ru(NH_3)_5(NO)]^{3+}$ to give $[Ru(NH_3)_5(N_2)]^{2+}$ (i.e. they do not just act as bases deprotonating $[Ru(NH_3)_5(NO)]^{3+}$).[168] The driving force for these reactions seems to be the remarkable thermodynamic stability of $[Ru(NH_3)_5(N_2)]^{2+}$ rather than the electrophilicity of $[Ru(NH_3)_5(NO)]^{3+}$.

It is finally appropriate to discuss "non-reactions" of nitrosyl with nucleophiles which did not occur despite purposeful attempts to carry them out. These negative results, though frustrating to those people performing them, have contributed to our understanding of nitrosyl reactivity. Even those nitrosyls which do react with some nucleophiles often do not react with all; N_3^- reacts with $[RuCl(pdma)_2(NO)]^{2+}$ [184] but not with $[OsCl(pdma)_2(NO)]^{2+}$ [186] for instance. Other than such scattered observations there has been no research into the nucleophilicity side of the equation, and if this area of nitrosyl reactivity is to expand some detailed work is necessary.

Purposeful attempts to induce nucleophilic attack at $[Os(NH_3)_5(NO)],^{3+ \; 203}$ trans-$[RuCl(NH_3)_4(NO)],^{2+ \; 164}$ $[ReCl_2(PR_3)_n(NO)]$ (n = 2 or 3),^{204} and $[MoCl_n(PR_3)_m(NO)]^{p-}$ ($n + m$ = 5, p = 3 - n)^{205} were unsuccessful. All these complexes have $\nu(NO) < 1860 \; cm^{-1}$.

Nucleophilic attack at mixed carbonyl-nitrosyls occurs usually at the CO, though sometimes at another ligand e.g. an aromatic ring, or at the metal. No cases are known where the nitrosyl is attacked. Examples include: $[CpMn(CO)(L)(NO)]^+$ (L = CO or PR$_3$);$^{206-210,225,417}$ $[M(PR_3)_2(CO)_2(NO)]^+$ (M = Fe, Os);^{213} and $[CpRe(CO)_2(NO)]^+.^{149,215}$ For the latter cations an extensive chemistry of the carbonyl ligand has been explored (see Volume I). The reactivity of the carbonyl ligand towards nucleophiles implies that the $\pi^*\{MCO\}$ orbital lies at lower energy than $\pi^*\{MNO\}$. If so, nucleophilic attack at mixed carbonyl-nitrosyls is never likely to occur at the nitrosyl, and if it occurs at the metal rather than the carbonyl, then a carbonyl ligand, not a nitrosyl, will be replaced. Examples of this process are the reactions of $[Fe(R_3P)_2(CO)_2(NO)]^+$ with nucleophiles (X) such as N_3^-, SCN$^-$ or CN$^-$ to give $[FeX(R_3P)_2(CO)(NO)]^{216}$ and the replacement of one CO ligand in $[Mn(CO)(NO)_3]$ or $[Co(CO)_3(NO)]$ by a wide variety of phosphines.$^{217-218}$

Nucleophilic attack of CH_3O^- or OH$^-$ on the mixed nitrosyl-isocyanide $[CpCr(CH_3NC)(NO)]^+$ occurs at the CH_3NC ligand;^{219} of CH_3O^- on the mixed carbene-nitrosyl $[CpRe(Ph_3P)(CHPh)(NO)]^+$ at the carbene ligand;^{220} and of R$_3$P on $[\eta^4\text{-}C_4H_4)Fe(CO)_2(NO)]^+$ at the aromatic ring.$^{221-4}$ Hence the LUMO in all these cases must contain a major contribution from a ligand other than the nitrosyl.

5.   REACTIONS OF NITROSYLS WITH ELECTROPHILES

As discussed in Section 4, nitrosyls with a high $\nu(NO)$ frequency react with nucleophiles. Conversely, nitrosyls with a very low $\nu(NO)$ frequency might be expected to react with electrophiles. Since very low γ(NO) frequencies are associated with bent {MNO} groups, one is tempted to associate electrophilic attack with bent nitrosyls. Although such an association is broadly true, the relations between electrophilic attack and either γ(NO) frequency or the valence bond description of the nitrosyl ligand are much less useful for electrophilic than they are for nucleophilic attack. This is because the valence bond description of a bent {MNO} group assigns lone pairs of electrons to both N and O, so that the site of electrophilic attack is no longer predictable. Nitrosyls with low γ(NO) frequencies are not necessarily bent,[226] and linear nitrosyls can be attacked by electrophiles (the valence bond description of a linear nitrosyl assigns a lone pair to the oxygen atom). Finally, whereas considering nucleophilic attack as occurring at coordinated NO^+ has the advantage that the chemistry observed does actually parallel that of uncoordinated NO^+, considering electrophilic attack as occurring at coordinated NO^- has no such advantage. The NO^- species has at best a fleeting existence,[227-228] and its chemistry is for practical purposes unknown.

Electrophilic attack also suffers from the fact that electrophiles are often oxidizing agents, and nitrosyls are usually unstable with respect to oxidation. There is competition between reaction of the nitrosyl ligand and metal oxidation.

The simplest electrophile is the proton. The simultaneous addition of electrons and protons to nitrosyls has been discussed in Section 3. Reactions in which a proton is added to a non-reduced isolable nitrosyl attached to a single metal are also known, and are listed in Table III. Of the complexes listed, the structures of $[CoBr(pdma)_2(NO)]^+$ (CoNO 132°),[232] $[Ir(Ph_3P)_3(NO)]$ (IrNO linear),[231] and $[Os(Ph_3P)_2(NO)_2]$ (OsNO 179, 174°)[226] have been determined. The Ir analogue of $[OsCl(CO)(Ph_3P)_2(NO)]$ is $[IrCl(CO)(Ph_3P)_2(NO)]^+$ and has an IrNO angle of 124°,[11] and the CoNO angle in $[CoCl(en)_2(NO)]^+$ (c.f. $[Co(OH)(en)_2(NO)]^+$ in Table III) is also 124°.[93]

In the cases of protonation of $[CoBr(pdma)_2(NO)]^+$ and $[Co(OH)(en)_2(NO)]^+$ it is clear that simple addition of a proton has occurred, with no other changes in the coordination sphere. Infrared and

TABLE III

Protonation of Nitrosyls bound to a Single Metal

Complex	ν(NO) (cm^{-1})	Acid	Product	Ref.
[OsCl(CO)(ph$_3$P)$_2$(NO)]	1560[a]	HCl	[OsCl$_2$(CO)(Ph$_3$P)$_2$(HNO)]	229
[Os(Ph$_3$P)$_2$(NO)$_2$]	1615, 1655	2HCl	[OsCl$_2$(NHOH)(Ph$_3$P)(NO)]^r	229
[Ir(Ph$_3$P)$_3$(NO)]	1600	3HX	[IrX$_3$(Ph$_3$P)$_2$(NH$_2$OH)]^r (X = Cl, Br)	230
[Rh(Ph$_3$P)$_3$(NO)]	1612	3HCl	[RhCl$_3$(Ph$_3$P)$_2$(NH$_2$OH)]^r	231
[CoBr(pdma)$_2$(NO)]$^+$	1550, 1565	HBr	[CoBr(pdma)$_2$(HNO)]$^{2+}$	232
[ReCl(Ph$_3$P)$_2$(NO)$_2$]	1600, 1650	HCl	[ReCl$_2$(Ph$_3$P)$_2$(HNO)(NO)]^r	233
trans-[MoF(cppe)$_2$(NO)]	1528	HPF$_6$	[MoF(dppe)$_2$(HNO)]$^{+ r}$	234
trans-[MoF(dppe)$_2$(NO)]		HX	[MoX$_2$(dppe)$_2$(HNO)] (X = Cl, Br)	234
[V(THA)(H$_2$O)(NO)]$^-$	1502	H$^+$	[(V(THA)(NH$_2$O)$_2$OH]$^{+ r}$	103b
[Co(OH)(en)$_2$(NO)]$^{2+}$		H$^+$	[Co(H$_2$O)(en)$_2$(NOH)]$^{3+ r}$	236

[a] weak to medium bands appear at 1769 and 1629 cm^{-1} also.[235]   [b] ν(THA) = $\overline{\text{N}(\text{CHMeCH}_2\text{O-})_3\text{V}}$

r = reversible

nmr evidence indicate that the site of protonation in $[CoBr(pdma)_2(NO)]^+$ is the nitrogen atom, and molecular orbital calculations indicate that the highest occupied molecular orbital (HOMO) in the nitrosyl is an antibonding π-type molecular orbital which has a large $\pi^*(NO)$ component.[232] There is little transfer of electron density from cobalt to the nitrosyl on protonation; in the simplest valence bond model the reaction:

$$\left[(pdma)_2ClCo{-}\ddot{N}\underset{\ddot{\ddot{O}}\cdot}{}\right]^+ + H^+ \longrightarrow \left[(pdma)_2ClCo{-}N\overset{H}{\underset{\ddot{\ddot{O}\cdot}}{\diagup}}\right]^{2+} \qquad (62)$$

has occurred.

Simple protonation apparently occurs for $[MoF(dppe)_2(NO)]$ also.[234] This is surprising, since this 18 electron $\{MNO\}^6$ complex would be expected to have a linear $\{MNO\}^6$ group and require addition of a seventh ligand before protonation at the nitrogen atom could occur.

It has been suggested that the site of protonation of $[Co(OH)(en)_2(NO)]^+$ is the oxygen atom of the nitrosyl ligand.[236] This suggestion must be doubted in view of: (i) the results for $[CoCl(pdma)_2(HNO)]^{2+}$;[232] (ii) the structure of $[OsCl_2(CO)(Ph_3P)_2(HNO)]$, where crystallographic evidence for protonation at the N atom exists;[235] (iii) the existence of several crystallographically certified complexes containing the HNO, RNO, H_2NO or R(H)NO ligands (References 93, 98-100, 239, 240-242, 468, 475, 481-482, 485-486).

Note that whereas $[OsCl_2(CO)(Ph_3P)_2(HNO)]$ contains unidentate N-coordinated HNO,[235] the other complexes contain bidentate N,O-coordinated HNO, H_2NO or R(H)NO. Protonation at the nitrogen atom of NO must therefore be assumed when the nitrosyl ligand is bound to a single metal. However nitrosyl bound to clusters as μ_2-NO or μ_3-NO reacts at the oxygen atom. Protonation of $[Ru_3(CO)_{10}(NO)]^-$ with CF_3SO_3H gives $[Ru_3(CO)_{10}(NOH)]$.[483] With the weaker acid CF_3CO_2H the hydride $[Ru_3(H)(CO)_{10}(NO)]$ is formed.[488] Unfortunately

$[Ru_3(CO)_{10}(NOH)]$ is an oil, but the analogous reaction of $CF_3SO_3CH_3$ with $[Ru_3(CO)_{10}(NO)]^-$ gives $[Ru_3(CO)_{10}(NOCH_3)]$ which has been structurally characterized. It contains a μ_3-$NOCH_3$ ligand having an NOC angle of 112.1°.[483] Protonation of $[MecpMn(NO)]_3(\mu_3$-$NO)$ (Mecp $= \eta^5$-$C_5H_4CH_3$) with one equivalent of $HBF_4 \cdot O(CH_3)_2$ or aqueous HPF_6 gives $[\{MecpMn(NO)\}_3(\mu_3$-$NOH)]^+$, which in turn reacts with a further two equivalents of acid to give $[\{MecpMn(NO)\}_3(\mu_3$-$NH)]^+$. Both reactant and product have been structurally characterized (the NOH angle is 107°).[484] It appears that the nitrosyl ligand behaves like an amine oxide when attached to a cluster but not when attached to a single metal.

The reaction[229]

$$[OsCl(CO)(Ph_3P)_2(NO)] + HCl \rightleftharpoons [OsCl_2(CO)(Ph_3P)_2(HNO)] \qquad (63)$$

could also involve initial protonation of the nitrosyl ligand (the {OsNO} group may be bent) followed by subsequent addition of Cl^- to the osmium coordination sphere. The reverse order of addition, a concerted reaction, or initial metal hydride formation and subsequent hydride transfer (the proton in HNO is distinctly hydridic according to the nmr spectrum[235]) are also possible mechanisms; attempts to obtain definitive information were not successful.[235] The probability of an initial oxidative addition of HX followed by hydride shift is quite likely for the five-coordinate dinitrosyl $[ReCl(Ph_3P)_2(NO)_2]$ and even more likely for the four-coordinate $[M(Ph_3P)_3(NO)]$ (M = Rh, Ir) and $[Os(Ph_3P)_2(NO)_2]$. The reaction of $[Ru(Ph_3P)_2(NO)_2]$ with HCl in fact gives initially $[Ru(H)(Ph_3P)_2(NO)_2]^+$ [243] but ultimately $[RuCl_3(Ph_3P)_2(NO)]$ is formed and not a complex containing (HNO). Treatment of $[Ir(Ph_3P)_3(NO)]$ with a non-complexing acid ($HClO_4$, HBF_4 or HPF_6) gives $[Ir(H)(Ph_3P)_3(NO)]^+$, which is converted to $[IrCl_3(Ph_3P)_2(NH_2OH)]$ by HCl.[230,267] However, hydride transfer to NO in $[Ir(H)(Ph_3P)_3(NO)]^+$ could not be induced by adding X^-, only $[Ir(H)(X)(Ph_3P)_2(NO)]$ being obtained, and this latter complex does not give any NH_2OH on treatment with HCl. The structure of $[Ir(H)(Ph_3P)_3(NO)]^+$ shows it to be a trigonal bipyramid with an axial linear IrNO group trans to the hydride,[237,245] although fluxional behavior involving a bent {IrNO} group may be expected in solution, since the isoelectronic and isostructural $[Ru(H)(Ph_3P)_3(NO)]$ is fluxional.[246] The related $[IrCl_2(Ph_3P)_2(NO)]$, which is formed when two instead of three moles of HCl are added to $[Ir(Ph_3P)_3(NO)]$,[230] and is a square pyramidal molecule with an apical bent (123°) {IrNO} group,[244] does not accept protons. Neither does $[Co(Ph_3P)_3(NO)]$.

The above conglomeration of facts clearly indicates that simple protonation, even of a bent nitrosyl, does not occur in these systems. However, the precise mechanism leading to NHOH or NH_2OH is not at all obvious. The formation of NHOH or H_2NOH complexes may indicate that simultaneous protonation and hydride shift to NO from the metal is necessary, and such a suggestion is at least not in disagreement with any of the above facts.

Protonation of the nitrosyl ligand followed by intramolecular condensation between an NHOH and a CF_3COO ligand is proposed to explain the formation of $[Os(ON = C(O)CF_3)(O_2CCF_3)(Ph_3P)(NO)]$ from $[Os(Ph_3P)_2(NO)_2]$ and CF_3COOH.[487] This mechanism seems very probable since $[Os(Ph_3P)_2(NO)_2]$ is known to give $[OsCl_2(NHOH)(Ph_3P)_2(NO)]$ on reaction with HCl.[229] The CF_3COOH reaction also gives a hydride $[Os(H)(O_2CCF_3)(Ph_3P)_2(NO)]$,[252] presumably by loss of NH_2OH. The ruthenium analogue $[Ru(Ph_3P)_2(NO)_2]$ gives $[Ru(O_2CCF_3)_2(Ph_3P)_2(NO)]$, no condensation taking place.[252,487]

As do carbonyls, nitrosyls interact with Lewis acids, particularly $[Cp_3Ln]$ (Ln = lanthanide). Crease and Legzdins first showed that the $\nu(NO)$ frequency of $[CpCrCl(NO)_2]$ decreases on interaction with Cp_3Er, Cp_3Yb and Cp_3Sm, implying adduct formation in which the oxygen atom of the nitrosyl donates a pair of electrons to Cp_3Ln.[247] For Ln = Sm an equilibrium between complexed and uncomplexed $[CpCrCl(NO)_2]$ is evident. Extension of this work to other lanthanide acceptors and to $[CpM(CO)_2(NO)]$ (M = Cr, Mo, W) and $[CpMn(CO)(NO)]_2$ gave the surprising result that $[CpMn(CO)(NO)]_2$ formed adducts in which bridging CO and terminal NO ligands are complexed.[248] In solution $[CpMn(CO)(NO)]_2$ is a fluxional molecule having both bridging and terminal CO and NO groups:[249]

$$\begin{array}{ccccc} & & O & & O \\ & & \| & & \| \\ Cp & & X & & X \\ \diagdown & \diagup & \diagdown & \diagup & \\ & Mn & & Mn & & \qquad X = C, N \\ \diagup & \diagdown & \diagup & \diagdown & \\ X & & X & & Cp \\ \| & & \| & \\ O & & O & \end{array}$$

Apparently the terminal nitrosyl oxygen is more basic than the bridging, but the reverse is true for the carbonyl. In $[CpM(CO)_2(NO)]$ (M = Cr, Mo, W) only the nitrosyl forms an adduct; the $\nu(NO)$ frequency goes to lower energy and $\nu(CO)$ to higher, because transfer of metal electron density into the nitrosyl to compensate for that removed by Cp_3M leaves less electrons to be donated to the carbonyl. Again, the terminal nitrosyl is a better Lewis base than the terminal carbonyl. Other nitrosyls such as $[Fe(CO)_2(NO)_2]$ and $[RuCl_3(Ph_3P)_2(NO)]$ do not form adducts with Cp_3Ln.[248] A comparative study of adduct formation between Cp_3Sm and a terminal nitrosyl (in $[CpCr(CO)_2(NO)]$), a doubly bridging nitrosyl (in $[MecpMn(CO)(NO)]_2$), and a triply bridging nitrosyl (in $[(MecpMn)_3(NO)_4]$) shows that there is no significant difference between the Lewis basicity of doubly bridging and triply bridging nitrosyls, but that terminal nitrosyls have an equal or slightly greater Lewis basicity than bridging.[138] It is interesting that adduct formation has only been observed with nitrosyls of the earlier transition metals; the polarization of the {MNO} group due to the π-interaction is in fact less for these metals than for the later members of the series.

Addition of BF_3 to $[Fe(CN)_5(NO)]^{2-}$ increases the frequency of $\nu(NO)$, consistent with adduct formation between a CN^- ligand and BF_3 rather than a nitrosyl-BF_3 interaction.[250] It has however been suggested that FeNOM bridging occurs in a variety of M^{2+} - $[Fe(CN)_5(NO)]$ salts.[46]

After Brønsted and Lewis acids the next obvious electrophile is dioxygen, and several detailed studies of the oxidation of nitrosyls have been undertaken. The reaction between O_2 and $[CoL_4(NO)]$:[253]

$$2[CoL_4(NO)] + 2B + O_2 \longrightarrow 2[Co(B)L_4(NO_2)] \qquad (64)$$

and between O_2 and $[Co(en)_2(NO)]^{2+}$:[253]

$$2[Co(en)_2(NO)]^{2+} + 2CH_3CN + O_2 \longrightarrow 2[Co(CH_3CN)(en)_2(NO_2)]^{2+} \quad (65)$$

where L_4 is one of the planar tetradentate ligands

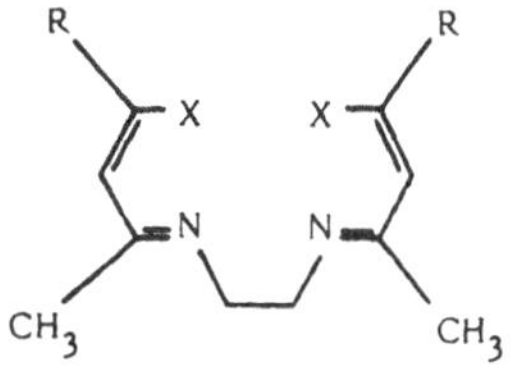

R = CH$_3$, X = O

R = C$_6$H$_5$, X = O   or

R = CH$_3$, X = S

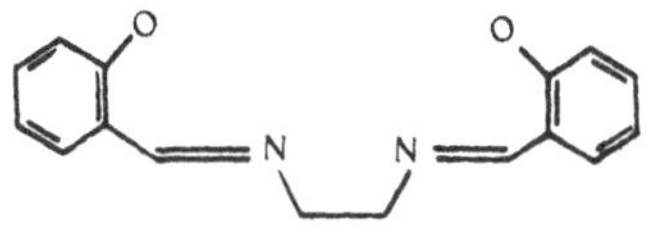

or twice the bidentate ligand $Me_2NC\overset{S}{\underset{S}{\diagdown}}$ and B is a base (R_3P or pyridine)

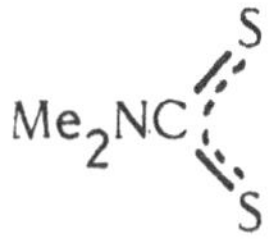

have been studied kinetically. All complexes have CoNO angles close to 120°. The base B is necessary to promote the reaction, the rate of which increases with increasing basicity of B. The mechanism proposed is:[253]

$$[CoL_4(NO)] + B \overset{fast}{\rightleftharpoons} [Co(B)L_4(NO)] \quad (66)$$

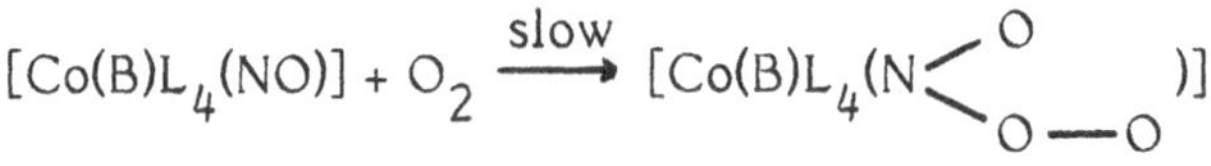

$$[Co(B)L_4(NO)] + O_2 \overset{slow}{\longrightarrow} [Co(B)L_4(N\overset{O}{\underset{O-O}{\diagdown}})] \quad (67)$$

$$[Co(B)L_4(N\overset{O}{\underset{O-O}{\diagdown}})] + [Co(B)L_4(NO)] \overset{fast}{\longrightarrow}$$

$$[Co(B)L_4(N\overset{O\quad O}{\underset{O-O}{\diagup\diagdown}}N)CoL_4(B)] \quad (68)$$

$$[Co(B)L_4(N\overset{O\quad O}{\underset{O-O}{\diagup\diagdown}}N)CoL_4(B)] \overset{fast}{\longrightarrow} 2[Co(B)L_4(NO_2)] \quad (69)$$

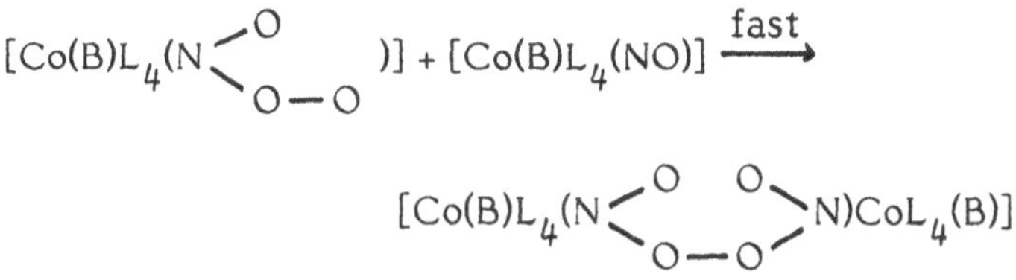

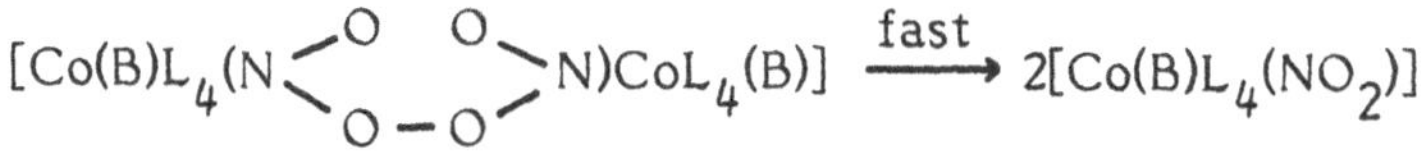

In a reaction which probably proceeds via the same mechanism $[Co(dmgH)_2(NO)]$ is oxidized by O_2 in the presence of H_2O:[254]

$$2[Co(dmgH)_2(NO)] + 2H_2O + O_2 \longrightarrow 2[Co(H_2O)(dmgH)_2(NO_2)] \quad (70)$$

where $dmgH_2$ = dimethylglyoxime. However, in CH_2Cl_2 in the presence of Ph_3P, $4\text{-Bu}^t\text{-}C_5H_4N$, $4\text{-(NC)}C_5H_4N$ or 1-methylimidazole as base B, the nitro product $[Co(B)(dmgH)_2(NO_2)]$ is less than 50% of the total, and the major product is the nitrato complex $[Co(B)(dmgH)_2(NO_3)]$.[255] Except for $[Os(Ph_3P)_2(CO)_2(NO)]$, which is oxidized to NO_2:[256]

$$[Os(Ph_3P)_2(CO)_2(NO)] + 2RNC + O_2 \rightarrow [Os(Ph_3P)_2(RNC)_2(CO)(NO_2)]$$

(71)

and photochemical oxidation:

$$2[NiCl(dppe)(NO)] + O_2 \xrightarrow{h\nu} 2[NiCl(dppe)(NO_2)] \qquad (72)^{[257,500]}$$

$$2[Ir(Ph_3P)_2(NO)_2]^+ + O_2 \xrightarrow{h\nu} 2[Ir(Ph_3P)_2(NO_2)(NO)] \qquad (73)^{[258]}$$

the usual products of O_2 oxidation of nitrosyls are nitrato complexes. Examples include:

$$[RuCl(Ph_3P)_2(O_2)(NO)] + CO \rightarrow [RuCl(Ph_3P)_2(CO)_2(NO_3)] \quad (74)^{[259]}$$

$$[Ru(Ph_3P)_2(NO)_2] + 2O_2 \rightarrow [Ru(Ph_3P)_2(O_2)(NO_3)(NO)] \quad (75)^{[243]}$$

$$[Ru(Ph_3P)_2(O_2)(NO_3)(NO)] + 2CO \rightarrow [Ru(Ph_3P)_2(CO)_2(NO_3)_2]$$

$$(76)^{[243]}$$

$$[IrCl(X)(Ph_3P)_2(CO)(NO)] + O_2 \rightarrow [IrCl(X)(Ph_3P)_2(CO)(NO_3)]$$

$$(77)^{[260]}$$

$(X = Cl, Br, I, NCS, NCO$ or $N_3)$

$$[Pt(Ph_3P)_2(NO_3)(NO)] + O_2 \rightarrow \underline{cis}\text{-}[Pt(Ph_3P)_2(NO_3)_2] \qquad (78)^{[268]}$$

There has been much speculation as to why NO_2 is formed in some oxidations, NO_3 in others, and both products can be obtained from some reactions; In fact adding pyridine to the $[IrCl(X)(Ph_3P)(CO)(NO)]/O_2$ reaction (equation 77) gives products containing NO_2, NO_3 and ONO.[260] The necessity of an added base in most reactions, and the clear evidence that O_2 itself acts as a base (ligand) in other reactions (e.g. reaction (75)), support the idea that the key reaction is electrophilic attack of O_2 at a bent {MNO} group, even when the starting material has linear {MNO} groups, as in $[Ru(Ph_3P)_2(NO)_2]$.[261] The rearrangement of the M—N$\overset{O}{\underset{O-O}{<}}$

intermediate (reaction (67)) to a nitrato ligand, instead of dimerization (reaction 68), has been proposed as a determining factor for the distribution of products,[260] but it is not clear why or when rearrangement or dimerization occurs. It was suggested also that nitrosyls with relatively high ν(NO) frequencies give nitrato and those with low ν(NO) frequencies give nitro,[255] but this suggestion has not been borne out by subsequent work; e.g. both $[Co(dmgH)_2(NO)]$ (ν(NO) 1710 cm^{-1}) and $[IrCl(X)(Ph_3P)_2(CO)(NO)]$ (ν(NO) 1520 cm^{-1}) give nitrato. It has been claimed that nitro products are obtained by nucleophilic attack of a radical, formed by attack of O_2 at the coordinated nitrosyl, on a second formally NO$^+$ ligand. Nitro formation therefore requires both NO$^+$ and NO$^-$ ligands, nitrato only NO$^-$.[320] It is noteworth that the two photochemical oxidations occur with complexes which have linear {MNO} groups and do not require a base; presumably excitation of an electron into the LUMO π*(MNO) orbital bends the {MNO} group. Also noteworthy is that these two reactions give NO$_2$ as the sole product. The possibility of photochemical excitation does not appear to have been considered in the other reactions given above, and before further debate on the products takes place a careful experimental reinvestigation with photochemistry in mind is necessary.

The very weakly coordinated nitrosyl in

and related complexes can be oxidized to NO$_2$ by O$_2$ without addition of a base.[262] The complexes are diamagnetic and have ν(NO) at 1715 cm^{-1}. It is therefore surprising that they react with O$_2$, and since the NO is readily removed thermally the possibility of initial NO loss, gas phase oxidation and re-coordination of NO$_2$ cannot be discounted.

It is interesting that $[Rh(Ph_3P)_2(SO_2)(NO)]$ which has an RhNO angle of $140°$[263] and contains two potential sites of oxidation (NO and SO_2) reacts with O_2 to give $[Rh(Ph_3P)_2(SO_4)(NO)]$[265] not $[Rh(Ph_3P)_2(SO_2)(NO_3)]$. Electrophiles add to the formyl-nitrosyl $[CpRe(Ph_3P)(CHO)(NO)]$ at the formyl, not the nitrosyl group.[264]

Nitrogen monoxide itself can act as an electrophile towards nitrosyls. The formation of red $[(Co(NH_3)_5)_2(N_2O_2)]^{4+}$ and $[Co(NH_3)_5(NO)]^{2+}$ from $[Co(H_2O)_6]^{2+}$, NH_3 and NO takes place via:[266]

$$[Co(H_2O)_6]^{2+} + 6NH_3 \rightleftharpoons [Co(NH_3)_6]^{2+} + 6H_2O \tag{79}$$

$$[Co(NH_3)_6]^{2+} + NO \rightleftharpoons [Co(NH_3)_5(NO)]^{2+} + NH_3 \tag{80}$$

$$[Co(NH_3)_5(NO)]^{2+} + NO \rightarrow [Co(NH_3)_5(N_2O_2)]^{2+} \tag{81}$$

$$[Co(NH_3)_5(N_2O_2)]^{2+} + [Co(NH_3)_6]^{2+} \rightarrow [(Co(NH_3)_5)_2(N_2O_2)]^{4+} + NH_3 \tag{82}$$

The CoNO angle in $[Co(NH_3)_5(NO)]^{2+}$ is $119°$.[94] This reaction can also proceed further, by NO attack on $[(Co(NH_3)_5)_2(N_2O_2)]^{4+}$, giving N_2O and NO_2^-. Most reactions of this type do proceed further and they are therefore discussed as disproportionation reactions in Section 7 below.

6.   FORMATION OF CARBON-NITROGEN BONDS

Carbon-nitrogen bonds are formed by nucleophilic attack of carbanions or enols on nitrosyls (Sections 2.1 and 4) and on reduction of nitrosyls in the presence of organic radicals (Section 3). Compounds containing carbon-nitrogen bonds are also formed by other reactions of nitrosyls whose mechanisms are speculative, and by insertion of NO itself into metal-carbon bonds. The latter reactions would not normally be considered as reactions of coordinated nitrosyl, but for the sake of completeness, and since NO coordination prior to insertion may occur, we include them in this section.

Benzyl bromide, $PhCH_2Br$, reacts slowly with $[Ru(Ph_3P)_2(NO)_2]$ under CO to give $PhCH=NOH$ (9%), PhCN (17%), $PhCONH_2$ (77%), PhCHO (5%) and a quantitative yield of $[RuBr_2(Ph_3P)_2(CO)_2]$.[269] Under N_2 or C_2H_4 the yield of free organonitrogen compounds is reduced and

$[RuBr_2(Ph_3P)_2(NCPh)_2]$ is obtained. Control experiments show that $PhCH_2Br$ and NO react inefficiently to give $PhCH_2NO_2$. The proposed mechanism involves reaction of $PhCH_2Br$ at the linear {RuNO} group[261,271] to give a complex containing {Ru(N(O)CH_2Ph)}. The tautomer of this, {Ru(N(OH) = CHPh)}, can be replaced by CO, dehydrated to PhCN (coordinated or uncoordinated), and the PhCN can be subsequently hydrolyzed to $PhCONH_2$. A radical mechanism was eliminated by trapping experiments, and initial oxidative addition of $PhCH_2Br$ at $[Ru(Ph_3P)_2(NO)_2]$ was also experimentally disproved. Lower yields of organonitrogen compounds are obtained using $[Ru(P(OPh)_3)_2(NO)_2]$, $[Ru(dppe)_2(NO)_2]$, $[M(PR_3)_3(NO)]$ (R = Ph, OC_2H_5; M = Co, Rh),[269] and $[RhCl_2(Ph_3P)_2(NO)]$.

Addition of $PhCH_2Cl$ across the nitrosyl in $[CoX(CO)_2(NO)]^-$ (X = Cl, Br, I, CN; the complex can be presumed to have a linear {CoNO}[10] group as does $[Co(Ph_3P)_2(CO)(NO)]$[289]) gives PhCH=NOH, $CoCl_2$ and a small quantity of $p\text{-}C_6H_5(C_6H_4)COOH$.[290] Initial oxidative addition of $PhCH_2Cl$ at $[CoX(CO)_2(NO)]$ is proposed as the key step, but no mechanistic investigation has been made, and the reaction may well proceed as in the case of $[Ru(Ph_3P)_2(NO)_2]$ discussed above.

A remarkable reaction between the two three-electron donor ligands NO and C_3H_5 occurs on treatment of $[M(Ph_3P)_2(\eta^3\text{-}C_3H_5)(NO)]^+$ (M = Rh, Ir) with CO. The products are $[M(Ph_3P)_2(CO)_3]^+$ and $CH_2 = CHCH = NOH$.[270] The {IrNO}[8] complex $[Ir(Ph_3P)_2(\eta^3\text{-}C_3H_5)(NO)]^+$ has the structure:

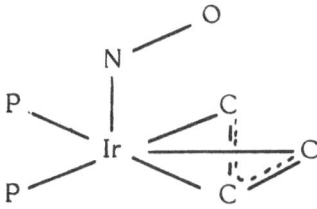

with an {IrNO} angle of 129° and formally 16 electrons around the iridium. There is clear ir and nmr evidence that the {MNO} angle straightens dynamically in solution with no change in the $\eta^3\text{-}C_3H_5$ coordination mode. The coupling reaction takes place via the sequence:

$$[M(Ph_3P)_2(\eta^3\text{-}C_3H_5)(NO)]^+ + CO \rightarrow$$

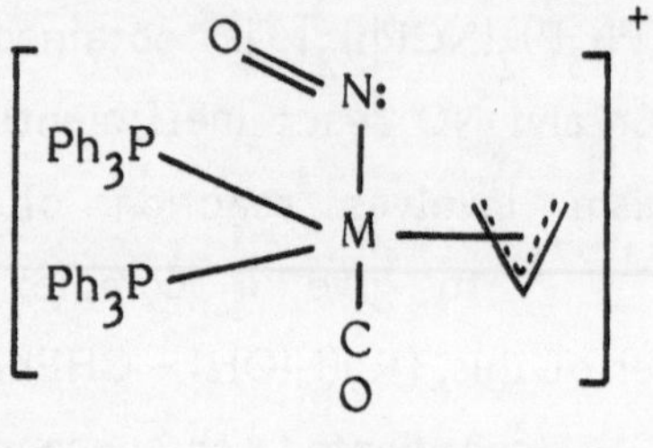

$$\left[\begin{array}{c} O \\ \| \\ N: \\ Ph_3P \diagdown \; | \\ \;\;\;\;\; M \text{---} \diagup\diagdown \\ Ph_3P \diagup \; | \\ C \\ \| \\ O \end{array}\right]^+ \qquad (83)$$

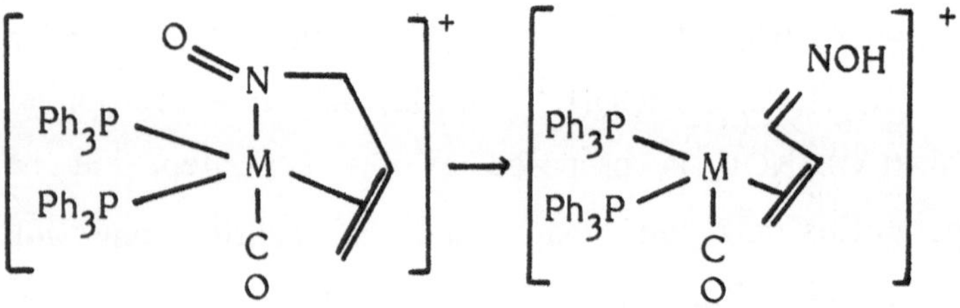

(A)

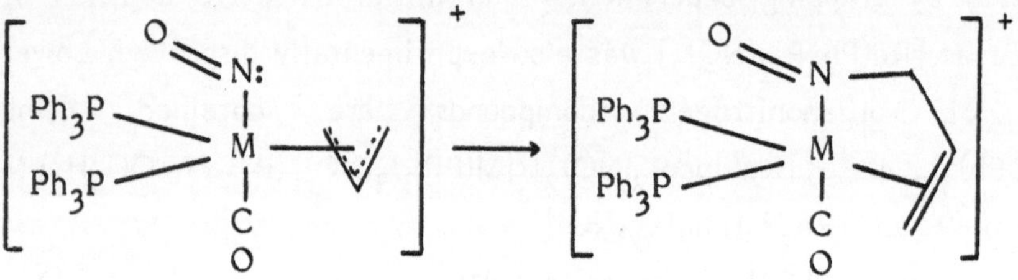

$$\left[\begin{array}{c} O \\ \| \\ N: \\ Ph_3P \diagdown \; | \\ \;\;\;\;\;\; M \text{---} \diagup\diagdown \\ Ph_3P \diagup \; | \\ C \\ \| \\ O \end{array}\right]^+ \longrightarrow \left[\begin{array}{c} O \\ \| \\ N \\ Ph_3P \diagdown \; | \\ \;\;\;\;\;\; M \\ Ph_3P \diagup \; | \\ C \\ \| \\ O \end{array}\right]^+ \qquad (84)$$

$$\left[\begin{array}{c} O \\ \| \\ N \\ Ph_3P \diagdown \; | \\ \;\;\;\;\;\; M \\ Ph_3P \diagup \; | \\ C \\ \| \\ O \end{array}\right]^+ \longrightarrow \left[\begin{array}{c} NOH \\ Ph_3P \diagdown \;\;\; \\ \;\;\;\;\;\; M \\ Ph_3P \diagup \; | \\ C \\ \| \\ O \end{array}\right]^+ \qquad (85)$$

(B)

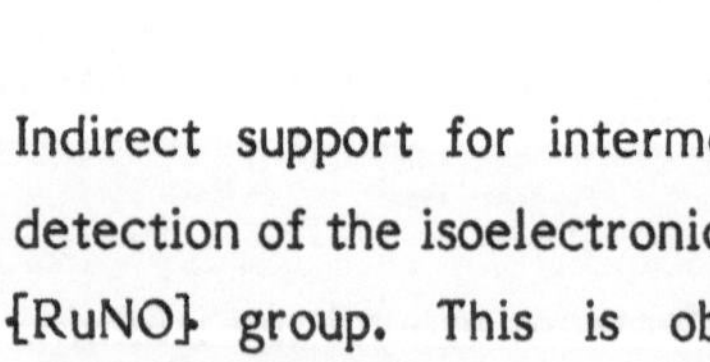

$$\left[\begin{array}{c} NOH \\ Ph_3P \diagdown \;\;\; \diagup \\ \;\;\;\;\;\; M \\ Ph_3P \diagup \;\; \diagdown \end{array}\right]^+ + 2CO \rightarrow [M(Ph_3P)_2(CO)_3]^+ +$$
$$Ph_3PCH_2 = CHCH = NOH \qquad (86)$$

Indirect support for intermediate (A) is provided by the spectroscopic detection of the isoelectronic $[Ru(Ph_3P)_2(CO)(\eta^3\text{-}C_3H_5)(NO)]$ with a bent {RuNO} group. This is obtained by adding CO to $[Ru(Ph_3P)_2(\eta^3\text{-}C_3H_5)(NO)]$, which like its analogue $[Ru(Ph_3P)_2(NO)_2]$, has a linear {RuNO} group.[272-273] The linear–bent fluxionality in these complexes is surprisingly facile; between –50° and 0° both linear and bent {IrNO} isomers of $[Ir(Ph_3P)_2(\eta^3-C_3H_5)(NO)]^+$ are observed in solution, but the allyl ligand remains fixed as a $\eta^3-C_3H_5$. Only at higher temperatures does $\eta^3 \rightleftharpoons \eta^1-C_3H_5$ and <u>syn</u> $\rightleftharpoons$ <u>anti</u> interconversion occur, though allyl rotation may be taking place even at low temperatures.[270]

The electronic structure of the allyl-nitrosyl is clearly important in the above reaction and in the fluxionality of the complexes, since the reaction is not general. The $\{MNO\}^4$ complexes of formula $[CpML(\eta^3-C_3H_5)(NO)]^{n+}$ (M = Mo, W; n = 1 for a neutral L, n = 0 for a uninegative L) add ligands to M with concomitant changes in the allyl (η^3 to η^1 [274] or distortion of the allyl to an asymmetric bonding mode[275-278]) or undergo nucleophilic attack at the allyl.[274] There is no evidence of bending of the $\{MNO\}$ group or of any reaction of the nitrosyl. Such nitrosyls are in fact extremely stable with respect to alterations in the linear $\{MNO\}$ group; the fluxionality of $[Cp_2Mo(I)(NO)]$ and related molecules proceeds with alteration in the hapticity of the C_5H_5 rings in preference to $\{MoNO\}$ bending.[279-282] In the $\{FeNO\}^8$ complex $[Fe(CO)_2(\eta^3-C_3H_5)(NO)]$ the CO and, surprisingly, NO ligands can be replaced by R_3P,[283] and basic phosphines also attack the allyl ligand,[284-286] but C - N bond formation does not occur. This may be because the Fe atom has insufficient electron density to bend the $\{FeNO\}$ group. The same reason can be advanced to explain why the allyl ligand in $[Fe(Ph_3P)(\eta^3-C_3H_5)(NO)_2]^+$ is fluxional, and the complex loses Ph_3P on reduction or treatment with H^- but the nitrosyl ligand remains unaffected.[287]

The existence of intermediate B, equation (85), also receives indirect support from the reaction:[288]

$$[(\eta^3-C_3H_5)Ni(\mu-Br)]_2 + 4NO \rightarrow 2 \left[\begin{array}{c} CH_2 \\ \| \\ C \\ CH \\ \| \\ CH \\ \| \\ NOH \end{array} \; NiBr(NO) \right] \quad (87)$$

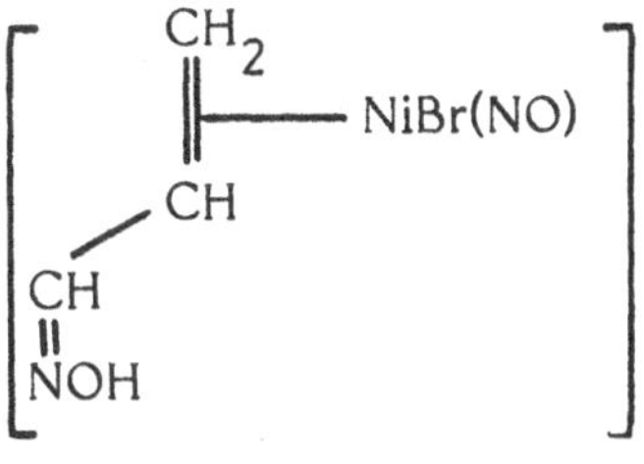

where the nickel-oxime complex can be isolated. The product arises <u>via</u> a reaction sequence similar to that in equations (83) - (86) after coordination of NO with bridge splitting (see also Section 10).

Brunner and Loskot first observed the reactions between $[CpCo(\mu\text{-}NO)]_2$ and several strained olefins in the presence of NO to give nitroso compounds:[291-292]

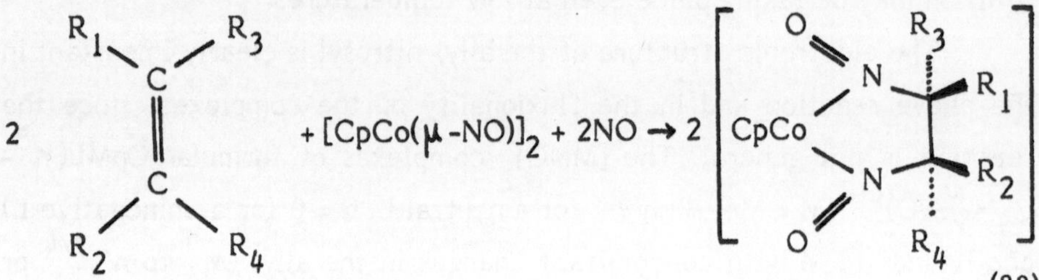

$$2 \quad \underset{R_2}{\overset{R_1}{\diagdown}} \underset{\Vert}{\overset{C}{\diagup}} \underset{R_4}{\overset{R_3}{\diagup}} + [CpCo(\mu\text{-}NO)]_2 + 2NO \rightarrow 2 \left[CpCo \underset{O}{\overset{O}{\diagdown}} \underset{N}{\overset{N}{\diagdown}} \underset{R_4}{\overset{R_3}{\underset{R_1}{\diagup}}} \underset{R_2}{\diagup} \right]$$

(88)

The structure of the product was first obtained by spectroscopic methods and later confirmed crystallographically for a derivative of the olefin norbornadiene which gave:[293]

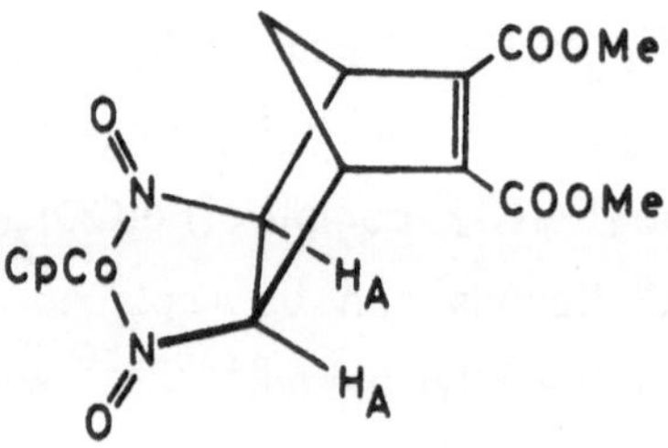

with the H_A protons in the <u>endo</u> position. Similar reactions with a variety of olefins take place, but the products are not always isolable. It has since been found that <u>in situ</u> reduction of these products with $LiAlH_4$ gives excellent yields of the vicinal diamines:[294,498]

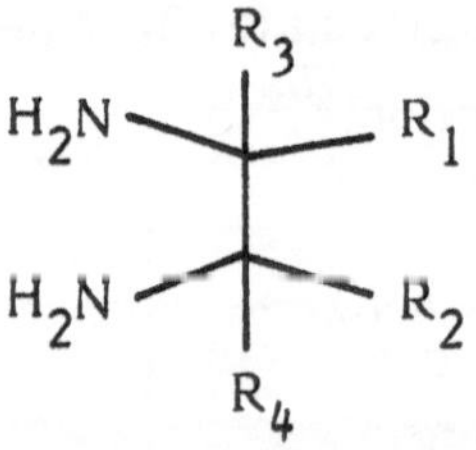

and the reaction (which has been extended to a variety of olefins[498]) is the first general method for the conversion of olefins into such diamines. Kinetic and mechanistic studies have shown that reaction (88) proceeds by initial formation of the highly reactive $[CpCo(NO)_2]$ monomer (hence the need for excess NO):[497]

TABLE IV

Insertion Reactions of Nitrogen Monoxide

Reactant	Product	Ref.
$[CpTi(CH_3)_3]$	$[CpTi(CH_3)(ON(CH_3)NO)_2]$	295
$[Cp_2M(CH_3)_2]$ M = Ti, Zr	$[Cp_2M(CH_3)(ON(CH_3)NO)]$	296
$[Cp_2ZrPh_2]$	$[Cp_2Zr(Ph)(O(NPh)NO)]$	518
$[Cp_2Nb(CH_3)_2]$	$[Cp_2Nb(O)(ON(CH_3)NO)]_n$	297
$[MR_4]$ R = $CH_2Si(CH_3)_3$, M = Ti, Zr $CH_2C(CH_3)_3$, M = Zr	$[M(ON(R)NO)_4]$	297
$[MCl_2R_2]$ M = Ti, Zr R = $CH_2Si(CH_3)_3$ $CH_2C(CH_3)_3$	$[MCl_2(ON(R)NO)_2]$	297
$[Nb(CH_3)_5]$	$[Nb(CH_3)_2(ON(CH_3)NO)_3]$	297
$[NbCl_2(CH_3)_3]$	$[NbCl_2(CH_3)(ON(CH_3)NO)_2]$	297
$[Ta(CH_3)_5]$	$[Ta(CH_3)(ON(CH_3)NO)_2]_2$	297
$[MCl_3(CH_3)_2]$ M = Nb, Ta	$[MCl_3(ON(CH_3)NO)_2]$	298
$[MCl_2(CH_3)_3]$ M = Nb, Ta	$[MCl_2(CH_3)(ON(CH_3)NO)_2]$	298
$[W(CH_3)_6]$	$[W(CH_3)(ON(CH_3)NO)_2]$	299-300
$[Re_3Cl_3R_6]$ R = $CH_2Si(CH_3)_3$	$[Re_3Cl_3R_5(ON(R)NO)]$	301
$[Al(CH_3)_3]$	$[Al(CH_3)_2(ON(CH_3)NO)Al(CH_3)_3]$	302
$[Ga(CH_3)_3]$	$[Ga(CH_3)_2(ON(CH_3NO)]$	302
$[Cu(CH_3)]_n$	$[Cu(ON(CH_3)NO)_2]$	304
$[R_2Zn]$ R = CH_3, C_2H_5	$[Zn(ON(CH_3)NO)_2]$	305
$[Cd(CH_3)_2]$	$[Cd(ON(CH_3NO)_2]$	306
$[B(C_2H_5)_3]$	$[B(C_2H_5)_2(ON(C_2H_5)(B(C_2H_5)_2)]$ + $[B(C_2H_5)_2(ON(C_2H_5)_2]$	307

Attempts to insert NO into $[Cp_2TiCl(CH_3)]$ were unsuccessful,[296] and various $[ZrR_4]$ complexes have given products as yet uncharacterized.[296,308]

$$[CpCo(\mu\text{-}NO)]_2 + 2NO \longrightarrow 2[CpCO(NO)_2] \tag{89}$$

This monomer then adds the olefin in a concerted fashion. Remarkably, this addition is <u>reversible</u>, so that exchange reactions between the dinitroso adduct and a free olefin take place with $[CpCo(NO)_2]$ as an intermediate.[497]

Insertion of NO into diamagnetic alkyls of the early transition metals or aluminium alkyls leads quite generally to <u>N</u>-alkyl-<u>N</u>-nitroso-hydroxylaminato compounds:

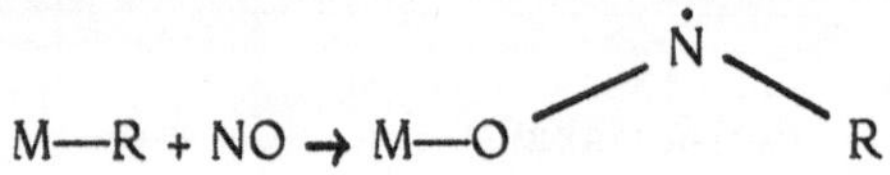

$$(90)$$

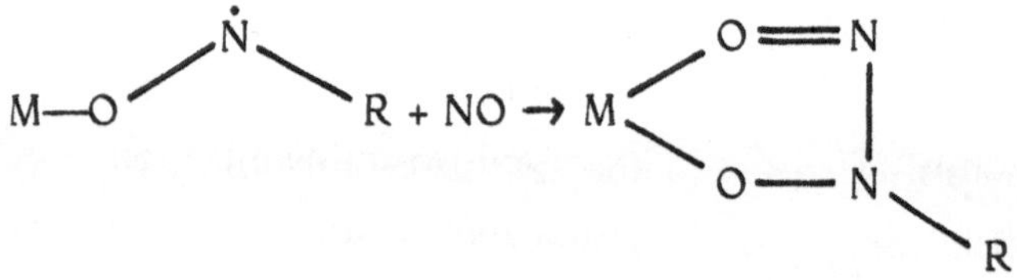

$$(91)$$

The known examples of this reaction are listed in Table IV; note that the earliest example, insertion of NO into $(CH_3)_2Zn$ or $(C_2H_5)_2Zn$:

$$[R_2Zn] + 2NO \longrightarrow [Zn(ON(R)NO)_2] \tag{92}$$

dates back to 1856,[305] and reactions of Grignard reagents with NO giving, after hydrolysis, free N-alkyl-N-nitrosohydroxylamines were also known very early.[309-310] The {ON(R)NO} ligand is bidentate, except where indicated in Table IV. Insertion of NO into the M - R bonds continues stepwise until the maximum coordination number of M is reached. Any remaining M - R bonds are not further attacked.

The structure of the M unit has been confirmed

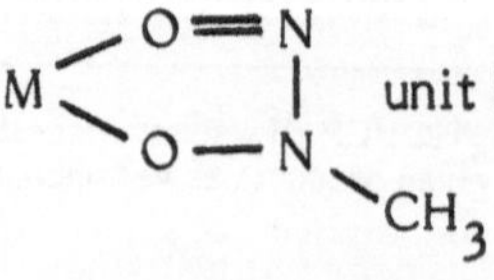

crystallographically for the complexes $[Ta(CH_3)Cl_2(ON(CH_3)NO)_2]$,[298] $[W(CH_3)_4(ON(CH_3)NO)_2]$,[299,309] $[Re_3Cl_3(CH_2Si(CH_3)_3)_5(ON(CH_2Si(CH_3)_3)NO)]$[301] and $[Al(CH_3)_2(ON(CH_3)NO)Al(CH_3)_3]$.[302] The structure of the last complex is:

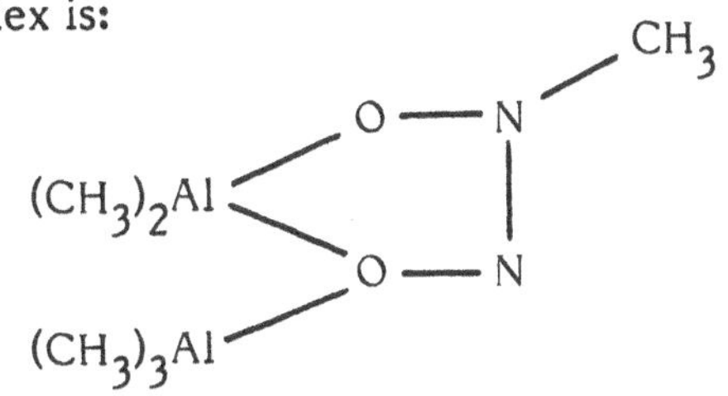

$$(CH_3)_2Al \quad \begin{array}{c} O-N-CH_3 \\ | \\ O-N \end{array} \quad (CH_3)_3Al$$

Although some of the reactant alkyls in Table IV are paramagnetic, for example $Cp_2Nb(CH_3)_2$, interaction of NO with paramagnetic metal alkyls does not generally lead to complexes containing the $\{ON(CH_3)NO\}$ ligand, but to oxo complexes. For $Cp_2Nb(CH_3)_2$, which at room temperature gives the polymeric oxo-hydroxylaminato complex $[Cp_2NbO(ON(CH_3)NO)]_n$, performing the reaction at low temperature shows that the oxo product is formed first:[297]

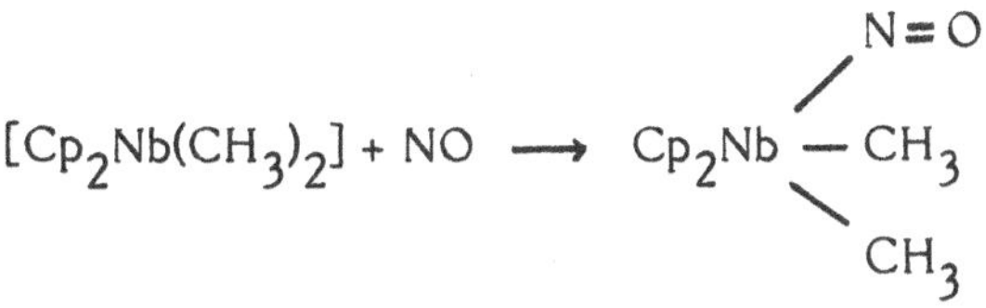

$$[Cp_2Nb(CH_3)_2] + NO \longrightarrow Cp_2Nb \begin{array}{c} N=O \\ CH_3 \\ CH_3 \end{array} \tag{93}$$

$$Cp_2Nb \begin{array}{c} N=O \\ CH_3 \\ CH_3 \end{array} \longrightarrow Cp_2Nb \begin{array}{c} O \\ | \\ N \\ CH_3 \quad CH_3 \end{array} \tag{94}$$

$$Cp_2Nb \begin{array}{c} O \\ | \\ N \\ CH_3 \quad CH_3 \end{array} \longrightarrow Cp_2Nb \begin{array}{c} O \\ \| \\ CH_3 \end{array} + CH_3N \tag{95}$$

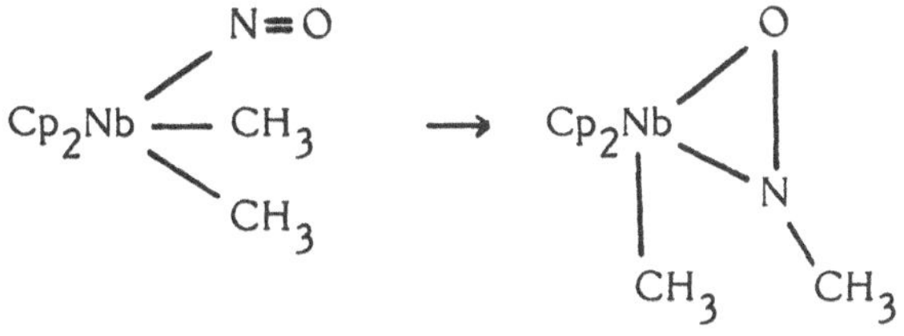

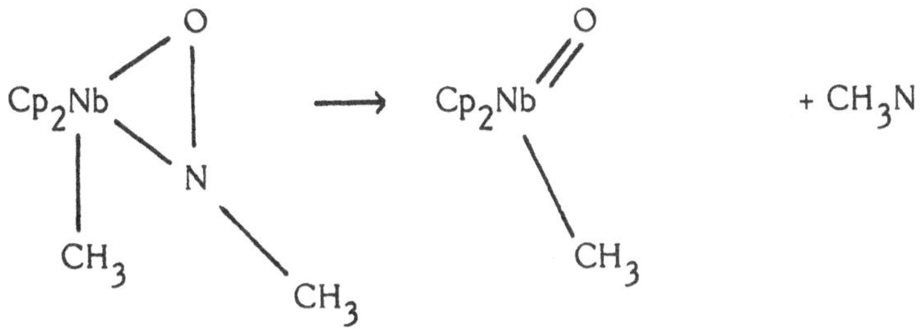

$$2CH_3N \longrightarrow (CH_3)_2N_2 \tag{96}$$

Insertion of NO into the $Nb - CH_3$ bond then follows.

The d^1 alkyls $[Re(CH_3)_6]$ and $[Re(O)(CH_3)_4]$ react with NO even at -78° to give $[Re(O)_2(CH_3)_3]$ and $(CH_3)_2N_2$;[311-312] $[Re(O)(CH_2Si(CH_3)_3)_4]$ and NO give $[Re_2O_3(CH_2Si(CH_3)_3)_6]$,[313] for which the structure

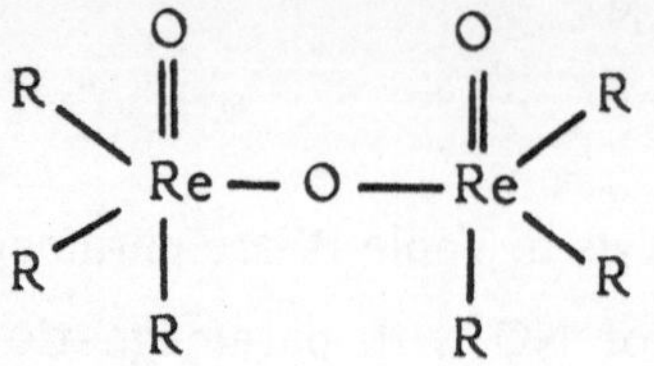

$(R = CH_2Si(CH_3)_3)$ is proposed. It is remarkable that NO is cleaved so readily at -78°. As with the reaction between $[Cp_2Nb(CH_3)_2]$ and NO discussed above, the initial reaction is believed to be formation of a CH_3NO complex on coordination of NO to the coordinatively unsaturated rhenium alkyl complex:[311]

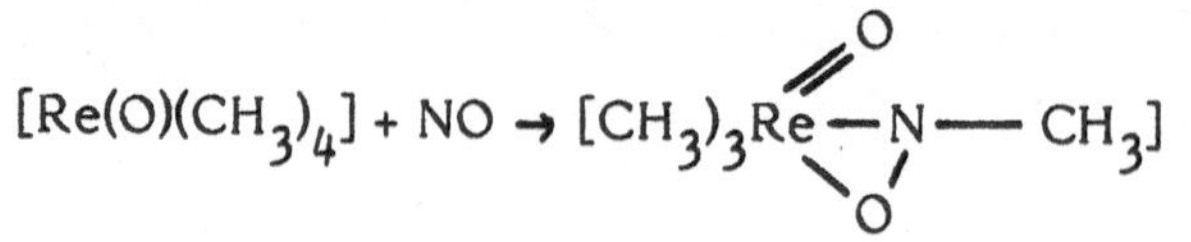

$$[Re(O)(CH_3)_4] + NO \rightarrow [(CH_3)_3Re-N-CH_3] \tag{97}$$

Other d^1 complexes react similarly; $[V(CH_2Si(CH_3)_3)_4]$ gives $[V(O)(CH_2Si(CH_3)_3)_3]$ and Cp_2TiR (R = Ph or $C_6H_5CH_2$) give $[Cp_3Ti_3O_4(NO)]$, for which the structure below is proposed:[297]

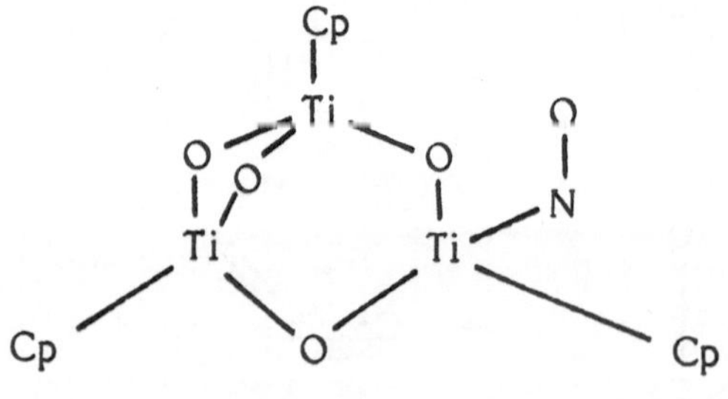

The intermediates containing the $M \diagdown \genfrac{}{}{0pt}{}{N-R}{O}$ moiety are interesting

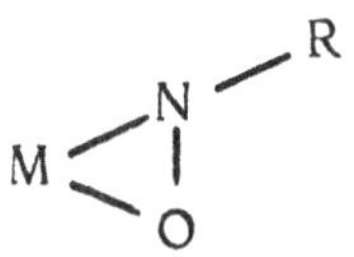

in that complexes containing such a grouping have been structurally characterized for early transition metals (vanadium and molybdenum) but they have not been reported to react further to give oxo complexes.

Whereas with the early transition metals insertion of NO into the metal-alkyl bond gives complexes of the (ON(R)NO) ligand if the complex is diamagnetic and oxo-complexes if it is paramagnetic, insertion reactions with alkyls of the later transition metals are rare and do not give either of the above products. The reaction between $[Co(CH_3)_2(PMe_3)_3]$ and NO gives $[Co(CH_3)_2(PMe_3)_2(NO)]$ at low temperature, and this product undergoes an insertion reaction on warming to produce $[Co(CH_3)(PMe_3)_2(CH_3NO)_2]_2$.[314] On the basis of spectroscopy the structure of the insertion product is proposed as

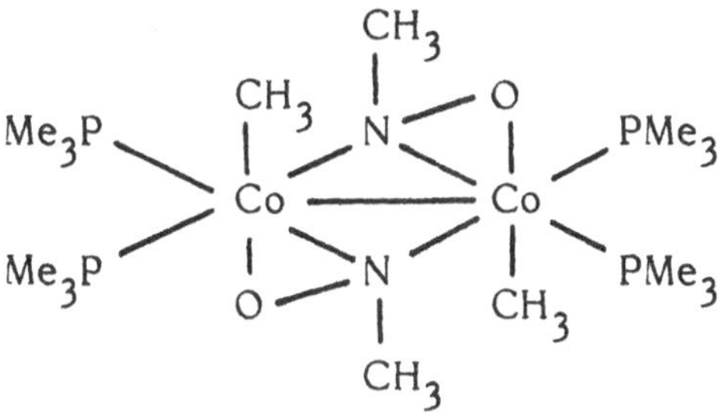

The initial insertion reaction

$$[Co(CH_3)_2(PMe_3)_2(NO)] \longrightarrow [Co(CH_3)(PMe_3)_2(CH_3NO)] \qquad (98)$$

can be followed by infrared spectroscopy, and is preceded by a shift in $\gamma(NO)$ from 1712 to 1657 cm^{-1}, considered to be indicative of bending of the $\{CoNO\}^8$ group prior to insertion in the Co - CH_3 bond.[314]

A very clean insertion reaction has been achieved with $[CpCo(NO)]^-$, which reacts with RI (R = CH_3, C_2H_5, $(CH_3)_2CH$ or p-$CH_2C_6H_4CH_3$) at low temperature to form [CpCo(R)(NO)]$^-$. On warming in the presence of a tertiary phosphine the insertion product [CpCo(R$_3$P)(N(O)R)] is formed.[481-482] The structure of [CpCo(Ph$_3$P)(N(O)C$_2$H$_5$)] has been determined by X-ray crystallography. Kinetic studies show that the insertion reaction takes place at [CpCO(R)(NO)], the phosphine actually inhibiting the insertion.[482] It was later shown that reduction of [CpFe(μ-NO)]$_2$ followed by methylation in the presence of [CpFe(μ-NO)]$_2$ and PMe$_3$ gives the insertion product [CpFe(CH$_3$)(PMe$_3$)(N(O)CH$_3$)]. An intermediate, [CpFe(CH$_3$)$_2$(NO)], has been isolated.[499]

Acetylacetonato complexes of cobalt,[315] nickel[316] and copper[316] undergo an insertion by NO whereby the O,O-chelate is converted into an N,O ring:

$$[Co(acac)_3] + 3NO \xrightarrow{h\nu} \qquad \qquad \qquad (99)$$

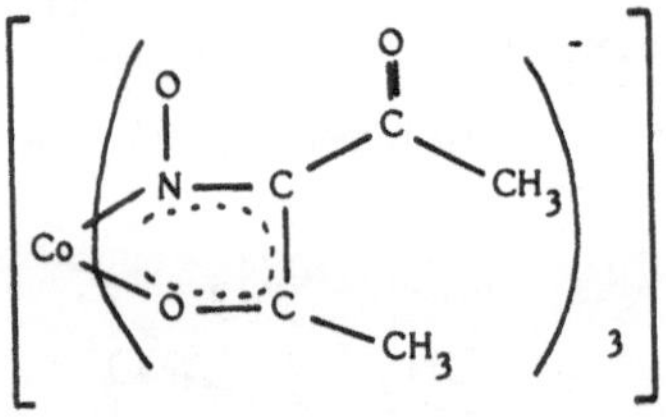

Interestingly the Co(II) complex, [Co$_4$(acac)$_8$], does not undergo insertion:[315]

$$[Co_4(acac)_8] + 4NO \longrightarrow 4[Co(acac)_2NO)] \qquad (100)$$

Reaction (99) does appear to be an actual insertion at the metal ligand bond since the yields are high. The analogous reaction of palladium:

$$[Pd(acac)_2] + NO \longrightarrow \qquad \qquad \qquad (101)$$

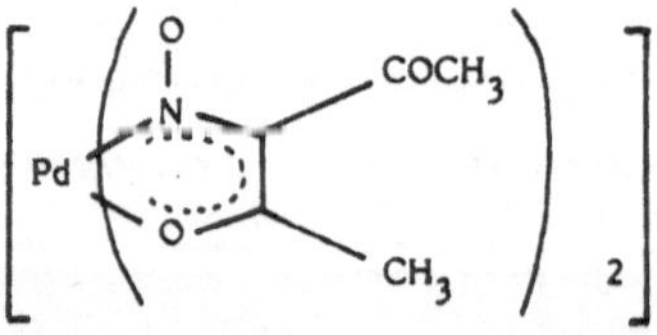

is on the other hand not strictly a reaction of a coordinated ligand since it proceeds via acid catalyzed loss of $CH_3COCH_2COCH_3$ from [Pd(acac)$_2$], subsequent nitrosation of the free ligand by free NO, and recoordination of the resultant 3-hydroxy-imino-pentane-2,4-dionato ligand.[317] Whether the use of photochemistry as in reaction (99) is the key to a direct insertion cannot be determined on present evidence.

Although many mixed carbonyl-nitrosyl complexes are known, reactions between these two ligands are rare. The first report of such a reaction was on irradiation of a benzene solution of $[CpMo(CO)_2(NO)]$ in the presence of Ph_3P:[318]

$$2[CpMo(CO)_2(NO)] + 3Ph_3P \xrightarrow{h\nu} [CpMo(CO)(Ph_3P)(NO)] + CO +$$
$$[CpMo(CO)(Ph_3P)(NCO)] + Ph_3PO \qquad (102)$$

Since irradiation of $[CpMo(CO)_n(Ph_3P)_{3-n}(N_3)]$ (n = 1 or 2) gives $[CpMo(CO)_n(Ph_3P)_{3-n}(NCO)]$ it was concluded that a metal nitrene (imido) intermediate, $[CpMo(CO)_n(Ph_3P)_{3-n}(N)]$, was involved. The $\nu(NO)$ frequencies for the analogous complexes fall in the order Cr > W > Mo, and the yield of NCO product is zero for the Cr complex and low for the W analogue. Since the N - O bond strength will decrease with a decrease in $\nu(NO)$, and since this is the bond which must be cleaved photochemically to form the nitrene (no reaction occurs in the dark), the observed yield also supports a nitrene mechanism.[318] A brief report of the conversion of $[Ir(Ph_3P)_2(CO)(NO)]$ into $[Ir(Ph_3P)_3(NCO)]$ and Ph_3PO on irradiation in the presence of Ph_3P[319-320] appears to be in error; cleavage of the Ir - NO bond occurs and the NO thus set free oxidizes Ph_3P to Ph_3PO, being itself reduced to N_2O.[504] A nitrene mechanism is favored in the reaction between $[Pt(PEt_3)_2(NO_2)(NO)]$ and CO to give $[Pt(PEt_3)_2(NO_2)(NCO)]$ and CO_2.[320] However coupling of CO and NO to form the intermediate

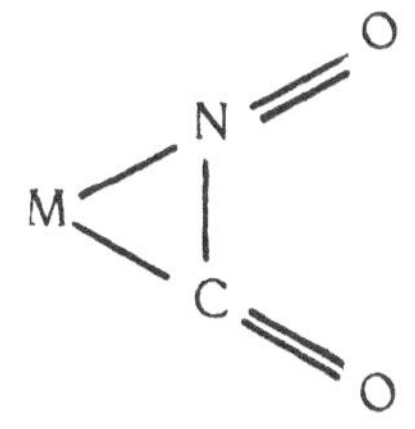

and capture of O by a second CO or Ph_3P molecule is an alternative mechanism which seems more likely since the reaction proceeds in the dark.

The coupling mechanism must be operative in the reaction:[321-322]

$$2[Cp_2Ti(CO)_2] + NO \longrightarrow [Cp_2Ti(NCO)] + [Cp_2TiO] + 3CO \qquad (103)$$

which takes place in the dark. A nitrene would have the unacceptable formula $[Cp_2Ti(N)]$ (formally a Ti(V) complex). Substitution of one NO for one CO in the Ti(II) complex $[Cp_2Ti(CO)_2]$ would certainly result in a bent TiNO group, and hence the electrophile-nucleophile reaction

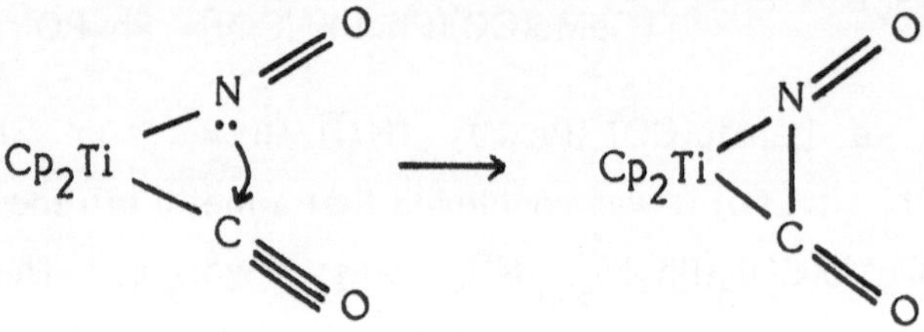

$$\qquad\qquad (104)$$

is facilitated. An analogous reaction occurs with $[Cp_2V(CO)]$,[323] and these reactions may generally be expected for carbonyl-nitrosyls in which the nitrosyl is strongly bent.

7.   COUPLING, DIMERIZATION AND DISPROPORTIONATION OF NITROGEN MONOXIDE IN NITROSYL COMPLEXES

Two reactions of nitrogen monoxide are of considerable interest because of their apparent simplicity and also because of their environmental importance. These are:

$$3NO \rightarrow N_2O + NO_2 \qquad (105)$$

and

$$2NO + CO \rightarrow N_2O + CO_2 \qquad (106)$$

Both reaction (105) and (106) involve coupling of two NO molecules, and both are examples of the general reduction of NO according to:

$$A + 2NO \rightarrow AO + N_2O \qquad (107)$$

which in aqueous solution becomes the reduction:

$$2H^+ + 2e + 2NO \rightarrow N_2O + H_2O \qquad (108)$$

Two possible mechanisms for these reactions can readily be envisaged. In the first there is initial coordination of NO to the metal center, after which the species A (equation (107)) captures O from NO forming AO and leaving a metal nitrene (imide). This nitrene then reacts with a second molecule of NO, giving the weak ligand N_2O which leaves the inner coordination sphere allowing further NO coordination. In the second mechanism initial coordination of NO to the metal is followed by immediate attack of a second NO. This may attack the coordinated NO directly forming the {ONNO} ligand (which is presumed to have an N - N bond but whose point(s) of attachment to the metal are not specified) or may coordinate to the metal forming a dinitrosyl. The substrate A then abstracts O from {(ONNO)M} or induces coupling of the two NO ligands in the dinitrosyl to give an {M(A)(N_2O_2)} complex. Oxygen atom transfer from {N_2O_2} to A gives AO and N_2O.

Early work on the mechanism of the reduction of NO to N_2O by Sn(II) in the presence of Cu(I) suffered from technical difficulties, and more extensive studies could still only suggest the possibility of an {M(N_2O_2)} intermediate,[324] partly by analogy with the reaction between NO and alkaline SO_3^{2-} which gives $[O_3SN(O)NO]^{2-}$. Definitive proof for {M(N_2O_2)} was lacking, but the evidence of these studies was not compatible with the nitrene mechanism.

Recent work on more tractable systems was anticipated by Hieber and his coworkers who observed N_2O, NO_2 and AO (A = CO, R_3P) as products of a number of reaction during the period 1938-1970, and suggested, without proof, that dinitrosyls were responsible for them.[4,326] The first investigations of the reaction of CO with $[Ir(Ph_3P)_2(NO)_2]^+$:

$$[Ir(Ph_3P)_2(NO)_2]^+ + 4CO \rightarrow [Ir(Ph_3P)_2(CO)_3]^+ + CO_2 + N_2O \qquad (109)$$

suggested that the nitrene mechanism was operative,[258] but later work clearly favors the coupling mechanism.[319,327] In fact all reactions of type (105)-(108) so far investigated appear to follow some variation of this route.

Reaction (106) takes place continuously when catalyzed by $[M(Ph_3P)_2(NO)]^{n+}$ complexes (M = Co, Rh or Ir for n = 1; M = Ru, Os for n = 0) and precursors to them,[258,319,327-329] and also by $[RhCl_2(CO)_2]^{[24,330-333]}$ which reacts with NO to give the very unstable active catalyst $[RhCl_2(CO)(NO)_2]^-$.[332] The complex $[Fe(Ph_3P)_2(NO)_2]$ is catalytically inactive. In order to discuss the important features of the reactions catalyzed by $[M(Ph_3P)_2(NO)_2]^{n+}$ it is most convenient to give first the proposed steps in the reactions, not all of which occur with every catalyst:[329]

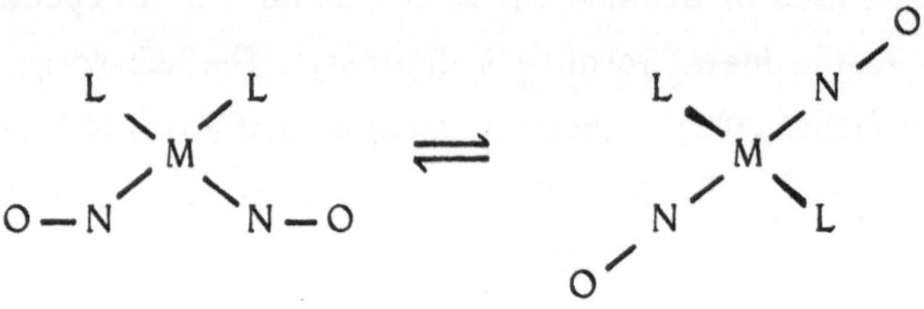

(110)

Square Planar   Tetrahedral

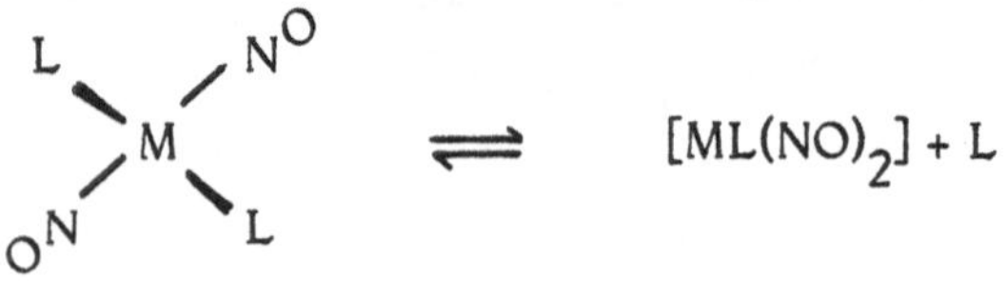

(111)

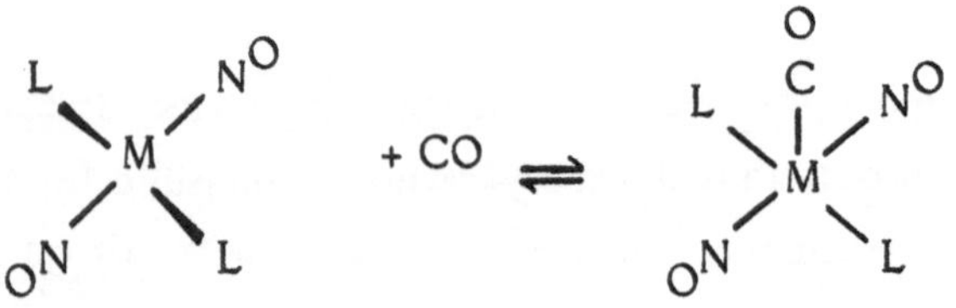

(112)

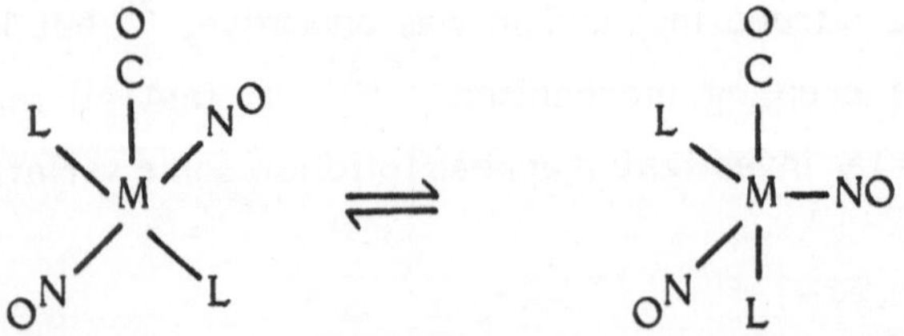

(113)

$$[ML_2(CO)(NO)_2] \rightleftharpoons [ML_2(CO)(NO)] + NO \tag{114}$$

$$(115)$$

$$+ CO \rightleftharpoons \tag{116}$$

$$+ CO \rightarrow [ML_2(CO_2)] + N_2O + CO_2 \tag{117}$$

(L = Ph_3P; charges omitted for clarity). The catalytic activity of the $[ML_2(NO)_2]^{n+}$ complexes lies in the order M = Rh > Ir > Ru > Os > Co > Fe (though there is some disagreement about this),[328-329] the {MNO} angles in the order Fe > Ru = Os > Co > Ir > Rh,[328] and the {NMN} angles in the order Rh > Ir > Ru = Os > Co > Fe.[328,334] These facts can be interpreted in at least two ways. One is that the increased bending of the {MNO} group in the repulso conformation,[334] which will be markedly accentuated by addition of a CO ligand (reaction 112), facilitates the coupling of NO (reaction 115).[327-328] The second is that the bending of {MNO} and increase in the {NMN} angle facilitates addition of CO to the $[ML_2(NO)_2]^{n+}$ complex.[329] In effect this latter explanation means that the idealized 18-electron tetrahedral $[ML_2(NO)_2]^{n+}$ complex on the right hand side of equation (110) is converted to the 16-electron square planar $[ML_2(NO)_2]^{n+}$ complex on the left hand side; the 16 electron complex can more easily add CO. The second explanation also notes that catalytically inactive $[Fe(Ph_3P)_2(NO)_2]$ loses a Ph_3P ligand rather easily, and hence the reaction

$$[Fe(Ph_3P)_2(NO)_2] + CO \rightleftharpoons [Fe(Ph_3P)(CO)(NO)_2] + Ph_3P \quad (118)$$

can take place, leaving two essentially linear {MNO} groups which will not couple.[329]

Although there are disagreements on the details of the CO addition and coupling reactions (110)-(115), there is general agreement on the necessity of the coupling reaction, and it is natural to assume that such coupling gives a complex of the type

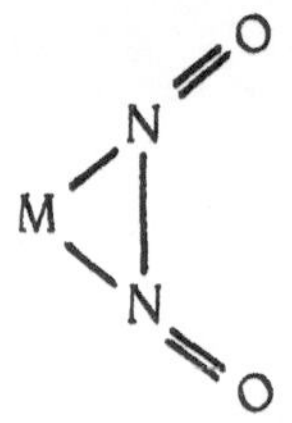

depicted in equation (115). It has however been pointed out that there is no clear evidence for such a species in these reactions, and there is not even any evidence that such a species is a necessity.[331] An alternative mechanism involving direct attack of NO can be postulated:

$$[ML_2(NO)_2]^{n+} + NO \rightarrow [ML_2(NO_2)]^{n+} + N_2O \quad (119)$$

$$[ML_2(NO_2)]^{n+} + CO \rightarrow [ML_2(CO)(NO_2)]^{n+} \quad (120)$$

$$[ML_2(CO)(NO_2)]^{n+} \rightarrow [ML_2(NO)]^{n+} + CO_2 \quad (121)$$

$$[ML_2(NO)]^{n+} + NO \rightarrow [ML_2(NO)_2]^{n+} \quad (122)$$

Only one $[ML_2(N_2O_2)]^{n+}$ complex is known. This is $[Pt(Ph_3P)_2(N_2O_2)]$ obtained from $[Pt(Ph_3P)_n]$ (n = 3 or 4) and NO[335] and it has the structure[336]

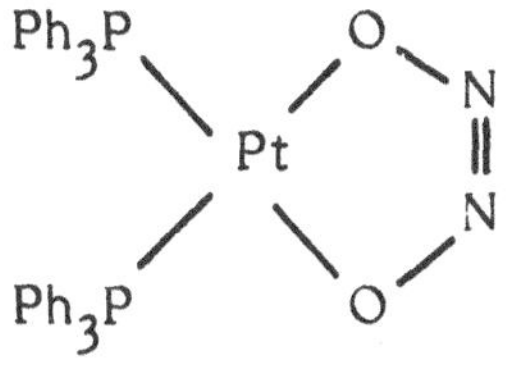

and reacts with CO:[319]

$$[Pt(Ph_3P)_2(N_2O_2)] + 3CO \rightarrow [Pt(Ph_3P)_2(CO)_2] + N_2O + CO_2 \qquad (123)$$

The reaction is not catalytic with CO and NO. Precisely how an N-bonded dinitrosyl could rearrange to give an O-bonded dinitrogen dioxide ligand is unclear. The $[Pt(Ph_3P)_2(ONNO)]$ complex undergoes an interesting reaction with SO_2 giving $[Pt(Ph_3P)_2(SO_3)]$ and $[Pt(Ph_3P)_2(N_2O_2)(SO_2)]$. The latter complex, which can also be obtained from the reaction of NO with $[Pt(Ph_3P)_3(SO_2)]$, appears to contain SO_2 bonded both to Pt and to the N_2O_2 ligand.[337]

The steps represented by equations (115) to (117) remain very uncertain but in any event the final gaseous products are CO_2 and N_2O. When $[Ir(Ph_3P)_2(NO)_2]^+$ is reacted stoichiometrically with CO the final iridium-containing product is $[Ir(Ph_3P)_2(CO)_3]^+$. This ion is known to readily lose CO and therefore one may propose the reaction sequence:[329]

$$[M(Ph_3P)_2(CO)_3]^{n+} \rightleftharpoons [M(Ph_3P)_2(CO)_2]^{n+} + CO \qquad (124)$$

$$[M(Ph_3P)_2(CO)_2]^{n+} + NO \rightleftharpoons [M(Ph_3P)_2(CO)_2(NO)]^{n+} \qquad (125)$$

$$[M(Ph_3P)_2(CO)_2(NO)]^{n+} \rightleftharpoons [M(Ph_3P)_2(CO)(NO)]^{n+} + CO \qquad (126)$$

$$[M(Ph_3P)_2(CO)(NO)]^{n+} + NO \rightleftharpoons [M(Ph_3P)_2(CO)(NO)_2]^{n+} \qquad (127)$$

by which the iridium complex returns to equation (112) to restart the catalytic cycle.

The catalysis of reaction (106) by $[RhCl_2(CO)_2]^-$ definitely follows a different mechanism from that for $[M(Ph_3P)_2(NO)_2]^{n+}$, since water is a necessary cocatalyst. The scheme proposed for the reaction is:[331]

$$[RhCl_2(CO)_2]^- + 2NO \rightleftharpoons [RhCl_2(CO)_2(NO)_2]^- \qquad (128)$$

$$[RhCl_2(CO)_2(NO)_2]^- + 2H^+ \rightleftharpoons \left[\begin{array}{c} \\ \text{Cl} \overset{\overset{\displaystyle O}{\overset{\displaystyle \|}{C}}}{\underset{\overset{\displaystyle C}{\underset{\displaystyle O}{}}}{\underset{\displaystyle Rh}{}}} \overset{\text{Cl}}{\underset{\underset{\displaystyle N}{\underset{\|}{N}}}{}} OH \\ H \overset{O}{} \end{array}\right]^+ \qquad (129)$$

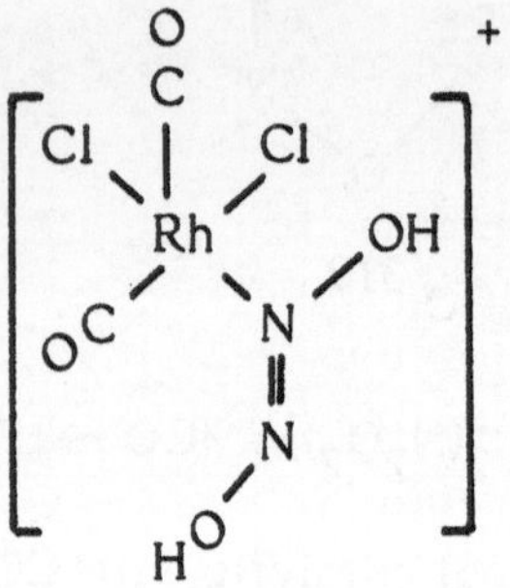

$$\left[\begin{array}{c} \text{Cl} \; Rh \; \text{Cl} \; OH \\ N=N \\ H O \end{array}\right]^+ + H_2O \rightleftharpoons \left[\begin{array}{c} \text{Cl} \; Rh \; \text{Cl} \; OH \\ O=C \\ H O \quad N=N \\ H O \end{array}\right]^+ \qquad (130)$$

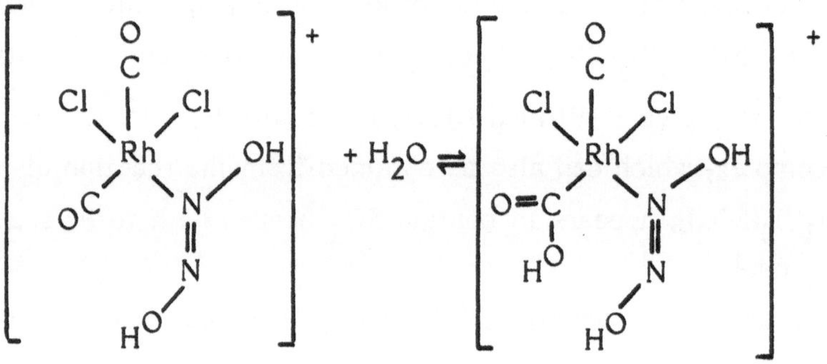

$$\left[\begin{array}{c} \text{Cl} \; Rh \; \text{Cl} \; OH \\ O=C \\ OH \quad N=N \\ H O \end{array}\right]^+ \longrightarrow [RhCl_2(CO)]^- + N_2O + CO_2 + 2H^+ + H_2O \qquad (131)$$

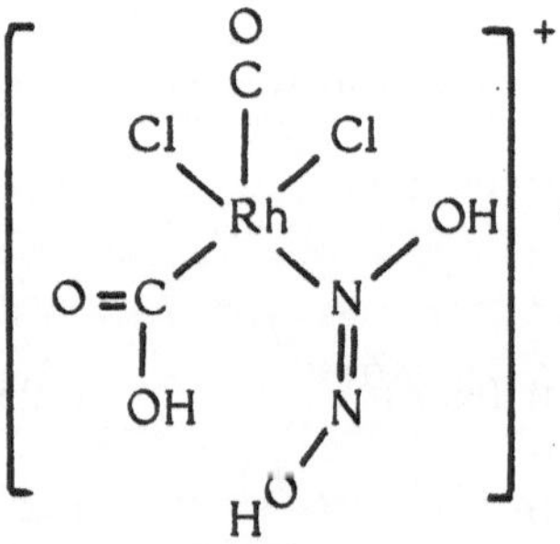

$$[RhCl_2(CO)]^- + 2NO \rightleftharpoons [RhCl_2(CO)(NO)_2]^- \qquad (132)$$

$$[RhCl_2(CO)(NO)_2]^- + CO \rightleftharpoons [RhCl_2(CO)_2(NO)_2]^- \qquad (133)$$

Convincing evidence has been presented that the active catalyst is $[RhCl_2(CO)(NO)_2]^-$, in which both {RhNO} groups are bent and therefore the rhodium has formally 16 electrons.[332] Addition of a second molecule of CO is accompanied, or closely followed, by protonation to give $[RhCl_2(CO)_2(HON_2OH)]^+$. The next step has been shown by labeling studies not to be the expected oxygen or hydroxyl transfer to CO but rather intermolecular nucleophilic attack of H_2O on coordinated CO. The oxygen in the CO_2 liberated in the overall reaction is derived solely from solvent H_2O and not at all from NO.[331] The details of the collapse of the remarkable intermediate $[RhCl_2(CO)(C(O)OH)(HON_2OH)]^+$ to give N_2O and CO_2 (equation 131) are not known.

A second mechanism for the reaction is possible since under NO alone, in the absence of added CO, $[RhCl_2(CO)_2]^-$ reacts according to[331]

$$[RhCl_2(CO)_2]^- + 6NO \rightarrow [RhCl_2(NO)_2]^- + 2N_2O + 2CO_2 \qquad (134)$$

This can be explained by the sequence

$$[RhCl_2(CO)_2]^- + 2NO \rightleftharpoons [RhCl_2(CO)_2(NO)_2]^- \qquad (135)$$

$$[RhCl_2(CO)_2(NO)_2]^- \rightleftharpoons [RhCl_2(CO)(NO)_2]^- + CO \qquad (136)$$

$$[RhCl_2(CO)(NO)_2]^- + 2NO \rightarrow [RhCl_2(CO)(NO_2)(NO)]^- + N_2O \qquad (137)$$

$$[RhCl_2(CO)(NO_2)(NO)]^- + OH^- \rightleftharpoons [RhCl_2(C(O)OH)(NO_2)(NO)]^{2-} \qquad (138)$$

$$[RhCl_2(C(O)OH)(NO_2)(NO)]^{2-} \rightarrow [RhCl_2(NO)_2]^- + CO_2 + OH^- \qquad (139)$$

Here formation of NO_2 occurs, but labeling experiments again show that the O of NO_2 does not transfer to CO but to the co-catalyst, water.

Catalysis of reaction (106) by $[Pt_2(\mu\text{-}Ph_2P)(CH_2)PPh_2)_3]$ has been observed.[338] Because NO slowly but irreversibly reacts with the catalyst to give an inactive nitrosyl complex the rate declines with time. The mechanism of the reaction is unknown.

The use of a bimetallic Wacker-type catalyst for simultaneous oxidation of CO and reduction of NO has been reported. The essential steps are:[339]

$$CO + Pd^{2+} + H_2O \rightarrow Pd^0 + CO_2 + 2H^+ \qquad (140)$$

$$Pd^0 + 2CuCl_2 \rightarrow Pd^{2+} + 2CuCl_2^- \qquad (141)$$

$$Pd^0 + 2NO + 2H^+ \rightarrow Pd^{2+} + N_2O + H_2O \qquad (142)$$

$$2CuCl_2^- + 2NO + 2H^+ \rightarrow 2CuCl_2 + H_2O + N_2O \qquad (143)$$

Although none of these reactions is novel and reaction (142) is rather slow, the combination is effective because Pd^{2+} catalyzes reaction (143), just as Cu^+ catalyzes NO reduction by Sn^{2+}. The mechanism of reaction (142) may also be similar to the Sn^{2+}/Cu^+ system, but once again though an N_2O_2 intermediate is suggested evidence is lacking.

An interesting system which is not catalytic, but which may throw light on the mechanism of the catalytic reactions discussed above is that obtained by reacting $AgNO_2$ with cis-$[Mo(dmpe)_2(CO)_2]$ (dmpe = $Me_2PCH_2CH_2PMe_2$). The overall reaction is:[340]

$$\text{cis-}[Mo(dmpe)_2(CO)_2] + 2AgNO_2 \rightarrow 2CO_2 + N_2O + 2Ag + [Mo(dmpe)_2(O)] +$$
$$\text{molybdenum products} \qquad (144)$$

and the mechanistic scheme proposed is:

$$\text{cis-}[Mo(dmpe)_2(CO)_2] + 2AgNO_2 \rightarrow \text{cis-}[Mo(dmpe)_2(CO)_2(ONO)](NO_2) + 2Ag$$
$$(145)$$

$$\text{cis-}[Mo(dmpe)_2(CO)_2(ONO)](NO_2) \rightarrow \text{cis-}[Mo(dmpe)_2(CO)_2(NO_2)](NO_2) \quad (146)$$

$$\text{cis-}[Mo(dmpe)_2(CO)_2(NO_2)](NO_2) \rightarrow \text{cis-}[Mo(dmpe)_2(CO)(NO)](NO_2) + CO_2$$
$$(147)$$

$$\text{cis-}[Mo(dmpe)_2(CO)(NO)](NO_2) \rightarrow [Mo(dmpe)_2(CO)(NO)(ONO)] \qquad (148)$$

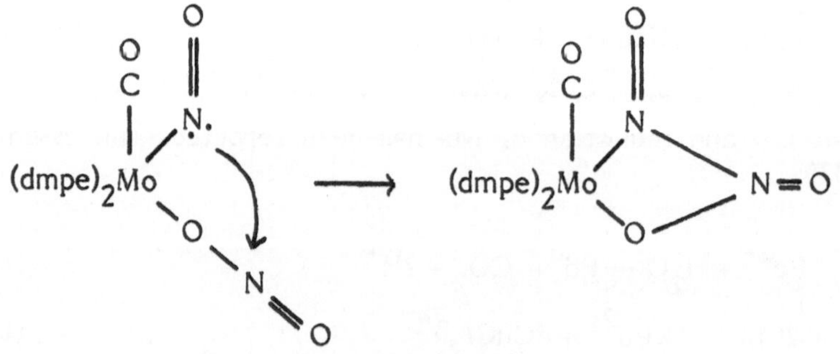

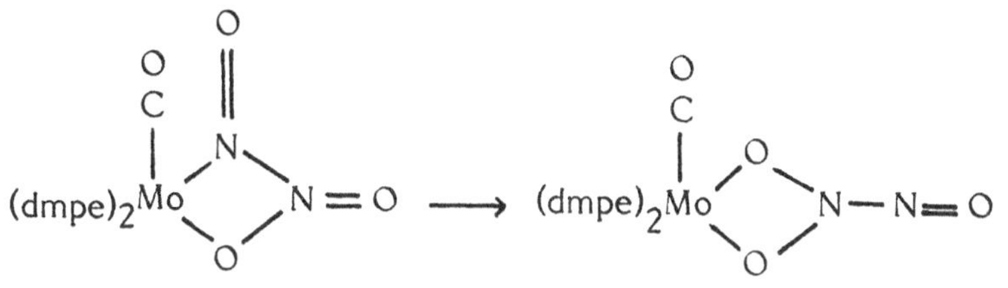

$$(dmpe)_2Mo \cdots N=O \longrightarrow (dmpe)_2Mo \cdots N-N\equiv O \qquad (150)$$

$$(dmpe)_2Mo \cdots N-N\equiv O \rightarrow [\,(dmpe)_2Mo(O)\,] + CO_2 + N_2O \qquad (151)$$

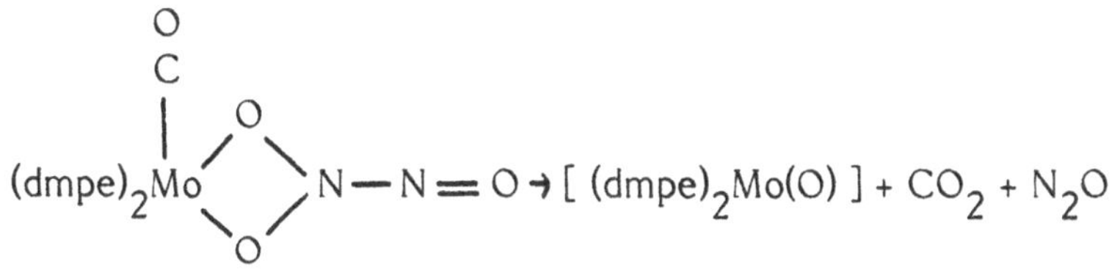

([(dmpe)$_2$Mo(O)] → unknown products)

Though this scheme is by no means proven, it does provide an alternative

to the formation of N_2O via $M \overset{N}{\underset{N}{<}}$ complexes. Oxygen atom

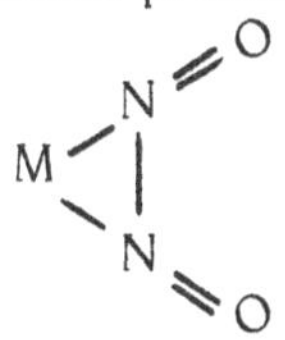

transfer to nitrosyls is a well-documented process (see Section 9), and NO_2 complexes could be obtained by NO or CO attack on a dinitrosyl. Much more work on these intriguing systems is needed before the question of their mechanisms will be resolved.

The formation of $[Pt(Ph_3P)_2(N_2O_2)]$ from $[Pt(Ph_3P)_3]$ or $[Pt(Ph_3P)_4]$ and NO has been referred to above. This complex appears to be the only monomeric complex containing the $\{N_2O_2\}$ ligand which has been adequately characterized. Bidentate bridging $\{N_2O_2\}$ occurs in $[(Co(NH_3)_5)_2(N_2O_2)]^{4+}$, produced by the reaction between NO and ammonical Co^{2+}.[266] Although first suggested to be a complex of $N_2O_2^{2-}$ by Werner in 1918,[341] this was not confirmed until 1969, when X-ray crystallography showed that the structure is[342]

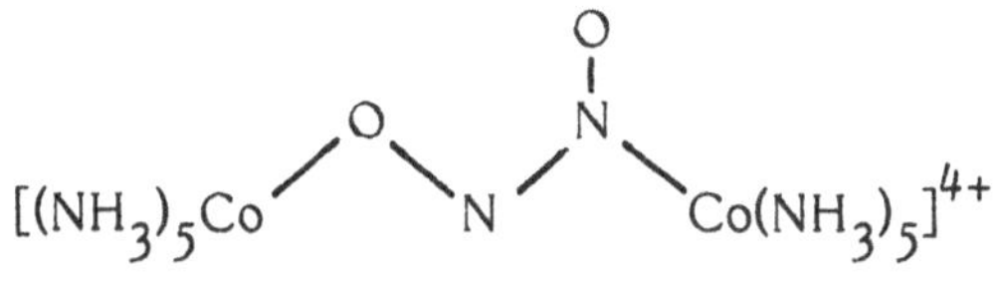

$$[(NH_3)_5Co \cdots N \cdots N(O) \cdots Co(NH_3)_5]^{4+}$$

It is formed only at room temperature (see Section 5). At low temperature the $N_2O_2^{2-}$ ligand is oxidized by NO to N_2O and NO_2^-. Tetradentate N_2O_2 occurs in $[Co_4(NO_2)_2(N_2O_2)(NO)_8]$, produced (along with N_2O) by the reaction between NO and $[Co(CO)_3(NO)]$; the structure is[343]

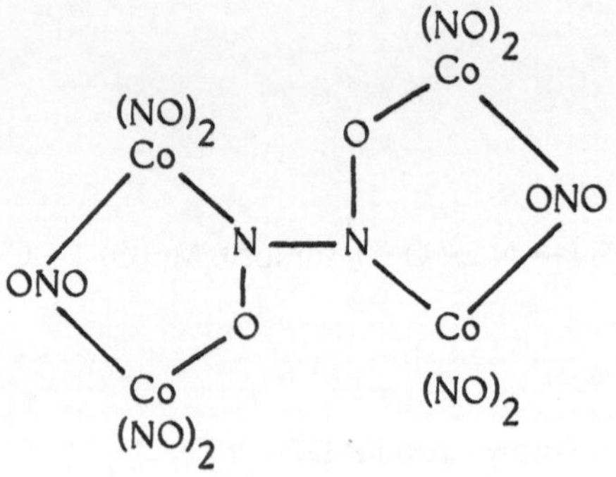

Structural evidence is lacking for three other complexes believed to contain $N_2O_2^{2-}$ as a ligand: $[Cp_2Ti(N_2O_2)]_n$,[322,344] $[Co(TPP)(N_2O_2)(NO)]$ (TPP = tetraphenylporphyrin)[345] and $[(MoCl(L-L)(NO)_2(N_2O_2)]^{2+}$ (the last complex being the result of the dimerization, with Cl^- elimination, of $[MoCl_2(L-L)(NO)_2]$ (L-L = ethylendiamine or pdma).[346]

In general N_2O_2 complexes appear to be reactive intermediates, either in the reaction of CO with NO (equation 106, see above) or more often in the disproportionation of NO (equation 105).[347-348] One probable reason for their comparative rarity is precisely that they are intermediates in reaction (105); reactions of transition metal complexes with NO are most often performed using a large excess of NO, which favors the completion of reaction (105). Small quantities of nitro (NO_2^-) complexes in the products cannot be taken as evidence of reaction (105) however. The use of unpurified NO, or the inadvertent admission of traces of O_2 during the experiment, can often account for the presence of nitro complexes. This is particularly true in earlier work where the practical difficulties of purifying and handling NO were greater than nowadays. Even when pure NO is used, and all precautions to avoid contamination by higher oxides of nitrogen are taken, the products obtained will be very dependent on the amount of NO used. This fact, probably more important in nitrosyl chemistry than for any other ligand, is often ignored. It probably explains the controversy over the products of the reaction between $[Rh(Ph_3P)_3Cl]$ and NO, from which N_2O, $[RhCl(Ph_3P)_2(NO_2)(NO)]$,[319,349-352] $[RhCl(Ph_3P)_2(NO)]$,[352] $[RhCl(Ph_3P)_2(NO_2)_2]$[351] and polymeric nitrosyls[350] have been obtained by different workers. A few reactions in which carefully purified NO has been used and a high yield of a nitro complex obtained are in the literature. Though disproportionation of NO can reasonably be assumed in such cases, nothing can be said about the course of the reactions until gas analysis is performed. Examples are

$$\underline{mer}\text{-}[OsCl_3(Me_2PhP)_3] + NO \xrightarrow{Zn} [OsCl(Me_2PhP)_3(NO_2)(NO)]^+ \quad (152)^{353}$$

$$[Ru(sal_2en)(Ph_3P)_2] + NO \rightarrow [Ru(sal_2en)(NO_2)(NO)] \quad (153)^{354}$$

$$[CpCr(CO)(THF)(NO)] + NO \rightarrow [CpCr(NO)(\mu\text{-}NO)]_2 + [CpCr(NO_2)(NO)_2] \quad (154)^{135}$$

$$n[Co_2(CO)_8] + 3nNO \rightarrow [Co(NO_2)(NO)_2]_n \quad (155)^{356}$$

$$[(CpCr)_2(\mu\text{-}SC(CH_3)_3)_2(\mu\text{-}S)] + NO \longrightarrow$$

$$[(CpCr(NO))_2(\mu\text{-}SC(CH_3)_3)(\mu\text{-}S\text{-}SC(CH_3)_3)] + [CpCr(NO)_2(ONO)] \quad (156)^{493}$$

$$\underline{cis}\text{-}[Pt(Me_2PhP)_2(CH_3)_2] + NO \rightarrow [Pt(CH_3)_2(Me_2PhP)_2(NO_2)_2] \quad (157)^{357}$$

$$[Pt(H)(I)(Me_2PhP)_2] + NO \rightarrow [PtI(Me_2PhP)_2(NO_2)] \quad (158)^{357}$$

$(sal_2en = \underline{N},\underline{N}'\text{-ethylenebis(salicyclideneiminato)}; THF = tetrahydrofuran).$ Note the lack of insertion of NO into the Pt - CH_3 bond. Other reactions have been more closely examined and the overall stoichiometry determined:

$$[Ni(CO)_4] + 4NO \rightarrow [Ni(NO_2)(NO)] + 4CO + N_2O \quad (159)^{358}$$

$$[Co(8\text{-}Q)_2(NO)] + 8NO \rightarrow [NO^+][Co(8\text{-}Q)_2(NO_3)(NO_2)] + 3N_2O \quad (160)^{359}$$

$(8\text{-}Q = 8\text{-quinolinolato})$

In the latter reaction labeling studies show that the coordinated NO in the starting complex is oxidized to coordinated NO_2^- by free NO.[359] No detailed mechanism can be suggested.

The remarkably complicated reaction:

$$2[Co(Ph_3P)_3(NO)] + 14NO \rightarrow 2[Co(Ph_3P)(NO_2)(NO)_2] + 4Ph_3PO +$$
$$4N_2O + N_2 \qquad (161)$$

has been the subject of a very detailed kinetic and mechanistic investigation[347-348] and the following steps suggested:

$$[CoL_3(NO)] \rightleftharpoons [CoL_2(NO)] + L \qquad (162)$$

$$L + 2NO \rightarrow LO + N_2O \qquad (163)$$

$$[CoL_2(NO)] + NO \rightarrow [CoL_2(NO)_2] \qquad (164)$$

$$[CoL_2(NO)_2] + NO \rightarrow [CoL_2(N_2O_2)(NO)] \qquad (165)$$

$$[CoL_2(N_2O_2)(NO)] \rightarrow [CoL(NO)] + LO + N_2O \qquad (166)$$

$$[CoL(NO)] + 2NO \rightarrow [CoL(NO)_3] \qquad (167)$$

$$2[CoL(NO)_3] \rightarrow [(CoL(NO)_2)_2(N_2O_2)] \qquad (168)$$

$$[(CoL(NO)_2)_2(N_2O_2)] \rightarrow 2[CoL(NO_2)(NO)] + N_2 \qquad (169)$$

$$[CoL(NO_2)(NO)] + NO \rightarrow [CoL(NO_2)(NO)_2] \qquad (170)$$

$L = Ph_3P$

A side-reaction produces the cluster $[Co_3(Ph_3P)_3(NO)_7]$ of unknown structure. Reactions (162) - (170) do give the correct overall stoichiometry of equation (161), and some of the proposed reactions, e.g. (162) and (163), have been independently proven. Moreover, some of the intermediates, e.g. $[(Co(Ph_3P)_2(NO)_2)_2(N_2O_2)]$, have been isolated from the reaction. Nevertheless intermediates such as $[Co(Ph_3P)(NO)]$ appear rather unlikely, and the decomposition of $[(Co(Ph_3P)(NO)_2)_2(N_2O_2)]$ to give N_2 and not N_2O is also very unusual. In general disproportionation or reduction of NO gives N_2O, and N_2 is only observed when excess reductant allows further reduction of N_2O.

The simplest and best understood case of NO disproportionation is the reaction[360]

$$[Co(B)(L-L)_2(NO)]^{n+} + 2NO \rightarrow [Co(B)(L-L)_2(NO_2)]^{n+} + N_2O \qquad (171)$$

where L-L = ethylenediamine for n = 2 and dimethylglyoximato for n = 0; B is a base (e.g. methanol or pyridine). This reaction can also be regarded as oxidation of coordinated NO by free NO, and as such should be compared with the reaction of $[Co(B)(L-L)_2(NO)]^{n+}$ with O_2 (Section 5). Oxidation of the bent $\{CoNO\}$ group by NO with retention of the Co - N bond gives the radical intermediate

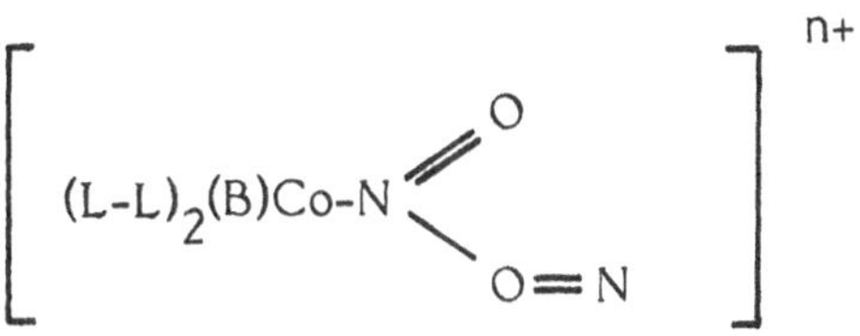

$$\left[(L-L)_2(B)Co-N{\overset{\displaystyle O}{\underset{\displaystyle O=N}{\Big\langle}}} \right]^{n+}$$

from which a second molecule of NO can abstract a nitrogen atom. This mechanism is considerably different from those previously postulated, in which coupling of two coordinated nitrosyls gives coordinated N_2O_2. Formation of a dinitrosyl, $[Co(L-L)_2(NO)_2]^{n+}$, seems unlikely because the reactions require a base B to proceed; dinitrosyl formation would be inhibited by base. Formation of a μ-N_2O_2 species such as $[(Co(B)(L-L)_2)_2(N_2O_2)]^{2n+}$ can be discounted because $[Co(B)(L-L)_2(NO)]^{n+}$ shows no tendency to dimerize in solution.

A reaction which does take place via a bridged N_2O_2 ligand is the base (N_3^-, CN^- or NCO^-) promoted loss of N_2O from $[M(L-L)_2(NO)_2]$ (M = Mo, W; L-L = N,N-diethyldithio(carbamato)):[361-362]

$$2[M(L-L)_2(NO)_2] + 2B \rightarrow 2[M(B)_2(L-L)_2(NO)] + N_2O \qquad (172)$$

The oxygen atom which is unaccounted for in this equation appears to become attached to M, i.e. metal oxides of unknown stoichiometry are formed. Reaction (172) proceeds via the sequence:

$$[M(L-L)_2(NO)_2] + B \rightleftharpoons \left[(L-L)_2(B)M{\overset{\displaystyle N^O}{\underset{\displaystyle N}{\Big\langle}}} O \right] \qquad (173)$$

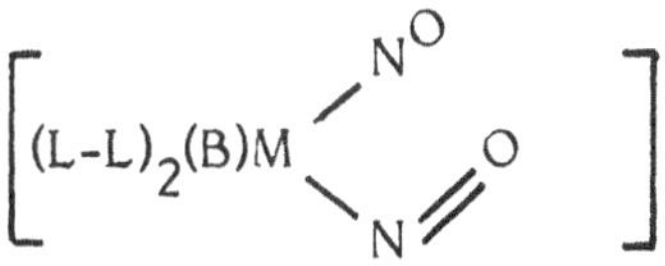

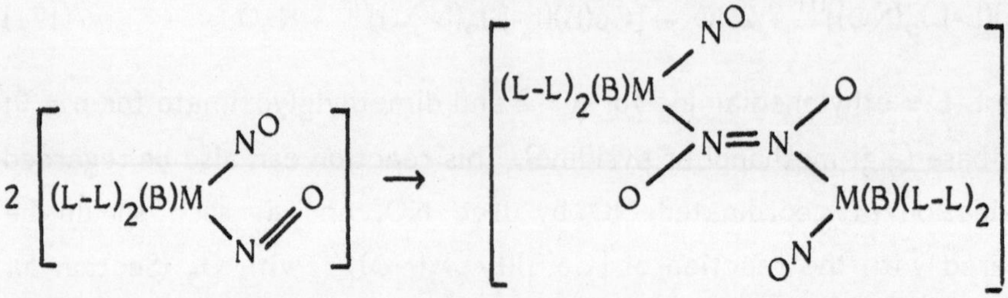

$$2 \left[(L\text{-}L)_2(B)M \overset{N\overset{O}{\diagup}}{\underset{N\diagdown}{\diagdown}} O \right] \rightarrow \left[(L\text{-}L)_2(B)M \overset{N\overset{O}{\diagdown}}{\diagdown} \overset{O}{\underset{O}{\diagdown N\!=\!N}} \overset{O}{\underset{O^{N}}{\diagdown}} M(B)(L\text{-}L)_2 \right] \quad (174)$$

Inter-rather than intra-molecular coupling is favored because only one of the {MNO} groups is bent. This is based on the interesting idea that dinitrosyls will couple intramolecularly if they have the repulso configuration, but intermolecularly if they have the attracto:[334]

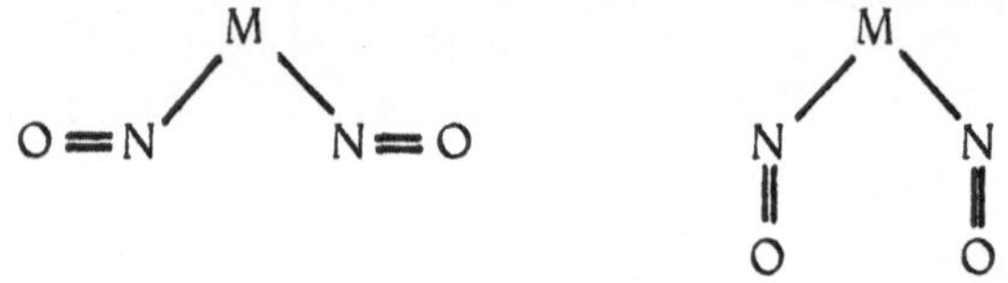

$$O\!=\!N \overset{M}{\diagup \diagdown} N\!=\!O \qquad\qquad \overset{M}{\underset{\overset{\|}{O}}{N}} \overset{}{\diagup \diagdown} \overset{}{\underset{\overset{\|}{O}}{N}}$$

                    repulso   attracto

8.   EXCHANGE, SUBSTITUTION, AND TRANSFER OF THE NITROSYL LIGAND

8.1   Introduction

There are three groups of reactions of nitrosyls which are not reactions of the coordinated NO in the sense that NO is converted into some new species $N(A)_n O(B)_m$, but which are sufficiently unusual to require discussion. These are:

Exchange of NO:

$$[L_n M(NO)] + \overset{*}{N}O \rightleftharpoons [L_n M(\overset{*}{N}O)] + NO \qquad\qquad (175)$$

Substitution of NO:

$$[L_n M(NO)] + L' \rightleftharpoons [L_n ML'] + NO \qquad\qquad (176)$$

Transfer of NO:

$$[L_n M(NO)] + L'_m M' \rightarrow [L'_m M'(NO)] + L_n M \qquad\qquad (177)$$

For metal carbonyls these reactions are very well established; the importance of carbonyls as catalysts is based on just these properties. For nitrosyls they are rare. The reason for this in the cases of substitution (175) and exchange (176) is obvious. Whatever the nature of the bonds between M and NO in a nitrosyl, it is clear that the $^2\Pi$ ground state of free NO has been modified almost irrevocably on complexation. Loss of NO from a complex requires far more electronic rearrangement than does loss of CO and in the case of reaction (176) a three-electron donor is being replaced by a two-electron donor. Loss of NO^- from those complexes in which it formally exists (a minority of nitrosyls) is complicated by the necessity of dimerization at the very least, and does not seem to have been observed. Loss of NO^+ is possible, and is observed in the reaction[122]

$$[Ir(Ph_3P)_2(NO)_2]^+ + Ph_3P \rightleftharpoons [Ir(Ph_3P)_3(NO)] + NO^+ \qquad (178)$$

but such reactions are very rare. Nikol'skii has calculated that the processes

$$[Fe(CN)_5(NO)]^{2-} + 2H_2O \rightleftharpoons [Fe(CN)_5(H_2O)]^{3-} + H^+ + HNO_2 \qquad (179)$$

and

$$[Fe(CN)_5(NO)]^{2-} + H_2O \rightleftharpoons [Fe(CN)_5(H_2O)]^{2-} + NO \qquad (180)$$

have ΔG°_{298} (179) = +125.5 kJ mol^{-1} and ΔG°_{298} (180) = +97.1 kJ mol^{-1},[364] both markedly unfavorable.

Assuming that reactions of type (177) take place _via_ bridging of a nitrosyl between two metal centres, and assuming that the accepting metal complex can accommodate three electrons (with or without loss of an existing ligand), there is no obvious reason why such reactions should not be observed. Bridging nitrosyls, while not as common as bridging carbonyls, are crystallographically and spectroscopically well established. Nitrosyl transfer reactions are nevertheless not common.

8.2   Exchange of Coordinated Nitrosyl for free Nitrogen Monoxide

The reactions of type (175) are potentially more facile than those of type (176) if bimolecular replacement of the three-electron donor NO by itself takes place. This does occur thermally, and by the simple associative mechanism, in the gas phase:[365]

$$[Fe(CO)_2(NO)_2] + {}^{15}NO \rightleftharpoons [Fe(CO)_2({}^{15}NO)({}^{14}NO)] + {}^{14}NO \qquad (181)$$

$$[Co(CO)_3(NO)] + {}^{15}NO \rightleftharpoons [Co(CO)_3({}^{15}NO)] + {}^{14}NO \qquad (182)$$

The reactions are both very slow ($k_f(181) = 18 \times 10^{-6}$ M^{-1} s^{-1} and $k_f(182) = 4 \times 10^{-6}$ M^{-1} s^{-1}), and [CpNi(NO)] does not show measurable exchange after ten days at 120°. The nitrosyl group confers stability with respect to thermal exchange not only on itself but also on the carbonyl ligands in these complexes; both CO exchange and substitution in $[Fe(CO)_2(NO)_2]$ and $[Co(CO)_3(NO)]$ are several orders of magnitude slower than in the isoelectronic $[Ni(CO)_4]$ or in $[Fe(CO)_5]$.[366-371]

The rate constants for NO exchange in $[M(Ph_3A)I(NO)_2]$ (M = Fe, Co; A = P, As) are close to 1×10^{-3} s^{-1} and both unimolecular and bimolecular mechanisms are observed.[372]

Exchange of ${}^{15}NO$ for ${}^{14}NO$ in $[M(CN)_5(NO)]^{2-}$ (M = Fe, Ru) and $[RuX_5(NO)]^{2-}$ (X = Cl, Br) shows that the rate of exchange falls in the order $[Fe(CN)_5(NO)]^{2-}$ $\gg$ $[Ru(CN)_5(NO)]^{2-}$ > $[RuCl_5(NO)]^{2-}$ > $[RuBr_5(NO)]^{2-}$ which is also the order of electrophilicity of these complexes.[373] The exchange process is accelerated as the pH increases (and by irradiation, see Section 8.5) indicating that nucleophilic attack by H_2O at the nitrosyl giving the more labile NO_2^- ligand may play a role in the reaction.

Slow NO exchange is also observed for the 18-electron complex $[FeI(S_2CN(CH_3)_2)_2(NO)]$ ($k = 1.22 \times 10^{-4}$ s^{-1}), but it is believed that the rate determining step is not NO exchange but homolytic Fe - I bond scission, and bimolecular NO exchange takes place at the 17 electron complex $[Fe(S_2CN(CH_3)_2)_2(NO)]$ (this being demonstrated in independent experiments). The Fe - I bond reforms after exchange.[374] An Fe - I bond scission, rather than nitrosyl transfer, also seems to be responsible for the reaction[374]

$$[Fe(S_2CN(CH_3)_2)_2(^{14}NO)] + [FeI(S_2CN(CH_3)_2)_2(^{15}NO)] \rightleftharpoons \qquad (183)$$
$$[Fe(S_2CN(CH_3)_2)_2(^{15}NO)] + (FeI(S_2CN(CH_3)_2)_2(^{14}NO)]$$

Reversible uptake of NO (a minor variation on reaction (175)) has been observed, most notably in the "brown-ring test" reaction

$$Fe^{2+}(aq) + NO(aq) \rightleftharpoons FeNO^{2+}(aq) \qquad (184)$$

for which an equilibrium constant of 4.8×10^2 mol^{-1} l^{-1} has been calculated.[375] For the analogous reaction of $[Fe(edta)]^{2-}$:

$$[Fe(edta)]^{2-} + NO \rightleftharpoons [Fe(edta)NO]^{2-} \qquad (185)$$

(edta = ethylenediaminetetraacetate)

values of 8.59×10^5 mol^{-1} l^{-1} for the equilibrium constant at 55°C, -66.1 kJ mol^{-1} for ΔH° and -86.6 J $°K^{-1}$ mol^{-1} for ΔS° were obtained.[376] It is noteworthy that the small value of ΔH° only just compensates for the decrease in entropy, thus making the reaction favorable (ΔG° = -37.7 kJ mol^{-1}). The apparently reversible uptake of NO by Fe(III) porphyrins:[377]

$$[FeCl(TPP)] + NO \rightleftharpoons [FeCl(TPP)(NO)] \qquad (186)$$

(TPP = tetraphenylporphyrin) is in dispute. Covincing evidence has been given that the actual product of reaction (186) is the Fe(II) complex $[Fe(TPP)(NO)]$[378] and it is suggested that this is produced by nucleophilic attack of solvent CH_3OH at the nitrosyl of $[FeCl(TPP)(NO)]$ (formally Fe(II) and NO^+; $\nu(NO) = 1880$ cm^{-1}) with subsequent loss of CH_3ONO and HCl and coordination of excess NO to the $[Fe(TPP)]$ so formed.[375,379] No evidence for CH_3ONO was given, and hence an outer-sphere reduction of $[FeCl(TPP)]$ to $[Fe(TPP)]$ by NO (or by CH_3OH) are equally likely

TABLE V

Substitution of Nitrogen Monoxide

Complex	Incoming Ligand	Product	Remarks	Ref.
$[PdCl_2(NO)_2]$	NH_3 CN^-	$[PdCl_2(NH_3)_3]$ $[PdCl_2(CN)_2]^{2-}$	NO evolved	392
$[Co(PR_3)_3(NO)]^+$ $R = OCH_3, OC_2H_5$	$R'NC$ CO	$[Co(PR_3)_2(CNR')_3]^+$ $[Co(PR_3)_3(CO)]^+$ $[Co(PR_3)_2(CO)_2]$		393
$[VCl_2(NO)_3]_2$	PH_3PO CH_3CN $\underline{o}$-phenanthroline	$[VCl(Ph_3PO)_4(NO)]^+$ $[VCl(CH_3CN)_4(NO)]^+$ $[VCl(\underline{o}\text{-phen})_2(NO)]^+$		394
$[CpCrCl(NO)_2]$	$C_5H_{10}NH$ HNC_4H_8NH	$[CpCrCl(NHC_5H_{10})(NO)]$ $[(CpCrCl(NO))_2(\mu\text{-}NHC_4H_8NH)]$	$[CpMCl(NO)_2]$ (M = Mo, W) have labile NO.	397 395–6
$[M(Ph_3P)_2(CH_3CN)_2(NO)]^{2+}$ M = Rh, Ir	CO S_2CNR_2	$[Rh(CH_3CN)_2(CO)_2]^+$ $[Rh(Ph_3P)_2(S_2CNR_2)_2]^+$		398–9 400
$[RuCl(Ph_3P)_2(NO)]$	Electron-rich olefins (L')	$[RuCl(Ph_3P)_2(L')]$		401–2

Complex	Incoming Ligand	Product	Remarks	Ref.
[Fe(LH)(NO)₂] (LH = S(CH₂)₂NHMe(CH₂)₃NMe(CH₂)₂S)	Solvent	[FeL(NO)]		403
[FeBr(NO)₂]₂	Ph₃P	[FeBr₂(Ph₃P)₂(NO)] [Fe(Ph₃P)₂(NO)₂]		4, 404,406
[FeBr(NO)₃]	None	[FeBr₂(NO)₂]		407

mechanisms. Similar uncertainty reigns over the mechanism of formation of $[Fe(heme)(NO)]^{2+}$ from $[Fe(heme)]^{3+}$ and $NO.$[380] Neither $[Fe(heme)(NO)]^{2+}$ nor $[Fe(TPP)(NO)]$ loses NO reversibly,[495-496] though the latter adds a second NO under pressure.[495] Both $[CrCl(TPP)]$ and $[MnCl(TPP)]$ absorb NO reversibly, though details are not known.[381] Recently convincing evidence for the reversible coordination of NO by $MnX_2(PR_3)$ (X = Cl, Br, I; R_3 = Pr^n_3, Bu^n_3, $PhMe_2$, $PhEt_2$) has been presented.[507]

Cobalt(II) complexes with imidazole,[382-383] amino acid and histidine[384] ligands also coordinate NO reversibly. The most detailed investigation has been made of $[CoCl(bpy)_2(NO)]^+$, which shows a $\nu(NO)$ vibration at 1642 cm^{-1} (cf. $[CoCl(en)_2(NO)]^+$ with $\nu(NO)$ 1635 cm^{-1} and a $\{CoNO\}$ angle of $136°$[93]) reversibly can be isolated as a crystalline perchlorate salt, but loses NO reversibly in solution.[385]

Anhydrous CuX_2 (X = Cl, Br) absorb NO reversibly in non-aqueous solution:[386]

$$2n[CuX_2] + 2nNO \rightleftharpoons n[CuX_2(NO)]_2 \qquad\qquad (187)$$

These weakly coordinated copper nitrosyls (K_{equil} = 3.7 x 10^{-3}, $\Delta H°$ = 37.6 kJ mol^{-1}, $\Delta S°$ = 79.5 J °K^{-1} mol^{-1}[386]) are useful for the preparation of organic halides and N-nitroso compounds (Section 9). Vanadium(IV) complexes of 3,5-di-tert-butylcatecholate also reversibly bind NO, O_2 and CO; these complexes are considered as models for the vanodocyte cells found in the blood of Ascidians.[387] It appears that $[RhX(Ph_3P)_2(NO)_2]$ (X = Cl, Br) reversibly lose NO in solution, yielding $[RhX(Ph_3P)_2(NO)]$.[388] However, the fate of the lost NO is not known, and ready reduction of one of the NO ligands may be occurring.[405] Another possibility is disproportionation, since $[IrI(Ph_3P)_2(NO)_2]$ gives $[IrI_2(Ph_3P)_2(NO)]$ on recrystallisation.[389] These reactions, along with the reported "substitution" of one NO in $[M(Ph_3P)_2(NO)_2]^+$ (M = Rh, Ir) by Ph_3P to give $[M(Ph_3P)_3(NO)]$[122,390] and of NO and NO_2 in $[RhCl(Ph_3P)_2(NO_2)(NO)]$ by

Ph_3P^{391} (actually all substitutions of NO^+, as in equation 178), would bear further investigation.

8.3   Substitution of the Nitrosyl Ligand, Including Photochemical Substitution

In Table V are listed reactions in which an NO ligand is replaced by another ligand (equation 176). It is stressed that in most of the reactions listed the overall stoichiometry of the reaction has not been established, and that in many the fate of the replaced NO is not known. It cannot simply be assumed that NO is evolved. It is probable that other "substitution" reactions are lurking in the literature and require investigation.

Photochemical reactions of nitrosyls are more common than thermal ones, and much better understood. The LUMO orbital of nitrosyls which do not contain CO as a co-ligand is mainly the $\pi*${MNO}, or in some cases the $\sigma*$ {MNO} orbital. Occupation of either of these orbitals causes bending of the {MNO} group and interaction of the two orbitals, so the precise ordering is not important. These orbitals also have significant contributions from co-ligand orbitals. Since the LUMO is antibonding with respect to the M - NO and/or M - L interaction, population of this orbital may be expected to favor loss of NO or L. The HOMO of nitrosyls mainly contains co-ligand orbitals, with the {MNO} σ and π bonding orbitals lying at lower energy. Irradiation of nitrosyls at long wavelengths ($>$ 450 nm) tends to produce little reaction, or changes in the co-ligands. Higher energy irradiation on the other hand often results in loss of the NO ligand. Examples are:

$$[Fe(CN)_5(NO)]^{2-} + H_2O \xrightarrow{h\nu} [Fe(CN)_5(H_2O)]^{2-} + NO \qquad (188)^{408}$$

$$[RuCl(bpy)_2(NO)]^{2+} + CH_3CN \xrightarrow{h\nu} [RuCl(bpy)_2(CH_3CN)]^{2+} + NO$$
$$(189)^{80}$$

$$[FeCl(pdma)_2(NO)]^{2+} + S \xrightarrow{h\nu} [FeCl(pdma)_2S]^{2+} + NO \qquad (190)^{409}$$

$$[RuCl(\eta^2\text{-V})_2(NO)] + H_2O \xrightarrow{h\nu} [RuCl(\eta^2\text{-V})_2(H_2O)] + NO \quad (191)^{157}$$

$$[RuX_5(NO)]^{2-} + H_2O \xrightleftharpoons{h\nu} [RuX_5(H_2O)]^{2-} + NO \quad (192)^{410-411}$$

X = Cl, Br, I; S = solvent; V = violurato

The changes taking place on irradiation of $[Fe(CN)_5(NO)]^{2-}$ at various wavelengths and under various conditions have been subjects of controversy for some time. It is now certain that the primary process on irradiation at 400 nm or shorter wavelengths is loss of NO.[408,412] However the mechanism of the apparently simple general reaction

$$[ML_5(NO)] + H_2O \xrightarrow{h\nu} [ML_5(H_2O)] + NO \quad (193)$$

is much more complicated than equation (193) implies. For a variety of complexes of M = Fe, Ru, Os, Cr, V, Pt, Re and Co and L = Cl, Br, I and CN, reaction (193), when performed with coordinated $N^{16}O$ in $H_2^{18}O$ as solvent, produces equal amounts of $N^{18}O$ and $N^{16}O$. This can only occur if loss of NO is assisted by nucleophilic attack of H_2O:[412]

$$[ML_5(NO)] + H_2^{18}O \xrightarrow{h\nu} L_5MN{\overset{\displaystyle {}^{16}O\text{---}}{\underset{\displaystyle {}^{18}O\text{---}}{}}}\begin{matrix} \text{H} \\ \text{H} \end{matrix} \quad (194)$$

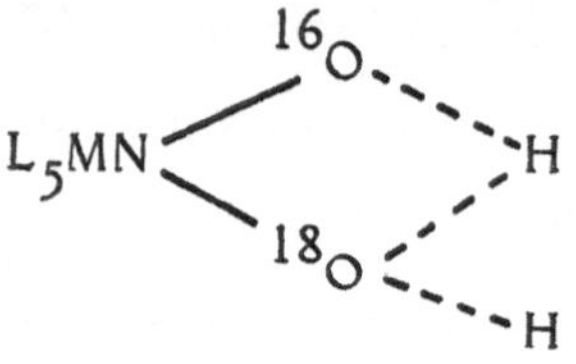

$$L_5MN{\overset{\displaystyle {}^{16}O\text{---}}{\underset{\displaystyle {}^{18}O\text{---}}{}}}\begin{matrix} \text{H} \\ \text{H} \end{matrix} \longrightarrow L_5MN{\overset{\displaystyle {}^{16}OH}{\underset{\displaystyle {}^{18}OH}{}}} \quad (195)$$

$$L_5MN{\overset{\displaystyle {}^{16}OH}{\underset{\displaystyle {}^{18}OH}{}}} \longrightarrow [ML_5] + 1/2\,(N^{16}O + N^{18}O) + 1/2(H_2^{16}O + H_2^{18}O) \quad (196)$$

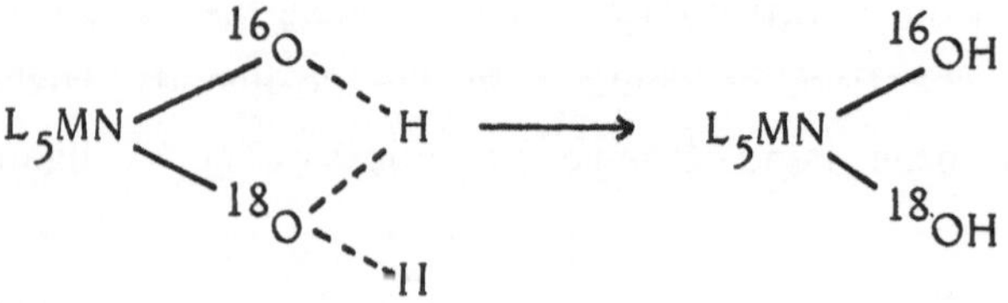

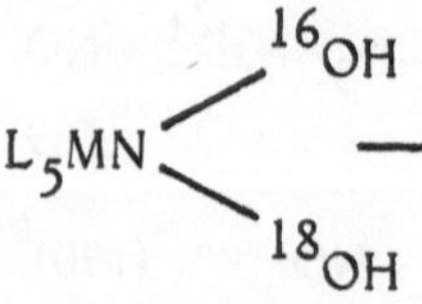

Note that free $N^{16}O$ does slowly exchange an oxygen atom with $^{18}OH_2$ to give $N^{18}O$ and $H_2^{16}O$.[122,456] For $[Fe(CN)_5(NO)]^{2-}$ and $[Ru(CN)_5(NO)]^{2-}$ the processes (193) - (196) are accompanied by a reaction reminiscent of the attack of $[RuNH_2(NH_3)_4(NO)]^{2+}$ on $[Ru(NH_3)_5(NO)]^{3+}$ (equations 59-61):[412]

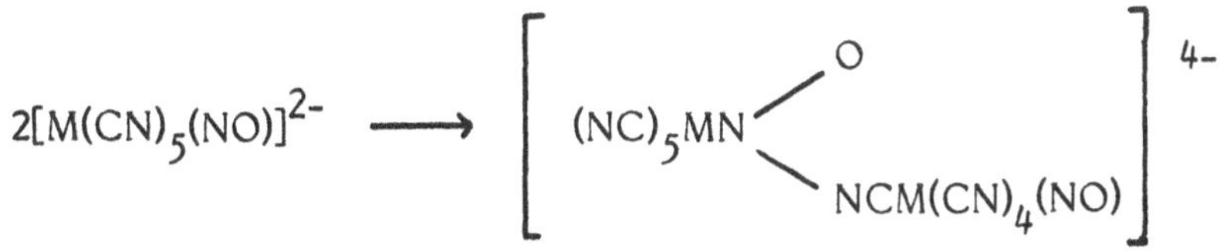

$$2[M(CN)_5(NO)]^{2-} \longrightarrow \left[(NC)_5MN \begin{array}{c} O \\ \diagup \\ \diagdown \\ NCM(CN)_4(NO) \end{array} \right]^{4-} \quad (197)$$

$$\left[(NC)_5MN \begin{array}{c} O \\ \diagup \\ \diagdown \\ NCM(CN)_4(NO) \end{array} \right]^{4-} + 3H_2O \quad \xrightarrow{h\nu}$$

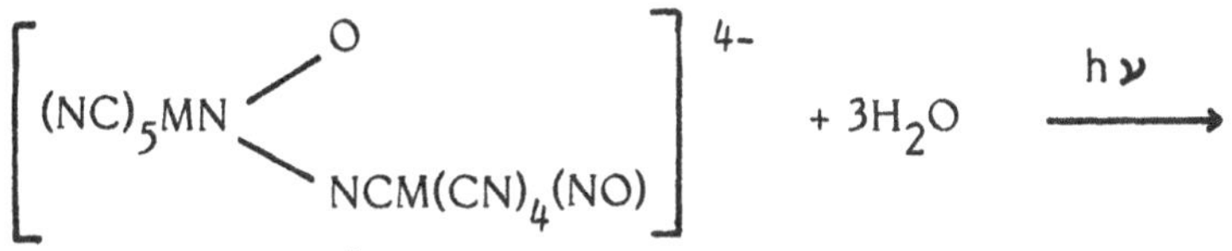

$$[M(CN)_5(H_2O)]^{3-} + [M(CN)_4(H_2O)(NO)]^{-} + N_2 + HCO_2H \quad (198)$$

Ultraviolet irradiation of a mixture of $[Co(CO)_3(NO)]$ and CF_2Br_2 or $(CF_2Br)_2$ gives $[Co_2(CO)_6(\mu_2\text{-}CF_2)_2]$[413] or $[CF_2BrCF_2Co(CO)_4]$[413-414] respectively. The primary reaction is not Co - NO but C - Br bond fission:

$$RBr \rightarrow R\cdot + Br\cdot \quad (R = CF_2Br \text{ or } CF_2BrCF_2) \quad (199)$$

and both of the product radicals abstract NO from $[Co(CO)_3(NO)]$:[414]

$$[Co(CO)_3(NO)] + CF_2Br \rightarrow NOBr + [Co(CO)_3CF_2] \quad (200)$$

$$[Co(CO)_3(NO)] + Br \rightarrow NOBr + [Co(CO)_3] \quad (201)$$

These reactions, which are not strictly photochemical reactions but rather radical reactions of nitrosyls, represent the only case of a reaction in a mixed carbonyl-nitrosyl which involves the nitrosyl ligand (with the possible exception of some partial replacement of NO in $[Fe(CO)_2(NO)_2]$ and $[Co(CO)_3(NO)]$ when these are photolyzed in a CO matrix at $20°K$[415-416,420]). Photolysis of $[Co(CO)_3(NO)]$ in the presence of HCl gas gives $CoCl_2$, CO and, from NO, N_2O and H_2O! The reaction is believed to occur via a CoN(H)O intermediate.

As noted in Section 4, the LUMO for a mixed carbonyl-nitrosyl is largely π^*{MCO} in character, and therefore excitation labilizes a CO, not an NO ligand. This fact has been exploited for the synthesis of nitrosyl complexes, particularly in the case of $[CpCr(CO)_2(NO)]$:

$$[CpCr(CO)_2(NO)] + L \xrightarrow{h\nu} [CpCr(CO)(L)(NO)] + CO \qquad (202)$$

where L is for instance Ph_3P, THF or cyclooctene.[135,418-425] This last ligand is thermally very labile, allowing many further ligands to be introduced. When reaction (202) is performed with L = NO both $[CpCr(NO_2)(NO)_2]$ and $[CpCr(\mu-NO)(NO)]_2$ are obtained.[135,502] This suggests that irradiation of carbonyl-nitrosyls in the presence of NO might be a viable route to elusive highly nitrosylated complexes, and in fact $[Mn(CO)(NO)_3]$ and $[Cr(NO)_4]$ have been so prepared.[426-429]

8.4   Transfer of the Nitrosyl Ligand

Two types of nitrosyl transfer are known. The first is intramolecular, and occurs in the fluxional behavior of nitrosyl-containing dimers or clusters. The best studied examples are $[CpCr(NO)_2]_2$ and $[CpMn(CO)(NO)]_2$.[249,430] For $[CpCr(NO)_2]_2$ cis-trans isomerization occurs:

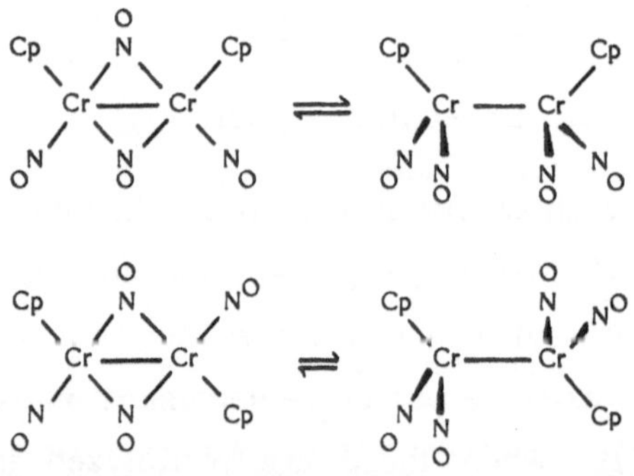

(203)

The cis $\rightleftharpoons$ trans equilibrium constant decreases with the polarity of the solvent, as would be expected since the cis isomer is the more polar of the two. The thermodynamic parameters for the cis $\rightleftharpoons$ trans conversion in toluere are ΔH^o = -8.8 kJ mol^{-1}, ΔS^o = -7.1 J °K^{-1} mol^{-1} and ΔG^o = -6.7

kJ mol^{-1} at 298°K. For [CpMn(CO)(NO)]$_2$ both <u>cis</u> and <u>trans</u> bridged isomers are disymmetric, and this introduces asymmetric forms of nitrosyl transfer:

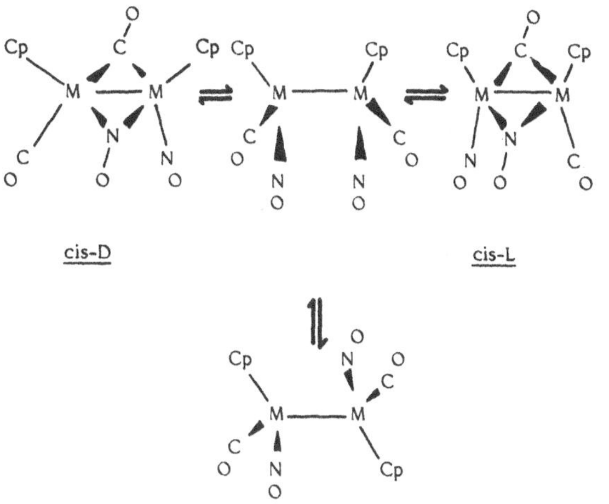

<u>cis-D</u>   <u>cis-L</u>   (204)

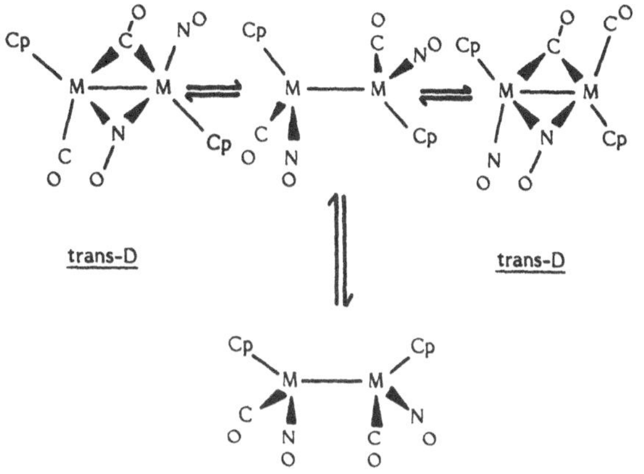

<u>trans-D</u>   <u>trans-D</u>   (205)

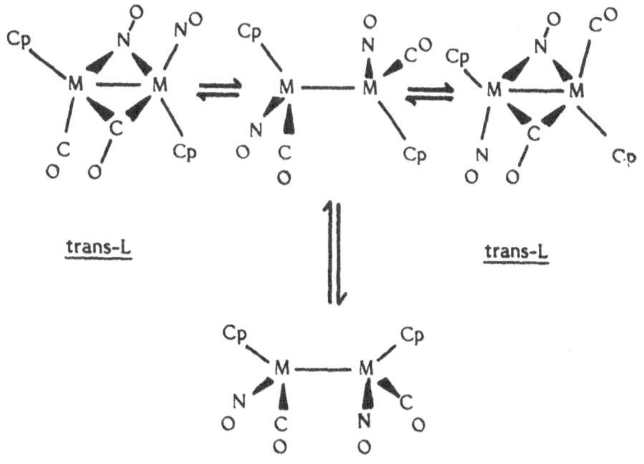

<u>trans-L</u>   <u>trans-L</u>   (206)

In addition an unsymmetrical transfer may be occurring via the intermediate[249]

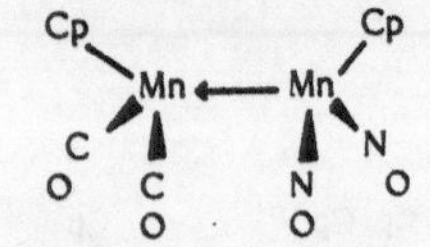

The clusters $[Os_3(CO)_{10}(NO)_2]$ and $[Os_3(CO)_9(NO)_2]$ and their derivatives, which have the basic structures[431-432]

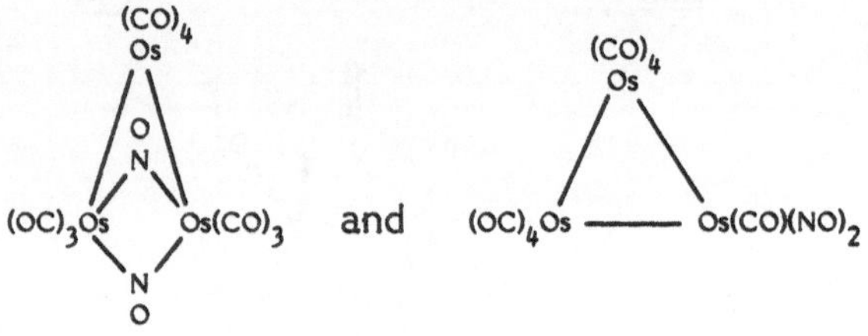

and

undergo both CO and NO migration. It seems likely that complexes having pairs of bridging nitrosyls will generally be fluxional, as is often the case for carbonyls.

The second type of nitrosyl transfer, intermolecular, was first observed in the reduction of $[Co(NH_3)_5(NO)]^{2+}$ by Cr^{2+}:

$$[Co(NH_3)_5(NO)]^{2+} + Cr^{2+} \rightarrow [Cr(H_2O)_5(NO)]^{2+} + Co^{2+} + 5NH_3 \quad (207)$$

though kinetic details were not given.[87] From a subsequent study of the related transfers[236]

$$[Co(H_2O)(en)_2(NO)]^{2+} + Cr^{2+} \rightarrow [Cr(H_2O)_5(NO)]^{2+} + 2en + Co^{2+} \quad (208)$$

$$[Cr(H_2O)_5(NO)]^{2+} + {}^*Cr^{2+} \rightleftharpoons [{}^*Cr(H_2O)_5(NO)]^{2+} + Cr^{2+} \quad (209)$$

it was suggested that Cr^{2+} effectively acts as a Lewis acid towards the nitrosyl ligand; $[Co(H_2O)(en)_2(NO)]^{2+}$ can in fact be protonated in aqueous solution. Although it was suggested that the site of protonation and Cr^{2+} attack is the oxygen atom[236] this must be doubted (see Section 5). In any event reactions (207) - (209) involve net transfer of NO (i.e. an N and an O nucleus plus 15 electrons), and since the starting nitrosyl contains a bent {MNO} group (formally coordinated NO⁻), transfer of one electron must take place in the {Co(NO)Cr} or {Cr(NO)Cr} intermediate.

Three extensive investigations of the transfer of NO from $[Co(dmgH)_2(NO)]^{433,478}$ and $[M(Ph_3P)_3(NO)]^{434}$ (M = Co or Rh) to other complexes have been made. The important results are several. Firstly, two types of nitrosyl transfer are observed:

$$M(NO) + M' \rightarrow M'(NO) + M \tag{210}$$

$$M(NO) + M'X \rightarrow MX + M'(NO) \tag{211}$$

Reaction (210) is simple nitrosyl transfer, exemplified by[433]

$$[Co(dmgH)_2(NO)] + [CoCl_2(Ph_3P)_2] \rightarrow [CoCl_2(Ph_3P)_2(NO)] + [Co(dmgH)_2] \tag{212}$$

and reaction (211) is nitrosyl/halide interchange, exemplified by[434]

$$[Rh(Ph_3P)_3(NO)] + [CoCl(Ph_3P)_3] \rightleftharpoons [RhCl(Ph_3P)_3] + [Co(Ph_3P)_3(NO)] \tag{213}$$

Secondly, NO does not dissociate from $[Co(dmgH)_2(NO)]$ under the conditions of reaction (212), and some of the products of nitrosyl transfer are not obtainable by the reaction of the receiving complex with free NO[433] (see however the reaction of $[Co(dmgH)_2(NO)]$ with hemoglobin A discussed below). Thirdly, in a formal sense NO may be transferred as NO (i.e. an N and an O nucleus plus 15 electrons) or as NO⁻ (i.e. an N and an O nucleus plus 16 electrons). Reactions (212) and (213) are examples of transfer of NO, and

$$[Co(dmgH)_2(NO)] + [Fe(dppe)_2(NO)]^+ \rightarrow [Fe(dppe)_2(NO)_2] + [Co(dmgH)_2]^+ \tag{214}$$

is transfer of NO⁻ [433]. Fourthly, the {CoNO} group is bent in

$[Co(dmgH)_2(NO)]$[433] but linear in $[M(Ph_3P)_3(NO)]$;[435] in both cases nitrosyl transfer requires formation of a $M(\mu\text{-}NO)M'$ bridge by electrophilic attack of M' at coordinated NO. Because of the necessity of bridge formation, nitrosyl transfer occurs only to complexes which are unsaturated or can lose a ligand in solution. Fifthly, where interchange of NO and X occurs, it appears that neutral NO and X, not charged versions of these ligands, are interchanged. A sequential interchange (NO first) occurs for the stereochemically rigid $[Co(dmgH)_2(NO)]$, but concerted double NO and Cl bridging may be possible where this would be stereochemically acceptable.

Nitrosyl transfer from $[Co(dmgH)_2(NO)]$, $[Co(tmgH)_2(NO)]$ ($tmgH_2$ = tetramethyleneglyoxime), and $[CoCl(en)_2(NO)]^+$ to hemoglobin A and myoglobin has also been studied in detail.[436,478] Kinetic results establish the mechanism:

$$[Co(dmgH)_2(NO)] + L \rightleftharpoons [LCo(dmgH)_2(NO)] \tag{215}$$

$$[Co(dmgH)_2(NO)] \rightleftharpoons [Co(dmgH)_2] + NO \tag{216}$$

$$[Co(dmgH)_2] + B \rightleftharpoons [BCo(dmgH)_2] \tag{217}$$

$$[Hb] + NO \rightarrow [Hb(NO)] \tag{218}$$

where L are accessible cysteine sulfhydryl groups or basic amino acids on hemoglobin (Hb) and B are basic substrates including cysteine residues. Note that the important step in the transfer of NO is the dissociation reaction (215) (cf. reaction (212) above, and its interpretation). Doyle and coworkers suggest that all nitrosyl transfer reactions proceed via a dissociation step and not via a bridge.[478] However it is possible that a bridging mechanism can operate when the steric constraints of hemoglobin are removed. Note however that free NO is observed in the rearrangement reaction

$$[CpMn(CO)I(NO)] + L \rightarrow [Cp_2Mn_2I(NO)_3] \tag{219}$$

where L is a Lewis base.[494] Nitrosylheme complexes act as nitrosyl transfer reagents.[437] The rather remarkable reaction[438]

$$2[Co(Sacsac)_2(NO)] \rightarrow [Co(Sacsac)(NO)_2] + [Co(Sacsac)_3] \qquad (220)$$

(Sacsac = dithioacetylacetonato), which must involve nitrosyl transfer, has been reported, as has the disproportionation of $[Fe(Ph_3P)_2(CO)(NO)]^+$ to $[Fe(Ph_3P)_2(NO)_2]$.[439]

Photochemically induced nitrosyl transfer appears to be a promising synthetic tool for the preparation of nitrosyls.[442, 509-12] An interesting example is the irradiation of a mixture of $[CpCo(\mu\text{-}NO)]_2$ and $[CpCo(CO)_2]$ which gives as the major product $[(CpCo)_2(\mu\text{-}CO)(\mu\text{-}NO)]$,[137] clearly showing that the bridging of NO is necessary for nitrosyl transfer. Photo-induced nitrosyl transfer between $[CoBr(NO)_2]_2$ and $[Cr(CO)_6]$ in THF gives $[CrBr_2(THF)_n(NO)]$;[441] between $[CoBr(NO)_2]_2$ and $[CpV(CO)_3L]$ (L = Ph_3P or CO) gives $[CpVL(NO)_2]$; between $[CoBr(Ph_3P)(NO)_2]$ and $[CpV(CO)_4]$ both NO and Ph_3P are exchanged, thus giving $[CpV(Ph_3P)(NO)_2]$.[442]

9.   MISCELLANEOUS REACTIONS OF NITROSYLS

When NO is passed through an alkaline solution (pH > 8.5) of $[Ru(NH_3)_6]^{3+}$ [443] or $[Ru(en)_3]^{3+}$ [444] high yields of $[Ru(NH_3)_5(N_2)]^{2+}$ and $[Ru(H_2O)(en)_2(N_2)]^{2+}$ respectively are obtained. Mechanistic investigation of the reaction involving $[Ru(NH_3)_6]^{3+}$ shows that the crucial intermediate is the seven-coordinate amido complex $[Ru(NH_3)_5(NH_2)(NO)]^{2+}$, with NO and NH_2 cis to one another.[445] Support for this intermediate is given by the known bimolecular attack of NO at $[Ru(NH_3)_6]^{3+}$ in acid solution[446] and the existence of $[Ru(NH_3)_4(NH_2)(NO)]^{2+}$.[164] Intramolecular coupling of the formally NO^+ complex ($\{RuNO\}^6$ but 20 electrons) with NH_2 leads to the N_2 ligand; an analogous intermolecular reaction[165] has been discussed previously (Section 4). The mechanism is supported by kinetic measurements. In the case of $[Ru(en)_3]^{3+}$ a similar mechanism would require C–N bond cleavage, driven by the thermodynamic stability of the N_2 ligand. An alternative mechanism is however available since O_2 oxidises $[Ru(en)_3]^{3+}$ to

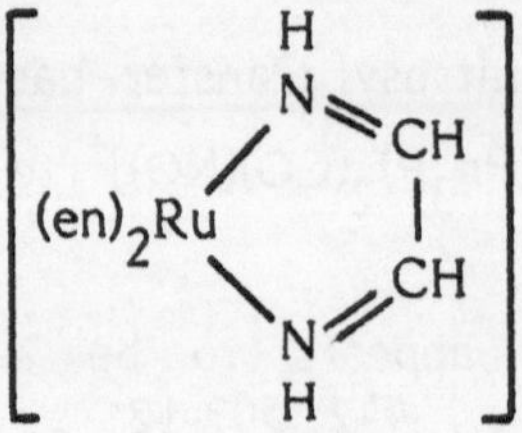

$$\left[(en)_2Ru \underset{\underset{H}{N}}{\overset{\overset{H}{N}}{\underset{\diagdown CH}{\diagup}}} \parallel \underset{\diagdown CH}{\diagup} \right]^{2+} \quad .447$$

A similar oxidation by NO, followed by coordination of further NO to the α-di-imine complex and nucleophilic attack of this ligand, rather than deprotonated ethylenediamine, at the NO^+ is therefore a viable route. The catalytic production of N_2O and N_2 from NO using $[Co(en)_2(NO_2)_2]^+$ in the presence of amines[448] may proceed via a similar mechanism, though metal-catalyzed decomposition of $R_2NN_2O_2^-$ formed by simple attack of free NO at the amine[449-451] must also be considered. Whether the formation of PhNNO in the reactions[452,453]

$$2[Fe(NPh_2)_2(dioxan)_2] + 6NO \rightarrow [Fe(NPh_2)_2(NO)_2]_2 + 2Ph_2NNO \qquad (221)$$

and

$$[Co(NPh_2)_2]_2 + 6NO \rightarrow [Co(NPh_2)_2(NO)_2]_2 + 2Ph_2NNO \qquad (222)$$

takes place by reaction of free or coordinated NO requires further investigation.

Oxygen atom interchange between coordinated NO_2 and NO is an interesting reaction which can be regarded as nucleophilic attack of an oxygen atom at a linearly coordinated NO. It was first observed in cis-$[Fe(S_2CN(CH_3)_2)_2(NO_2)(NO)]$, and proceeds via a symmetrical intermediate:

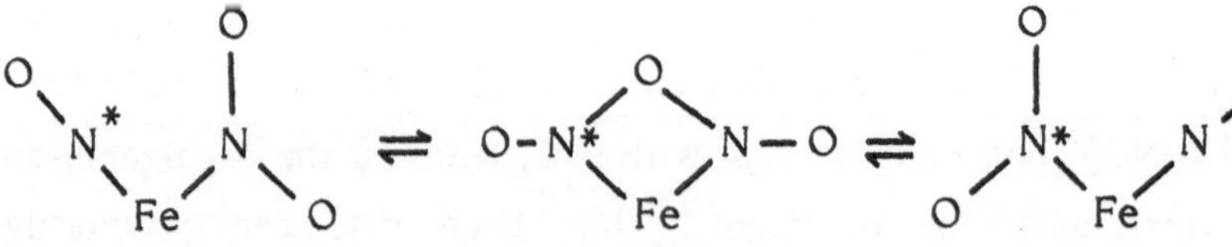

$$\underset{Fe}{\overset{O}{\underset{\diagup\quad\diagdown}{N^* \quad N}}} \overset{\overset{O}{|}}{\underset{\diagdown O}{}} \rightleftharpoons O-N^* \overset{O}{\underset{Fe}{\diagdown\diagup}} N-O \rightleftharpoons \underset{O}{\overset{O}{\underset{Fe}{\diagup \quad \diagdown}}} N^* \quad N \overset{O}{\diagdown O} \qquad (223)$$

As this implies, the reaction occurs only for the cis and not the trans isomer, and is independent of added NO or NO_2.[454-455] In the solid state the plane of the NO_2 ligand is perpendicular to the FeNO bond, and therefore exchange only takes place in solution.[455] The analogous Ru complex does not interchange oxygen; ν(NO) is at 1805 cm^{-1} versus 1857 cm^{-1} for the Fe analogue, indicated a higher electron density at the nitrosyl attached to ruthenium.[465]

Considering the reaction as nucleophilic attack of oxygen at coordinated nitrosyl suggests that electrophiles other than NO may be used, and the reaction of trans-[Ni(PEt$_3$)$_2$(NO$_2$)$_2$] with CO to give [Ni(PEt$_3$)$_2$(NO$_2$)(NO)] and CO_2 proceeds via the intermediate[457-9]

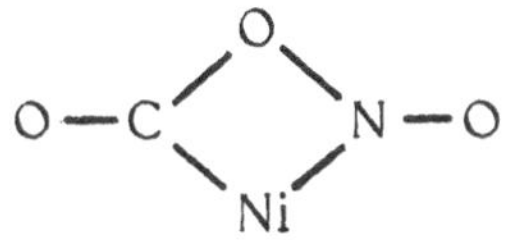

Intermolecular oxygen atom transfers from coordinated NO_2 to free CO,[460-461] coordinated CO,[214,243,251] coordinated CS,[251] coordinated NO_2[440] and free Ph$_3$P[243] have been used to prepare nitrosyls. The oxidation of olefins to aldehydes or alcohols to ketones by oxygen atom transfer from the coordinated NO_2 in [Co(py)(Saloph)(NO$_2$)] (Saloph = N,N'-bisalicylidene-o-phenylenediamino) or [Co(py)(TPP)(NO$_2$)] (Tpp = tetraphenylporphyin) can be made catalytic since the nitrosyl product can be reoxidised to the nitro complex by O_2.[146,363,422,479] This reoxidation is facile in the presence of Lewis acids, which are a necessity for a successful catalytic cycle.[479]

Reactions of considerable utility in organic synthesis are those of the weakly coordinated copper nitrosyls (formed in situ by reaction (187)) to give geminal dihalides:[355]

$$2RCH_2NH_2 + [CuX_2(NO)]_2 \rightarrow 2RCHX_2 + 4[CuX] + N_2 + 2HCl + 2H_2O \quad (224)$$

or N-nitrosamines:[83,238]

$$2R_2NH + 4NO + Cu^{2+} \rightarrow 2RNNO + N_2O + H_2O \quad (225)$$

$$2ArNH_2 + [CuCl_2(NO)]_2 \rightarrow [CuCl]_2 + 2ArCl + N_2 + H_2O \qquad (226)$$

Attack by alcohols also occurs giving alkyl nitrites, RONO.[83] The observations that N_3^- gives N_2 and N_2O, and NH_3 gives N_2, indicate that nucleophilic attack of the organic amine at the weakly coordinated formally NO^+ ligand (ν(NO) for these copper nitrosyls ranges from 1837 to 1865 cm^{-1} [355]) is responsible for the reactions. Silver(I) also catalyzes the formation of N-nitrosamines from NO and amines.[147]

It has been reported that $[Cp_2VI(NO)]$ decomposes in vacuo in tetrahydrofuran solution to give $\{(CpVI)_2[CpV(NO)]_2(\mu\text{-}O)_4\}$.[469,505] The structure of the vanadium-containing product was determined by X-ray diffraction, but the fate of the NO is unknown. No N_2O was produced. It has been suggested that the NO and Cp groups in $[CpV(CO)(NO)_2]$ can form pyridine under electron impact.[470]

The remarkable formation of the oxo complexes $[CpW(C_2H_4)(CH_3)(O)]$ and $[CpW(C_2H_4)(C(O)CH_3)(O)]$ on reaction of $[CpW(CO)_2(O=C(CH_3)CH=CH)$ with NO has been reported.[517]

A reaction related to the coupling of NO and $\eta^3\text{-}C_3H_5$ to form an oxime on nickel has recently been observed on ruthenium:[514]

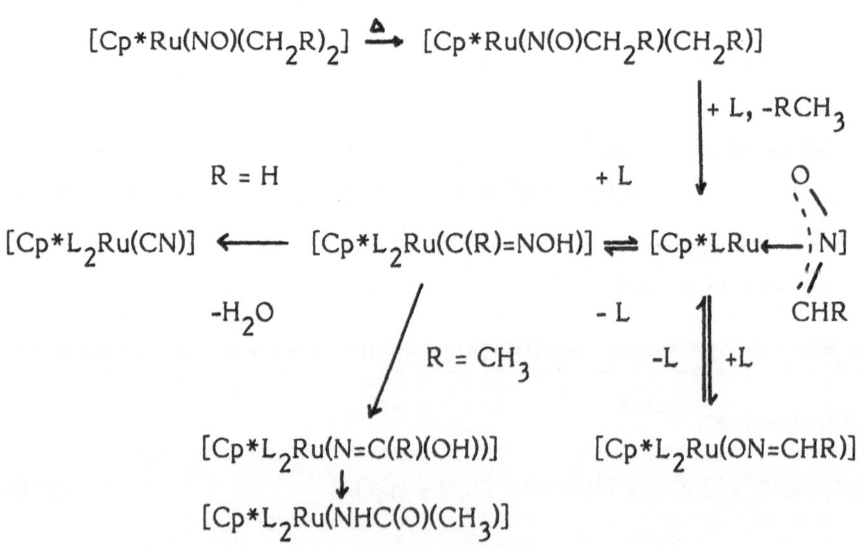

L = PMe$_3$, Cp* = $\eta^5\text{-}C_5(CH_3)_5$, R = H, CH$_3$

Acknowledgement

I wish to thank Carol Fairley for assistance in preparing and typing the manuscript for this Chapter.

REFERENCES

1. L. Playfair, Proc. Roy. Soc. 5:846 (1849).

2. C. Boedeker, Liebigs Annalen, 117:193 (1861).

3. R. L. Mond and A. E. Wallis, J. Chem. Soc. 121:32 (1922).

4. See for instance the reaction between $[CoBr(NO)_2]$ or $[FeI(NO)_2]_2$ and pyridine:
   W. Hieber and R. Màrin, Z. Anorg. Allg. Chem. 240:241 (1939);
   W. Hieber and R. Nast, Z. Anorg. Allg. Chem. 244:231 (1940);
   W. Hieber and K. Heinicke, Z. Naturforsch. 14B:819A (1959);
   W. Hieber and R. Kramolowsky, Z. Anorg. Allg. Chem. 321:94 (1963).

5. C. Claus, J. Prakt. Chem. 79:57 (1860); see also A. Werner, Chem. Ber. 40:2614 (1907).

6. F. Bottomley, Coord. Chem. Rev. 26:7 (1978).

7. J. H. Swinehart, Coord. Chem. Rev. 2:385 (1967).

8. J. Mašek, Inorg. Chim. Acta Rev. 3:99 (1969).

9. J. Wei, Adv. Catal. 24:57 (1975).

10. N. V. Sidgwick and R. W. Bailey, Proc. Roy. Soc. A144:521 (1934).

11. D. J. Hodgson and J. A. Ibers, Inorg. Chem. 7:2345 (1968).

12. J. H. Enemark and R. D. Feltham, Coord. Chem. Rev. 13:339 (1974).

13. R. Hoffman, M. M.-L. Chen, M. Elian, A. R. Rossi and D. M. P. Mingos, Inorg. Chem. 13:2666 (1974).

14. R. Hoffman, M. M.-L. Chen and D. L. Thorn, Inorg. Chem. 16:503 (1977).

15. N. H. Guiha, J. N. Cohn, E. Mikulic, J. A. Franciosa and C. J. Limas, New Engl. J. Med. 291:587 (1974).

16. T. Taylor, M. Styles and A. Lamming, Brit. J. Anaesth. 42:859 (1970).

17. J. H. Enemark and R. D. Feltham, Proc. Nat. Acad. Sci. USA, 69:3534 (1972).

18. B. F. G. Johnson and J. A. McCleverty, Prog. Inorg. Chem. 7:277 (1966).

19. N. G. Connelly, Inorg. Chim. Acta Rev. 47 (1972).

20. J. A. McGinnety, M.T.P. Int. Rev. Sci. Inorg. Chem. Ser. Cne 5:229 (1972).

21. B. A. Frenz and J. A. Ibers, M.T.P. Int. Rev. Sci. Phys. Chem. Ser. One 11:33 (1972).

22. K. G. Caulton, Coord. Chem. Rev. 14:317 (1975).

23. W. P. Griffith, Adv. Organomet. Chem. 7:211 (1968).

24. R. Eisenberg and C. D. Meyer, Acc. Chem. Res. 8:26 (1975).

25. F. Bottomley, Acc. Chem. Res. 11:158 (1978).

26. J. A. McCleverty, Chem. Rev. 79:53 (1979).

27. Yu. N. Kukushkin, L. I. Danilina and N. S. Patina, Soviet. J. Coord. Chem. 3:1127 (1977).

28. F. Bottomley and P. S. White, Acta Cryst. B35:2193 (1979).

29. J. H. Swinehart and P. A. Rock, Inorg. Chem. 5:573 (1966).

30. J. Mašek and H. Wendt, Inorg. Chim. Acta. 3:455 (1969).

31. C. Andrade and J. H. Swinehart, Inorg. Chim. Acta. 5:207 (1971).

32. A. N. Sergeeva, Z. I. Tkachenko and D. I. Semenishin, Russ. J. Inorg. Chem. 20:739 (1975).

33. N. E. Katz, M. A. Blesa, J. A. Olabe and P. J. Aymonino, J. Inorg. Nucl. Chem. 42:581 (1980).

34. H. Maltz, M. A. Grant and M. C. Navaroli, J. Org. Chem. 36:363 (1971).

35. R. Nast and J. Schmidt, Z. Naturforsch. 32B:469 (1977).

36. S. K. Wolfe, C. Andrade and J. H. Swinehart, Inorg. Chem. 13:2567 (1974).

37. S. Luňák and J. Vepřek-Šiška, Collect. Czech. Chem. Commun. 39:2719 (1974).

38. P. A. Rock and J. H. Swinehart, Inorg. Chem. 5:1078 (1966).

39. A. Müller and E. J. Baran, Angew. Chem. Int. Ed. Engl. 8:890 (1969).

40. D. Mulvey and W. A. Waters, J.C.S. Dalton Trans. 951 (1975).

41. J. H. Swinehart and W. G. Schmidt, Inorg. Chem. 6:232 (1967).

42. A. Ishigaki, M. Oue, Y. Matsushita, I. Masuda and T. Shono, Bull. Chem. Soc. Jpn. 50:726 (1977).

43. C. Andrade and J. H. Swinehart, Inorg. Chem. 11:648 (1972).

44. P. J. Morando, E. B. Borghi, L. M. de Schteingart and M. A. Blesa, J.C.S. Dalton Trans. 435 (1981).

45. A. B. Nikol'skii, Soviet J. Coord. Chem. 3:879 (1976).

46.   D. B. Brown, Inorg. Chem. 14:2582 (1975).

47.   H. Kruszyna, R. Kruszyna, J. Hurst and R. P. Smith, J. Toxicol. Environ. Health, 6:757 (1980).

48.   For instance: W. Wiegrebe and M. Villig, Z. Naturforsch. 36B:129 (1981).

49.   R. P. Smith and H. Kruszyna, J. Pharmacol. Exp. Ther. 191:557 (1974).

50.   F. Bottomley and F. Grein, J.C.S. Dalton Trans. 1359 (1980).

51.   P. T. Manoharan and H. B. Gray, J. Am. Chem. Soc. 87:3340 (1965).

52.   R. P. Cheney, M. G. Simic, M. Z. Hoffman, I. A. Taub and K.-D. Asmus, Inorg. Chem. 16:2187 (1977).

53.   W. L. Bowden, P. Bonnar, D. B. Brown and W. E. Geiger, Inorg. Chem. 16:41 (1977).

54.   M. C. R. Symons, D. X. West and J. G. Wilkinson, Inorg. Chem. 15:1022 (1976).

55.   J. D. W. van Voorst and P. Hemmerich, J. Chem. Phys. 45:3914 (1966).

56.   R. P. Cheney, S. D. Pell and M. Z. Hoffman, J. Inorg. Nucl. Chem. 41:489 (1979).

57.   J. Schmidt, H. Kühr, N. L. Dorn and J. Kopf, Inorg. Nucl. Chem. Lett. 10:55 (1974).

58.   J. Kopf and J. Schmidt, Z. Naturforsch. B32:275 (1977).

59.   J. Schmidt and W. Dorn, Inorg. Chim. Acta 16:223 (1976).

60.   T. W. Hawkins and M. B. Hall, Inorg. Chem. 19:1735 (1980).

61.   R. Nast and J. Schmidt, Angew. Chem. Int. Ed. Engl. 8:383 (1969).

62.   R. Nast and J. Schmidt, Z. Anorg. Allg. Chem 421:15 (1976).

63.   J. Mašek and J. Dempir, Collect. Czech. Chem. Commun. 34:727 (1969).

64.   J. Mašek, M. G. Bapat, B. Ćosović and J. Dempir, Collect. Czech. Chem. Commun. 34:485 (1969).

65.   J. Mašek and E. Mášlová, Collect. Czech. Chem. Commun. 39:2141 (1974)

66.   N.-G. Vannerberg and S. Jagner, Chem. Scripta 6:19 (1974).

67.   E. J. Baran and A. Müller, Chem. Ber. 102:3915 (1969).

68.   E. J. Baran and A. Müller, Z. Anorg. Allgem. Chem. 370:283 (1969).

69.   S. Jagner and N.-G. Vannerberg, Acta Chem. Scand. 24:1988 (1970).

70.   S. Jagner and E. Ljungström, Acta Cryst. B34:653 (1978).

71. D. I. Bustin and A. A. Vlček. Collect. Czech. Chem. Commun. 31:2374 (1966).

72. D. I. Bustin and A. A. Vlček, Collect. Czech. Chem. Commun. 32:1655 (1967).

73. S. Sarkar and A. Müller. Z. Naturforsch. 33B:1053 (1978).

74. J. Schmidt, Z. Anorg. Allg. Chem. 431:284 (1977).

75. J. Mašek and J. Dempir, Inorg. Chim. Acta 2:443 (1968).

76. W. M. Latimer, "Oxidation Potentials", 2nd Ed., p. 101, Prentice-Hall Inc., New Jersey (1952).

77. R. P. Cheney, M. Z. Hoffman and J. A. Lust, Inorg. Chem. 17:1177 (1978).

78. J. N. Armor and M. Z. Hoffman, Inorg. Chem. 14:444 (1975).

79. R. W. Callahan, G. M. Brown and T. J. Meyer, J. Am. Chem. Soc. 97:894 (1975).

80. R. W. Callahan and T. J. Meyer, Inorg. Chem. 16:574 (1977).

81. J. A. McCleverty, N. M. Atherton, J. Locke, E. J. Wharton and C. J. Winscom, J. Am. Chem. Soc. 89:6082 (1967).

82. J. N. Armor, R. Furman and M. Z. Hoffman, J. Am. Chem. Soc. 97:1737 (1975).

83. W. Brackman and P. J. Smit, Rec. Trav. Chim. Pays-Bas 84:372 (1965).

84. W. Brackman and P. J. Smit, Rec. Trav. Chim. Pays-Bas 85:857 (1966).

85. S. Clamp, N. G. Connelly, G. E. Taylor and T. S. Lauttit, J.C.S. Dalton Trans. 2162 (1980).

86. W. P. Griffith, J. Chem. Soc. 3286 (1963).

87. J. Armor, Inorg. Chem. 12:1959 (1973).

88. R. P. Cheney and J. N. Armor, Inorg. Chem. 16:3338 (1977).

89. M. J. Cleare and W. P. Griffith, J. Chem. Soc. (A). 1117 (1970).

90. M. Mukaida, Bull. Chem. Soc. Jpn. 43:3805 (1970).

91. J. N. Armor and M. Buchbinder, Inorg. Chem. 12:1086 (1973).

92. F. Bottomley, J.C.S. Dalton Trans. 1600 (1974).

93. D. A. Snyder and D. L. Weaver, Inorg. Chem. 9:2760 (1970).

94. C. S. Pratt, B. A. Coyle and J. A. Ibers, J. Chem. Soc. (A). 2146 (1971).

95. R. L. Roberts, D. W. Carlyle and G. L. Blackmer, Inorg. Chem. 14:2739 (1975).

96.   W. Hieber, R. Nast and E. Proeschel, Z. Anorg. Chem. 256:159 (1948).

97.   R. Nast and J. Föppl, Z. Anorg. Chem. 263:310 (1950).

98.   K. Wieghardt, W. Holzbach, B. Nuber and J. Weiss, Chem. Ber. 113:629 (1980).

99.   K. Wieghardt and W. Holzbach, Angew. Chem. Int. Ed. Engl. 18:549 (1979).

100. K. Wieghardt, U. Quilitzsch, B. Nuber and J. Weiss, Angew. Chem. Int. Ed. Engl. 17:351 (1978).

101. U. Quilitzsch and K. Wieghardt, Z. Naturforsch. 34B:640 (1979).

102. K. Wieghardt and U. Quilitzsch, Z. Anorg. Allg. Chem. 457:75 (1979).

103. K. Wieghardt and U. Quilitzsch, Z. Naturforsch. 36B:683 (1981).

104. H. Ogino, K. Tsukahara and N. Tanaka, Bull. Chem. Soc. Jpn. 48:3401 (1975).

105. R. Přibil, J. Mašek and A. A. Vlček, Inorg. Chim. Acta 5:57 (1971).

106. R. E. Dessy, J. C. Charkoudian and A. L. Rheingold, J. Am. Chem. Soc. 94:738 (1972).

107. G. Piazza, A. Foffani and G. Paliani, Z. Phys. Chem. N.F. 60:167 (1968).

108. R. Seeber, G. A. Mazzocchin, G. Albertin and E. Bordingnon, J. C.S. Dalton Trans. 979 (1979).

109. G. Paliani, S. M. Murgia and G. Cardaci, J. Organomet. Chem. 30:221 (1971).

110. B. F. G. Johnson, S. Bhaduri and N. G. Connelly, J. Organomet. Chem. 40:C36 (1972).

111. C. Coutoure, J. R. Morton, K. F. Preston and S. Strach, J. Mag. Res. 41:88 (1980).

112. J. R. Morton, K. F. Preston and S. J. Strach, J. Phys. Chem. 84:2478 (1980).

113. D. Ballivet-Tkatchenko, M. Riveccie and N. El Murr, J. Am. Chem. Soc. 101:2763 (1979).

114. D. Ballivet, C. Billard and I. Tkatchenko, Inorg. Chim. Acta. 25:L58 (1977).

115. D. Huchette, B. Thery and F. Petit, J. Mol. Catal. 4:433 (1978).

116. E. Leroy, D. Huchette, A. Mortreux and F. Petit, Nouv. J. Chim. 4:173 (1980).

117. A. Mortreux, J. C. Bavay and F. Petit, Nouv. J. Chim, 4:671 (1980).

118. D. Huchette, J. Nicole and F. Petit, Tetrahed. Lett. 1035 (1979).

119. E. LeRoy, F. Petit, J. Hennion and J. Nicole, Tetrahed. Lett. 2403 (1978).

120. I. Tkatchenko, J. Mol. Catal. 4:163 (1978).

121. G. Piazza and G. Paliani, Z. Phys. Chem. N. F. 71:91 (1970).

122. S. Bhaduri, K. Grundy and B. F. G. Johnson, J.C.S. Dalton Trans. 2085 (1977).

123. B. W. S. Kolthammer and P. Legzdins, J.C.S. Dalton Trans. 31 (1978).

124. N. Flitcroft, J. Organomet. Chem 15:254 (1968).

125. C. Y. Y. Chan and F. W. B. Einstein, Acta Cryst. B26:1899 (1970).

126. G. Hoch, H.-E. Sasse and M. L. Ziegler, Z. Naturforsch. 30B:704 (1975).

127. B. W. Hames, P. Legzdins and J. C. Oxley, Inorg. Chem. 19:1565 (1980).

128. R. G. Ball, B. W. Hames, P. Legzdins and J. Trotter, Inorg. Chem. 19:3626 (1980).

129. J. Müller and S. Schmitt, J. Organomet. Chem. 160:109 (1978).

130. J. L. Calderón, S. Fontana, E. Frauendorfer and V. W. Day, J. Organomet. Chem. 64:C10 (1974).

131. P. Legzdins and D. T. Martin, Inorg. Chem. 18:1250 (1979).

132. J. K. Hoyano, P. Legzdins and J. T. Malito, J.C.S. Dalton Trans. 1022 (1975).

133. W. P. Griffith, J. Lewis and G. Wilkinson, J. Chem. Soc. 872 (1959).

134. M. K. Lloyd and J. A. McCleverty, J. Organomet. Chem. 61:261 (1973).

135. B. W. Hames, P. Legzdins and D. T. Martin, Inorg. Chem. 17:3644 (1978).

136. J. Müller; quoted by J. A. McCleverty, ref. 26, p. 65.

137. J. Müller and S. Schmitt, J. Organomet. Chem. 97:C54 (1975).

138. S. Onaka, Inorg. Chem. 19:2132 (1980).

139. J. Müller, H. Dorner and F. H. Köhler, Chem Ber. 106:1122 (1973).

140. J. Müller, H. Dorner, G. Huttner and H. Lorenz, Angew. Chem. Int. Ed. Engl. 12:1005 (1973).

141. J. Muller and H. Dorner, Angew. Chem. Int. Ed. Engl. 12:834 (1973).

142. R. C. Elder, Inorg. Chem 13:1037 (1974).

143. R. A. Armstrong and H. Taube, Inorg. Chem. 15:1904 (1976).

144. K. Nakamoto, "Infrared Spectra of Inorganic and Coordination Compounds," 2nd Edn., Wiley-Interscience, Toronto, 1970, p. 78.

145. L. S. Liebeskind, K. B. Sharpless, R. D. Wilson and J. A. Ibers, J. Am. Chem. Soc. 100:7061 (1978).

146. J. V. Dubrawski and R. D. Feltham, Inorg. Chem. 19:355 (1980).

147. B. C. Challis and J. R. Outram, J.C.S. Chem. Commun. 707 (1978).

148. J. Muller and S. Schmitt, Z. Anorg. Allg. Chem. 426:77 (1976).

149. C. P. Casey, M. A. Andrews, D. R. McAlister and J. E. Rinz, J. Am. Chem. Soc. 102:1927 (1980).

150. N. G. Connelly and R. L. Kelly, J.C.S. Dalton Trans. 2334 (1974).

151. F. Bottomley, S. G. Clarkson and S.-B. Tong, J.C.S. Dalton Trans. 2344 (1974).

152. C. A. Reed and W. R. Roper, J.C.S. Dalton Trans. 1243 (1972).

153. S. Sueur, C. Bremard and G. Nowogrocki, C. R. Acad. Sci. Paris 281C:401 (1975).

154. S. Sueur and C. Bremard, Bull. Soc. Chim. France 961 (1975).

155. C. Bremard, M. Muller, G. Nowogrocki and S. Sueur, J.C.S. Dalton Trans. 2307 (1977).

156. C. Bremard, G. Nowogrocki and S. Sueur, Inorg. Chem. 18:1549 (1979).

157. C. Bremard, G. Nowogrocki and S. Sueur, Inorg. Chem. 17:3220 (1978).

158. S. Sueur, C. Bremard and G. Nowogrocki, J. Inorg. Nucl. Chem. 38:2037 (1976).

159. F. Abraham, G. Nowogrocki, S. Sueur and C. Bremard, Acta Cryst. B34:1466 (1978).

160. F. Abraham, G. Nowogrocki, S. Sueur and C. Bremard, Acta Cryst. B36:799 (1980).

161. F. Bottomley, J.C.S. Dalton Trans. 2148 (1972).

162. F. Bottomley and J. R. Crawford, Chem. Commun. 200 (1971).

163. F. Bottomley and J. R. Crawford, J. Am. Chem. Soc. 94:9092 (1972).

164. F. Bottomley and J. R. Crawford, J.C.S. Dalton Trans. 2145 (1972).

165. F. Bottomley, E. M. R. Kiremire and S. G. Clarkson, J.C.S. Dalton Trans. 1909 (1975).

166. F. Bottomley, S. G. Clarkson and E. M. R. Kiremire, J.C.S. Chem. Commun. 91 (1975).

167. K. Schug and C. P. Guengerich, J. Am. Chem. Soc. 101:235 (1979).

168.   C. P. Guengerich and K. Schug, Inorg. Chem. 17:1378 (1978).

169.   T. J. Meyer, J. B. Godwin and N. Winterton, Chem. Commun. 872 (1970). .

170.   J. B. Godwin and T. J. Meyer, Inorg. Chem. 10:2150 (1971).

171.   F. J. Miller and T. J. Meyer, J. Am. Chem. Soc. 93:1294 (1971).

172.   S. A. Adeyemi, F. J. Miller and T. J. Meyer, Inorg. Chem. 11:994 (1972).

173.   S. A. Adeyemi, E. C. Johnson, F. J. Miller and T. J. Meyer, Inorg. Chem. 12:2371 (1973).

174.   W. L. Bowden, W. F. Little and T. J. Meyer, J. Am. Chem. Soc. 95:5084 (1973).

175.   W. L. Bowden, W. F. Little and T. J. Meyer, J. Am. Chem. Soc. 99:4340 (1977).

176.   W. L. Bowden, W. F. Little and T. J. Meyer, J. Am. Chem. Soc. 96:5605 (1974).

177.   W. L. Bowden, W. F. Little and T. J. Meyer, J. Am. Chem. Soc. 98:444 (1976).

178.   J. L. Walsh, R. M. Bullock and T. J. Meyer, Inorg. Chem. 19:865 (1980).

179.   S. A. Adeyemi, J. N. Braddock, G. M. Brown, J. A. Ferguson, F. J. Miller and T. J. Meyer, J. Am. Chem. Soc. 94:300 (1972).

180.   F. Bottomley, W. V. F. Brooks, S. G. Clarkson and S.-B. Tong, J.C.S. Chem. Commun. 919 (1973).

181.   M. Mukaida, M. Yoneda and T. Nomura, Bull. Chem. Soc. Jpn. 50:3053 (1977).

182.   M. Mukaida, T. Nomura and T. Ishimari, Bull Chem. Soc. Jpn. 48:1443 (1975).

183.   P. G. Douglas, R. D. Feltham and H. G. Metzger, J. Am. Chem. Soc. 93:84 (1971).

184.   P. G. Douglas and R. D. Feltham, J. Am. Chem. Soc. 94:5254 (1972).

185.   P. G. Douglas, R. D. Feltham and H. G. Metzger, Chem. Commun. 889 (1970).

186.   F. Bottomley and E. M. R. Kiremire, J.C.S. Dalton Trans. 1125 (1977).

187.   F. Bottomley, J.C.S. Dalton Trans. 2538 (1975).

188.   S. H. Simonson and M. H. Mueller, J. Inorg. Nucl. Chem. 27:309 (1965).

189. J. Chatt, C. M. Elson, N. E. Hooper and G. J. Leigh, J.C.S. Dalton Trans. 2392 (1975).

190. C. M. Elson, J.C.S. Dalton Trans. 2401 (1975).

191. V. I. Nefedov, N. M. Sinitsyn, Ya. V. Salyn and L. Baier, Soviet J. Coord. Chem. 1:1332 (1975).

192. W. B. Hughes and B. A. Baldwin, Inorg. Chem. 13:1531 (1974).

193. B. Folkesson, Acta Chem. Scand. A28:491 (1974).

194. V. I. Nefedov, Ya. V. Salyn', I. B. Baranovskii and A. B. Nikol'skii, Russ. J. Inorg. Chem. 22:931 (1977).

195. Ya. V. Salyn', V. I. Nefedov, R. Taube and K. Seyferth, Soviet J. Coord. Chem. 3:1192 (1977).

196. P. Finn and W. L. Jolly, Inorg. Chem. 11:893 (1972).

197. M. N. Hughes and G. Stedman, J. Chem. Soc. 2824 (1963).

198. F. Bottomley and W. V. F. Brooks, Inorg. Chem. 16:501 (1977).

199. J. R. Perrott, G. Stedman and N. Uysal, J.C.S. Dalton Trans. 2058 (1976).

200 J. H. Boyer in "The Chemistry of the Nitro and Nitroso Groups", Part 1, H. Feuer, Ed., Interscience, New York, Chap. 5 (1968).

201. J. N. Armor, J. Inorg. Nucl. Chem. 35:2067 (1973).

202. I. P. Evans, G. W. Everett and A. M. Sargeson, J.C.S. Chem. Commun. 139 (1975).

203. F. Bottomley and S.-B. Tong, J.C.S. Dalton Trans. 217 (1973).

204. R. W. Adams, J. Chatt, N. E. Hooper and G. J. Leigh, J.C.S. Dalton Trans. 1075 (1974).

205. F. King and G. J. Leigh, J.C.S. Dalton Trans. 429 (1977).

206. L. Busetto, A. Palazzi, D. Pietropaolo and G. Dolcetti, J. Organomet. Chem. 66:453 (1974).

207. H. Brunner, Z. Anorg. Allg. Chem. 368:120 (1969).

208. E. O. Fischer, F. R. Kreissl, E. Winkler and C. G. Kreiter, Chem. Ber. 105:588 (1972).

209. R. B. King, M. B. Bisnette and A. Fronzaglia, J. Organomet. Chem. 4:256 (1965); idem, ibid. 5:341 (1966).

210. D. Messer, G. Landgraf and H. Behrens, J. Organomet. Chem. 172:349 (1979).

211. B. F. G. Johnson and J. A. Segal, J.C.S. Dalton Trans. 1268 (1972).

212. B. F. G. Johnson and J. A. Segal, J. Organomet. Chem. 31:C79 (1971).

213.   B. F. G. Johnson and J. A. Segal, J.C.S. Dalton Trans. 478 (1973).

214.   P. K. Baker, K. Broadley and N. G. Connelly, J.C.S. Chem. Commun. 775 (1980).

215.   C. P. Casey, M. A. Andrews and J. E. Rinz, J. Am. Chem. Soc. 101:741 (1979).

216   G. Dolcetti, L. Busetto and A. Palazzi, Inorg. Chem. 13:222 (1974).

217.   H. Schumann and M. Meissner, Z. Naturforsch. 35B:863 (1980).

218.   H. Schumann and M. Meissner, Z. Naturforsch. 35B:594 (1980).

219.   F. J. Regina and A. Wojcicki, Inorg. Chem. 19:3803 (1980).

220.   A. G. Constable and J. A. Gladysz, J. Organomet. Chem. 202:C21 (1980).

221.   D. M. Birney, A. M. Crane and D. A. Sweigart, J. Organomet. Chem. 152:187 (1978).

222.   A. Efraty, S. S. Sandhu, R. Bystrek and D. Z. Denney, Inorg. Chem. 16:2522 (1977).

223.   A. Efraty, D. Liebman, J. Sikora and D. Z. Denney, Inorg. Chem. 15:886 (1976).

224.   J. A. Potenza, R. Johnson, D. Williams, B. H. Toby, R. A. Lalancette and A. Efraty, Acta Cryst. B37:442 (1981).

225.   E. O. Fischer and H.-J. Beck, Chem. Ber. 104:3101 (1971).

226.   B. L. Haymore and J. A. Ibers, Inorg. Chem. 14:2610 (1975).

227.   K. Jäger and A. Henglein, Z. Naturforsch. 22A:700 (1967).

228.   J. A. D. Stockdale, R. N. Compton, G. S. Hurst and P. W. Reinhardt, J. Chem. Phys. 50:2176 (1969).

229.   K. R. Grundy, C. A. Reed and W. R. Roper, Chem. Commun. 1501 (1970).

230.   C. A. Reed and W. R. Roper, J. Chem. Soc. (A) 3054 (1970).

231.   V. G. Albano, P. Bellon and M. Sansoni, J. Chem. Soc. (A) 2420 (1971).

232.   J. H. Enemark, R. D. Feltham, J. Riker-Nappier and K. F. Bizot, Inorg. Chem. 14:624 (1975).

233.   G. LaMonica, M. Freni and S. Cenini, J. Organomet. Chem. 71:57 (1974).

234.   T. Tatsumi, K. Sekizawa and H. Tominaga, Bull. Chem. Soc. Jpn. 53:2297 (1980).

235.   R. D. Wilson and J. A. Ibers, Inorg. Chem. 18:336 (1978).

236.   R. L. Roberts, D. W. Carlyle and G. L. Blackmer, Inorg. Chem. 14:2739 (1975).

237. D. M. P. Mingos and J. A. Ibers, Inorg. Chem. 10:1479 (1971).

238. W. Brackman and P. J. Smit, Rec. Trav. Chim. Pays-Bas 84:357 (1965).

239. D. A. Muccigrosso, S. E. Jacobson, P. A. Apgar and F. Mares, J. Am. Chem. Soc. 100:7063 (1978).

240   K. Wieghardt, W. Holzbach, J. Weiss, B. Nuber and B. Prikner, Angew. Chem. Int. Ed. Engl. 18:548 (1979).

241. K. Wieghardt, E. Hofer, W. Holzbach, B. Nuber and J. Weiss, Inorg. Chem. 19:2927 (1980).

242. K. Wieghardt, W. Holzbach, E. Hofer and J. Weiss, Inorg. Chem. 20:343 (1981).

243. K. R. Grundy, K. R. Laing and W. R. Roper, J.C.S Chem. Commun. 1500 (1970).

244. D. M. P. Mingos and J. A. Ibers, Inorg. Chem. 10:1035 (1971).

245. G. R. Clark, J. M. Waters and K. R. Whittle, Inorg. Chem. 13:1628 (1974).

246. S. T. Wilson and J. A. Osborn, J. Am. Chem. Soc. 93:3068 (1971).

247. A. E. Crease and P. Legzdins, J.C.S. Chem. Commun. 268 (1972).

248. A. E. Crease and P. Legzdins, J.C.S. Dalton Trans. 1501 (1973).

249. R. M. Kirchner, T. J. Marks, J. S. Kristoff and J. A. Ibers, J. Am. Chem. Soc. 95:6602 (1973).

250. D. F. Shriver and J. Posner, J. Am. Chem. Soc. 88:1672 (1966).

251. A. Efraty, R. Arneri and J. Sikora, J. Organomet. Chem. 91:65 (1975).

252. A. Dobson and S. D. Robinson, J. Organomet. Chem. 99:C63 (1975).

253. S. G. Clarkson and F. Basolo, Inorg. Chem. 12:1528 (1973).

254. M. Tamaki, I. Masuda and K. Shinra, Bull. Chem. Soc. Jpn. 45:171 (1972).

255. W. Trogler and L. G. Marzilli, Inorg. Chem. 13:1008 (1974).

256. K. R. Laing and W. R. Roper, Chem. Commun. 1568 (1968).

257. R. Ugo, S. Bhaduri, B. F. G. Johnson, A. Khair, A. Pickard and Y. Benn-Taarit, J.C.S. Chem. Commun. 694 (1976).

258. B. F. G. Johnson and S. Bhaduri, J.C.S. Chem. Commun. 650 (1973).

259. K. R. Laing and W. R. Roper, Chem. Commun. 1556 (1968).

260. M. Kubota and D. A. Phillips, J. Am. Chem. Soc. 97:5637 (1975).

261. A. P. Gaughan, B. J. Corden, R. Eisenberg and J. A. Ibers, Inorg. Chem. 13:786 (1974).

262. W. M. Coleman and L. T. Taylor, J. Am. Chem. Soc. 100:1705 (1978).

263. D. C. Moody and R. R. Ryan, J.C.S. Chem. Commun. 503 (1976).

264. W.-K. Wong, W. Tam and J. A. Gladysz, J. Am. Chem. Soc. 101:5440 (1979).

265. J. Valentine, D. Valentine and J. P. Collman, Inorg. Chem. 10:219 (1971).

266. P. Gans, J. Chem. Soc. (A). 943 (1967).

267. C. A. Reed and W. R. Roper, Chem. Commun. 155 (1969).

268. M. Kubota, C. A. Koerntgen and G. W. McDonald, Inorg. Chim. Acta 30:119 (1978).

269. J. A. McCleverty, C. W. Ninnes and I. Wolochowicz, J.C.S. Chem. Commun. 1061 (1976).

270. M. W. Schoonover, E. C. Baker and R. Eisenberg, J. Am. Chem. Soc. 101:1880 (1978).

271. S. Bhaduri and G. M. Sheldrick, Acta Cryst. B31:897 (1975).

272. M. W. Schoonover and R. Eisenberg, J. Am. Chem. Soc. 99:8371 (1977).

273. M. W. Schoonover, C. P. Kubiak and R. Eisenberg, Inorg. Chem. 17:3050 (1978).

274. N. A. Bailey, W. G. Kita, J. A. McCleverty, A. J. Murray, B. E. Mann and N. W. J. Walker, J.C.S. Chem. Commun. 592 (1974).

275. J. W. Faller and A. M. Rosan, J. Am. Chem. Soc. 98:3388 (1976).

276. R. D. Adams, D. F. Chodosh, J. W. Faller and A. M. Rosan, J. Am. Chem. Soc. 101:2570 (1979).

277. J. W. Faller, D. F. Chodosh and D. Katahira, J. Organomet. Chem. 187:227 (1980).

278. T. J. Greenhough, P. Legzdins, D. T. Martin and J. Trotter, Inorg. Chem. 18:3268 (1979).

279. J. A. McCleverty and A. J. Murray, J.C.S. Dalton Trans. 1424 (1979).

280. M. M. Hunt, W. G. Kita, B. E. Mann and J. A. McCleverty, J.C.S. Dalton Trans. 467 (1978).

281. M. M. Hunt, W. G. Kita and J. A. McCleverty, J.C.S. Dalton Trans. 474 (1978).

282. M. M. Hunt and J. A. McCleverty, J.C.S. Dalton Trans. 480 (1978).

283. F. M. Chaudhari, G. R. Knox and P. L. Pauson, J. Chem. Soc. (C). 2255 (1967).

284. G. Cardaci, J. Organomet. Chem. 202:C81 (1980).

285. G. Cardaci, J.C.S. Dalton Trans. 1808 (1974).

286. G. Cardaci, J.C.S. Dalton Trans. 2452 (1974).

287. P. K. Baker and N. G. Connelly, J. Organomet. Chem. 179:C33 (1979).

288. R. A. Clement, U. Klabunde and G. W. Parshall, J. Mol. Catal. 4:87 (1978).

289. V. Albano, P. L. Bellon and G. Ciani, J. Organomet. Chem. 38:155 (1972).

290. M. Foà and L. Cassar, J. Organomet. Chem. 30:123 (1971).

291. H. Brunner and S. Loskot, Angew. Chem. Int. Ed. Engl. 10:515 (1971).

292. H. Brunner and S. Loskot, J. Organomet. Chem. 61:401 (1973).

293. G. Evrard, R. Thomas, B. R. Davis and I. Bernal, J. Organomet. Chem. 124:59 (1977).

294. P. N. Becker, M. A. White and R. G. Bergman, J. Am. Chem. Soc. 102:5676 (1980).

295. R. J. H. Clark, J. A. Stockwell and J. D. Wilkins, J.C.S. Dalton Trans. 120 (1976).

296. P. C. Wailes, H. Weigold and A. P. Bell, J. Organomet. Chem. 34:135 (1972).

297. R. A. Middleton and G. Wilkinson, J.C.S. Dalton Trans. 1888 (1980).

298. J. D. Wilkins and M. G. B. Drew, J. Organomet. Chem. 69:111 (1974).

299. S. R. Fletcher, A. Shortland, A. C. Skapski and G. Wilkinson, J.C.S. Chem. Commun. 922 (1972).

300. A. Shortland and G. Wilkinson, J.C.S. Dalton Trans. 872 (1973).

301. P. Edwards, K. Mertis, G. Wilkinson, M. B. Hursthouse and K. M. A. Malik, J.C.S. Dalton Trans. 334 (1980).

302. S. Amirkhalili, P. B. Hitchcock, J. D. Smith and J. G. Stamper, J.C.S. Dalton Trans. 2493 (1980).

303. S. Amirkhalili, A. J. Conway and J. D. Smith, J. Organomet. Chem. 149:407 (1978).

304. J. Sand and F. Singer, Justus Liebigs Ann. Chem. 329:190 (1903)

305. E. Frankland, Justus Liebigs Ann. Chem. 99:345 (1856).

306. M. H. Abraham, J. H. N. Garland, J. A. Hill and L. F. Larkworthy, Chem. Ind. 1615 (1962).

307.   S. J. Brois, Tet. Lett. 345 (1964).

308.   R. F. Clarke, G. W. A. Fowles and D. A. Rice, J. Organomet. Chem.
        74:417 (1974).

309.   S. R. Fletcher and A. C. Skapski, J. Organomet. Chem. 59:299
        (1973).

310.   E. Müller and H. Metzger, Chem. Ber. 89:396 (1956).

311.   K. Mertis and G. Wilkinson, J.C.S. Dalton Trans. 1488 (1976).

312.   L. Galyer, K. Mertis and G. Wilkinson, J. Organomet. Chem. 85:C37
        (1975).

313.   K. Mertis, D. H. Williamson and G. Wilkinson, J.C.S. Dalton Trans.
        607 (1975).

314.   H.-F. Klein and H. H. Karsch, Chem. Ber. 109:1453 (1976).

315.   M. Herberhold and H. Kratzer, Z. Naturforsch. 32B:1263 (1977).

316.   I. Masuda, M. Tamaki and K. Shinra, Bull. Chem. Soc. Jpn. 42:157
        (1969).

317.   D. A. White, J. Chem. Soc. (A) 233 (1971).

318.   A. T. McPhail, G. R. Knox, D. G. Robertson and G. A. Sim,
        J. Chem. Soc. (A) 205 (1971).

319.   S. Bhaduri, B. F. G. Johnson, C. J. Savory, J. A. Segal and R. H.
        Walters, J.C.S. Chem. Commun. 809 (1974).

320.   S. A. Bhaduri, I. Bratt, B. F. G. Johnson, A. Khair, J. A. Segal, R.
        Walters and C. Zuccaro, J.C.S. Dalton Trans. 234 (1981).

321.   F. Bottomley and H. H. Brintzinger, J.C.S. Chem. Commun. 234
        (1978).

322.   F. Bottomley and I. J. B. Lin, J.C.S Dalton Trans. 271 (1981).

323.   F. Bottomley and J. Darkwa, J.C.S. Dalton Trans. 1435 (1985).

324.   T. L. Nunes and R. E. Powell, Inorg. Chem. 9:1912 (1970).

325.   T. L. Nunes and R. E. Powell, Inorg. Chem. 9:1916 (1970).

326.   W. Hieber and K. Heinicke, Z. Anorg. Allg. Chem. 316:321 (1962).

327.   B. L. Haymore and J. A. Ibers, J. Am. Chem. Soc. 96:3325 (1974).

328.   J. A. Kaduk and J. A. Ibers, Inorg. Chem. 14:3070 (1975).

329.   S. Bhaduri and B. F. G. Johnson, Transition Met. Chem. 3:156 (1978).

330.   J. Reed and R. Eisenberg, Science 184:568 (1974).

331.   C. D. Meyer and R. Eisenberg, J. Am. Chem. Soc. 98:1364 (1976).

332.   D. E. Hendriksen, C. D. Meyer and R. Eisenberg, Inorg. Chem.
        16:970 (1977).

333. D. E. Hendriksen and R. Eisenberg, J. Am. Chem. Soc. 98:4662 (1976).

334. R. L. Martin and D. Taylor, Inorg. Chem. 15:2970 (1976).

335. S. Cenini, R. Ugo, G. LaMonica and S. D. Robinson, Inorg. Chim. Acta 6:182 (1972).

336. S. Bhaduri, B. F. G. Johnson, A. Pickard, P. R. Raithby, G. M. Sheldrick and C. I. Zuccaro, J.C.S. Chem. Commun. 354 (1977).

337. S. Bhaduri, B. F. G. Johnson, A. Khair, I. Ghatak and D. M. P. Mingos, J.C.S. Dalton Trans. 1572 (1980).

338. C.-S. Chin, M. S. Sennett, P. T. Wier and L. Vaska, Inorg. Chim. Acta 31:L443 (1978).

339. M. Kubota, K. J. Evans, C. A. Koerntgen and J. C. Marsters, J. Am. Chem. Soc. 100:342 (1978).

340. J. A. Connor and P. I. Riley, J.C.S. Dalton Trans. 1231 (1979).

341. A. Werner and P. Karrer, Helv. Chim. Acta 1:54 (1918).

342. B. F. Hoskins, F. D. Whillans, D. H. Dale and D. C. Hodgkin, J.C.S. Chem. Commun. 69 (1969).

343. R. Bau, I. H. Sabherwal and A. B. Burg. J. Am. Chem. Soc. 93:4926 (1971).

344. J.-J. Salzmann, Helv. Chim. Acta 51:903 (1968).

345. B. B. Wayland and J. V. Minkiewicz, J.C.S. Chem. Commun. 1015 (1976).

346. R. D. Feltham, W. Silverthorn and G. McPherson, Inorg. Chem. 8:344 (1969).

347. M. Gargano, P. Giannoccaro, M. Rossi, A. Sacco and G. Vasapollo, Gazz. Chim. Ital. 105:1279 (1975).

348. M. Rossi and A. Sacco, J.C.S. Chem. Commun. 694 (1971).

349. W. B. Hughes, J.C.S. Chem. Commun. 1126 (1969).

350. J. Kiji, S. Yoshikawa and J. Furukawa, Bull. Chem. Soc. Jpn. 43:3614 (1970).

351. E. Miki, K. Mizumachi and T. Ishimori, Bull. Chem. Soc. Jpn. 48:2975 (1975).

352. Yu. N. Kukushkin, L. I. Danilina and M. M. Singh, Russ. J. Inorg. Chem. 16:1449 (1971).

353. J. Chatt, D. P. Melville and R. L. Richards, J. Chem. Soc. (A) 1169 (1971).

354. M. A. A. F. de C. T. Carrondo, P. R. Rudolf, A. C. Skapski, J. R. Thornback and G. Wilkinson, Inorg. Chim. Acta 24:L95 (1977).

355.   M. P. Doyle, B. Siegfried and J. J. Hammond, J. Am. Chem. Soc. 98:1627 (1976).

356.   C. E. Strouse and B. I. Swanson, Chem. Commun. 55 (1971).

357.   R. J. Puddephatt and P. J. Thompson, J.C.S. Dalton Trans. 2091 (1976).

358.   R. D. Feltham and J. T. Carriel, Inorg. Chem. 3:121 (1964).

359.   E. Miki, Chem. Lett. 835 (1980).

360.   D. Gwost and K. G. Caulton, Inorg. Chem. 13:414 (1974).

361.   J. A. Broomhead and J. R. Budge, Inorg. Chem. 17:2414 (1978).

362.   J. A. Broomhead and J. R. Budge, Aust. J. Chem. 32:1187 (1979).

363.   B. S. Tovrog, F. Mares and S. E. Diamond, J. Am. Chem. Soc. 102:6616 (1980).

364.   A. B. Nikol'skii, Soviet J. Coord. Chem. 3:879 (1977).

365.   F. A. Palocsay and J. V. Rund, Inorg. Chem. 8:696 (1969).

366.   D. F. Keeley and R. E. Johnson, J. Inorg. Nucl. Chem. 11:33 (1959).

367.   A. Cardaci, A. Foffani, G. Distefano and G. Innorta, Inorg. Chim. Acta 1:340 (1967).

368.   E. M. Thorsteinson and F. Basolo, J. Am. Chem. Soc. 88:3929 (1966).

369.   A. Foffani, S. Pignatoro, G. Distefano and G. Innorta, J. Organomet. Chem. 7:473 (1967).

370.   D. E. Morris and F. Basolo, J. Am. Chem. Soc. 90:2531 (1968).

371.   B. J. Plankey and J. V. Rund, Inorg. Chem. 18:957 (1979).

372.   G. Innorta, S. Torroni and A. Foffani, J. Organomet. Chem. 66:459 (1974).

373.   A. B. Nikol'skii and A. Yu. Ershov, Russ. J. Phys. Chem. 51:1069 (1977).

374.   O. A. Ileperuma and R. D. Feltham, Inorg. Chem. 16:1876 (1977).

375.   K. Kustin, I. A. Taub and E. Weinstock, Inorg. Chem. 5:1079 (1966).

376.   Y. Hishinuma, R. Kaji, H. Akinoto, F. Nakajina, T. Mori, T. Kamo, Y. Arikawa and S. Nozawa, Bull. Chem. Soc. Jpn. 52:2863 (1979).

377.   L. Vaska and H. Nakai, J. Am. Chem. Soc. 95:5431 (1973).

378.   B. B. Wayland and L. W. Olson, J. Am. Chem. Soc. 96:6037 (1974).

379.   W. R. Scheidt and M. E. Frisse, J. Am. Chem. Soc. 97:17 (1975).

380.   J. C. W. Chien, J. Am. Chem. Soc. 91:2166 (1969).

381.   B. B. Wayland, L. W. Olson and Z. U. Siddiqui, J. Am. Chem. Soc. 98:94 (1976).

382. B. Jeżowska-Trzebiatowska, K. Gerega and A. Vogt, Inorg. Chim. Acta 31:183 (1978).

383. B. Jeżowska-Trzebiatowska, K. Gerega and G. Formicka-Kozłowska, Inorg. Chim. Acta 40:187 (1980).

384. P. Silvestroni and L. Ceciarelli, J. Am. Chem. Soc. 83:3905 (1961).

385. A. Vlček and A. A. Vlček, Inorg. Chim. Acta 9:165 (1974).

386. R. T. M. Fraser and W. E. Dasent, J. Am. Chem. Soc. 82:348 (1960).

387. J. P. Wilshire and D. T. Sawyer, J. Am. Chem. Soc. 100:3972 (1978).

388. Yu. N. Kukushkin and M. M. Singh, Russ. J. Inorg. Chem. 14:1670 (1969).

389. D. M. P. Mingos and J. A. Ibers, Inorg. Chem. 9:1105 (1970).

390. Yu. N. Kukushkin, L. I. Danilina and A. I. Osokin, Soviet J. Coord. Chem. 4:322 (1978).

391. J. Kiji, S. Yoshikawa and J. Furukawa, Bull. Chem. Soc. Jpn. 43:3614 (1970).

392. J. Smidt and R. Jira, Chem. Ber. 93:162 (1960).

393. G. Albertin, E. Bordingnon, L. Canovese and A. A. Oro, Inorg. Chim. Acta 38:77 (1980).

394. W. Beck, H. G. Fick, K. Lottes and K. H. Schmidtner, Z. Anorg. Allg. Chem. 416:97 (1975).

395. R. B. King, Inorg. Chem. 7:90 (1968).

396. R. P. Stewart and G. T. Moore, Inorg. Chem. 14:2699 (1975).

397. E. O. Fischer and H. Strametz, J. Organomet. Chem. 10:323 (1967).

398. N. G. Connelly, P. T. Draggett, M. Green and T. A. Kuc, J.C.S. Dalton Trans. 70 (1977).

399. N. G. Connelly, M. Green and T. A. Kuc, J.C.S. Chem. Commun. 542 (1974).

400. M. Ghedini, G. Dolcetti, O. Gandolfi and B. Giovannitti, Inorg. Chem. 15:2385 (1976).

401. P. B. Hitchcock, M. F. Lappert, P. L. Pye and S. Thomas, J.C.S. Dalton Trans. 1929 (1979).

402. M. F. Lappert and P. L. Pye, J.C. S. Dalton Trans. 837 (1978).

403. L. M. Baltusis, K. D. Karlin, H. N. Rabinowitz and J. C. Dewan, Inorg. Chem. 19:2627 (1980).

404. W. Hieber and K. Heinicke, Z. Anorg. Allg. Chem. 316:305 (1962).

405. K. K. Pandey and U. C. Agarwala, J. Inorg. Nucl. Chem. 42:293 (1980).

406.   W. Hieber and K. Heinicke, Z. Anorg. Allg. Chem. 316:321 (1962).

407.   W. Hieber and W. Beck, Z. Naturforsch. 13B:194 (1958).

408.   S. K. Wolfe and J. H. Swinehart, Inorg. Chem. 14:1049 (1975).

409.   P.-H. Liu and J. I. Zink, Inorg. Chem. 16:3165 (1977).

410.   A. B. Nikol'skii, A. M. Popov and I. V. Vasilevskii, Soviet J. Coord. Chem. 2:508 (1976).

411.   O. V. Sizova, N. G. Antonov, A. B. Nikol'skii and V. I. Baranovskii, Soviet J. Coord. Chem. 2:853 (1976).

412.   A. B. Nikol'skii and A. M. Popov, Dokl. Akad. Nauk SSSR 250:105 (1980).

413.   F. Seel and G.-V. Röschenthaler, Angew. Chem. Int. Ed. Engl. 9:166 (1970).

414.   F. Seel and G.-V. Röschenthaler, Z. Anorg. Allg. Chem. 386:297 (1971).

415.   O. Crichton, M. Poliakoff, A. J. Rest and J. J. Turner, J.C.S. Dalton Trans. 1321 (1973).

416.   O. Crichton and A. J. Rest, J. C. S. Dalton Trans. 986 (1977).

417.   H. Brunner, Chem Ber. 102:305 (1969).

418.   M. Herberhold and P. D. Smith, Angew. Chem. Int. Ed. Engl. 18:631 (1979).

419.   M. Herberhold and H. Alt, J. Organomet. Chem. 42:407 (1972).

420.   O. Crichton and A. J Rest, J. C. S. Dalton Trans. 656 (1977).

421.   H. Brunner, J. Organomet. Chem. 16:119 (1969).

422.   B. S. Tovrog, S. E. Diamond and F. Mares, J. Am. Chem. Soc. 101:270 (1979).

423.   M. Herberhold and H. Alt, J. Organomet. Chem. 42:407 (1972).

424.   M. Herberhold and H. Alt, Liebigs Ann. Chem. 292 (1976).

425.   M. Herberhold, H. Alt and C. G. Kreiter, Liebigs Ann. Chem. 300 (1976).

426.   M. Herberhold and A. Razavi, J. Organomet. Chem. 67:81 (1974).

427.   M. Herberhold and A. Razavi, Angew. Chem. Int. Ed. Engl. 11:1092 (1972).

428.   B. I. Swanson and S. K. Satija, J. C. S. Chem. Commun. 40 (1973).

429.   M. Herberhold and L. Haumaier, J. Organomet. Chem. 160:101 (1978).

430.   T. J. Marks and J. S. Kristoff, J. Organomet. Chem. 42:C91 (1972).

431.   J. R. Norton, J. P. Collman, G. Dolcetti and W. T. Robinson, Inorg. Chem. 11:382 (1972).

432.   S. Bhaduri, B. F. G. Johnson, J. Lewis, D. J. Watson and C. Zuccaro, J. C. S. Dalton Trans. 557 (1979).

433.   C. B. Ungermann and K. G. Caulton, J. Am. Chem. Soc. 98:3862 (1976).

434.   A. Sacco, G. Vasapollo and P. Giannoccaro, Inorg. Chim. Acta 32:171 (1979).

435.   B. L. Haymore and J. A. Ibers, Inorg. Chem. 14:3060 (1975).

436.   M. P. Doyle, F. J. Van Doornik and C. L. Funckes, Inorg. Chim. Acta 46:L111 (1980).

437.   R. Bonnett, A. A. Charalambides, R. A. Martin, K. D. Sales and B. W. Fitzsimmons, J. C. S. Chem. Commun. 884 (1975).

438.   A. R. Hendrickson, R. K. Y. Ho and R. L. Martin, Inorg. Chem. 13:1279 (1974).

439.   J. L. A. Roustan, J. Y. Merour and A. Forgues, J. Organomet. Chem. 186:C23 (1980).

440.   F. R. Keene, D. J. Salmon and T. J. Meyer, J. Am. Chem. Soc. 99:2384 (1970).

441.   J. Schmidt and D. Rehder, Chem. Lett. 933 (1976).

442.   F. Näumann and D. Rehder, J. Organomet. Chem. 204:411 (1981).

443.   S. Pell and J. N. Armor, J. Am. Chem. Soc. 94:686 (1972).

444.   S. Pell and J. N. Armor, J. C. S. Chem. Commun. 259 (1974).

445.   S. D. Pell and J. N. Armor, J. Am. Chem. Soc. 95:7625 (1973).

446.   J. N. Armor, H. A. Scheidegger and H. Taube, J. Am. Chem. Soc. 90:5928 (1968).

447.   B. C. Lane, J. E. Lester and F. Basolo, J. C. S. Chem. Commun. 1618 (1971).

448.   S. Naito, J. C. S. Chem. Commun. 175 (1978).

449.   R. Longhi, R. O. Ragsdale and R. S. Drago, Inorg. Chem. 1:768 (1962).

450   R. S. Drago and F. E. Paulik, J. Am. Chem. Soc. 82:96 (1961).

451.   R. S. Drago and B. R. Karstetter, J. Am. Chem. Soc. 83:1819 (1961).

452.   H.-O. Fröhlich and W. Römhild, Z. Chem. 19:414 (1979).

453.   H.-O. Fröhlich and W. Römhild, Z. Chem. 20:154 (1980).

454.   O. A. Ileperuma and R. D. Feltham, J. Am. Chem. Soc. 98:6039 (1976).

455.   O. A. Ileperuma and R. D. Feltham, Inorg. Chem. 16:1876 (1977).

456.   T. I. Taylor and J. C. Clarke, J. Chem. Phys. 31:277 (1959).

457.   D. T. Doughty, G. Gordon and R. P. Stewart, J. Am. Chem. Soc. 101:2645 (1979).

458.   R. D. Feltham and J. C. Kriege, J. Am. Chem. Soc. 101:5064 (1979).

459.   J. Kriege-Simondsen, G. Elbaze, M. Dartiguenave, R. D. Feltham and Y. Dartiguenave, Inorg. Chem. 21:230 (1982).

460.   S. Bhaduri, B. F. G. Johnson and T. W. Matheson, J.C.S. Dalton Trans. 561 (1977).

461.   G. Booth and J. Chatt, J. Chem. Soc. 2099 (1962).

462.   F. Bottomley and M. Mukaida, J.C.S. Dalton Trans. 1933 (1982).

463.   F. Bottomley, W. V. F. Brooks, D. E. Paez, P. S. White and M. Mukaida, J.C.S. Dalton Trans. 2465 (1983).

464.   K. K. Pandey, Coord. Chem. Rev. 51:69 (1983).

465.   R. Bhattacharyya, G. P. Bhattacharjee and A. M. Saha, Trans. Met. Chem. 8:255 (1983).

466.   R. Bhattacharyya and A. M. Saha, Inorg. Chim. Acta 77:L81 (1983).

467.   R. Bhattacharyya and P. S. Roy, Trans. Met. Chem. 7:285 (1982).

468.   K. Wieghardt, G. Backes-Dahmann, W. Swiridoff and J. Weiss, Inorg. Chem. 22:1221 (1983).

469.   F. Bottomley, J. Darkwa and P. S. White, J.C.S. Chem. Commun. 1039 (1982).

470.   J. Müller, J. Organomet. Chem. 23:C38 (1970).

471.   W. R. Murphy, K. D. Taheuchi and T. J. Meyer, J. Am. Chem. Soc. 104:5817 (1982).

472.   M. S. Thompson and T. J. Meyer, J. Am. Chem. Soc. 103:5577 (1981).

473.   L. Dózsa, V. Kormos and M. T. Beck, Inorg. Chim. Acta 82:69 (1984).

474.   M. T. Beck, A. Kathó and L. Dózsa, Inorg. Chim. Acta 55:L55 (1981).

475.   A. Müller and N. Mohar, Z. Anorg. Allg. Chem. 480:157 (1981).

476.   A. R. Chakravarty and A. Chakravarty, J.C.S. Dalton Trans. 961 (1983).

477.   A. R. Chakravarty and A. Chakravarty, J.C.S. Dalton Trans. 1765 (1982).

478.   M. P. Doyle, R. A. Pickering, R. L. Dykstra and B. R. Cook, J. Am. Chem. Soc. 104:3392 (1982).

479.   B. S. Tovrog, S. E. Diamond, F. Mares and A. Szalkiewicz, J. Am. Chem. Soc. 103:3522 (1981).

480.   B. Haymore, J. C. Huffman, A. Dobson and S. D. Robinson, Inorg. Chim. Acta 65:L231 (1982).

481.   W. P. Weines, M. A. White and R. G. Bergman, J. Am. Chem. Soc. 103:3612 (1981).

482.   W. P. Weiner and R. G. Bergman, J. Am. Chem. Soc. 105:3922 (1983).

483.   R. E. Stevens and W. L. Gladfelter, J. Am. Chem. Soc. 104:6454 (1982).

484.   P. Legzdins, C. R. Nurse and S. J. Rettig, J. Am. Chem. Soc. 105:3727 (1983).

485.   B. Nuber and J. Weiss, Acta Cryst. B37:947 (1981).

486.   H. W. W. Adrian and A. van Tets, Acta Cryst. B35:153 (1979).

487.   B. Haymore, J. C. Huffman, A. Dobson and S. D. Robinson, Inorg. Chim. Acta 65:L231 (1982).

488.   R. E. Stevens, T. J. Yanta and W. L. Gladfelter, J. Am. Chem. Soc. 103:4981 (1981).

489.   D. Braga, B. F. G. Johnson, J. Lewis, J. M. Mace, M. McPartlin, J. Puga, W. J. H. Nelson, P. R. Raithly and K. H. Whitmire, J.C.S. Chem. Commun. 1081 (1982).

490.   M. A. Collins, B. F. G. Johnson, J. Lewis, J. M. Mace, J. Morris, M. McPartlin, W. J. H. Nelson, J. Puga and P. R. Raithley, J.C.S. Chem. Commun. 689 (1983).

491.   B. F. G. Johnson, J. Lewis and J. M. Mace, J.C.S. Chem. Commun. 186 (1984).

492.   D. E. Fjare and W L. Gladfelter, J. Am. Chem. Soc. 103:1572 (1981).

493.   I. L. Eremenko, A. A. Pasynskii, V. T. Kalinnikov, Yu. T. Struchkov and G. G. Aleksandrov, Inorg. Chim. Acta 52:107 (1981).

494.   B. W. Hames, B. W. S. Kolthammer and P. Legzdins, Inorg. Chem. 20:650 (1981).

495.   D. Lancon and K. M. Kadish, J. Am. Chem. Soc. 105:5610 (1983).

496.   E. Fujita and J. Fajer, J. Am. Chem. Soc. 105:6743 (1983).

497.   P. N. Becker and R. G. Bergman, J. Am. Chem. Soc. 105:2985 (1983).

498.   P. N. Becker and R. G. Bergman, Organometallics 2:787 (1983).

499.   M. D. Seidler and R. G. Bergman, Organometallics 2:1897 (1983).

500. S. Bhaduri, K. R. Sarma and B. A. Narayan, Trans. Met. Chem. 6:206 (1981).

501. W. Evans and J. I. Zirk, J. Am. Chem. Soc. 103:2635 (1981).

502. M. Herberhold, W. Kremnitz, H. Trampisch, R. B. Hitam, A. J. Rest and D. J. Taylor, J.C.S. Dalton Trans. 1261 (1982).

503. S. Sakar and P. Subramanian, J. Inorg. Nucl. Chem. 43:202 (1981).

504. M. Kubota, M. K. Chan and L. K. Woo, Inorg. Chem. 23:1639 (1984).

505. F. Bottomley, J. Darkwa and P. S. White, J.C.S. Dalton Trans. 1435 (1985).

506. K. Wieghardt, U. Quilitzsch and J. Weiss, Inorg. Chim. Acta 89:L43 (1984).

507. D. S. Barratt and C. A. McAuliffe, J.C.S. Chem. Commun. 594 (1984).

508. J. A. Olabe, L. A. Gentil, G. Rigotti and A. Navaza, Inorg. Chem. 23:4297 (1984).

509. P. Oltmanns and D. Rehder, J. Organomet. Chem. 281:263 (1985).

510. J. Schiemann, E. Weiss, F. Näumann and D. Rehder, J. Organomet. Chem. 232:219 (1982).

511. F. Näumann and D. Rehder, Inorg. Chim. Acta 84:117 (1984).

512. F. Näumann and D. Rehder, Z. Naturforsch. 39B:1647 (1984).

513. M. D. Johnson and R. G. Wilkins, Inorg. Chem. 23:231 (1984).

514. M. D. Seidler and R. G. Bergman, J. Am. Chem. Soc. 106:6110 (1984).

515. D. W. Pipes and T. J. Meyer, Inorg. Chem. 23:2466 (1984).

516. H. D. Abruña, J. L. Walsh, T. J. Meyer and R. W. Murray, J. Am. Chem. Soc. 102:3272 (1980).

517. H. G. Alt and H. I. Hayen, Angew. Chem. Int. Ed. Engl. 24:497 (1985).

518. C. J. Jones, J. A. McCleverty and A. S. Rothin, J.C.S. Dalton Trans. 405 (1985).

HYDROLYSIS AND CONDENSATION REACTIONS

OF O- AND N-BOUND LIGANDS

Robert W. Hay

Department of Chemistry, University of Stirling

1. INTRODUCTION

Although scattered references to metal ion promoted reactions are to be found in the early literature it was not until the late 1950's that the significance of these reactions was fully recognized. A strong driving force has been the realization that some 30% of enzymes are metalloenzymes or require metal ions for activity. Many of the reactions dealt with in this chapter have been studied in an attempt to mimic the biochemical systems. Since Kroll's 1953 observation[1] that the hydrolysis of α-amino-acid esters was catalyzed by a number of transition metal ions, a large research effort has been devoted to this area, and the associated field of peptide and amide hydrolysis. In many ways the results obtained have been quite remarkable. Rate enhancements of 10^9-10^{11} have been noted in some reactions, values similar to those observed in enzymic reactions. In addition, many advances have occurred in the more general field of inorganic reaction mechanisms.

On reading this Chapter, the reader will become aware that much kinetic and mechanistic work remains to be done on the role of metal ions in areas such as phosphate ester hydrolysis, sulfate ester hydrolysis, glycoside hydrolysis, and many others. It is hoped that this Chapter will stimulate interest in some of these areas. The compartmentalization of chemistry into organic and inorganic sectors has tended to hinder developments. Inorganic chemists have neglected such fields as the hydrolysis of phosphate and sulfate esters, regarding this as the domain of the physical-organic chemist. Fortunately this view is slowly changing, mainly as a result of the growing importance of the new developing area of biological inorganic chemistry.

In spite of the extensive high resolution data available on metalloenzymes, for example carboxypeptidase A and carbonic anhydrase, many mechanistic problems remain. Thus some six mechanisms have been suggested for carbonic anhydrase although X-ray data at 2Å resolution has been available for several years. The crystallographic data can only give a static picture of an enzyme or an enzyme-substrate complex, and although of immense importance the data have their limitations in the area of reaction dynamics.

Studies on model systems (or biomimetic investigations) have proved invaluable in delineating possible mechanisms and in assessing their merits. Biomimetic studies have, for example, demonstrated the high reactivity of M-OH in intramolecular reactions[23].

For any proton-catalyzed reaction that occurs, an analogous Lewis acid-catalyzed reaction involving a metal ion will exist under the appropriate conditions. In biological reactions, zinc(II) is the main Lewis acid catalyst[21], but other metals such as manganese(II) can be involved. In metal ion catalyzed reactions it is normally necessary to provide at least two bonding sites on the substrate molecule for substantial rate accelerations to occur. However, for reactions involving intramolecular attack by coordinated nucleophiles such as M-OH or M-NH$_2^-$ this is not a necessary requirements.

Several book and review articles are available dealing with various aspects of metal ion catalysis[2-18] and the role of metal ions in enzymatic catalysis[19-21].

2.   HYDROLYSIS OF AMINO ACID ESTERS AND AMIDES, PEPTIDES

2.1.   Labile Metal Systems

The metal ion-catalyzed (or, more correctly, promoted) hydrolysis of amino-acid esters and peptides has been a subject of continuing interest over the past three decades. In the early 1930's the hydrolysis of peptides was found to be subject to metal-ion catalysis, but the discovery by Kroll[1] in 1952 that the hydrolysis of α-amino-acid esters was catalyzed by metal ions stimulated considerable interest in the area. Many of these reactions can be considered as simple model systems for such metalloenzymes as carboxypeptidase A, leucine aminopeptidase, and glycylglycine dipeptidase[19].

Work in this area has been reviewed by Hay and Morris[22] and by Buckingham[23]. The latter review deals specifically with M-OH systems and their ability to hydrolyze such biologically important substrates as amino-esters and peptides. The general area of the interaction of metal ions and metal complexes with amino acid derivatives and peptides is the subject of a continuing biannual review[24]. A major problem in studying the hydrolysis of α-amino-acid esters in the presence of labile metal ions has been the determination of the ligand binding sites; and the identification of the catalytically active species in solution. Many of the early kinetic studies were carried out with monoamino esters where the formation constants are quite low. Solutions containing, for example, copper(II) and the ester ligand (E), will contain a large variety of complexes such as CuE^{2+}, CuE_2^{2+}, CuE_3^{2+} and CuE_4^{2+}. The most abundant species will depend on the copper(II): ester ratio, and inevitably there is more than one complex present. In addition, there will be mixed species $CuEA^+$, CuE_2A^+ etc. formed as the hydrolysis of E to A^- (the anion of the amino acid) occurs. As a result it can be extremely difficult to establish unequivocally the nature of the hydrolytically most active species.

A number of these difficulties have been overcome by using polydentate amino acid derivatives containing additional functional groups capable of interacting with a metal ion or by using mixed-ligand complexes. There has been considerable success in interpreting the kinetic data for such systems. Many of the initial investigations (as with all studies of metal catalyzed reactions) were complicated by the use of complexing buffers which led to additional problems due to metal ion-buffer interactions. The advent of pH-stats, the use of non-complexing buffers such as Hepes and Pipes[25], and the development of Elias buffers,[541] has essentially resolved this problem.

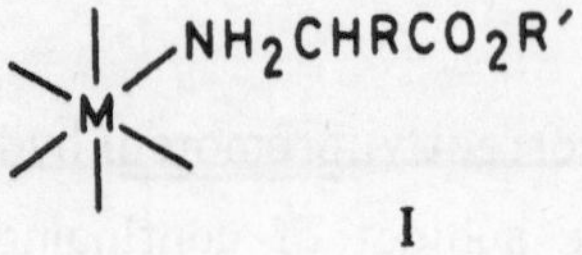

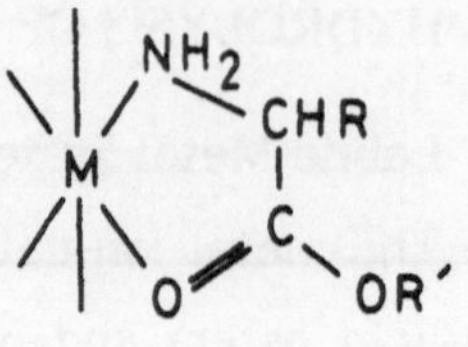

Amino-acid esters can be monodentate (I) or bidentate (II), and examples of both types of complex have been isolated and their solid state infrared spectra studied[26-30].

Many investigations have shown that formation of the monodentate ester species has similar effects to protonation of the α-amino group. Thus the pK_a values of $M-NH_2CHRCO_2H$ and $\overset{+}{N}H_3CH_2CO_2H$ (carboxyl ionisation) are usually quite similar[31-33].

Formation of the chelated ester species leads to a considerable reduction[27] in the carbonyl stretching frequency, e.g. from 1740 cm^{-1} to 1600 cm^{-1} in the case of copper(II), indicating significant polarization of the carbonyl bond by the metal ion. Kroll[1] initially suggested that such chelated species were involved in the metal-ion-catalyzed hydrolysis of glycine methyl ester and this view is now supported by much experimental evidence. Polarization of the carbonyl group by the metal ion assists nucleophilic attack by reagents such as OH$^-$ and H_2O (Scheme 1).

These reactions involve the formation of the tetrahedral intermediate (III), and such an intermediate has been detected spectroscopically in the hydrolysis of some cobalt(III) derivatives[34]. Oxygen-exchange data are also fully consistent with this view[35].

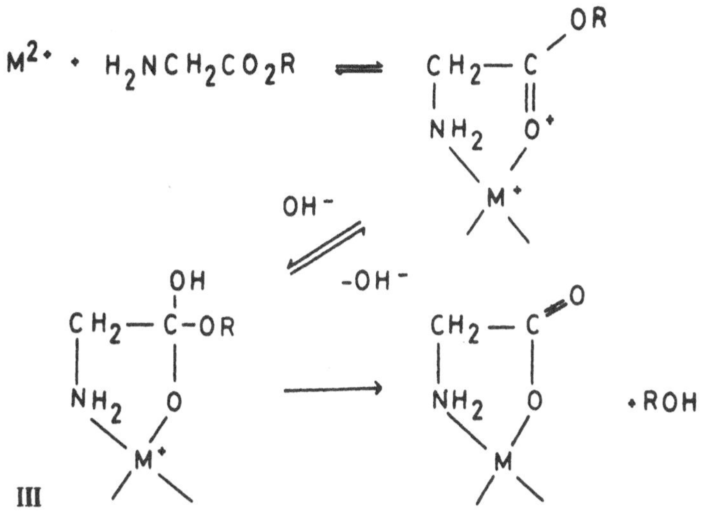

Scheme 1. Metal-ion catalyzed hydrolysis of glycine esters.

2.1.1. Hydrolysis of the Free Ligands

In order to provide information on the magnitude of the catalytic effect of metal ions on amino-acid ester hydrolysis it is necessary to have data on the base hydrolysis of amino-acid esters (hydroxide ion is a more important nucleophile than water). The reaction scheme for the base hydrolysis of amino-acid esters can be summarized by equations (1) - (4).

$$\overset{+}{H_3}NCHRCO_2R' + H_2O \overset{K_E}{\rightleftharpoons} H_2NCH(R)CO_2R' + H_3O^+ \qquad (1)$$

$$H_2NCHRCO_2R' + OH^- \overset{k_E}{\longrightarrow} H_2NCHRCO_2^- + R'OH \qquad (2)$$

$$\overset{+}{H_3}NCHRCO_2R' + OH^- \overset{k_{EH^+}}{\longrightarrow} \overset{+}{H_3}NCHRCO_2^- + R'OH \qquad (3)$$

$$\overset{+}{H_3}NCHRCO_2^- + H_2O \overset{K_A}{\rightleftharpoons} H_2NCHRCO_2^- + H_3O^+ \qquad (4)$$

For most α-amino acid methyl and ethyl esters the values of pK_E fall within the range 7 to 7.6[22]. At physiological pH significant amounts of both the protonated and unprotonated ester are present. The temperature dependence of the pK_E values has been studied and the resulting thermodynamic parameters discussed[36-38].

The base hydrolysis of an amino-acid ester follows the rate expression:

$$- \, d\,[\text{Total ester}]\,/dt = k_E[E][OH^-] + k_{EH+}[EH^+][OH^-] \qquad (5)$$

where E and EH^+ represent the protonated and unprotonated forms of the ester. Quite precise values of k_E are available for a large number of amino-acid esters[36-7,39] (Table 1).

Table 1. Rate Constants for the Base Hydrolysis of Amino Acid Esters at 25°C and $\mu = 0.1M$ (KCl) Data from ref. 22

Amino Ester	$k_E(M^{-1}s^{-1})$	$k_{EH+}(M^{-1}s^{-1})$
Methyl glycinate	1.28	28.3
Ethyl glycinate	0.64	22.9
Methyl α-alaninate	1.11	80.3
Methyl α-amino-n-butyrate	0.39	44.5
Methyl norvalinate	0.40	40.2
Methyl norleucinate	0.37	40.8
Methyl valinate	0.076	-
Methyl leucinate	0.455	-
Ethyl leucinate	0.187	-
Methyl isoleucinate	0.067	-
Methyl β-alaninate	0.136	6.87
Methyl serinate	0.99	-
Methyl β-phenylalaninate	0.55	-
Ethyl β-phenylalaninate	0.235	-
Methyl methioninate	0.77	-
Methyl tryptophanate	0.29	-
Methyl S-methylcysteinate		
Methyl tyrosinate		

Table 1 (Continued)

	k_E	k_{EH^+}	$k_{EH_2^{2+}}$
Methyl 2, 3-diamino-propionate	0.73	57.3	-
Methyl lysinate	0.46	1.26	73.5
Methyl histidinate	0.62	46.5	-
Methyl cysteinate [a]	0.07 (E⁻)	1.07 (EH)	3.83 (EH±)
Ethyl cysteinate [a]	0.04 (E⁻)	0.60 (EH)	6.67 (EH±)

Rate constants are not listed for amino acid esters such as methyl 4-amino-n-butyrate where lactamization occurs (see ref. 38)

(a)   In the case of the cysteine esters,

$E^- = {}^-SCH_2CH(NH_2)CO_2R$; $EH = HSCH_2CH(NH_2)CO_2R$;

$EH^{\pm} = {}^-SCH_2CH(NH_3^+)CO_2R$

Rate constants have also been reported for a number of N-acyl amino-acid esters[40]. Methyl esters undergo base hydrolysis at roughly twice the rate of ethyl esters. The rate constants k_E can be correlated with Taft's polar substituent constant σ^* with a slope $\rho^* = 0.65$[37]. The rate constants k_{EH^+} cannot be determined with high precision by pH-stat techniques; however, a number of rate constants have been reported (Table 1). The effect of charge on the rate of base hydrolysis of carboxylic esters has been extensively studied[41-44]. Generally, the hydrolysis rate of the 1+ charged form of the ester is ca 100 to 1000 times that of the neutral form. Values of k_{EH^+}/k_E are ca 100 for α-amino-acid esters. Withdrawal of the amino group to the β-position lowers k_{EH^+}/k_E to about 50 (methyl β-alaninate, methyl 2,3-diamino-propionate) and further withdrawal to the ε-position (methyl lysinate) reduces the ratio to 2.7.

Table 2. Rate Constants for the Metal Ion Promoted Hydrolysis of Methyl Glycinate and Ethyl Glycinate

Reaction	Rate Constant (Ionic strength; temp)	Ref.
$CuE^{2+} + OH^-$	7.6×10^4 M^{-1} s^{-1} (0.16M, 25°C)	46
$CuE^{2+} + H_2O$	4.3×10^{-5} s^{-1} (0.16M, 25°C)	46
$NiE^{2+} + OH^-$(a)	3.98×10^3 M^{-1} s^{-1} (1.0M, 30°C)	47
NiE^{2+} H$_2$O (a)	1.9×10^{-4} s^{-1} (1.0M, 30°C)	47
$CuE^{2+} + OH^-$ (a)	1.4×10^5 M^{-1} s^{-1} (?, 25°C)	48
$CuEA^+ + OH^-$ (a)	8.4×10^3 M^{-1} s^{-1} (?, 25°C)	48
$CuE^{2+} + OH^-$(a)	2.5×10^{-5} s^{-1} (?, 25°C)	48

E is $H_2NCH_2CO_2Me$ or $H_2NCH_2CO_2Et$;   A$^-$ is $H_2NCH_2CO_2^-$.

(a) These values refer to glycine ethyl ester.

2.1.2. Monoamino Esters

Following the initial work of Kroll[1], studies of amino acid ester hydrolysis were centered on simple mono-amino esters[1,35,45,49]. However, the results obtained were rather conflicting. More recent studies[46-48] (Table 2), have provided rate constants which show only order of magnitude agreement; however, uncertainty from the earlier studies regarding the nature of the nucelophile has been resolved. Hydroxide ion is the predominant nucleophile in spite of the reaction being studied at pH values of around 5. Higher pH values lead to precipitation of metal hydroxides, due to the rather low formation constants of the metal complexes. Evidence for nucleophilic attack by water has also been obtained[46-48]. The rate enhancement, in the copper(II)-glycine ester systems are large, ca 10^5-10^6. These large rate enhancements are similar to the accelerations of ca 10^6 observed in the inert cobalt(III) systems where direct metal-ester carbonyl

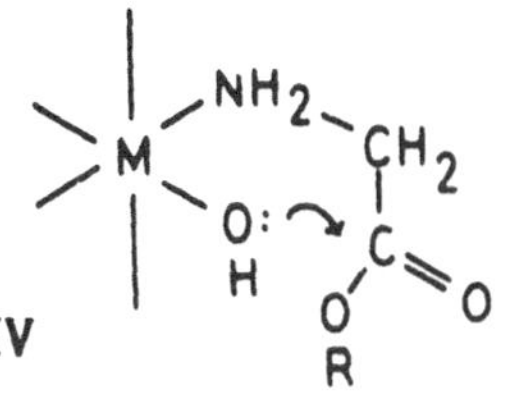

IV

bonding occurs. As a result it is probable that the glycine ester systems react as shown in Scheme 1. The question of the occurrence of coordinated nucleophile reactions of the type shown in **IV** is still unresolved, but the observed accelerations are more in accord with chelated ester intermediates.

Some recent studies have been carried out using palladium(II) complexes[50]. α-Amino-acid esters react with $[Pd(en)(OH_2)_2]^{2+}$ according to the equilibrium (6).

$$[Pd(en)(OH_2)_2]^{2+} + \overset{+}{N}H_3CH(R)CO_2R' \xrightarrow{K}$$

$$[Pd(en)\{NH_2CH(R)CO_2R'\}]^{2+} + H_3O^+ + H_2O \qquad (6)$$

The kinetics of hydrolysis of the ester ligand in the mixed ligand complexes have been studied in detail, and nucleophilic attack by both OH^- and OH_2 has been shown to occur. The hydrolysis data obtained are summarized in Table 3.

Substantial rate enhancements are observed for base hydrolysis (factors of 4×10^4 for glyOEt to 1.4×10^7 for ethyl picolinate) which involve carbonyl bonded intermediate complexes such as **V**. With esters of histidine (**VI**) and cysteine which form complexes with pendant ester groups only small rate enhancements of 20-100 occur. For those esters where there is a direct interaction between the alkoxy carbonyl group and palladium(II), the ratios k_{OH}/k_{H_2O} fall within the range 3.8×10^9 to 3.2×10^{11}. Such values for the relative nucleophilicity of hydroxide ion and water are quite comparable with those previously noted for copper(II) complexes[51].

Table 3. Hydrolysis data for [Pd (en) (NH$_2$CH(R)CO$_2$R')]$^{2+}$ Complexes at 25°C and μ = 0.1M (KNO$_3$) (k in M^{-1} s^{-1}). Data from ref. 50

Ester	k_{OH}(complex)	k_{H_2O}(complex)	k_{OH}(ester)
glyOEt	2.45×10^4	5.3×10^{-6}	0.64
glyOMe	6.25×10^4	4.9×10^{-6}	1.28
α-alaOEt	6.15×10^4	1.6×10^{-5}	0.55
pheOEt	11.75×10^4	1.04×10^{-5}	0.24
cysOEt	4.20	5.17×10^{-6}	0.04
hisOMe	12.76	5.48×10^{-6}	0.62
picOEt(a)	6.47×10^6	2.05×10^{-5}	0.46

(a) PicOEt = ethyl picolinate (2-carbethoxpyridine)

2.1.3. Mixed Ligand Complexes of Monoamino Esters

The palladium(II) system discussed above illustrates the useful data which can be obtained from mixed ligand complexes containing simple mono-amino acid ligands. Angelici and co-workers have studied the hydrolysis of α-amino-acid esters in mixed-ligand complexes of the type [CuL{NH$_2$CH(R)CO$_2$R'}]$^{n+}$ using ligands (L) such as iminodiacetate (imda)[52], nitrilotriacetate (nta)[53], 2,2',2"-tris(aminoethyl)amine (tren)[54], terpyridyl (terpy)[54], diethylenetriamine (dien)[55] and bis(2-pyridylmethyl-amine) (dpa)[56].

Further work by Hay and Banerjee has concerned mixed ligand complexes of nta[57], ethylenediaminemonoacetate (edma)[58], glycylglycine dianion[59], and imda[60]. To various degrees these studies can be said to mimic metalloenzyme-substrate complexes. These reactions can be illustrated by a specific example. Using a ligand such as ethylenediaminemonoacetate[58], ternary complexes with amino-acids can be formulated as either **VII** or **VIII**.

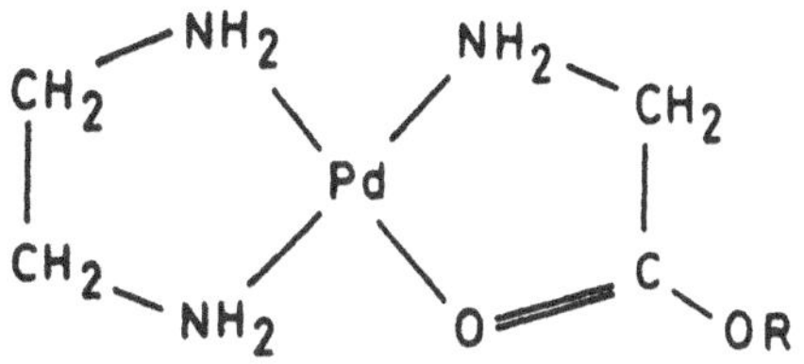

V

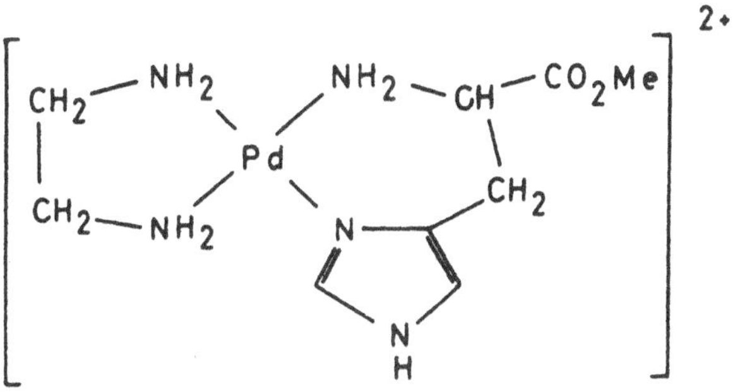

VI

VII

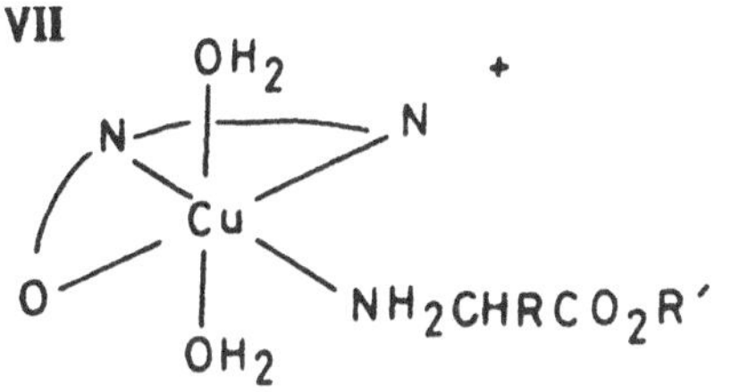

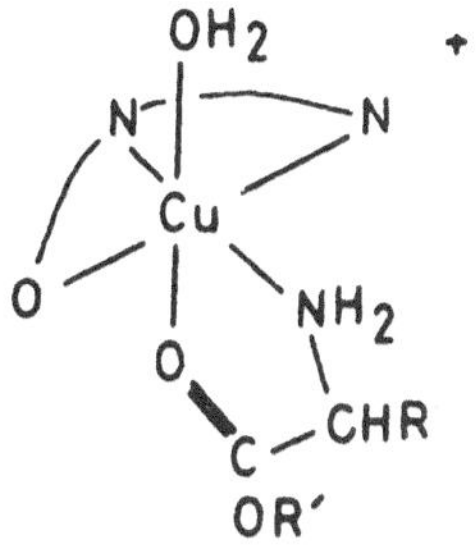

VIII

The ester ligands in the ternary complexes of edma⁻ undergo base hydrolysis some 10^3 times faster than the free unprotonated esters, although considerably lower effects occur with methyl-L-histidinate where the ratio is only 40. The magnitude of these effects suggest that the amino-acid ester ligand is probably bidentate with a rather weak alkoxycarbonyl-metal interaction. Intramolecular attack by coordinated hydroxide ion as shown in **IX** appears unlikely in this system in view of the magnitude of the accelerations and the fact that the reaction shows a first order dependence on the hydroxide ion concentration up to pH 8. Typical kinetic data[58] for this system are shown in Table 4.

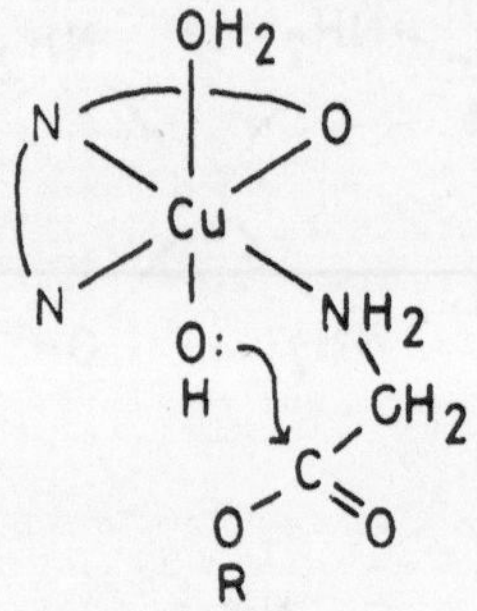

IX

Table 4. Catalytic Effects for the Hydrolysis of the Mixed Ligand Complexes [Cu (edma) $(NH_2CH(R)CO_2R')$]$^+$ at 25°C and μ = 0.1M (KNO_3) (k in M^{-1} s^{-1}). Data from ref. 58

Ester	10^{-3} k_{OH}(complex)	k_{OH}(ester)	Ratio
glyOMe	1.71	1.28	1.34 x 10^3
L-β-pheOMe	0.99	0.55	1.18 x 10^3
glyOEt	0.63	0.64	1.0 x 10^3
L-α-alaOEt	0.59	0.55	1.1 x 10^3
L-α-pheOEt	0.38	0.24	1.6 x 10^3
L-hisOMe	0.025	0.62	40

Angelici and co-workers[56] first noted a correlation between the formation constant K_{CuL} of the metal complex and the catalytic activity of the complex. Large values of K_{CuL} lead to

$$Cu^{2+} + L \; \underset{\longleftarrow}{\overset{K_{CuL}}{\longrightarrow}} \; CuL^{2+} \qquad\qquad (7)$$

lower base hydrolysis rates of the α-amino-acid esters in the ternary complex. The Lewis acidities of the CuL^{n+} complexes decrease as the formation constant for CuL^{n+} increases and values of k_{OH} decrease as log K_{CuL} increases (Table 5).[58] The effects of charge on these reactions do not appear to be of great importance; thus [Cu(edma)(glyOMe)]$^+$ undergoes slower base hydrolysis than [Cu(imda)(glyOMe)]^o, while the hydrolysis of

Table 5. Rate Constants (k_{OH}) and Equilibrium Constants (log K_{CuL}) Associated with the CuL Promoted Hydrolysis of Methyl Glycinate at 25°C

[CuL (glyOMe)]	k_{OH} (M^{-1} s^{-1})	log K_{CuL}	Ref.
[Cu(imda) (glyOMe)]^0	7.6 x 10^3(a)	10.63	54
[Cu(edma) (glyOMe)]$^+$	1.7 x 10^3(a)	13.29	58
[Cu(nta) (glyOMe)]$^-$	4.6 x 10^2(b)	13.10	53
[Cu(terpy) (glyOMe)]$^{2+}$	2.2 x 10^2(c)	13.40	54
[Cu(dpa) (glyOMe)]$^{2+}$	1.7 x 10^2(c)	14.4	56
[Cu(dien) (glyOMe)]$^{2+}$	1.4 x 10^2(a)	15.9	55
[Cu(tren) (glyOMe)]$^{2+}$	1.3(c)	18.8	54

(a) μ = 0.1M

(b) μ = 0.07M

(c) μ = 0.05M

[Cu(dpa)(glyOMe)]$^{2+}$ is some ten times slower than that of the edma derivative.

The results are of some significance as far as metalloenzymes are concerned, since the binding constants for these systems will presumably influence their effectiveness as Lewis-acid catalysts. For the binding of Zn(II) to apocarboxypeptidase a value of log K = 10.5[61] has been estimated and a similar value has been reported[62] for Zn(II) with apocarbonic anhydrase. Such considerations may be of importance in determining metal-ion specificities of metalloenzymes since metal ions with high binding constants for apoenzymes may be poor Lewis acid catalysts.

Table 6. Activation Parameters for the Hydrolysis of Methyl Glycinate in the Presence of Metal Nitriloacetates and Tetradentate Nickel (II) Chelates at $\mu = 0.1M$ (k in $M^{-1} s^{-1}$ at 25°C). Data from Table IV of ref. 63

Complex	log K_L	k_{OH}	$\Delta H^{\ddagger}$ (kcal mol^{-1})	$\Delta S^{\ddagger}$ (cal K^{-1}mol^{-1})
[Ni(tren)glyOMe]$^{2+}$	14.8	67.1	3.4 ± 0.8	-39 ± 2
[Ni(trien)(glyOMe]$^{2+}$	14.0	53.1	7.2 ± 0.7	-26 ± 1
[Ni(Edda)(glyOMe)]	13.5	41.2	4.8 ± 0.8	-35 ± 2
[Ni(nta)(glyOMe)]	11.47	52.3	0.9 ± 0.7	-47 ± 2
[Co(nta)(glyOMe)]	10.81	18.6	1.5 ± 0.8	-48 ± 2
[Cu(nta)(glyOMe)]	13.05	460	3.4 ± 1.2	-38 ± 3
[Zn(nta)(glyOMe)]	10.44	34.6	4.0 ± 0.7	-38 ± 2
[Cu(nta)(glyOEt)]	13.05	78.2	4.9	-33

Activation parameters have been determined[63] for the hydrolysis of α-amino-acid esters in mixed ligand complexes (Table 6). For base hydrolysis[53] of the complex [Cu(nta)-(NH$_2$CH$_2$CO$_2$Et)]$^-$, $\Delta H^{\ddagger}$ = 20.5 kJ mol^{-1} and $\Delta S^{\ddagger}$ = -138 J K^{-1} mol^{-1}. Catalysis in this system is primarily due to a substantial decrease in $\Delta H^{\ddagger}$ (by ca 21 kJ mol^{-1}) compared with the free ligand. A detailed discussion of the parameters is available[63].

2.1.4. Bidentate and Polydentate Esters

Detailed kinetic studies have been carried out on the base hydrolysis of a variety of bidentate esters such as methyl 2,3-diaminopropionate (**X**), methyl histidinate (**XI**), methyl cysteinate (**XII**) and the ethyl ester of ethylenediaminemonoacetate (**XIII**). For the first three

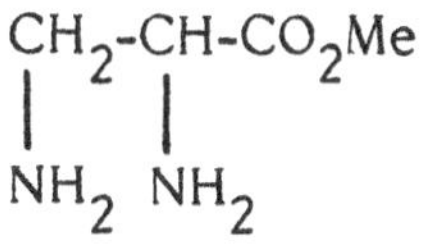

CH$_2$-CH-CO$_2$Me
   | |
NH$_2$ NH$_2$

X

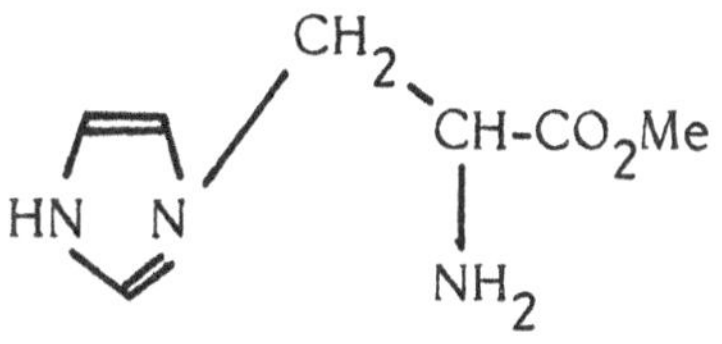

CH$_2$
   \
    CH-CO$_2$Me
    |
HN   N   NH$_2$

XI

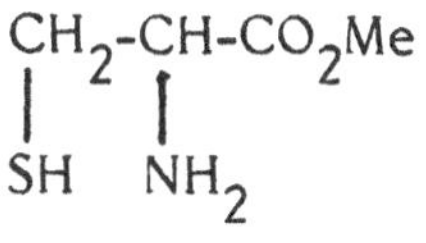

CH$_2$-CH-CO$_2$Me
   | |
SH   NH$_2$

XII

CH$_2$-CH$_2$
   | |
NH$_2$ NH-CH$_2$CO$_2$Et

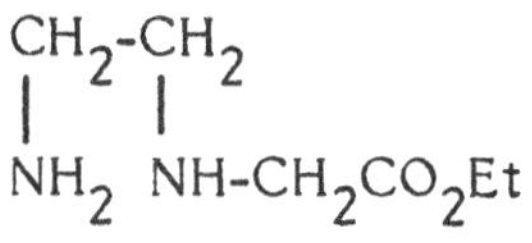

XIII

esters very stable metal complexes are formed with pendant ester functions[64-70]. The kinetic data obtained are summarised in Tables 7 and 8.

The order of decreasing reactivity of the various L-(+)-histidine methyl ester species[73] towards nucelophilic attack by hydroxide ion at 25°C and I = 0.1M is

$$CuE_2^{2+} > NiE_2^{2+} > CuE^{2+} > NiE^{2+} > EH > CuEA^+ > CuEOH^+ > NiEA > E.$$

A very similar situation occurs with methyl 2,3-diaminopropionate[74] where the reactivity order is $CuE^{2+} > CuE_2^{2+} > CuEOH^+ > CuE(en)^{2+} > HgE_2^{2+} > CuEA^+ > HgE^{2+} > EH^+ > HgEOH^+ > HgEA^+ > E$. The bis complex (CuE_2^{2+}) of methylhistidinate with copper(II) hydrolyzes 265 times more rapidly than the free ester (E) when statistical corrections are made, and EH^+ hydrolyzes 75 times more rapidly than E.

The rate enhancements are very much less than those observed in the glycine systems, in which metal-alkoxycarbonyl interactions occur.

Thermodynamic parameters $\Delta H^{\ddagger}$ and $\Delta S^{\ddagger}_{298}$ have been reported for the hydrolysis of CuE_2^{2+} and $CuEA^+$ (E = L-methylhistidinate[73] and methyl DL-2,3-diaminopropionate[74]). The metal complexes have $\Delta H^{\ddagger}$ values of 4 to 8 kJ mol^{-1} higher than for the free ligands, and the rate increases are due to more positive values of $\Delta S^{\ddagger}_{298}$.

Table 7. Rate Constants for the Metal Ion Promoted Base Hydrolysis of Some Bidentate Amino Acid Ester Species at 25°C

System	Complexes				
	ME_2^{2+}	MEA^+	ME^{2+}	$ME(OH)^+$	Ref.
Cu(II)-L-methyl histidinate	2.8(b)	-	-	-	66
	3.2	38.7	-	-	72
	198.8(c)	-	-	-	67
	328	42.7	175	33.7	73
Ni(II)-L-methyl histidinate	1.57(b)	-	-	-	66
	118	16.5	78.3	-	72
	37.5(c)	-	322.5	-	67
	188.3	28.3	87.7	-	73
Cu(II)-D,L-methyl 2,3-diaminopropionate	305	88.7	618	135	74
Hg(II)-D,L-methyl 2,3-diaminopropionate	115	24.3	82	50	74

(a) All rate constants refer to base hydrolysis at 25°C (k_{OH} in M^{-1} s^{-1}) and $\mu = 0.1M$ unless otherwise stated.

(b) μ not stated

(c) $\mu = 0.16M$

The practical difficulties arising due to the low formation constants of simple monoamino acid complexes with transition metal ions can be overcome by incorporating the ester moiety as part of a larger polydentate ligand, e.g. by using ligands such as ethyl glycinate-N,N-diacetic acid (XIV) which hydrolyzes to give nitrilotriacetic acid (XV):

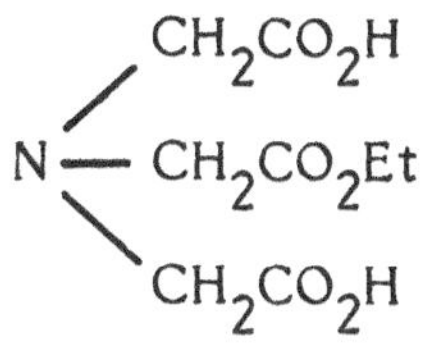

$$N \begin{array}{l} — CH_2CO_2H \\ — CH_2CO_2Et \\ — CH_2CO_2H \end{array} \quad \textbf{(XIV)}$$

$$N \begin{array}{l} — CH_2CO_2H \\ — CH_2CO_2H \\ — CH_2CO_2H \end{array} \quad \textbf{(XV)}$$

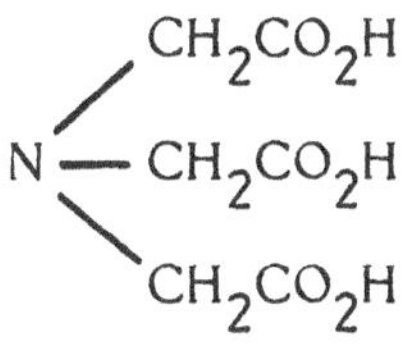

Kinetic measurements have been reported[75-6,78] for ligands involving ethyl glycinate (**XVIa**), ethyl alaninate (**XVIb**), ethyl valinate (**XVIc**), ethyl leucinate (**XVId**), butyl glycinate, ethyl-β-alaninate (**XVIIa**) and ethyl 4-aminobutyrate (**XVIIb**). The catalytic effects of the metal ions

Table 8. Rate Constants for the Metal Ion Promoted Base Hydrolysis of Some Bidentate Species of Cysteinate Esters at 25°C

System	Complexes		
	ME_2	MEA^-	Ref.
Ni(II)-L-methylcysteinate	10.8	3.6	71
Pb(II)-L-methyl cysteinate	5.7	0.97	71
Zn(II)-L-methyl cysteinate	4.7	1.38	71
Cd(II)-L-methyl cysteinate	3.5	0.7	71
Hg(II)-L-methyl cysteinate	2.3	0.77	71
Ni(II)-L-ethyl cysteinate	2.6(b)		67
Zn(II)-L-ethyl cysteinate	1.9(b)		67
Cd(II)-L-ethyl cysteinate	1.05(b)		67

(a) All rate constants refer to base hydrolysis at 25°C (k_{OH} in $M^{-1} s^{-1}$) and $\mu = 0.1M$ unless otherwise stated

(b) $\mu = 0.16M$

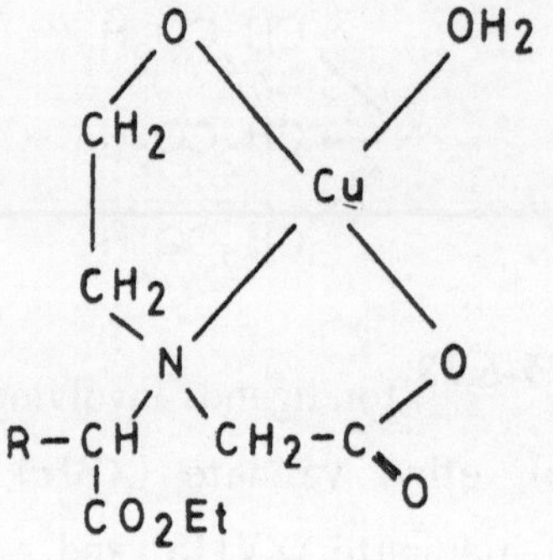

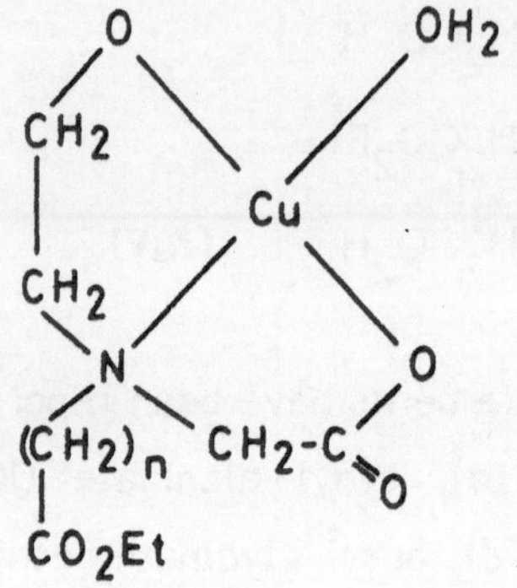

XVI

a, R = H

b, R = CH$_3$

c, R = CH(CH$_3$)$_2$

d, R = CH$_2$CH(CH$_3$)$_2$

XVII

a, n = 2

b, n = 3

Cd(II), Ni(II), Mn(II), Fe(II), Co(II), Cu(II), Zn(II), Pb(II), La(III), Nd(III), Gd(III), Dy(III), Er(III), Yb(III), Lu(III) and Sm(III) have been studied for a few of the ligands. The 1:1 complexes have large formation constants and display little tendency to add a second ligand. Hydroxocomplexes may, however, form by deprotonation of coordinated water molecules. For the ethyl glycinate derivative, which may be taken as an example of these reactions, pH-stat measurements have ben carried out in the pH range 5.0-7.0 for the reaction (8),

$$[Cu(EGDA)] + OH^- \longrightarrow [Cu(NTA)]^- + C_2H_5OH \tag{8}$$

where EGDA = ethyl glycinate-N,N-diacetate and NTA = nitrilotriacetate. The rate expression takes the form

rate = k[Cu(EGDA)] [OH$^-$]

with k_{OH} = 2.18 x 10^4 M^{-1} s^{-1} at 25°C [75]. The kinetic measurements themselves do not differentiate between bimolecular hydroxide ion attack on the 1:1 aquo chelate **XVIa** and intramolecular attack by coordinated hydroxide (ca 0.01 to 20% of [Cu(EGDA) (OH)]$^-$ exists in solution in the pH range 5.0-7.0). However, attack by coordinated hydroxide does not appear to

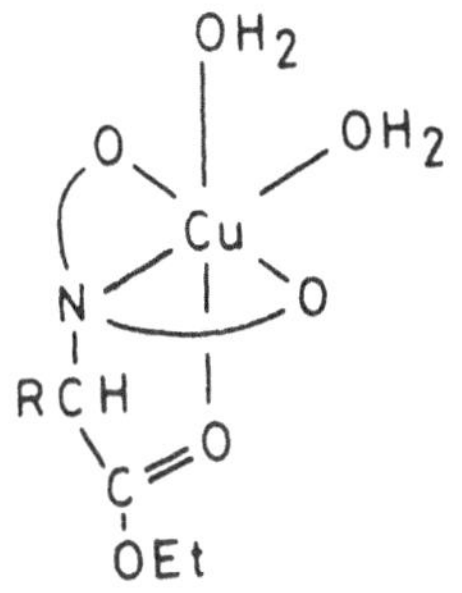

$$\text{XVIII}$$

be favorable in this type of system due to the relative dispositions of the hydroxide ion and the ester carbonyl group. The hydrolytic reactions are very rapid with rate enhancements of a similar magnitude to those observed for the glycinate esters (Table 2), suggesting that a direct interaction between the metal ion and the carbonyl group occurs in solution, as in **XVIII**. Kinetic data for the base hydrolysis of a variety of metal complexes of EGDA are summarized in Table 9. Rare earth ions such as Yb(III) and Lu(III)

Table 9. Rate Constants for the Base Hydrolysis of Metal Complexes of Ethyl Glycinate-N,N-Diacetate at 25°C. Data from ref. 78

Metal Ion	$10^{-2}k$ $(M^{-1} s^{-1})$	Metal Ion	$10^{-2}k$ $(M^{-1} s^{-1})$
Cd (II)	2.14	La(III)	115
Ni (II)	3.89	Nd(III)	347
Mn (II)	4.18	Sm(III)	447
Co (II)	10.1	Gd(III)	376
Fe (II)	38.6	Dy(III)	877
Zn (II)	66.2	Er(III)	1920
Cu (II)	218	Yb(III)	3460
Pb (II)	283	Lu(III)	3450

$\mu = 0.05M$ (KNO_3), $[M] = [EGDA] = 6.7 \times 10^{-4}M$

Table 10. Activation Parameters for the Base Hydrolysis of Ethyl Glycinate and Metal Complexes of Ethyl Glycinate-N,N-Diacetate. Data for the metal complexes from ref. 78

Ester	$\Delta H^{\ddagger}$ (kcal mol^{-1})	$\Delta S^{\ddagger}$ (cal K^{-1} M^{-1})
Ethyl glycinate	10.6 ± 0.5	-21.7 ± 1.0
Betaine ethyl ester	9.7 ± 0.3	-18.5 ± 1.1
[Cu(EGDA)]	5.0 ± 0.5	-21.9 ± 1.5
[Ni(EGDA)]	5.7 ± 0.4	-24.2 ± 1.2
[Pb(EGDA)]	4.4 ± 0.6	-23.5 ± 2.1
[Sm(EGDA)]$^+$	10.0 ± 0.4	-4.0 ± 1.2
[Lu(EGDA)]$^+$	10.2 ± 0.6	$+0.8 \pm 2.0$

have very marked catalytic effects. Activation parameters have been determined for some of these reactions (Table 10). In complexes of nickle(II), copper(II) and lead(II), the rate enhancements are due primarily to a lowered $\Delta H^{\ddagger}$ with little change in $\Delta S^{\ddagger}$. For rare earth ions quite the reverse is observed, with more positive (i.e. less negative) $\Delta S^{\ddagger}$ values leading to the rate enhancement, with no change in $\Delta H^{\ddagger}$. Clearly quite different mechanisms operate for the two systems. Nucleophilic catalysis (by acetate, HPO_4^{2-}, pyridine, 4-picoline and nitrite) of some of these reactions has been studied[76] using esters derived from ethyl valinate and ethyl leucinate which are sterically hindered to base hydrolysis.

Rodgers and Jacobson[79] have published the crystal structure of DL-(ethyl valinate-N,N-diacetato) diaquocopper(II), the copper(II) complex of EVDA **(XVIc)**. The stereochemistry around copper is distorted octahedral, the equatorial coordination consisting of the tertiary nitrogen, the two acetate oxygens, and water molecule. The second water molecule occupies an axial position at a greater distance from the metal, while the ether

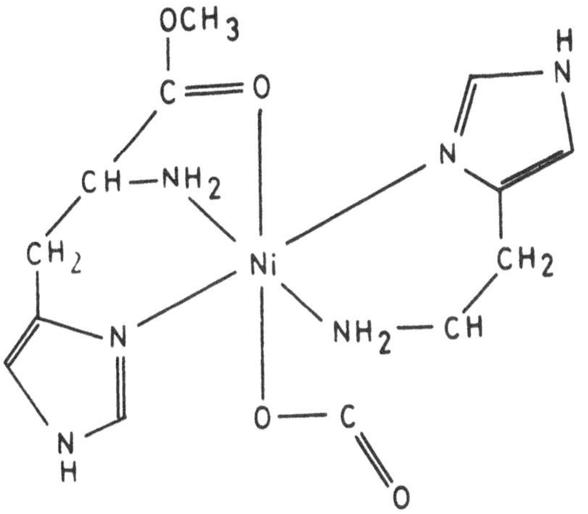

XIX

oxygen of the ester group occupies the opposite axial position, lying at 2.84 Å from the metal. The ester carbonyl group is not coordinated in the solid state.

2.1.5. Kinetic Stereoselectivity

The general topic of steroselectivity in metal complexes of amino-acids and their derivatives has been reviewed[80]. The rates of metal ion-promoted ester hydrolysis are usually independent of the configuration of the ligand (D or L); however, some stereoselective effects have been observed. The 1:1 nickel(II) complex of histidine promotes the hydrolysis of methyl histidinate more effectively if the two ligands are of opposite configurations[81]. [Ni(D-HisOMe) (L-His)]$^+$, for example, hydrolyzes some 40% faster than [Ni(D-HisOMe) (D-His)]$^+$. Stereoselectivity is not observed for the analogous copper(II) complexes,[82-3], but the nickel(II) result has been confirmed.[82-3] The NiEA$^+$ complex probably involves tridentate A$^-$ and primarily bidentate E (A$^-$ = the amino acid anion, E = the amino acid ester). The octahedral coordination of nickel(II) in these systems results in the possibility of a direct metal ion ester carbonyl interaction (XIX). Such an interaction can only occur when the two ligands are of opposite configurations, since a _trans_ arrangement of the bulky imidazole donors is preferred[84-6]. The interaction must be very weak in view of the small rate differences between the diastereoisomers. Such transient coordination is less likely for copper(II), which favors tetragonally distorted geometry. In $CuEA^+$ both E and A^- are probably bidentate with possibly weak apical A^- carboxylate coordination.

Metal ions can form three bis-complexes with optically active amino acids (aa), viz. $M(D-aa)_2$, $M(L-aa)_2$ and $M(D-aa)(L-aa)$. The first two complexes will have identical thermodynamic stabilities while the latter complex, containing ligands with opposite chiralities, is diastereoisomeric with them. Stereoselectivity is the preferential formation of one complex species rather than another and can be expressed quantitatively (eq.9):

$$\Delta\log \beta = \log \beta [M(D-aa)(L-aa)] - \log \beta [M(L/D-aa)_2] \qquad (9)$$

Very few cases of stereoselectivity in the formation of binary copper complexes of simple amino acids have been reported. With Ni(II), Co(II) and Zn(II) with histidine, stereoselectivity has been observed[84,87], but with Cu(II) the effect is negligible or very small. Stereoselectivity is present, however, in the formation of ternary complexes of Cu(II) with histidine and some simple amino acids containing aromatic substituents (e.g. tryptophan, where $\Delta\log \beta = 0.47$)[88].

Hix and Jones[81] have established that the more rapid hydrolysis of $[Ni(D-HisOMe)(L-His)]^+$ is a kinetic stereoselective effect since the formation constants of the diastereoisomeric complexes are identical. More detailed studies of these reactions have been reported[89-90]. Nickel(II) complexes of histidine and tryptophan provide stereoselectivity in the hydrolysis of histidine methyl ester, but not nickel(II) complexes of aspartic acid or methionine. Of the tridentate ligands, only those with a certain minimum steric bulk appear to be capable of exhibiting steroselectivity in such reactions.

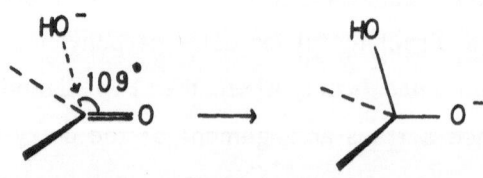

XX

2.1.6. Intramolecular Attack by Coordinated Hydroxide

The occurrence of intramolecular attack by coordinated hydroxide in labile metal complexes is not well confirmed, although it is now well established for the non-labile cobalt(II) complexes (Section 2.2). The pK_a values of coordinated water on copper(II) normally fall within the range 6-8. If coordinated hydroxide ion is an important nucleophile in copper(II) promoted reactions, the reaction would be expected to become independent of pH at ca. pH 8 when the bulk of the complex was converted to the active hydroxo-species. The pH-dependence of a number of copper(II)-promoted reactions has been studied at such pH levels, and no evidence for the production of catalytically active hydroxo compounds observed.

Baldwin[91-2] has discussed the importance of the "flight path" of the nucleophile at tetrahedral, trigonal and digonal centres. For attack at an sp^2-hybridized center the nucleophile should approach at an angle of 109° to the carbonyl bond, as shown in (**XX**). For many complexes it appears that the "flight path" of the nucleophile is unfavorable for intramolecular attack. It is presumably for this reason that no evidence has been obtained for the attack of "coordinated hydroxide" on <u>chelated</u> glycine esters in cobalt(III) chemistry.

Woolley[93] has observed that a number of hydroxometal complexes of the CR macrocycles **XXI** are active in the hydration of CO_2 and acetaldehyde. Electronic spectroscopy[94] and X-ray crystallography[95] suggest coordination numbers in aqueous solution of five for $CoCR^{2+}$ and $ZnCR^{2+}$, and six for $NiCR^{2+}$ and $CuCR^{2+}$. For the $ZnCR^{2+}$ system the pK_a of the coordinated water molecule is 8.69 at $25^\circ C$ and the hydroxo complex is a reasonable catalyst for CO_2 hydration, as is the complex $[Cu(glycylglycinate)(OH)]$[96].

Definitive evidence for intramolecular hydrolysis of the methyl ester (**XXII**) by metal hydroxide has been provided[97]. Molecular models of the metal complex **XXIII** indicate that when complexation with the imidazole nitrogen and the phenol hydroxyl occurs, it is not possible for coordination with the ester carbonyl to occur. This point taken in conjunction with the observed pH-rate profile, which shows that ionization of the $M-OH_2$ group is

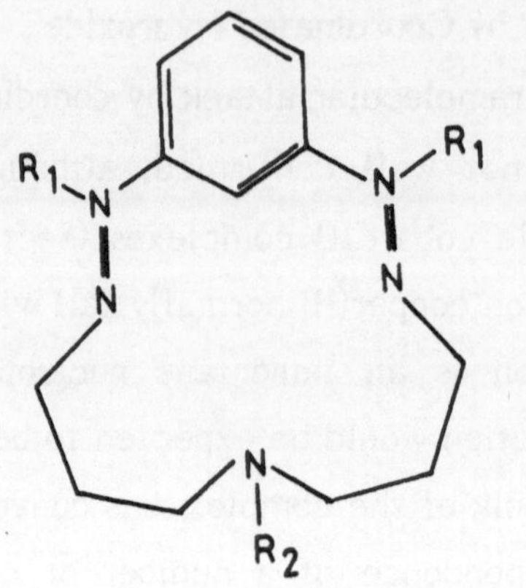

XXI

associated with catalysis, eliminates metal ion activation of the carbonyl bond to intermolecular attack by OH⁻ as a contributing factor. For base hydrolysis of **XXII**, $K_{OH}=2.7 \times 10^{-2} M^{-1}s^{-1}$ at 25°C. The specific rate constants for intramolecular hydrolysis by the M–OH species are $0.245 s^{-1}$ and $2 \times 10^{-2} s^{-1}$ for the cobalt (II) and nickle(II) derivatives.

The hydration of CO_2 by kinetically inert metal hydroxides of the type $(NH_3)_5MOH^{2+}$ (M = Co[98], Rh[99], Ir[99]) to give $(NH_3)_5MOCO_2H$, and the condensation of $[Co(en)_2OH(H_2O)]^{2+}$ with acetylacetone[100] to give $[Co(en)_2(acac)]^{2+}$ have demonstrated the ability of metal-bound hydroxide to add rapidly to carbonyl substrates. The cleavage of 4-nitrophenyl acetate

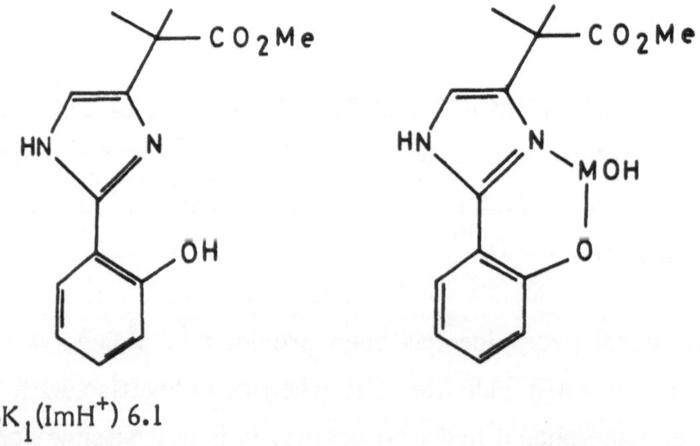

pK$_1$(ImH⁺) 6.1

pK$_{2}$(-OH) 9.6

(XXII)

(XXIII)

by simple metal complexes $[(NH_3)_5CoOH]^{2+}$ and $[(NH_3)_5CoIm)^{2+}$ (Im = N-deprotonated imidazole) has been studied[101] in water and dimethyl sulfoxide solution. In both solvents, the reactions are exclusively nucleophilic, as demonstrated by the detection of the acetylated reactants $[(NH_3)_5CoO_2CCH_3]^{2+}$ and $[(NH_3)_5Co(ImCOCH_3)]^{3+}$. The reaction of $[(NH_3)_5CoOH]^{2+}$ with chloral hydrate, diethyloxalate and succinic and maleic anhydrides in dimethyl sulfoxide gives the expected carboxylatopentammine-cobalt(III) complexes rapidly in high yield.

The pK_a values of $[Co(NH_3)_5OH_2]^{3+}$ and $[Co(NH_3)_5ImH]^{3+}$ are 6.4 and 10.0 respectively at 25°C and the large difference in nucleophilicity towards 4-nitrophenyl acetate (K_N= 9 M^{-1} s^{-1} for the imidazole compound, K_N = 1.5 x 10^{-3} M^{-1} s^{-1} for the hydroxo compound) closely parallels the difference in basicity.

Buckingham and Englehardt[102] have studied the intermolecular reaction of a variety of metal hydroxo complexes (M = Cr, Ru, Co, Ir, Rh) with propionic anhydride. Plots of log K_N versus the pK_a of the conjugate acid of the nucleophile (Bronsted plot) fall on a fairly smooth curve. Thus the rate constant K_N is simply related to the pK_a of the aquo complex. However, a metal bound nucleophile (M-N) is far superior in a kinetic sense to the free nucleophile (N) at pH values where the latter is largely protonated. The area of M-OH nucleophiles has been reviewed[23].

2.1.7 Ester Exchange Reactions

Metal ions catalyze ester exchange reactions of the type (10) using the appropriate ligand systems

$$R\ CO.OR' + R''OH \rightleftharpoons RCO.OR'' + R'OH \qquad (10)$$

A number of metal complexes of carboxylic esters undergo transesterification on refluxing with alcohols[103-4]. Thus copper(II) complexes of ethyl picolinate on refluxing in methanol give the analogous complexes of methyl picolinate[104]. In some cases methanolysis rather that transesterification occurs as with bis(ethyl acetoacetato)copper(II) which gives methoxy-bridged derivatives in refluxing methanol[105].

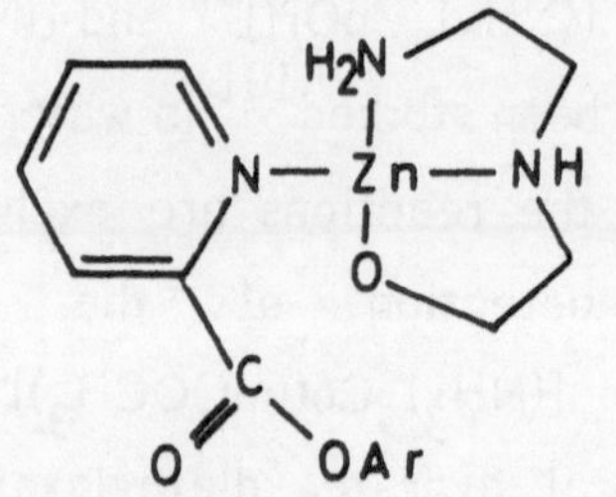

The complex bis(pentane-2,4-dionato)μ,μ'-dimethoxydicopper(II) has been found to be a very effective and selective catalyst[106] for the ester exchanges of 2-carbethoxypyridine (ethyl picolinate) and ethyl 2-pyridylacetate to give the corresponding methyl esters.

Sigman and Jorgenson[107] have found that zinc(II) catalyzes the transesterification reaction between 4-(β-hydroxyethyl) ethylenediamine and 4-nitrophenyl picolinate. The reaction involves a reactive mixed ligand complex (XXIV) in which the zinc(II) ion perturbs the pK_a of the hydroxyethyl group of N-(β-hydroxyethyl) ethylenediamine to provide a high effective concentration of the nucleophile. Intramolecular nucleophilic atack then occurs at the carbonyl group of p-nitrophenyl picolinate:

2.1.8 Peptides and Amides

Very limited data exist on the base hydrolysis of amino acids, amides and simple peptides. A number of papers by Long[108] and his co-workers do, however, provide data on the acid-catalyzed hydrolysis of peptides determined using an automatic amino acid analyser.

A number of discussions are available on the general topic of metal peptide complexes[109-114] and the subject is currently reviewed[24]. A variety of metal ions, nickel(II)[115-16] copper(II)[117] and palladium(II)[118-9] promote the ionization of peptide hydrogens. Promotion of peptide hydrogen ionization increases in the series Co(II) < Ni(II) < Cu(II) < Pd(II)[118]. A number of bisdipeptide-cobalt(III) complexes containing deprotonated amide groups have been investigated and their spectral properties studied[111,120-2]. Protonation in acidic media occurs on oxygen rather than nitrogen to give the iminol tautomer of the dipeptide ligand[122-3]. Many studies have shown that the oxygen atom is the site of both protonation and metal ion coordination to the neutral amide function, while the nitrogen atom becomes the metal ion binding site if ionization of an amide hydrogen occurs.

Hydrolytic Reactions

An extensive series of studies of the metal ion-catalyzed hydrolysis of peptides and related compounds, including both homogeneous and heterogeneous reactions, has been carried out by Bamann and his collaborators[124]. A comprehensive review of the work of Bamann's group is available[124]. Highly charged ions such as thorium(IV) catalyze the hydrolysis of leucylglycylglycine at pH values as low as 5. The thorium(IV) species is very extensively hydrolyzed at this pH and the reaction is presumably heterogeneous. Gel hydrolysis is effected at relatively low temperatures (37°C), while observable effects are only obtained with such ions as copper(II) at temperatures of ca 70°C.

The hydrolysis of di- and tri-peptides is also catalyzed by rare earth ions at pH 8.6[126]. Cerium(IV) and cerium(III) are particularly effective catalysts even at a temperature of 37°C. The same reactions with La(III) as a catalyst are much slower and only occur at an appreciable rate at 70°C.

Copper(II), nickel(II) and cobalt(II) promote the hydrolysis of glycine amide in the pH range 9.35 to 10.35 and temperatures of 65 to 75°C[127]. In the absence of metal ions, phenylalanyglycine amide undergoes ring closure to 3-benzyl-2,5-diketopiperazine, but in the presence of copper(II) at pH 5 hydrolysis at both amide and peptide bonds is competitive with ring closure.

The copper(II)-catalyzed hydrolysis of glycylglycine has been studied in some detail. Copper(II) ions catalyze the hydrolysis of glycylglycine in the pH range 3.5 to 6 at 85°C[128]. The pH-rate profile has a maximum at pH 4.2, consistent with the view that the catalytically active species in the reaction is the carbonyl bonded complex. This view has subsequently been confirmed by other workers[129], in conjunction with an ir investigation of the structures of the copper(II) and zinc(II) complexes in D_2O solution[130]. Catalysis by cobalt(II)[131], and zinc(II), nickel(II) and manganese(II) has also been studied[132-4]. For zinc(II) a decrease in rate at high pH is not observed since ionization of the amide proton does not occur.

$$\underset{\text{(XXV)}}{\overset{\displaystyle \overset{+}{\underset{}{}}}{\begin{array}{c} \text{Cl} \\ \text{H}_2\text{N} \diagdown | \diagup \text{NH}_2 \\ \text{(} \diagup \text{Co} \diagdown \text{)} \\ \text{H}_2\text{N} \diagup | \diagdown \text{NH}_2 \\ \text{Cl} \end{array}}}
\qquad\qquad
\underset{\text{(XXVI)}}{\overset{\displaystyle 2+}{\begin{array}{c} \text{NH}_2 \\ \text{H}_2\text{N} \diagdown | \diagup \text{NH}_2\text{CH}_2\text{CO}_2\text{R} \\ \text{(} \diagup \text{Co} \diagdown \\ \text{H}_2\text{N} \diagup | \diagdown \text{NH}_2 \\ \text{Cl} \end{array}}}$$

2.2   Non-Labile Cobalt(III) Complexes

The use of kinetically inert cobalt(III) complexes has led to very significant developments in our understanding of metal ion promoted hydrolysis of esters, amides and peptides. The very elegant work of Buckingham and Sargeson and their co-workers has greatly clarified the mechanistic pathways available in reactions of this type. Some of the work is summarized in two reviews[22].

It is convenient to discuss these processes under the headings of ester hydrolysis, peptide (amide) hydrolysis and peptide bond formation, but it should be remembered that similar intermediates occur in each case and the discussion overlaps the three sections.

2.2.1. Ester Hydrolysis

The preparation of complexes of the type cis-$[Co(en)_2(NH_2CH_2CO_2R)Cl]Cl_2$ (XXVI) was first decribed by Alexander and Busch[135] who reacted the appropriate amino acid ester hydrochloride with trans-dichlorobis(ethylenediamine)cobalt(III) chloride (XXV) in aqueous solution. The free amino acid ester was generated in situ by the presence of a weakly coordinating base such as diethylamine.

The N-coordinated amino acid derivatives can be prepared by hydrolysis of the ester function with 4M HCl[136].

Alexander and Busch[137] studied the mercury(II) promoted hydrolysis of cis-$[Co(en)_2X(glyOR)]^{2+}$ in acid solution:

$$[Co(en)_2X(glyOR)]^{2+} + Hg^{2+} + H_2O \longrightarrow [Co(en)_2gly]^{2+}$$
$$+ ROH + H^+ + HgX^+$$
$$(X = Cl, Br; R = CH_3, C_2H_5, i\text{-}C_3H_7)$$

and suggested that a chelated ester species $[Co(en)_2glyOR)]^{3+}$ (XXVII) was the reactive intermediate in the Hg^{2+}-promoted reaction.

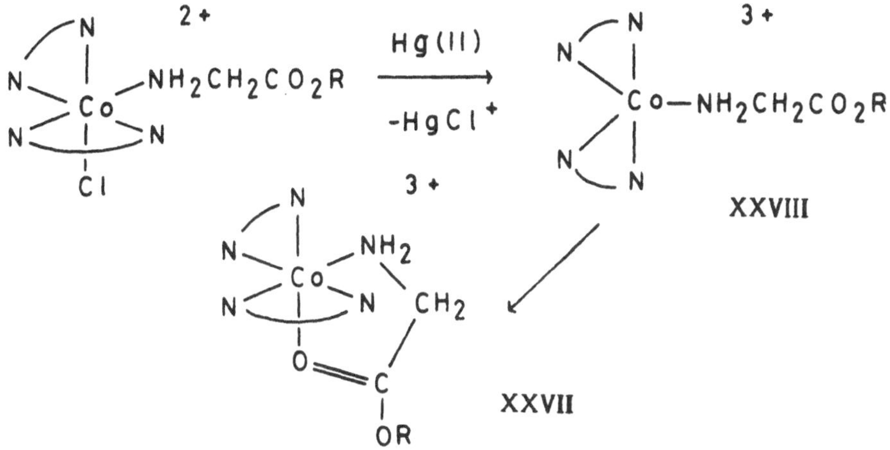

Evidence in support of the above scheme was provided by changes in the C=O stretching frequency as the monodentate ester (1735 cm^{-1}) was first chelated (1610 cm^{-1}) and then hydrolyzed to [Co(en)$_2$gly]$^{2+}$ (1640 cm^{-1}). The reaction presumably occurs via the 5-coordinate species (XXVIII), and the ester carbonyl oxygen competes so effectively with solvent water for the vacant coordination site that the chelated ester complex is formed exclusively. Wu and Busch[138] studied the analogous reactions of the t-butyl glycinate and obtained spectroscopic evidence for the chelated ester species. Hydrolysis is much slower than for the ethyl ester and occurs by alkyl-oxy rather than acyl-oxy fission.

Treatment of XXVI with AgClO$_4$ in acetone[139] allows the isolation of cis-[Co(en)$_2$(glyOR)] (ClO$_4$)$_3$ (XXIX). This complex can be used for the synthesis of peptide esters[139]; thus treatment of [Co(en)$_2$(glyOCH$_3$)] (ClO$_4$)$_3$ with amino acid or peptide esters in anhydrous sulfolane, dimethylsulfoxide or acetone solution gives the [Co(en)$_2$(peptide-OR)]$^{3+}$ ion (XXX).

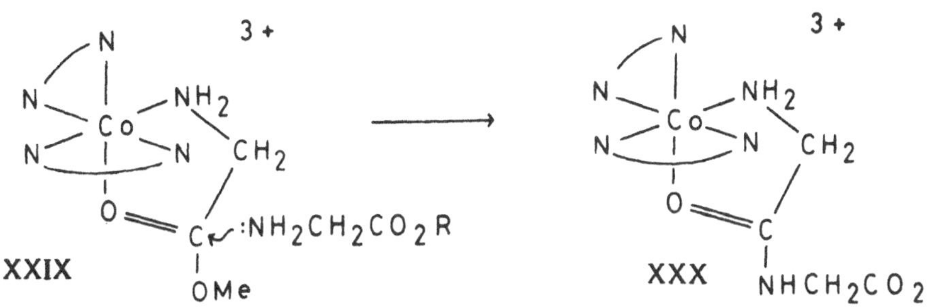

Kinetic data for the hydrolysis of $[Co(en)_2(glyOC_3H_7)](ClO_4)_3$ have been reported [140]. For the pH-independent attack by water $k = 1.15 \times 10^{-3}$ s^{-1} at 25°C, and evidence for OH$^-$ attack at pH 8.5 was obtained. The kinetic results confirmed the proposal that a chelated ester intermediate is formed following the Hg(II) promoted removal of halide ion from the chloropentammine complex. A similar intermediate[140] is generated following HOCl oxidation of the coordinated bromide in $[Co(en)_2Br(glyOR)]^{2+}$ but some of the aquo complex $[Co(en)_2(H_2O)(glyOR)]^{3+}$ is also produced in this reaction. Isotope studies using ^{18}O confirm that hydrolysis of the chelated ester occurs without rupture of the chelate ring.

It is of importance to extend the investigations to pH values of physiological significance, where base hydrolysis is expected to occur. The base hydrolysis of halopentammines of the type $[CoN_5X]$ (X=Cl,Br; N_5 = a series of nitrogen donors) normally occurs via an S_N1CB mechanism in which a dissociatively labile amido conjugate base is generated in a rapid pre-equilibrium step with hydroxide ion. The conjugate base loses halide ion in the slow rate-determining step to give a five-coordinate intermediate[141]. For ester complexes of the type $[Co(en)_2(glyOR)Cl]^{2+}$, the ester carbonyl group and solvent water can compete for the vacant site in the 5-coordinate intermediate (Scheme 2), giving rise to the chelated glycine ester species and the hydroxypentammine respectively.

Tracer experiments with ^{18}O [142] have established that the two main pathways to the chelated glycine complex (which is not the sole product) are (a) internal nucleophilic attack by bound hydroxide ion on the N-coordinated ester and (b) attack of "external" hydroxide on the chelated ester (Scheme 2).

The Hg(II)-promoted acid hydrolysis and base hydrolysis of the β_2-$[Co(trien)Cl(glyOEt)]^{2+}$ ion has also been studied[143]. Base hydrolysis of Cl$^-$ occurs some 10^3 times more rapidly than in the analogous bisethylenediamine system and only chelated glycine products are formed:

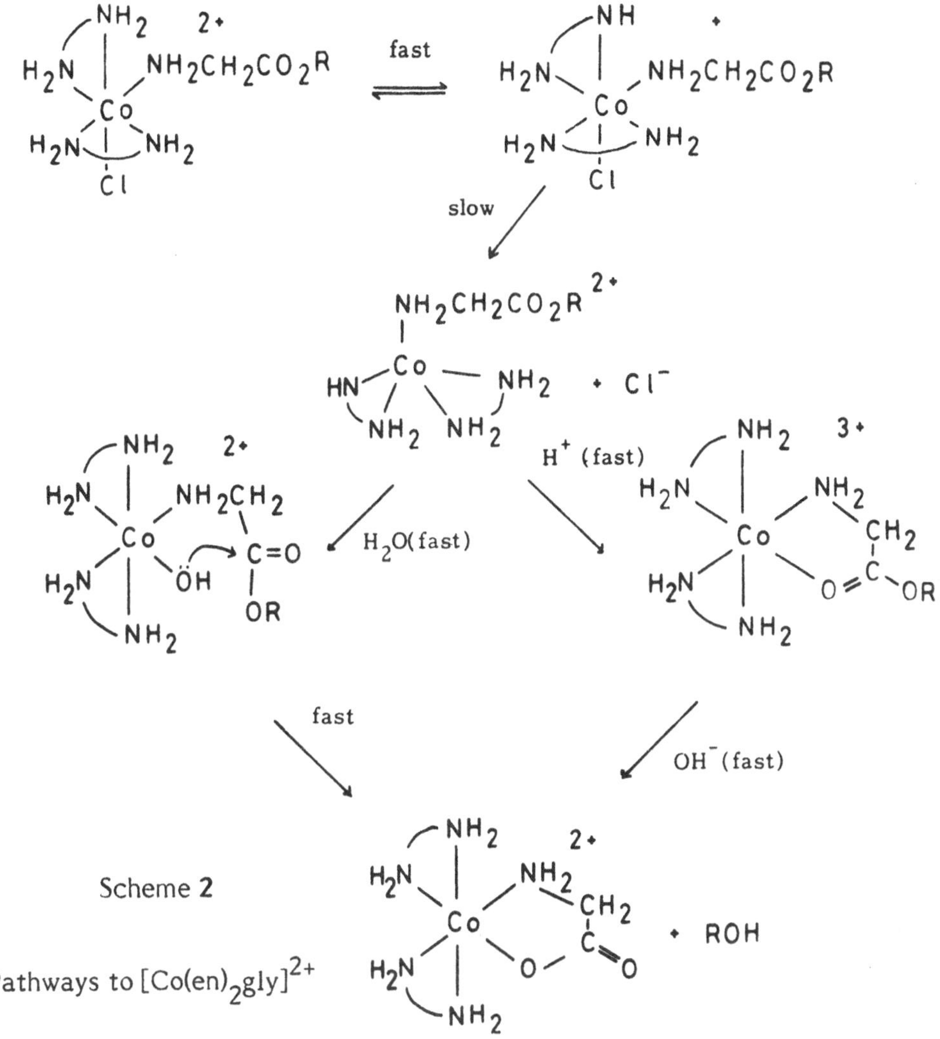

Scheme 2

Pathways to $[Co(en)_2gly]^{2+}$

$\beta_2\text{-}[Co(trien)Cl(glyOC_2H_5)]^{2+} + Hg^{2+} + H_2O \longrightarrow$

$\quad \beta_2\text{-}[Co(trien)gly]^{2+} + C_2H_5OH + HgCl^+ + H^+$

$\beta_2\text{-}[Co(trien)Cl(glyOC_2H_5)]^{2+} + OH^- \longrightarrow$

$\quad \beta_2\text{-}[Co(trien)gly]^{2+} + Cl^- + C_2H_5OH$

Base hydrolysis in the ^{18}O-labeled solvent established that some 84% of the $\beta_2\text{-}[Co(trien)gly]^{2+}$ product arises via the chelated ester and the remainder through intramolecular nucleophilic attack by coordinated hydroxide.

Detailed studies have now been made of the hydrolysis and aminolysis of the chelated ester species[144]:

$$
\underset{\textbf{XXXI}}{en_2Co\underset{O}{\overset{NH_2}{\diagdown}}\overset{CH_2}{\underset{C}{\diagup}}\overset{}{\diagdown}OC_3H_7}
\quad +H_2NR \quad
\underset{k_{-1}}{\overset{k_1\,(slow)}{\rightleftharpoons}}
\quad
\underset{\textbf{XXXII}}{en_2Co\underset{O}{\overset{NH_2}{\diagdown}}\overset{CH_2}{\underset{\underset{OC_3H_7}{|}}{\overset{|}{C-\overset{+}{N}H_2R}}}}
$$

$$
k_3 \diagup \text{ fast}
$$

$$
en_2Co\underset{O}{\overset{NH_2}{\diagdown}}\overset{CH_2}{\underset{C}{\diagup}}\diagdown NHR \quad +C_3H_7OH
$$

It has been found[144] that all oxygen and nitrogen bases act as specific nucleophiles rather than as general bases towards acyl-activated esters or amides of the general type (XXXI), irrespective of their basicity towards a proton.

Although the tetrahedral intermediate involved in the aminolysis of organic esters has not been observed directly, the tetrahedral intermediate (XXXII) has been detected spectrophotometrically[145] in reactions even though (XXXI) is highly activated towards nucleophilic attack and contains a "poor" leaving group.

The reaction of (XXXI) with glycine ethyl ester in DMSO solution has also been studied kinetically and the tetrahedral intermediate observed spectrophotometrically[145]. It has been suggested[146] that high concentrations of ClO_4^- alter the mechanism of the Hg(II)-promoted hydrolysis of the cis-[Co(en)$_2$X(glyOEt)]$^{2+}$ ion (X = Cl, Br), with high ClO_4^- concentrations (> 4M) giving a significant amount of [Co(en)$_2$(H$_2$O)-(glyOEt)]$^{3+}$ which only slowly ($t_{\frac{1}{2}} \sim 23$ min) reverted to the chelated ester intermediate with loss of bound water. Buckingham and Wein[147] have claimed that ClO_4^- does not alter the mechanism, but alters the rate and that in the presence of Hg(II) the reaction proceeeds exclusively via the chelated ester intermediate, a view consistent with previous work[140,143]. Quite marked variations have been observed in the ability of pendant ligand groups to become incorporated in five-coordinate intermediates generated

XXXIII

by Hg(II) and by OH⁻. Thus base hydrolysis of the complex cis-
[CoCl(en)$_2$(NH$_2$CH$_2$CH$_2$NH$_2$)]$^{2+}$ containing monodentate ethylenediamine
gives (ca 97%) the complex [Co(en)$_2$OH(NH$_2$CH$_2$CH$_2$NH$_2$)] containing a
pendant ethylenediamine[148]. Similar observations[149] have been made for
the complex **(XXXIII)** prepared from 1,1,1-tris(aminomethyl)ethane, where
base hydrolysis gives the hydroxo-complex in a clean reaction without any
formation of a hexammine species.

A number of kinetic studies of chloride loss from the N-coordinated
glycine complex cis-[CoCl(en)$_2$(NH$_2$CH$_2$CO$_2$H)]$^{2+}$ have also appeared[150-
151]. Base hydrolysis of cis-[CoCl(en)$_2${NH$_2$(CH$_2$)$_5$CO$_2$Me}]$^{2+}$ [152] and cis-
[CoCl(en)$_2${NH$_2$(CH$_2$)$_3$CO$_2$Me}]$^{2+}$ [153] have been studied. For the first
complex, following loss of chloride ion, a slower base hydrolysis of the
pendant ester function occurs. The final product is the hydroxypentammine
and the reaction is free from the intramolecular effects observed with the
glycine esters. Similar results were obtained[153] for the second complex. In
this case coordination to cobalt prevents lactamization of
NH$_2$(CH$_2$)$_3$CO$_2$Me to 2-pyrrolidinone from competing with base hydrolysis.
Base hydrolysis of the N-coordinated ester is only ca 7 times faster than for
NH$_2$(CH$_2$)$_3$CO$_2$Me. N-Coordination to cobalt has similar effects to
protonation of the amino group. Base hydrolysis of the aminoalcohol ester
complex cis-[Co(en)$_2$X(NH$_2$CH$_2$CH$_2$OCOMe)]$^{2+}$ (X = Cl or Br) has been
investigated in detail[154]. The complex hydrolyzes in two steps, the first
involving Cl⁻ or Br⁻ loss and the second ester hydrolysis. In basic solution 2-
aminoethyl acetate **(XXXIV)** undergoes rapid base catalyzed isomerization
to 2-acetylaminoethanol **(XXXVI)** via the tetrahedral intermediate **(XXXV)**.
Coordination to cobalt prevents this acyl transfer reaction from occurring.

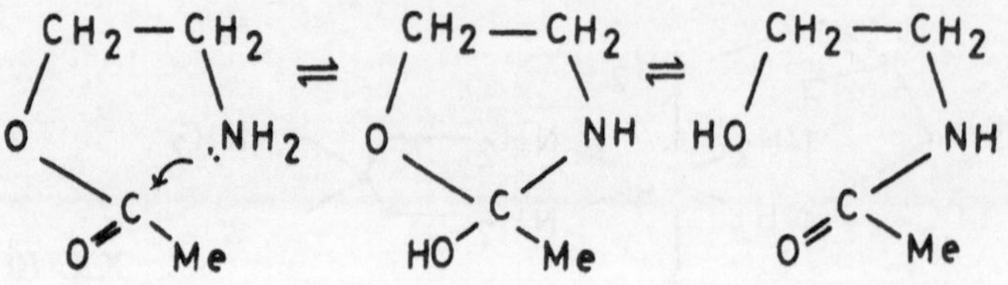

XXXIV   XXXV   XXXVI

In base hydrolysis reactions additional complexities may arise due to intramolecular attack by the amido conjugate base or displacement of the substrate itself in an S_N1CB process. Thus base hydrolysis of $[Co(NH_3)_5(NH_2CH_2CO_2R)]^{3+}$ **(XXXVII)** in the pH range 9-14 results in the formation of both $[Co(NH_3)_5(NH_2CH_2CO_2^-)]^{2+}$, containing the monodentate glycinate anion, and $[Co(NH_3)_4(NH_2CH_2CONH)]^{2+}$, containing glycine amide chelated via both nitrogen atoms[155]. This latter reaction can be rationalized by intramolecular attack in the conjugate base **(XXXVIII)**. N-coordinated aziridenes also occur in the base hydrolysis of $[Co(en)_2Br(NH_2CH_2Br)]^{2+}$ by a similar intramolecular pathway[156]. Base hydrolysis[157] of oxygen-bonded $[Co(NH_3)_5DMF]^{3+}$ (DMF = N,N-dimethylformamide) leads to two reaction pathways. The major pathway corresponds to hydroxide ion attack at the carbonyl center giving $[Co(NH_3)_5OOCH]^{2+}$ and $HN(CH_3)_2$. The minor path leads to $[Co(NH_3)_5OH]^{2+}$ and DMF and is the dissociative conjugate base process. Amide cleavage is accelerated $\geqslant 10^4$ times due solely to a more favorable entropy term.

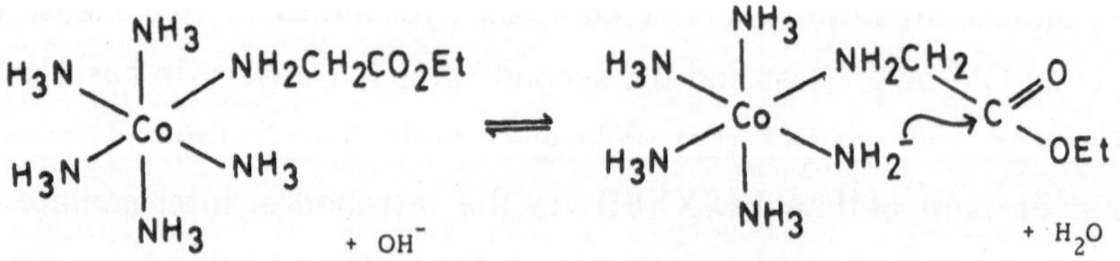

XXXVII   XXXVIII

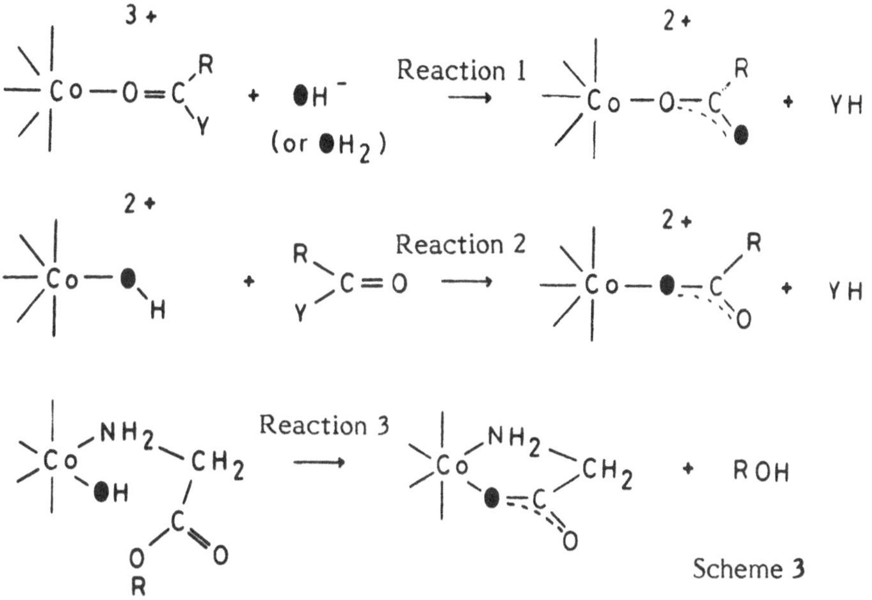

Scheme 3

It is appropriate to consider the "metal-carbonyl" and "metal-hydroxide" mechanisms in some detail (Scheme 3). Reaction 1 provides rate enhancements of 10^4-10^6 for all substrates independent of the leaving group Y.[137,140,143,158-9]. Reaction 2 is effective only with the more reactive species (CO_2, anhydrides, aldehydes and esters with good leaving groups)[23]. As a result the hydrolysis of amono acid esters, amides and simple peptides has not been observed in a bimolecular "metal hydroxide" process. In addition, the rate enhancement observed in Reaction 1 is due solely to a more favorable entropy term, whereas both $\Delta H^{\ddagger}$ and $\Delta S^{\ddagger}$ contribute in Reaction 2. Reaction 1 also leads to nucleophilic attack by species other than OH^- (NH_2R, ROH, H_2O) and general acid or general base catalysis can occur[139,144]. Only hydrolysis is observed with metal hydroxides and general acid or general base catalysis has not been observed[102].

Reaction 3, the intramolecular counterpart of Reaction 2, has been observed for amino acid esters[155], amides[160] and nitriles[161] where five-and six-membered chelate rings can result. For aminoacetonitrile a rate enhancement of $\sim 10^{11}$ occurs at pH 7, and this may be compared with an acceleration of 10^6 for the Reaction 1 analogue. For these reactions $\Delta H^{\ddagger}$ values become of great significance.

In the case of amino acid ester and amide complexes, the intramolecular hydrolysis reaction was not observed directly, but was deduced from the results of ^{18}O tracer studies. However, the cis-aqua complexes containing glycinamide, glycylglycine and isopropylglycylglycinate have been isolated and their subsequent cyclization studied over the pH range 0-14.[162-3].

2.2.2. Peptide and Amide Hydrolysis

A number of complexes of the type $[CoN_4(OH)(OH_2)]^{2+}$ (N_4 = a system of four nitrogen donors) stoichiometrically cleave the N-terminal amino acid from di- or tripeptides. Such reactions have been described for $N_4 = en_2$[164-5], trien[166-7] and tren[5]. Some of the difficulties in the use of β-$[Co(trien)(OH)(OH_2)]^{2+}$ for the sequential analysis of peptides have been discussed[169-172,174] and an investigation[171] of the relative effectiveness of the complexes cis-β-$[Co(trien)(OH)(OH_2)]^{2+}$, cis-$\alpha$-$[Co(trien)(OH)-(OH_2)]^{2+}$ and $[Co(NH_3)_4(OH)(OH_2)]^{2+}$ in the hydrolysis of glycyl dipeptides and dipeptide esters has shown that the cis-β-$[Co(trien)(OH)(OH_2)]^{2+}$ species is at least 50 times more effective than any of the other species.

The $[Co(EDDA)(OH)(OH_2)]$ ion (EDDA = ethylenediaminediacetate) has been reported[173] to be as effective as cis-β-$[Co(trien)(OH)(OH_2)]^{2+}$ in the hydrolysis of L-ala-L-Phe, so that the charge carried by the complex is of limited importance. A variety of other ligand system have been evaluated[174], $[Co(eee)(OH)(OH_2)]^{2+}$ (eee = 1, 8-diamino-3,6-dithia-octane) and $[Co(bpdah)(OH)(OH_2)]^{2+}$ (bpdah = 1,6-bis(α-pyridyl)-2,5-diazahexane). Both complexes are inactive towards glycylglycine at pH 8 and 65°C.

Cyclen complexes[175] and complexes of the diethylenetriamineacetic acids[176] have also been prepared as possible peptide cleaving reagents; however their potential in this area has not been assessed. The reasons for the marked dependence of these reactions on the "inert" ligand system employed is not fully clear and would bear detailed investigation.

Gillard[177] has shown that $[Co(dien)X_3]$ complexes react with glycylglycylglycine to give $[Co(dien)(glygly)]^+$ and free glycine, so that, in this case, the two N-terminal amino acids can be removed in a single step. Wu and Busch[178] have described the preparation of a number of complexes

$$\begin{array}{c}
X \\
HN{\underset{H_2N}{\overset{}{\diagup}}}\overset{|}{\underset{|}{Co}}{\overset{NH_2}{\underset{NH_2}{\diagdown}}} \\
O \diagdown \underset{C}{} \diagup CH_2 \\
\overset{|}{N}HCH_2CO_2R
\end{array}$$

XXXIX

of the type $[CoX(dien)(glyglyOR)]^{2+}$ and $[CoX(dien)(glyO)]^{2+}$ where X = Cl, NO_2. The formation of peptides using the three site moiety $Co(dien)^{3+}$ has been shown to occur[178] and $[Co(dien)OH(OH_2)_2]^{2+}$ cleaves the peptide bond of α-L-aspartylglycine[179]. Girgis and Legg[179] have studied the $[Co(trien)(OH)(OH_2)]^{2+}$ promoted hydrolysis of the diethyl ester of L-aspartic acid and dipeptides containing aspartic acid, glutamic acid and methionine. If the trifunctional amino-acid is N-terminal as in α-L-aspartylglycine and L-aspartic acid diethyl ester, essentially no hydrolysis occurs with the formation of $[Co(dien)(L-AspO)]^+$ where the amino-acid is coordinated as a tridentate. The $[Co(dien)(glyglyOR)X]^{2+}$ complexes[178] are believed to have the configuration **XXXIX**, which in the dien ligand adopts a mer-configuration and the ligand X lies trans to oxygen. A series of trans (O,X)-$[CoX(dien)(AA)]^+$ complexes (AA$^-$ = amino acid anion, X = Cl,NO_2,CN) have been prepared[180] and the endo- and exo-isomers which result from the asymmetry of the secondary nitrogen donor of the dien ligand identified for X = CN. Complexes such as **XXXIX** are likely intermediates in the reaction of $[Co(dien)(OH)(OH_2)_2]^{2+}$ with dipeptides [177,181]. Hay and Piplani[182] have studied the kinetics of base hydrolysis of a variety of complexes of type **XXXXIX** and have determined k_{OH} values for peptide bond hydrolysis, NO_2 and Cl hydrolysis (Table 11).

The rate constant k_{OH} for peptide bond hydrolysis falls within the range 0.67 - 0.88 M^{-1} s^{-1} at 25°C and base hydrolysis of the complexed peptide is ca 2 x 10^4 times faster than for the uncomplexed peptide. The pK_a of the amide proton of $[Co(dien)(glyglyO)NO_2]^+$ is > 11[178,182].

The peptide cleavage reactions are stoichiometric rather than catalytic, but in spite of this restriction they do provide interesting models

Table 11. Base hydrolysis of the peptide bond and the NO_2 and Cl ligands in Co(dien) complexes at $\mu = 0.1M$ and 25°C (k in M^{-1} s^{-1}). Data from ref. 182. For additional data, see ref. 542

Initial Complex	peptide k_{OH}	NO_2 k_{OH}	Cl k_{OH}
$[Co(dien)(glyglyO)NO_2]^+$	0.88		
$[Co(dien)(glyglyO)NO_2]^+$	0.67		
$[Co(dien)(glyglyOEt)NO_2]^{2+}$	0.68	2.5×10^{-2}	
$[CoCl(dien)(glyglyOEt)]^{2+}$	7.0		1.1×10^6
$[CoCl(dien)(glyO)]^+$	-		1.3×10^4
$[CoCl(dien)(glyNH_2)]^{2+}$	-		5.85×10^5
$[Co(dien)(glyO)NO_2]^+$		2.45×10^{-2}	

For base hydrolysis of glycylglycine, $k_{OH} \sim 4 \times 10^{-5}$ M^{-1} s^{-1} at 26°C.

for enzymes such as carboxypeptidase A where the X-ray data[183] indicate displacement of zinc-bound water by the carbonyl group of the peptide substrate. Lewis acid catalysis by zinc(II) plays an important role in catalysis and could account for a factor of ca 10^6 in the overall rate enhancement observed.

Since the first reports that cis-β-$[Co(trien)(OH)(H_2O)]^{2+}$ rapidly hydrolyzes amino acid esters and amides and acts selectively in the N-terminal hydrolysis of small peptides, many detailed studies of the reaction have appeared. Cobalt(III) complexes of triethylenetetramine, of the type $[Co(trien)X_2]^{n+}$, can exist in three geometric configurations designated cis-α, cis-β and trans (see structures below). The trans-isomer need not be considered as it does not allow cis-chelation of amino acid. As an amino acid contains two different donor atoms, the reaction of an amino acid with the cis-β isomer gives two isomeric complexes. These isomers are designated β_1 where the amino acid nitrogen is trans to a primary amino group of trien

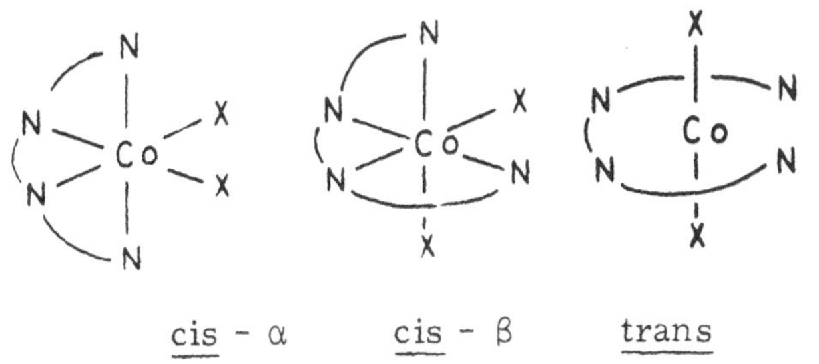

cis - α   cis - β   trans

Isomers of $[Co(III)(trien)X_2]^{n+}$

and β_2 where the amino acid nitrogen is _trans_ to a secondary nitrogen of trien[184-5].

The reaction of cis-β-$[Co(trien)(OH)(OH_2)]^{2+}$ with glycine, L-alanine, L-valine, L-leucine etc. in the pH range 7.5-8.0 and 60°C gives quantitative yields of β-$[Co(trien)AA]^{2+}$ (AA = amino acid anion)[164]. The same reagent also reacts with peptides, amino acid esters and amino-acid amides to give β-$[Co(trien)AA]^{2+}$, in which the N-terminal amino acid residue is cleaved from the amide or ester[166]. Cleavage is not observed with N-protected peptides such as N-carbobenzyloxyglycyl-L-phenylalanine[164]. Kinetic studies[166] on the hydrolysis of amino acid esters and glycinamide suggest a rate law first order in β-$[Co(trien)(OH)(H_2O)]^{2+}$ and first order in the free base form of the amino acid derivative. The proposed mechanism is shown in Scheme 4. Collman and Kimura[186] first isolated the carbonyl bonded complex **XXXIXb**. Detailed kinetic

Scheme 4. Proposed mechanisms for peptide hydrolysis.

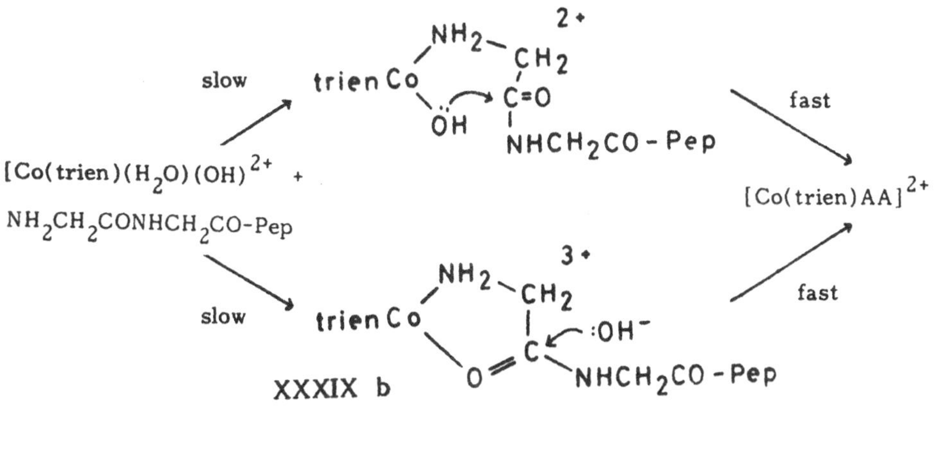

XXXIX b

Table 12. Rate Constants for the Base Hydrolysis of Ester Amide and Peptide Bonds in Various Cobalt(III) Complexes (25°C, μ = 1.0M). Data from ref. 158

Complex	$k/M^{-1} s^{-1}$
$[Co(en)_2(glyOC_3H_7)]^{3+}$	1.5×10^6
$[Co(en)_2(glyNH_2)]^{3+}$	25
$[Co(en)_2(glyNHCH_3)]^{3+}$	1.6
$[Co(en)_2(glyN(CH_3)_2)]^{3+}$	1.1
$[Co(en)_2(glyglyO)]^{2+}$	2.6
β_2-$[Co(trien)(glyNHCH_3)]^{3+}$	2
β_2-$[Co(trien)(glyglyOCH_3)]^{3+}$	5
β_2-$[Co(trien)(glyglyOCH_3)]^{3+}$	5
β_2-$[Co(trien)(glyglyOC_3H_7)]^{3+}$	3
β_2-$[Co(trien)(glyglyO)]^{2+}$	3

information is now available on the base hydrolysis rates of such carbonyl bonded complexes[158,187] (Table 12). At high pH (~11) deprotonation of the amide or peptide nitrogen occurs, and the deprotonated complexes are inert to base hydrolysis. For peptide bond cleavage k_{OH} values of 2-5 $M^{-1} s^{-1}$ are observed, giving rate accelerations of 10^4-10^6, values somewhat similar to those noted for the cobalt(III)-dien systems. The hydrolysis of carbonyl-bonded amides and peptides can be rationalized in terms of Scheme 5. A number of comparisons of intramolecular and intermolecular hydrolysis have been given.[162,160] Base hydrolysis[160] of the cis-$[Co(en)_2Br(glyNR_1R_2)]^{2+}$ ion over the pH range 9-14 results in two pathways for the production of $[Co(en)_2gly]^{2+}$. Following loss of bromide, competition for the 5-coordinate intermediate by solvent water and the amide carbonyl results in formation of the cis-$[Co(en)_2OH(glyNR_1NR_2)]^{2+}$ and $[Co(en)_2(glyNR_1R_2)]^{3+}$ (carbonyl chelated) in the ratio 54:46 (R_1=R_2=H). Intramolecular hydrolysis in the cis-$[Co(en)_2OH(glyNR_1NR_2)]^{2+}$ ion is at least 10^7 and possibly 10^{11} times faster

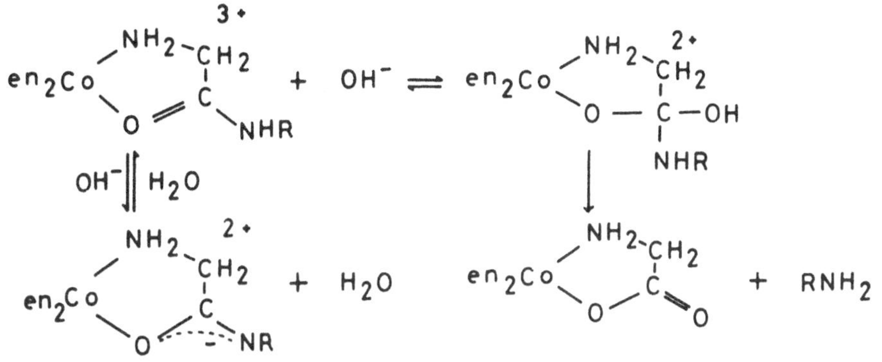

$$\text{en}_2\text{Co}\left[\begin{array}{c}\text{NH}_2-\text{CH}_2\\ \text{C}\\ \text{O}=\text{C}\\ \text{NHR}\end{array}\right]^{3+} + \text{OH}^- \rightleftharpoons \text{en}_2\text{Co}\left[\begin{array}{c}\text{NH}_2\\ \text{CH}_2\\ \text{O}-\text{C}-\text{OH}\\ \text{NHR}\end{array}\right]^{2+}$$

$$\text{OH}^-\big\|\,\text{H}_2\text{O}$$

$$\text{en}_2\text{Co}\left[\begin{array}{c}\text{NH}_2\\ \text{CH}_2\\ \text{O}\cdots\text{NR}\end{array}\right]^{2+} + \text{H}_2\text{O} \qquad \text{en}_2\text{Co}\left[\begin{array}{c}\text{NH}_2-\text{CH}_2\\ \text{C}\\ \text{O}\qquad\text{O}\end{array}\right] + \text{RNH}_2$$

Scheme 5. Base hydrolysis of carbonyl-bonded amides and peptides.

than base hydrolysis of uncoordinated glycine amide. Such a mechanism is markedly more efficient than attack by "external" hydroxide on the chelated species. Intramolecular hydrolysis of glycinamide and glycine dipeptides in complexes cis-[Co(en)$_2$Br(glyNHR)]$^{2+}$ (R = H, CH$_2$CO$_2$C$_3$H$_7$ and CH$_2$CO$_2^-$) has recently been studied in detail [162-3]. Mercury(II)-catalyzed removal of Br$^-$ from these complexes results in the immediate formation of the chelated amide or dipeptide and no aquo intermediate. There is also full retention of configuration about the cobalt(III) center. Similar results have been noted for other cis-[Co(en)$_2$X(NH$_2$R)]$^{2+}$ complexes where X = Cl, and NH$_2$R = glycinamide[160], glycine esters[137,140,143] or glycinate. The carbonyl or carboxylic acid function appears to be an excellent competitor for the five coordinate intermediate generated by mercury(II). HOCl-catalyzed oxidation of cis-[Co(en)$_2$Br(glyglyOC$_3$H$_7$)]$^{2+}$ results in the formation of some cis-[Co(en)$_2$(OH$_2$)(glyglyOC$_3$H$_7$)]$^{2+}$ in addition to [Co(en)$_2$(glyglyOC$_3$H$_7$)]$^{3+}$. Base hydrolysis (pH 9-14) results in the formation of some trans-[Co(en)$_2$(OH)(glyNHR)]$^{2+}$ ($\sim$6.2%) as well as cis-[Co(en)$_2$(OH)(glyNHR)]$^{2+}$ and [Co(en)$_2$(glyNHR)]$^{3+}$.

The reactions[163] of the cis-[Co(en)$_2$(OH$_2$/OH)(glyNHR)]$^{n+}$ (R = H, CH$_2$CO$_2$C$_3$H$_7$, CH$_2$CO$_2^-$) species have been investigated both in the presence and absence of buffers. For the dipeptide complex with R = CH$_2$CO$_2$C$_3$H$_7$, both the aquo and hydroxo species form [Co(en)$_2$(glyO)]$^{2+}$ but loss of hydroxide also occurs resulting in the chelated amide

$[Co(en)_2(glyNHR)]^{3+}$, as previously suggested by the other workers[188] for the glycine ester system.

A combination of rate and product analysis data suggests that the initial cyclization is rate-determining under all conditions. Buffer species act as general bases in this rate-determining process, but they also enhance the formation of the hydrolysis product. Coordinated water is more reactive than coordinated hydroxide owing largely to a more positive $\Delta S^{\ddagger}$. The aquo dipeptide complex hydrolyzes 10^{11} times more rapidly than the free dipeptide at pH 5, while a factor of 10^7 occurs for the hydroxo dipeptide complex at pH 8. (0.1M buffer). These simple "model" systems provide rate enhancements quite similar to those observed with metalloenzyme proteases.

The more recent work[162] also clarifies a previous report on the hydrolysis of the cis-$[Co(en)_2Br(glyNH_2)]^{2+}$ ion. In one study [160] it was concluded that the hydroxo glycinamide ion reacted rapidly under the conditions of base hydrolysis and that the subsequent process observed spectrophotometrically was hydrolysis of the chelated amide. Chan and Chan[190] had earlier reported that a relatively stable hydroxo-species was first formed, and this conclusion has now been confirmed. Base hydrolysis of cis-$[Co(en)_2Br(glyNH_2)]^{2+}$ gives considerable amounts of the hydroxo amide $[Co(en)_2(OH)(glyNH_2)]^{2+}$, in addition to the chelated amide.

Some further work has also been published on the $[Co(tren)(OH)(OH_2)]^{2+}$ promoted reactions. Hydrolysis of gly-L-phe at pH 7 gives the orange isomer of $[Co(tren)(glyO)]^{2+}$, and the orange isomer is also obtained when $[Co(tren)(glyOR)Cl]^{2+}$ is hydrolyzed in acid solution in the presence of Hg(II)[189]. The crystal structure[189] establishes that the glycine is coordinated with its nitrogen atom trans to the tertiary amino group of tren.

2.2.3 Peptide Bond Formation

The rapid condensation of glycine ethyl ester to form glycylglycine ethyl ester using the $Co(trien)^{3+}$ moiety as an N-terminal protecting group and an activating reagent was first established in 1967[191]. The rapid and quantitative formation of chelated peptide esters using the isolated intermediate $[Co(en)_2(glyOCH_3)]\ (ClO_4)_3$ has been shown to occur[139].

Treatment of $[Co(en)_2(glyOCH_3)]$ $(ClO_4)_3$ with amino acid or peptide esters in anhydrous sulfolane, dimethyl sulfoxide or acetone solutions results in the formation of the $[Co(en)_2(peptide-OR)]^{3+}$ ion. The reactions are rapid at $20°C$, being complete in minutes.

Peptide bond formation in the coordination sphere of cobalt(III) is also observed[186] in the reaction of cis-[Co(trien)Cl$_2$]Cl with glycine esters to give cis-β[Co(trien)(glyglyOR)]$^{3+}$. Kinetic and product analysis data for the reaction of $[Co(en)_2(glyOC_3H_7)]$ $(ClO_4)_3$ with H_2O, HO^-, NH_2CH_2CN and $CH_3CO_2^-$ have been reported[144]. For base hydrolysis $k_{OH} = 0.8 \times 10^6$ M^{-1} s^{-1} and for water attack $k_{H_2O} = 1.89 \times 10^{-5}$ M^{-1} s^{-1} at $25°C$ and $\mu = 1.0M$. The preparation of $[Co(en)_2(glyOC_3H_7)](ClO_4)_3$ has been described[145]. Attack of the ester at the carbonyl center occurs with $k = 14$ M^{-1} s^{-1}. At $25°C$, a second reaction is observed attributed to general acid catalyzed removal of isopropanol from the conjugate base of the tetrahedral intermediate.

The structure of the complex cis-[Co(trien)(glyglyOC$_2$H$_5$)] $(ClO_4)_3$. H_2O has been determined by X-ray diffraction[192]. The peptide ester is coordinated through the terminal amino group and the carbonyl oxygen of the same amino acid residue.

The reaction of $[Co(dien)X_3]$ $(X = Cl, NO_2)$ with glycine esters and glycylglycine esters gives $[Co(dien)(glyglyOR)X]^{2+}$ and $[Co(dien)(glyglyOR)]^{2+}$ respectively[178].

3.   CONDENSATION REACTIONS OF AMINO-ACIDS COORDINATED
      TO METAL IONS

Coordination of α-amino-acids to metal ions has long been known to increase the acidity of the C-H bond at the α-carbon atom, and various reviews of the topic are available[193-4]. This effect may be important in some biological transformations of amino acids in the presence of metal ions[195-6].

Since the pioneering work of Akabori and co-workers[197] a great deal of literature has appeared concerning the condensation reactions of metal-coordinated amino-acids[198-205] or small peptides[206] with various

$$
\begin{array}{c}
\diagup\!\!\!\!\diagdown \!\!\!\overset{NH_2-CH_2}{\underset{O-C=O}{Cu}} \quad +\quad CH_3CHO
\end{array}
\quad \xrightarrow[\;\;\;\;\;\;]{OH^-}\quad
\begin{array}{c}
\diagup\!\!\!\!\diagdown \!\!\!\overset{NH_2-CH-\overset{CH_3}{\underset{OH}{CH}}}{\underset{O-C=O}{Cu}}
\end{array}
$$

XL

aldehydes. The reaction has also been carried out starting from chiral octahedral cobalt(III) complexes containing glycine[208-210] with relatively high asymmetric yields. Later it was discovered that the use of an amino-acid Schiff-base complex instead of a simple amino-acid metal complex increases the nucleophilicity at the amino-acid α-carbon atom and at the same time prevents the occurence of N-alkylation[211-6].

Such an increase in reactivity has been exploited for reactions with aldehydes susceptible to self-condensation since it is possible to use mild basic conditions. Recently, chiral bis(salicylideneaminoacidato)cobalt(III) complexes have been used to perform a reasonably good asymmetric condensation of acetaldehyde with glycine with high overall yields[217].

3.1   Copper(II) Complexes

The original observation[197a] that copper(II) ions promote the base catalyzed condensation of acetaldehyde with glycine to give threonine (XL) was rapidly followed by a number of reports similar reactions[198,219], but little further attention was paid to simple copper(II)-amino acid systems until more recently[203-4]. The related Schiff base complexes did, however, continue to attract some attention, as described later, and it seems generally to have been presumed that the essential intermediate was the carbanion. It has been shown[220] that alanine specifically deuterated in the 2-position could be isolated from the reaction of $[Cu(ala)_2]$ with basic D_2O. The reaction was accompanied by loss of stereochemical integrity in the ligand, suggesting a carbanion intermediate. The reaction was thought to be steroselective since the rate of incorporation of deuteruim (determined by ^{1}H nmr of the ligand after removal of copper(II)) was apparently faster than racemization, a result which has been questioned by other workers[221]. This system has been re-examined[222], and while the carbanion mechanism

XLI

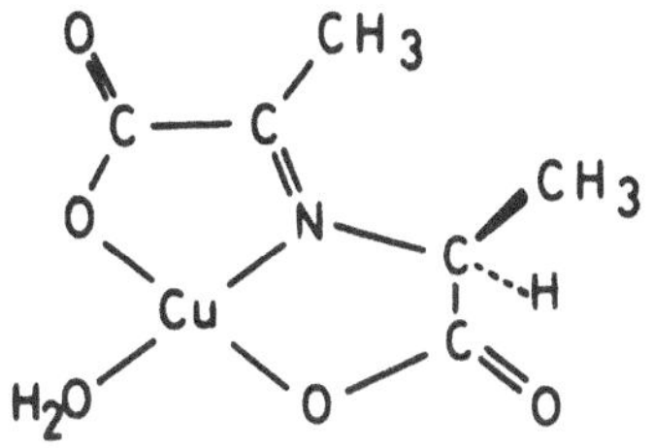

probably exists, it has been shown that in the case of alanine the key intermediate is the aqua(N-pyruvylidene-S-alaninato)copper(II) (**XLI**) formed by condensation of S-alanine with pyruvate which is generated in situ. It has been proposed that the reactions between acetaldehyde and metal-coordinated glycine in basic solution proceed via the intermediate formation of an N-hydroxyethyl glycine derivative **XLII** (Scheme 6). The bis(oxazolidine)copper(II) complex **XLIII** (R = CH_3) has been characterized by X-ray crystallography[198], and treatment of the complex with hydrogen sulfide in acid solution gives threonine.

The aldol condensation of formaldehyde with glycine complexes is

Scheme 6

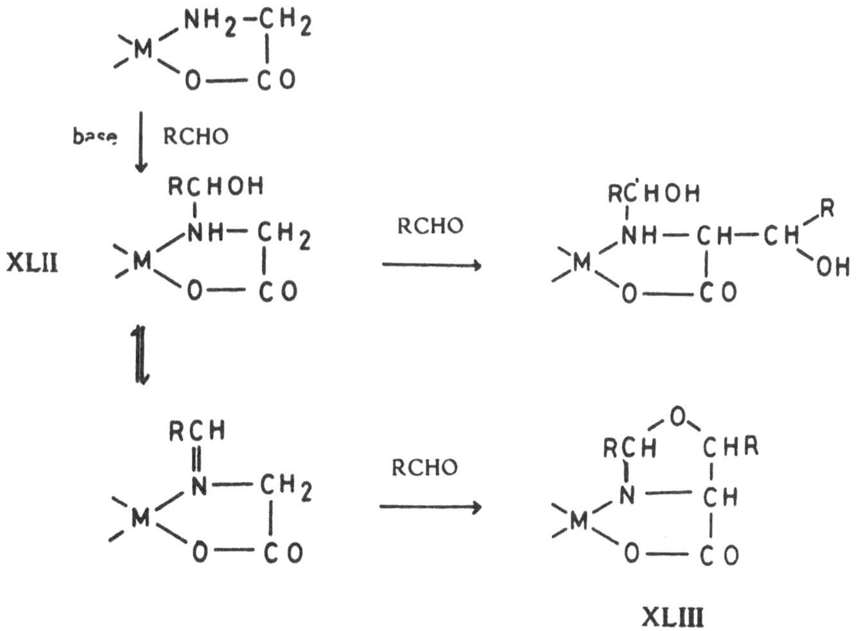

XLIII

much less well understood. Comparative studies show that formaldehyde is unable to form serine by reation of bis(glycinato)copper(II) under the conditions in which actaldehyde gives threonine in ca 40% yield[213]. Recent work has shown that condensation of formaldehyde at the α-carbon of [Cu(glyO)$_2$] (glyO = glycinate) is preceded by condensation on the amino group and is followed by cyclization to give on oxazolidine-type ring[223]. No evidence for such a mechanism was obtained in the case of [Co(en)$_2$(glyO)]$^{2+}$.

3.2   Cobalt(III) Complexes

The cobalt(III) complexes of amino acids, such as Co(gly)$_3$, are quite robust. Thus both isomers of this complex ion can be precipitated from 1M HCl or 1M NaOH, and they decompose only slowly over several days in boiling water[203]. The activation of amino-acids in cobalt(III) complexes was first reported using [Co(gly)$_3$][197b], and later similar results were obtained for other cobalt(III) glycinate complexes[200,203-4,208-9]. Activation has been studied by following the ^{1}H-^{2}H exchange of the methylene protons in basic solutions of D$_2$O[220,224-9], thus providing additional support for the carbanion mechanism. The reaction has been elegantly employed in the stereospecific synthesis of ^{2}H labeled amino-acids[230]. Stereoselectivity in a similar sense was noted by Murakami and Takahashi[208] who reported that the reaction of acetaldehyde with resolved [Co(en)$_2$(gly)]$^{2+}$ gave a small but detectable asymmetric yield of threonine. The threonine/allothreonine ratio was variable and depended on the conditions.

The reaction has been re-examined[209], and it was found that by careful control of the experimental conditions a higher yield of optically active threonine could be obtained. An explanation for the observed sterochemical preference has been given[210].

In chiral [Co(en)$_2$gly]$^{2+}$, the two hydrogens occupy diasteroisomeric positions, but Sargeson and co-workers[231] have shown that the rates of exchange of the pro-R and pro-S protons are indistinguishable. However, by using N-benzylglycine and thereby introducing a bulky substituent into the amino-acid nitrogen, it was possible to generate a distinct sterochemical difference between the two hydrogens, with the pro-S undergoing exchange more readily than the pro-R for the Λ-configuration on cobalt[231]. The presence of a bulky group attached to the nitrogen atom could therefore lead directly to the observed stereoselectivity at the 2-carbon. Such a substituent could be generated by the reaction of the aldehyde with the nitrogen of the amino acid prior to condensation of a further molecule of aldehyde at the 2-carbon. Such reactions leading to oxazolidine derivatives occur for amino acid complexes of copper(II)[201-2], and a similar reaction, although with different products, occurs with amino groups bound to cobalt(III)[204,232]. For a further discussion of this problem a review[193] should be consulted.

Although the reaction of $[Co(en)_2gly]^{2+}$ with acetaldehyde generally gives only threonine/allothreonine, a variety of products can be obtained by reaction with formaldehyde, depending upn the reaction conditions. Gillard and co-workers[203,200] have observed the expected formation of serine at pH 8. However, another group,[204] using a saturated aqueous solution of lithium carbonate as the base, observed the formation of the disubstituted product, 2-hydroxymethylserine, both as the $[Co(en)_2AA]^{2+}$ complex, and as [2-hydroxymethylserine-1,4,8,11-tetra-aza-6,13-dioxacyclotetradecane cobalt(III)]$^{2+}$ (XLIV), a complex of the macrocycle XLV. Somewhat similar reactions have been used to prepare caged metal ions[233] of the sepulchrate type.

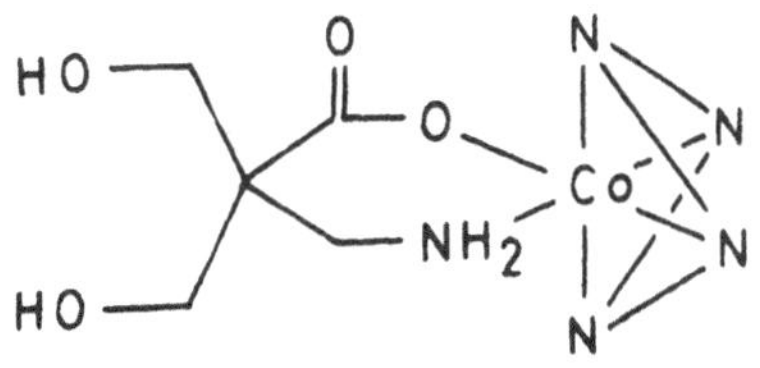

XLIV 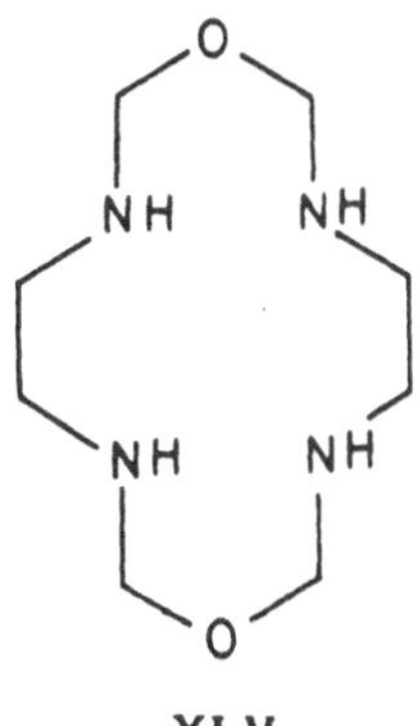   XLV

3.3   Schiff Base Complexes

The Schiff bases derived from amino acids and pyridoxal (vitamin B_6) have attracted considerable attention, principally as a result of the biological significance of vitamin B_6, and the realization that many of the reactions catalyzed by vitamin B_6 in the presence of enzymes could also be brought about in the absence of enzyme by using pyridoxal and various metal ions. In a classic series of papers[234-40], Snell and his co-workers reported the non-enzymic deamination[238], transamination[212] (with keto acid and pyridoxamine), racemization[237], decarboxylation[239], desulfydration of cysteine[238] and reversible cleavage[114] of 3-hydroxyamino acids and tryptophan. These reactions occur with lower rates and specificity than in vivo, and the activity of the metal ions roughly followed the order of stability of the Schiff base chelate[239]. Schiff base formation can have a considerable effect both on the position and degree of activation of a coordinated amino acid. The formation and stability of such Schiff bases and their metal complexes, especially with biologically important carbonyl compounds such as pyridoxal (or its crude model salicylaldehyde), glyoxalate and pyruvate have received considerable attention and the topic has been extensively reviewed[193-4,196a,b,241-5].

4.   HYDROLYSIS OF CARBOXYLIC ESTERS AND AMIDES

In addition to the large volume of work dealing with the metal ion promoted hydrolysis of α-amino-acid esters and amides a considerable amount of work has been reported on other esters and amides. Esters and amides of carboxylic acids, rather than undergoing solvolysis to give an acylium or oxo-carbonium ion (eq. 11), usually hydrolyze by way of an associative mechanism

$$R - \overset{\overset{\text{O}}{\|}}{C} - X \longrightarrow X^- + R - \overset{+}{C} = O \qquad (11)$$
$$X = OR, NR_2, SR$$

involving a tetrahedral intemediate **XLVI**. An S_N2 type of displacement process does not occur. Metal coordination of the carbonyl oxygen can accelerate the process by stabilizing the

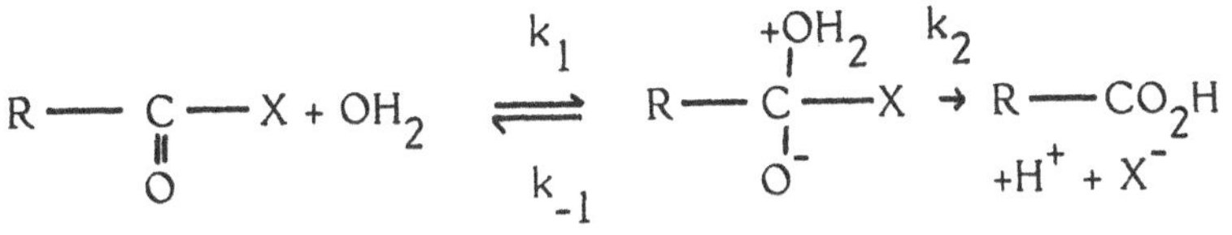

$$R \mathbin{-\!\!\!\underset{\underset{O}{\|}}{C}} \mathbin{-\!\!} X + OH_2 \underset{k_{-1}}{\overset{k_1}{\rightleftharpoons}} R \mathbin{-\!\!\!\underset{\underset{O^-}{|}}{\overset{\overset{+OH_2}{|}}{C}} \mathbin{-\!\!} X} \overset{k_2}{\rightarrow} R \mathbin{-\!\!} CO_2H$$
$$+H^+ + X^-$$

XLVI

tetrahedral intermediate and by increasing the susceptibility of the carbonyl oxygen to nucleophilic attack (general acid catalysis). Metal ion coordination of the leaving group X can accelerate the reaction in two ways. First, it can reduce the delocalization of lone pairs of electrons from X into the carbonyl group in the ground state, thus making the carbonyl group more susceptible to nucleophilic attack. Second, coordination of X can stabilize it during its departure from the tetrahedral intermediate. Examples of both effects may be seen in the hydrolysis of carboxylate esters of 8-hydroxyquinoline[246]. Metal ion promotion of the hydrolysis of 8-acetoxyquinoline (**XLVII**) has been studied in some detail[248]; however, the substrate binds rather weakly to metal ions and the system is not very amenable to detailed kinetic analysis. The ester 8-hydroxyquinoline-2-carboxylic acid (**XLVIII**) binds strongly to a variety of metal ions and detailed kinetic investigations are possible[249]. In the pH range 5-12 the hydrolysis of the carboxylic acid (HA) follows the rate law,

$$\text{rate} = k_o[A^-] + k_{OH}[A^-][OH^-]$$

where A^- is the anion, with $k_o = 1.67 \times 10^{-4}$ s^{-1} and $k_{OH} = 0.84$ $M^{-1}s^{-1}$ at $25°$. The 1:1 metal complexes of the ester MA^+ (M = Zn^{II} and Cu^{II}) undergo

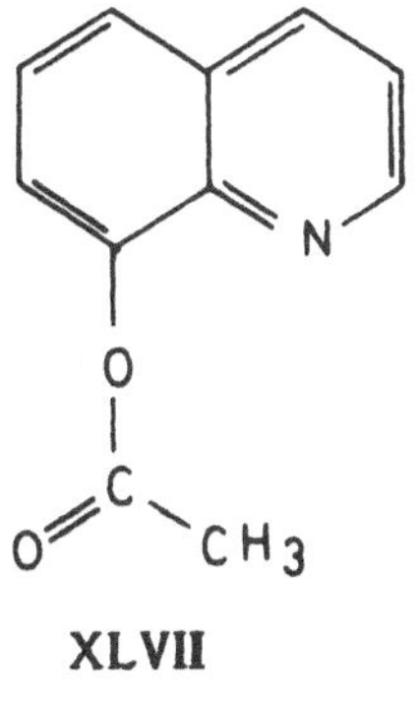

XLVII 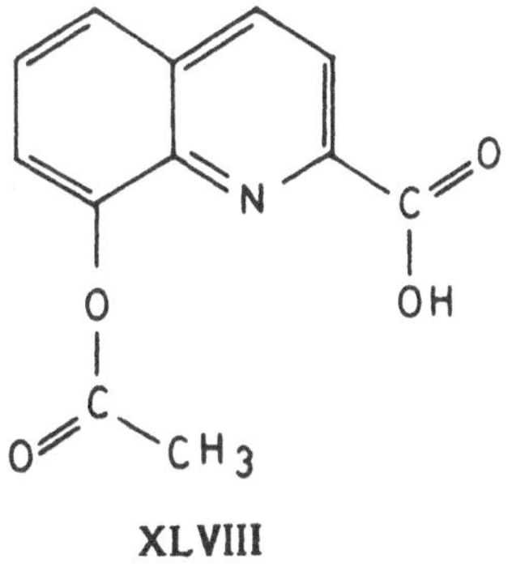   XLVIII

XLIX

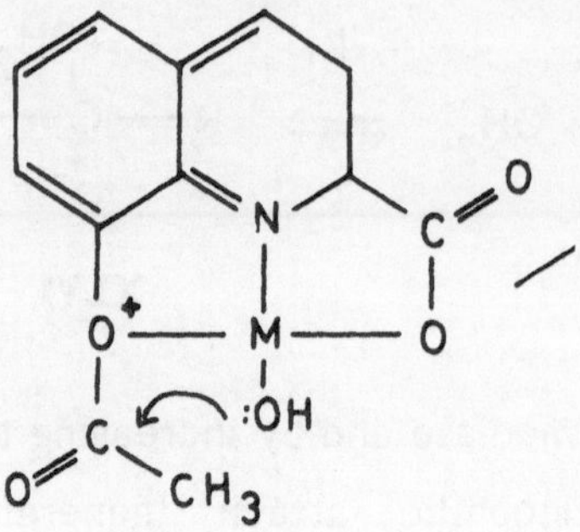

base hydrolysis ca 2 x 10^8 times faster than A^-. The metal ion promoted reactions are considered to involve attack by coordinated hydroxide in **XLIX** and this effect in conjunction with a perturbation of 6 pK units in the acidity of the conjugate acid of the leaving group leads to a rate of hydrolysis comparable with the reported values of k_{cat} for the hydrolysis of a good ester substrate by the zinc metalloenzyme carboxypeptidase A.

The methanolysis (Scheme **7**) of amides of N, N-di(2-picolyl)amine also illustrates the leaving group effect[250]. For example addition of the p-nitrophenyl derivative to a hot solution of $CuCl_2$ in methanol, resulted in the "almost instantaneous" formation of methyl p-nitrobenzoate and the deep blue color of the copper(II) complex of di(2-picolylamine). By contrast, no ester was formed on heating N,N-dibenzyl-p-nitrobenzamide with $CuCl_2$ in methanol for 24 hours, or on heating the ligand in methanol in the absence of copper. Chelation of the metal by the "leaving group" resulted in an enormous increase in the methanolysis rate. Increased hydrolysis rates in aqueous solution were also noted.

Effects of this type can also be of value in synthetic chemistry. Carboxylate esters of 8-hydroxyquinoline, are readily hydrolyzed in the presence of metal ions and this has led to the development[246] of the carbo(8-quinoloxy) substituent as an amino-protecting group for peptide synthesis (Scheme **8**).

Metal ions have marked catalytic effects on the hydrolysis of a variety of other esters of 8-hydroxyquinoline and o-phenanthroline. The metal-ion promoted hydrolysis of methyl 8-hydroxyquinoline-2-carboxylate **(L)** has been studied in detail[251], as has that of ethyl 1,10-phenanthroline-2-carboxylate **(LI)**[252]. Rate constants have been obtained for the base

Scheme 7

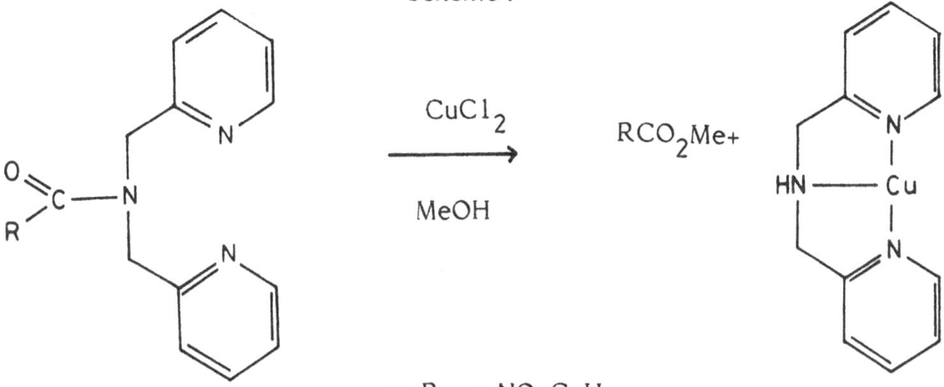
CuCl2
MeOH
RCO2Me+
HN
Cu

a, R = p-$NO_2C_6H_4$-
b, R = Me_3C-
c, R = CH_3CH=CH-

Scheme 8

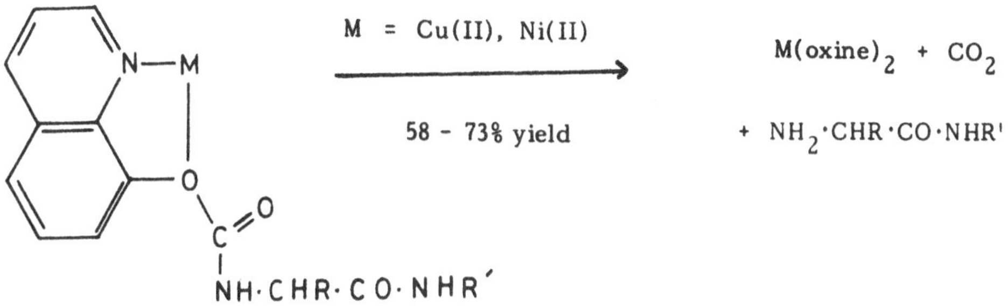
M = Cu(II), Ni(II)
58 - 73% yield
M(oxine)2 + CO2
+ NH2·CHR·CO·NHR'
NH·CHR·CO·NHR'

hydrolysis of **L** (=HA) and for the corresponding anion A^-. Formation constants K_{MA} have been determined at 25° for the equilibrium $M^{2+} + A^- \rightleftharpoons MA^+$ and rate constants k_{OH} for base hydrolysis of the MA^+ complexes obtained (Table 13).

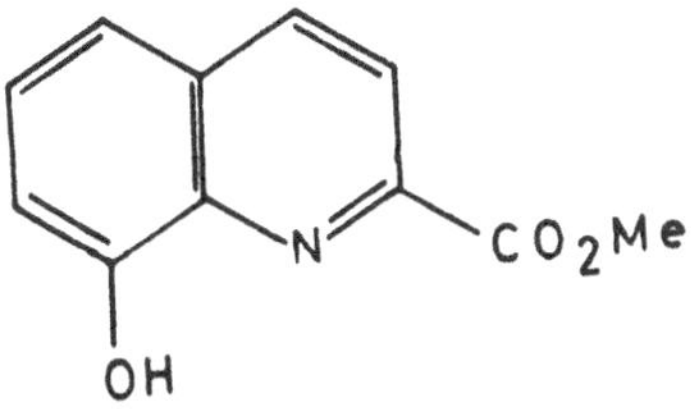
CO2Me
OH

L

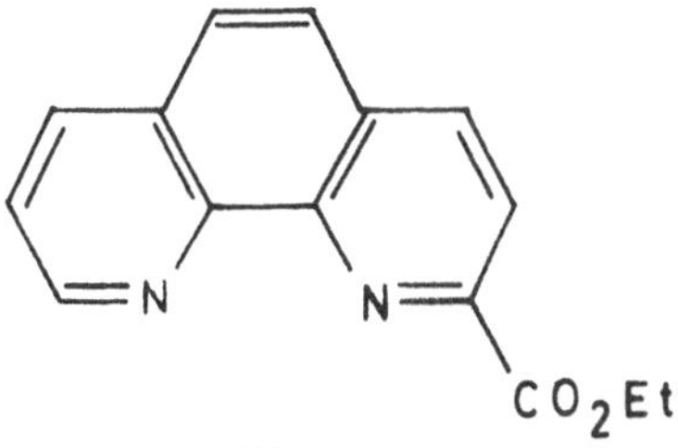
CO2Et

LI

Table 13. Rate Constants, Activation Parameters and Formation Constants for the Metal Complexes of Methyl 8-Hydroxquinoline-2-Carboxylate. Data from ref. 251

Complex	K_{MA}[a] (M^{-1})	k_{OH}[a] $(M^{-1} s^{-1})$	ΔH $(kJ\ mol^{-1})$	ΔS_{298} $(J\ K^{-1} mol^{-1})$
HA	-	4.0		
A^-	-	0.485	47.7	-90.8
MnA^+	4.9×10^4	4.9×10^2	47.7	-33.1
ZnA^+	3.8×10^6	2.6×10^3	54.4	2.9
NiA^+	5.6×10^5	1.01×10^4	30.5	-65.3
CoA^+	3.4×10^5	1.12×10^4	38.1	-39.7
CuA^+	$>2 \times 10^8$	6.3×10^5	33.1	-22.6

(a) Rate constants and formation constants at 25°C.

Quite large accelerations are observed (10^3-10^6) for the base hydrolysis of the MA^+ complexes when comparisons are made with the base hydrolysis of A^-. The total charge carrried by the complex does not appear to be a major factor in determining the hydrolysis rates. Thus at 25^0 the complex CuA^+ undergoes base hydrolysis (k_{OH} = 6.3 x 10^5 dm^3 mol^{-1} s^{-1}) at a very similar rate to the corresponding complex of L1 which carries a dipositive charge (k_{OH} = 5 x 10^5 dm^3 mol^{-1} s^{-1})[7]. The thermodynamic parameters for the hydrolysis of the various ester species (Table 13) show that the rate enhancements in the metal-ion-promoted reactions arise from more positive values of $\Delta S^{\ddagger}$ and, in general, lower enthalpies of activation. There is a close correspondence between log K_{ME} and values of $\Delta G^{\ddagger}$ (Figure 1). The similar trends of the two parameters is strong evidence in support of a pathway involving metal-ion-ester carbonyl binding in the ground state of the reaction. Bender[10] has discussed the importance of metal-ion

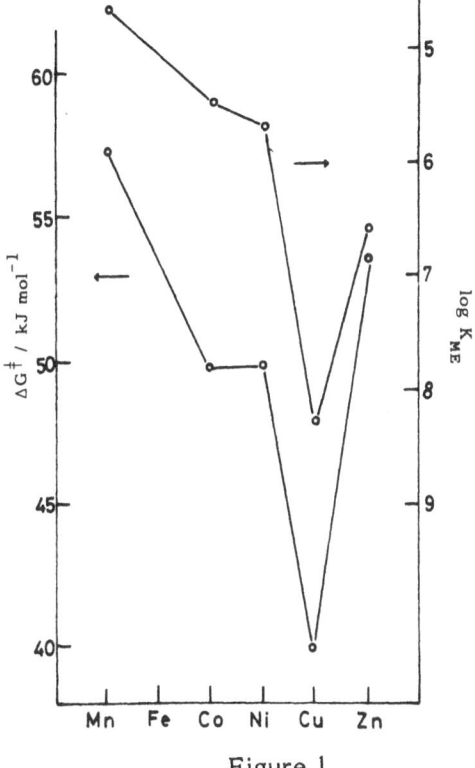

Figure 1

Relation between $\Delta G^{\ddagger}$ for the base hydrolysis

of $[ME]^+$ and log K_{ME}

stabilization of the transition state relative to ground state stabilization. In ester hydrolysis, stabilization of the transition state appears to be of primary importance when the metal ion can interact directly with the developing negative charge on the ester carbonyl group. The orientation of solvent molecules around charges or developing charges results in negative entropy changes and the effect may be as large as, or larger than, that resulting from the molecularity of the reaction. The metal ion can be regarded as "solvating" the transition state, thus leading to a more positive (i.e. less negative) entropy of activation. The bimolecular reaction of a positively charged metal complex with hydroxide ion gives rise to an overall neutral transition state where less solvent is contained than in the initial state. As a result a more positive entropy of activation for hydrolysis would be expected than occurs with the unpromoted reactions, and this is observed experimentally.

The general trends in $\Delta H^{\ddagger}$ and $\Delta S^{\ddagger}$ in the metal promoted reactions can be rationalized in terms of ground state and transition state interactions between the metal ion and the carbonyl group of the ester. Where a strong interaction occurs between the metal ion and the ester function in the ground state of the reaction, as appears to occur in the nickel(II)-promoted reaction, the value of $\Delta H^{\ddagger}$ is significantly lowered due to polarization of the carbonyl group, but $\Delta S^{\ddagger}$ remains quite negative since there is little change in solvation on moving to the transition state. If ground state interactions are weak, but a significant interaction occurs in the transition state, as appears to be the case with Zn(II), $\Delta H^{\ddagger}$ remains quite large but $\Delta S^{\ddagger}$ is significantly lowered since the metal ion is able to "solvate" the devloping negative charge on the carbonyl group and desolvation is more extensive.

Fife and Squillacote[253] have recently studied the influence of metal ions on the hydrolysis of N-(8-quinolyl)-phthalamic acid **(LII)** and 8-quinolyl hydrogen glutarate **(LIII)** in order to assess their effect on intramolecular nucleophilic carboxyl group participation in amide and ester hydrolysis. The divalent metal ions Cu(II), Ni(II), Co(II) and Zn(II) have little effect on the intramolecular nucleophilic reaction of either the amide or the ester. Above pH 6, the metal ions promote the base hydrolysis of the ester but not of the amide. The order of reactivity of the metal ion complexes is Zn(II) > Co(II) > Ni(II), with rate enhancements ranging from 10^2 to 2×10^4 although

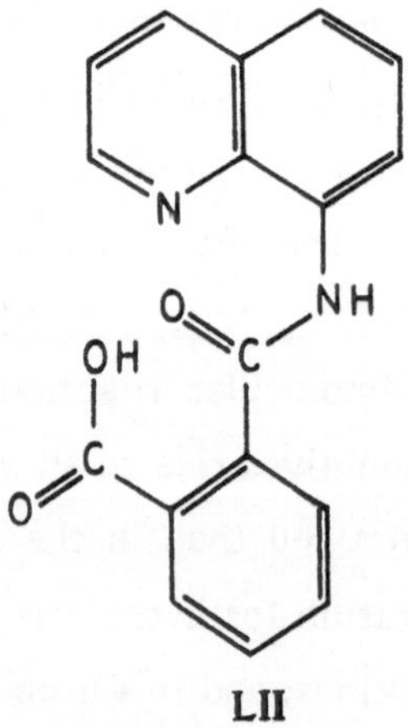

LII

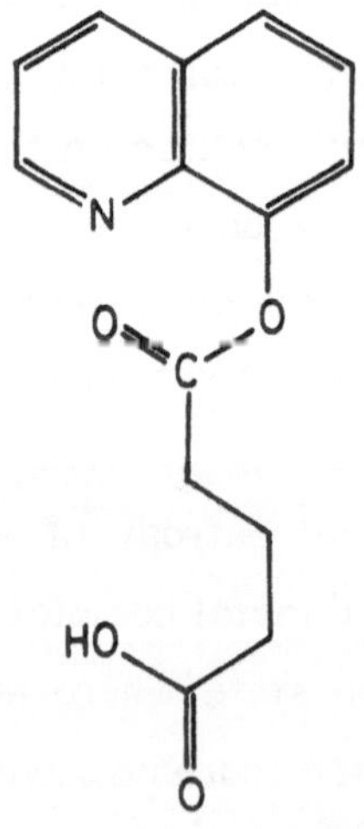

LIII

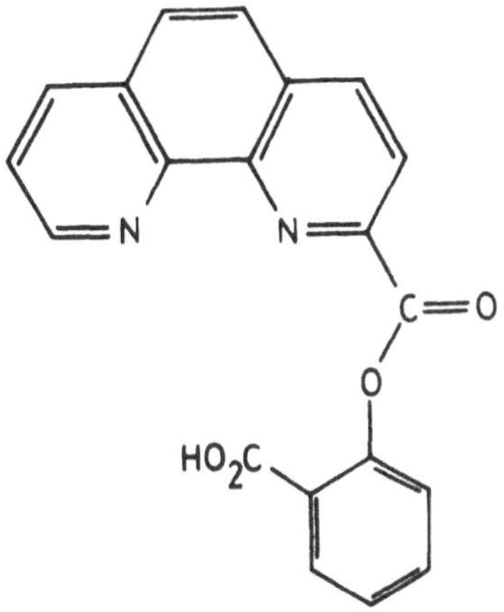

LV

LIV

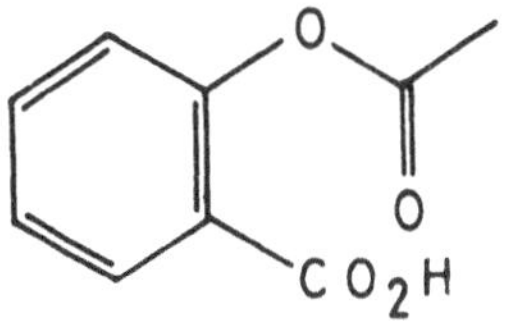

full complexation of the substrate was not achieved. The metal promoted base hydrolysis although capable of large rate enhancements cannot compete with intramolecular nucleophilic catalysis by the carboxylate group below pH 6. The pH-rate profile for the hydrolysis of 8-quinolyl hydrogen glutarate in the presence of zinc(II) is strikingly similar to the pH-k_{cat} profile for the carboxypeptidase A catalyzed hydrolysis of O-(trans-cinnamoyl)-L-β-phenyllactate.

The hydrolysis of salicyl phenanthroline-2-carboxylate (LIV)[254] involves general base catalysis by the neighboring carboxylate group analogous to the hydrolysis of aspirin (LV). However, in the presence of saturating concentrations of Cu(II), Ni(II), Co(II) or Zn(II) the plot of log k_{obs} versus pH is linear with a slope of 1.0 indicating a first order dependence on the hydroxide concentration and establishing that the carboxyl group is not participating in the presence of metal ion. Any rate-enhancing effect of metal ion bonding on intramolecular general base catalysis cannot compete with base hydrolysis of the metal complex. As a result, metal ion assisted general base catalysis in the carboxypeptidase A catalyzed hydrolysis of esters is unlikely.

A more recent study[254] of the metal ion promoted hydrolysis of 8-(2-carboxyquinoly)hydrogen glutarate has extended the earlier measurements[253]. Full complexation with Zn(II) and Cu(II) was achieved leading to rate enhancements of 4 x 10^7 for base hydrolysis of the copper (II) complex.

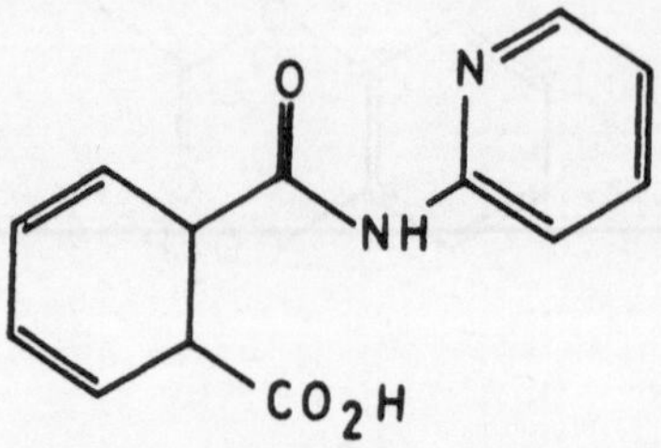

LVII

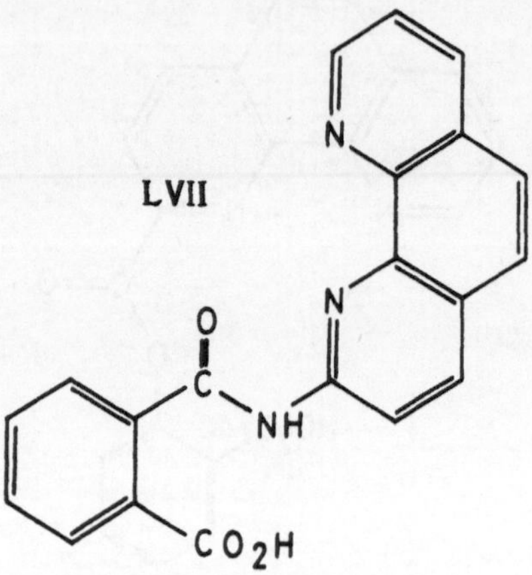

LVI

The metal ion inhibited hydrolysis of N-(2-pyridyl)-phthalamic acid (LVI) and N-(2-phenanthrolyl)phthalamic acid (LVII) has been studied in detail. It has been suggested that a metal ion chelated carbonyl group will not permit facile proton transfer to the leaving group, which is a requirement in the hydrolysis of amides[256].

A very rapid amide hydrolysis promoted by copper(II) and zinc(II) has been studied[257]. The lactam LVIII gives 1:1 complexes with Cu(II), Zn(II), Co(II) and Zn(II). For the copper(II) complex LIX (M = Cu) the pK_a of the coordinated water molecule is ca 7.6. Hydrolysis of the Cu(II) complex is 1.6 x 10^6 faster than that of the free lactam at pH 7.6, while a rate enhancement of 1.9 x 10^5 occurs with Zn(II). Nucleophilic attack by metal-bound hydroxide is considered to be the most plausible pathway for hydrolysis.

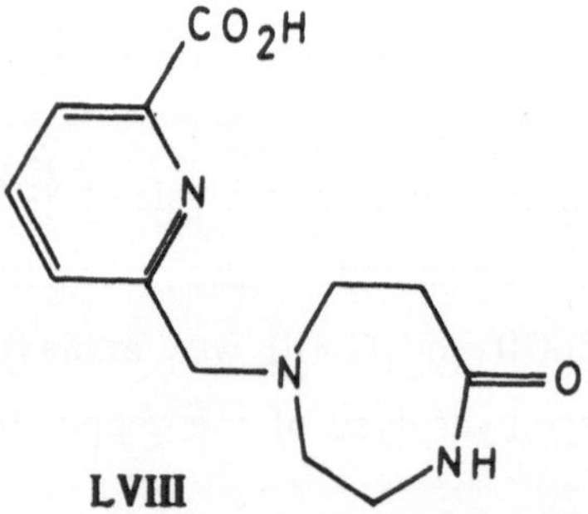

LVIII

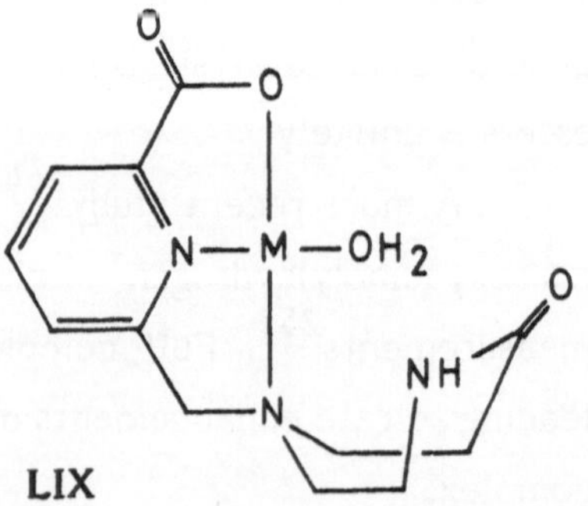

LIX

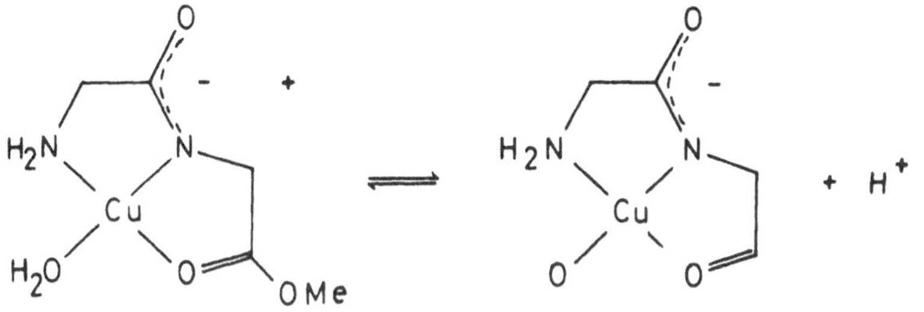

LX

Some kinetic work has also been carried out on the copper(II) catalyzed hydrolysis of peptide esters[258-9]. At pH values > 7.6 and 1:1 metal to ligand ratio, the peptide esters act as tridentate ligands, donation occurring as in LX via the terminal amino-group and the deprotonated amide nitrogen, with a weak interaction between the metal ion and the carbonyl group of the ester. The pK_a of the coordinated water molecule is in the range 8-9 so that an aquo-hydroxo equilibrium occurs in solution. Table 14 lists some of the kinetic data obtained with these systems. Rate enhancements of the order of 10^3 (compared with the unprotonated ligands) are observed.

Table 14. Rate Constants and pK_a Values for the Base Hydrolysis of Some Copper(II) Complexes of Dipeptide Esters at 25°C and μ = 0.01M. Data from ref. 259

Ligand	pK_a	$k_{OH}(a)$ $(M^{-1}s^{-1})$	$k_{OH}(b)$ $(M^{-1}s^{-1})$
Ethyl glyclylglycinate	8.0	2.87×10^3	1.98×10^2
Ethyl glycyl-β-alaninate	9.1	1.36×10^2	1.51×10^1
Ethyl glycyl-L-leucinate			1.68×10^2

(a) Base hydrolysis of the aquo complex

(b) Base hydrolysis of the hydroxo complex

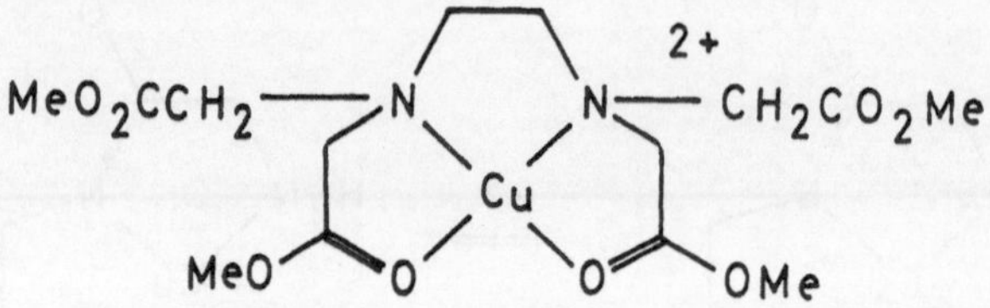

LXI

Tetramethyl ethylenediaminetetraacetate (Me$_4$edta) interacts with copper(II) to give the complex [Cu(Me$_4$edta)]$^{2+}$ **(LXI)** in which two ester groups of the ligand are bonded to the metal ion in conjunction with the two nitrogen donors[51]. A variety of metal complexes of the ester have been isolated and characterized[260]. Base hydrolysis of two ester groups occurs in the pH range 4.1-5.4 with k_{OH} = 3.23 x 10^5 and 7.02 x 10^4 M^{-1} s^{-1} at 25°C and I = 0.1M. Nucleophilic attack by water also occurs in this pH range. Hydrolysis of the second pair (presumably the pendant groups) takes place in the pH range 7.2-9.0 with k_{OH} = 1.54 x 10^3 and 74.7 M^{-1} s^{-1} (Table 15).

5.   HYDROLYSIS OF NITRILES

In recent years there has been considerable interest in the hydrolysis of nitriles in the coordination sphere of metal ions. Breslow, Fairweather and Keena[261] first reported that the hydrolysis of 2-cyano-1,10-

Table 15. Rate Constants for the Base Hydrolysis of EDTA Methyl Esters and their Copper(II) Complexes at 25°C and μ = 0.1M. Data from ref. 51

Ligand	k_{OH} (M^{-1} s^{-1})	Complex	k_{OH} (M^{-1} s^{-1})	Rate enhancement
Me$_4$edta	2.10	Cu(Me$_4$edta)$^{2+}$	3.2 x 10^5	1.6 x 10^5
Me$_3$edta$^-$	0.86	Cu(Me$_3$edta)$^+$	8.0 x 10^4	8.0 x 10^4
Me$_2$edta^{2-}	0.35	Cu(Me$_2$edta)	1.5 x 10^3	4.4 x 10^3
Meedta^{3-}	0.06	Cu(Meedta)$^-$	7.5 x 10^1	1.2 x 10^3

LXII

LXIII

Solvent ligands,
one coordinated OH⁻,
and one possible
CH_2OH-group interaction
with Ni omitted for clarity

phenanthroline to the corresponding amide is strongly promoted by metal ions such as copper(II), nickel(II) and zinc(II). Base hydrolysis of the nickel(II) complex was 10^7 times faster than that of the uncomplexed substrate. The entire rate acceleration was accounted for by the change in $\Delta S^{\ddagger}$ (more positive). Somewhat similar effects have been observed for base hydrolysis of 2-cyano-pyridine to the corresponding carboxamide. In this case rate enhancements of 10^9 occurred with the nickel(II) complex[262]. In the presence of tris(hydroxymethyl)aminomethane ("tris") and copper(II) or nickel(II) the product obtained was not the expected pyridine-2-carboxamide, but 2-(2'-pyridyl)-4,4-dihydroxymethyl-Δ^2-isooxazoline (LXII), which presumably arises via the mixed ligand complex LXIII. A number of studies have also been made of the hydrolysis of nitriles in the coordination sphere of cobalt(III). Pinnell et al[264] reported that aromatic nitriles coordinated to penta-ammine cobalt(III) are rapidly hydrolyzed by base to give the corresponding nitrogen bonded carboxamido complexes. Rate enhancements of 2 x 10^6 were observed in these reactions. Buckingham et al[265] found similar effects with aliphatic nitriles; thus base hydrolysis of acetonitrile to acetamide was promoted by a factor of 2 x 10^6 on coordination to $[Co(NH_3)_5]^{3+}$. A number of papers[266-9] dealing with the synthetic applications of metal-ion-promoted hydrolysis of nitriles have also been published.

Table 16. Base Hydrolysis of Nitrile Complexes. Data from refs. 270(b) and 265

Nitrile	$k_{OH}(a)$ $(M^{-1} s^{-1})$	Rel. rate (b)
CH_3CN	1.60×10^{-6}	1.0
$[(NH_3)_5Ru(CH_3CN)]^{3+}$	2.2×10^2	1.4×10^8
$[(NH_3)_5Rh(CH_3CN)]^{3+}$	1.0	6.2×10^5
$[(NH_3)_5Co(CH_3CN)]^{3+}$	3.40	2.12×10^6
$[(NH_3)_5Ru(CH_3CN)]^{2+}$	$< 6 \times 10^{-5}$	< 38
C_6H_5CN	7.2×10^{-6}	1.0
$[(NH_3)_5Ru(C_6H_5CN)]^{3+}$	2.0×10^3	2.8×10^8
$[(NH_3)_5Co(C_6H_5CN)]^{3+}$	18.2	2.53×10^6
2-cyanophenanthroline	2.6×10^{-3}	1.0
2-cyanophenanthroline-Ni^{2+}	2.4×10^4	10^7

(a) At $25^{\circ}C$, $\mu = 1.0M$ except for Ru(II)

(b) Relative rate $= k_{OH}(complex)/k_{OH}(free\ ligand)$

Particular emphasis has been directed towards comparisons of ligand properties and reactions between different organonitrile complexes of the general type $M(NH_3)_5(RCN)^{n+}$ where M = Ru(II), Ru(III), Rh(III) or Co(III) and RCN is an organonitrile such as acetonitrile or benzonitrile. Table 16 summarises some of the date obtained. Zanella and Ford[270] have found that specific rates of base hydrolysis of the penta-ammineruthenium(III) complexes of acetonitrile ($k_{OH} = 2.2 \times 10^2$ M^{-1} s^{-1}) and benzonitrile ($k_{OH} = 2.0 \times 10^3$ M^{-1} s^{-1}), $Ru(NH_3)_5(NCR)^{3+}$ (R = Me or Ph) are approximately 10^8 times faster than the free ligand at $25^{\circ}C$ and about 10^2 times faster than the analogous cobalt(III) complexes. The reaction products are the corresponding amido complexes **LXIV**, which reversibly protonate in acidic solution to give the amide complexes. Base hydrolysis of the pentamminerhodium(III) complex of acetonitrile (k_{OH} = 1.0 M^{-1} s^{-1}) is comparable to that of the analogous cobalt(III)

$$[(NH_3)_5M(NCR)]^{3+} + OH^- \longrightarrow [(NH_3)_5M(NHCOR)]^{2+}$$

(LXIV)

complex; the ruthenium(II) complex, however, is at least 10^6 less reactive than the corresponding ruthenium(III) complex. It has been suggested[270] that the d^5 Ru(III) species is significantly more reactive towards base hydrolysis to the amido complex than the d^6 Rh(III) and Co(III) analogue because of relative ability of the Ru(III) center to act as a π-acceptor and so stabilize the developing negative charge on the ligand which results from rate determining hydroxide ion attack on the nitrile carbon atom. The Ru(II) analogue is unreactive due to this center's ability to π back-bond into the nitrile ligand. It has also been observed that in HCO_3^-/CO_3^{2-} buffers the hydrolyses of the Ru(III) complexes are subject to general base catalysis, suggesting that in aqueous solution coordinated nitriles are subject to nucleophilic attack by nucleophiles other than hydroxide ion.

Activation parameters have been determined for the hydrolysis of some coordinated nitriles (Table 17). In the case of the $(NH_3)_5Co^{3+}$ moiety the acceleration is due to both more favorable $\Delta H^{\ddagger}$ and $\Delta S^{\ddagger}$ terms, while for the nickel(II) promoted reactions only $\Delta S^{\ddagger}$ appears to be significantly affected. This difference is at least qualitatively consistent with the arguments of Breslow et al[261] that the more favorable $\Delta S^{\ddagger}$ with nickel(II) is associated with bonding of the developing negative charge. In the cobalt(III) and the other metal(III) systems the imino group is already bonded to the metal center.

Metal ions have also been found to strongly promote the base hydrolysis of 2-cyano-8-hydroxyquinoline **(LXV)** to the corresponding carboxamide[271]. For the free ligand k_{OH} = 4.85 x 10^{-3} M^{-1} s^{-1} at 45°C and

Table 17. Rate Constants and Activation Parameters for Base Hydrolysis of Nitriles and Nitrile Complexes

Nitrile	k_{OH} (M^{-1} s^{-1})	$\Delta H^{\ddagger}$ (kcal mol^{-1})	$\Delta S^{\ddagger}$ (cal deg^{-1}mol^{-1}
Benzonitrile	8.2×10^{-6}(a)	19.9	-15.2
[(NH$_3$)$_5$Co(C$_6$H$_5$CN)]$^{3+}$	18.8(a)	16.5	2.7
2-cyano-1,10-phenanthroline	2.6×10^{-3}(a)	15.7	-20
[Ni(2-cyano-1,10-phenanthroline)]$^{2+}$	2.4×10^{4}(a)	15.1	14
[Ni(2-cyanopyridine)]$^{2+}$	6.3×10^{7}(a)	13.7	23
2-cyano-8-hydroxy quinolinate	4.85×10^{-3}(b)	13.4	-27
[Ni(2-cyano-8-hydroxy quinolinate)]$^{+}$	7.97×10^{2}(b)	14.2	-0.8
[Co(2-cyano-8-hydroxy quinolinate)]$^{+}$	2.74×10^{2}(b)	14.3	-3

(a) Data from ref. 264 (25°C)

(b) Data from ref. 271 (25°C)

μ = 1.0M. The hydrolysis is strongly promoted by nickel(II), cobalt(II) and copper(II). Thus the 1:1 complex of nickel(II) and the ligand undergoes base hydrolysis 2×10^{5} times faster than the free ligand at 45°C. The activation parameters obtained for the free ligand and the cobalt and nickel complexes (Table 17) show that the substantial rate accelerations are due to an entropy effect alone, as is also the case with 2-cyanophenanthroline.

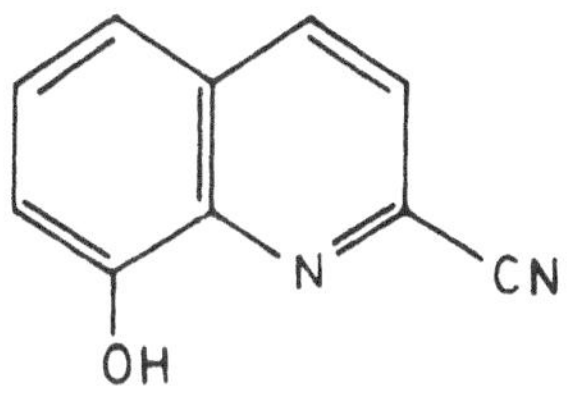

LXV

For steric reasons bonding of the cyano group to the metal ion cannot occur in the initial state **(LXVI)** of the reaction with this ligand. It therefore appears that the transition state must involve bonding of the developing iminoanion to the metal ion, possibly resulting in displacement of a coordinated water molecule (reaction A, Scheme 9) The metal ion can be regarded as "solvating" the developing negative charge on the nitrogen atom of the cyano-group. These effects would lead to more positive entropies of activation and thus account for the observed rate enhancements. A mechanism involving attack by coordinated hydroxide (reaction B) can also be considered. In the case of intramolecular hydroxide ion attack the function $k_{OH} = k_{obs}/[OH^-]$ will only remain constant provided that no more than ca 10% of the complex is in the hydroxo form. Only under these conditions will the concentration of the M-OH species **(LXVII)**

Scheme 9

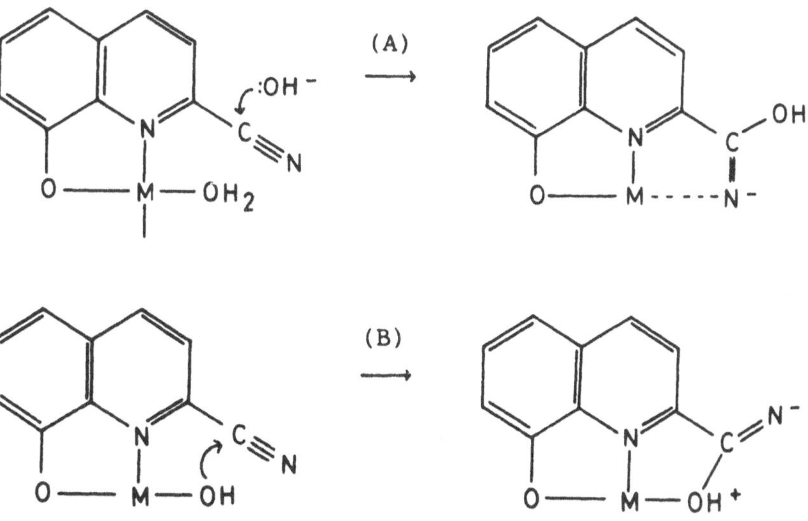

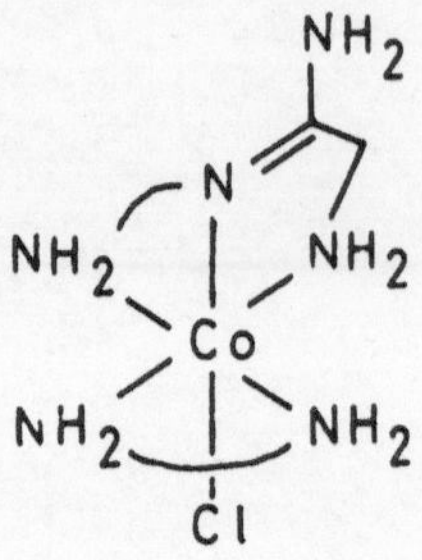

LXVIII

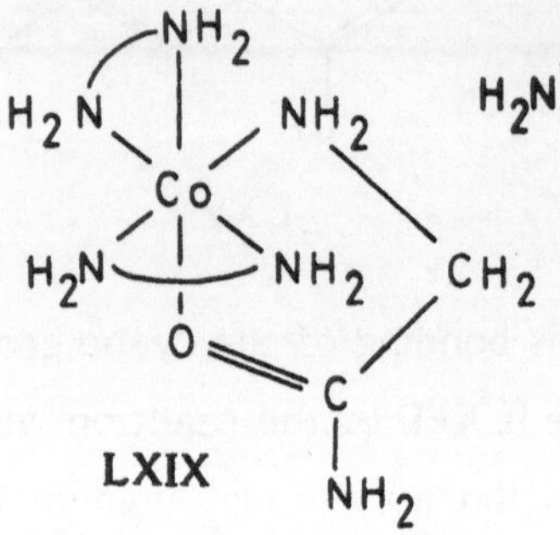

LXIX

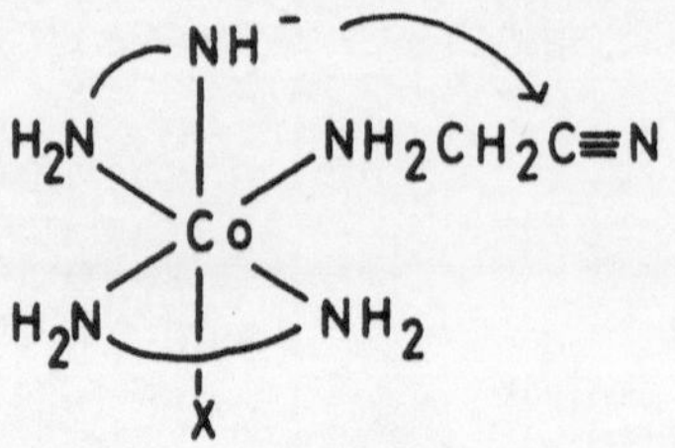

LXX

be proportional to [OH⁻]. It can be argued that similar catalytic effects are observed with both cobalt(II) and nickel(II), although since the ionization constants of the cobalt(II) and nickel(II) complexes of type **LXVI** differ considerably, hydrolysis by reaction (B) does not occur.

Chan and Chan[190] first reported the preparation of the complex cis-$[CoCl(en)_2NH_2CH_2CN]^{2+}$, by reaction of trans-$[CoCl_2(en)_2]^+$ with aminoacetonitrile. They observed that in near neutral solution the red cis-complex rapidly became violet and that $\nu(CN)$ at 2,260 cm^{-1} disappeared to be replaced by a band at 1680 cm^{-1}; this result was attributed to cis ⇌ trans isomerization. Buckingham and co-workers[266] determined the crystal structure of the violet complex and established its structure as **LXVIII**, and the same group[272] also observed that addition of mercury(II) to acidic solutions of the ions $[CoX(en)_2(NH_2CH_2CN)]^{2+}$ (X = Cl or Br) rapidly gave the chelated O-bonded glycine amide species **LXIX**. Detailed kinetic and mechanistic studies of these reactions have now been carried out[263,273]. Formation of **LXVIII** can be rationalized in terms of intramolecular attack by the amido species **LXX** on the -C≡N group.

A substantial variety of plant and bacterial species are known to possess enzymes, nitrilases, capable of catalyzing the hydrolysis of a variety of organic nitriles to the corresponding carboxylic acids[274-5]. In each case, reactions catalyzed by these enzymes are susceptible to inhibition by reagents which react with the thiol groups of enzymes. This behavior suggests that thiol groups of enzymes are involved at the active site of the nitrilases. This view is supported by model studies; thus 2-mercaptoethanol catalyzes the hydrolysis of N-benzyl-3-cyano-pyridinium bromide (LXXI) to the corresponding carboxamide[276]:-

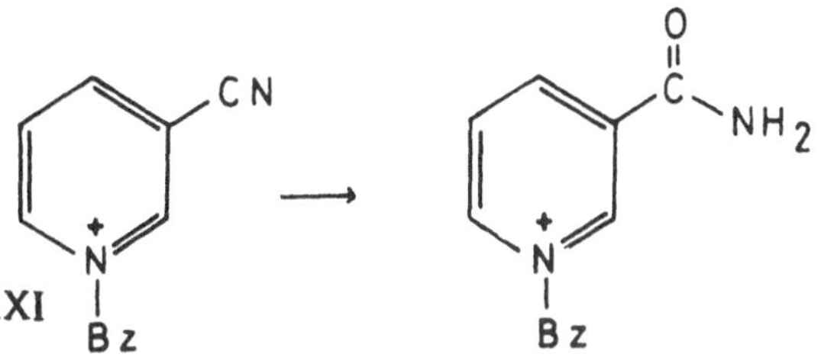

2-Mercaptoethanol increases the rate of hydrolysis of LXXI at pH 7 by a factor of 10^4 - 10^5. In view of the very effective catalysis of nitrile hydrolysis by metal ions and metal complexes it will be interesting to see if some metallonitrilases are subsequently discovered.

6.   HYDROLYSIS OF PHOSPHATE ESTERS

Phosphoryl and nucleotidyl transfer enzymes are extremely important and widespread in biology. They have in common the catalysis of nucleophilic reactions of phosphorus esters, and the general requirement for divalent metal ions, particularly Mg(II), for activity. This requirement has stimulated considerable interest in the catalytic roles of divalent metal ions in these reactions.

This section emphasizes work done in the last few years. The reader is referred to other sources for reviews of older work[277] or more general considerations of nucleophilic reactions at phosphorus[278-87].

General discussions of enzymic phosphoryl and nucleotidyl transfer are available[288-90], and the role of divalent metal ions has been reviewed[291-3].

6.1   Labile Metal Ions

Many interesting studies of metal ion catalyzed reactions of phosphate derivatives have been published; however, the mechanistic details are in many cases unclear. One example will illustrate this point. In an introduction to their paper on the metal ion catalyzed hydrolysis of acetyl phosphate, $CH_3COPO_4^{2-}$, Satchell and co-workers[294] state that "information[295] about the metal ion catalyzed hydrolysis of acetyl and other acyl phosphates in water is fragmentary and unsatisfactory. Catalysis by bivalent metals there certainly is[295-6], but the suggested[296] catalysis by univalent ions has been disputed[298], and that by bivalent ions has been claimed[298] to be a function of pH and buffer composition. Careful examination of the mechanism suggested[298] to explain the pH-dependence of the catalysis, shows that it does not, in fact, explain it; moreover, relevant rate and equilibrium constants have been wrongly calculated".

The pH-independent hydrolysis of acetyl phosphate is believed[297] to be a unimolecular decomposition of CH_3COOPO_3-, the substrate species which predominates at pH > 6 (eq 12)

$$\underset{\substack{| \\ O^-}}{CH_3\overset{\substack{O \\ ||}}{C}\text{-}O\text{-}\overset{\substack{O \\ ||}}{P}\text{-}O^-} \xrightarrow{\text{slow}} CH_3CO_2^- + PO_3^- \xrightarrow{\text{fast}} CH_3CO_2H + HPO_4^{2-} \qquad (12)$$

The reaction involves P-O bond cleavage. At higher pH a reaction involving attack by OH^- on the carbonyl carbon atom becomes increasingly important[295,297]. The detailed kinetic work of Satchell and co-workers[294] supports the catalytic pathways shown by the transition states **LXXII-LXXIV** [M = Ca(II) or Mg(II)]. There is an important contribution from a pathway involving reaction between MOH^+ and CH_3COPO_4M at high pH.

A very substantial acceleration has been observed for the hydrolysis of 2-(4(5)-imidazolyl)phenyl phosphate **(LXXVI)** in the presence of copper(II)[299]. At pH 6 a copper(II)/substrate ration of ca 2 leads to an acceleration of more than 10^4 compared to the non-catalyzed reaction. The effect is very much larger than occurs with salicyl phosphate **(LXXVII)**

LXXII   **LXXIII**   **LXXIV**

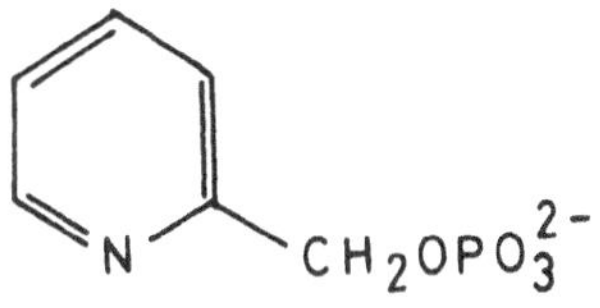

LXXV 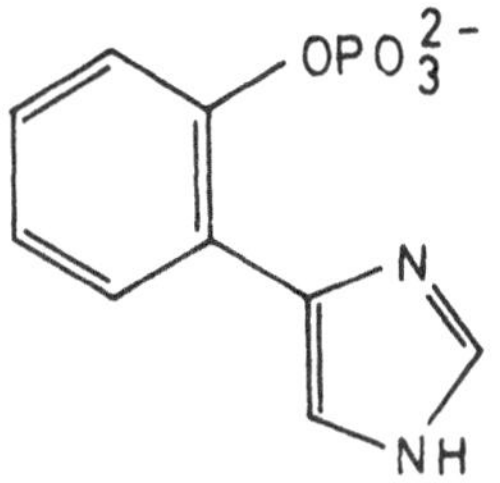   **LXXVI** 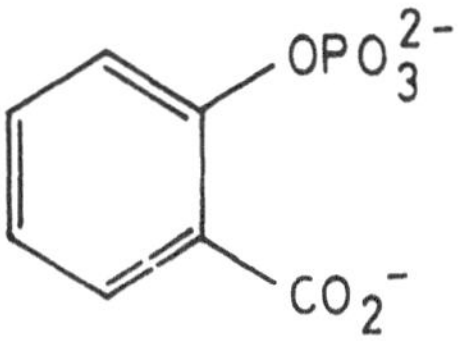   **LXXVII**

where at a copper/substrate ratio of unity k_{obs} is increased sevenfold at pH 5.1, while VO^{2+} increases k_{obs} by a factor of 28 at pH 4.5[301]. For the more analogous 2-pyridylmethyl phosphate system **(LXXV)** at a copper(II)/ substrate ratio of unity k_{obs} increases by a factor of 18 at pH 4.93[310]. The transition state in the copper(II)-promoted reaction of **LXXVI** has been formulated as **LXXVIII** or **LXXIX**. In **LXXVIII** the copper(II) ion acts as a more effective acid catalyst than a proton, lowering the pK_a of the leaving group so that facile hydrolysis of the dianion (generally observed only with leaving groups of pK_a < 7) may be observed. In **LXXIX** the copper(II) is expected to induce strain in the P-O bond and/or partially neutralize charge on the phosphate, leading to a nucleophilic displacement by solvent on phosphorus.

The hydrolysis of salicyl phosphate has also attracted considerable attention. Hydrolysis of the dianion **LXXX** at 39°C is 1.57×10^3 times faster than that of the monoanion of phenyl phosphate. Bromilow and Kirby[300] have concluded that all the experimental evidence regarding the hydrolysis can be rationalized in terms of the transition state **LXXXI**. Cleavage of the P-O bond which is well advanced is assisted by general acid catalysis by the

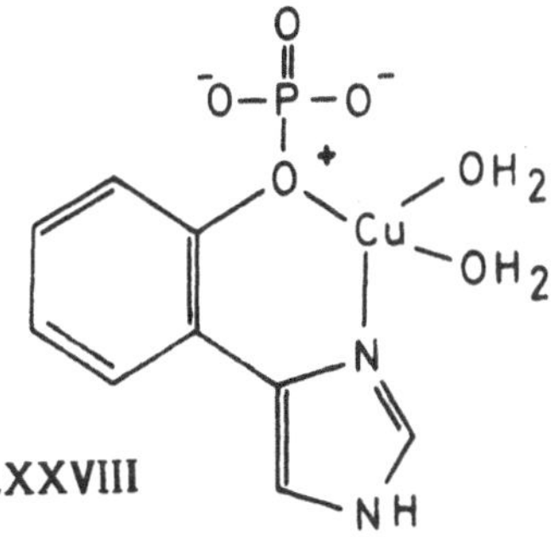

LXXVIII 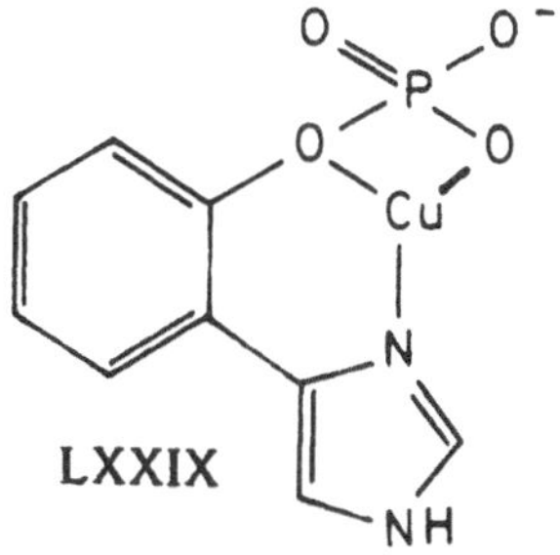   **LXXIX**

LXXX

LXXXI

LXXXIII

LXXXII

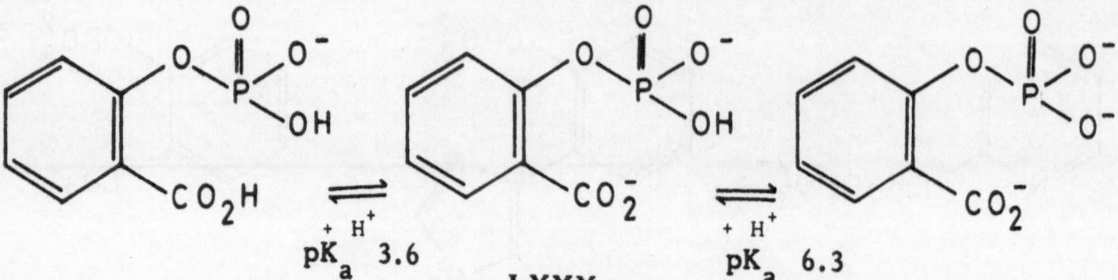

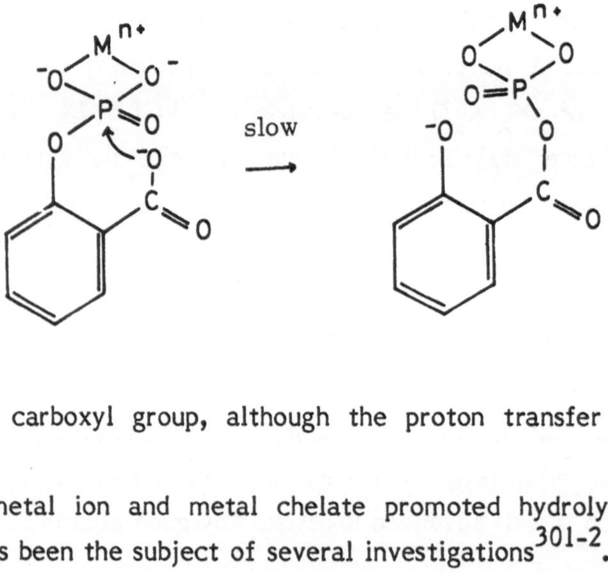

neighbouring carboxyl group, although the proton transfer has scarcely begun.

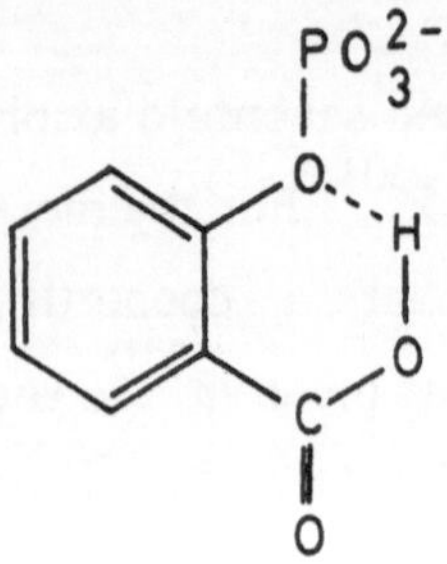

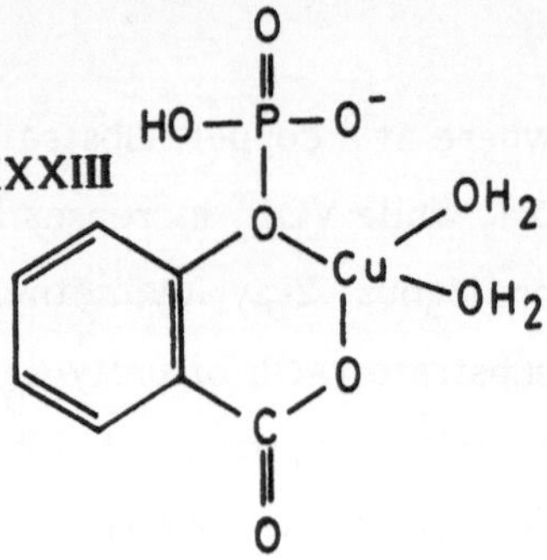

The metal ion and metal chelate promoted hydrolysis of salicyl phosphate has been the subject of several investigations[301-2]. The order of increasing activity of metal ions is Cu(II) < UO_2(VI) < VO(IV) < ZrO(IV) ~ Fe(III), while Ni(II), Co(II), Zn(II) and Cd(II) showed no activity. It was also found that some metal complexes, e.g. $Cu(en)^{2+}$, with vacant coordination sites exhibited some catalytic activity. Hofstetter et al [301] considered a species of type **LXXXII** to be responsible for catalysis. The mechanism involving intramolecular nucleophilic attack by the ionized carboxyl group to give a mixed anhydride or acyl phosphate is based on an early suggestion by Chanley et al[303] that the uncatalyzed hydrolysis involved an acyl phosphate intermediate. Subsequent work[304] has established that no such intermediate is involved. Murakami and Martell[302] have suggested intermediates or transition states of type **LXXXII** and **LXXXIII**. The intermediate **LXXXIII** would provide a similar mechanistic pathway to that provided in spontaneous hydrolysis (**LXXXI**) and is similar to the intermediate **LXXVIII** proposed for the hydrolysis of 2-[4(5)-imidazolyl]-phenyl phosphate in the presence of copper(II)[299]. For the spontaneous hydrolysis of salicyl phosphate at pH 5.1, k_{obs} = 3.46 x 10^{-5} s^{-1} at 30°C, while at a 2:1 ratio of copper(II):ligand at pH 5.1, the value of k_{obs} is 3.21 x 10^{-4} s^{-1}, a tenfold acceleration.

Metal complexes of adenosinediphosphoric and adenosine-monophosphoric acids have been studied[319] and the effect of divalent metal ions on the hydrolysis of adenosine di- and triphosphate (ADP and ATP) investigated[320]. The coupling of ATP hydrolysis to a simple inorganic redox system (VO^{2+} + H_2O_2) has been observed[321].

A summary of the effects of metal ions and metal complexes on phosphate ester hydrolysis is given in Table 18. Much remains to be done to define the mechanism of these reactions (see also Section 14.3.2).

6.2   Cobalt(III) Complexes

In recent years there has been considerable interest in studying the hydrolytic reactivity of cobalt(III) complexes of phosphate esters. This procedure has been adopted in order to avoid some of the mechanistic complexities which can arise with kinetically labile metal centres.

Farrell et al[328] appreciated that chelation of $ROPO_3^-$ to cobalt(III) to give a four-membered chelate ring should activate the phosphorus atom toward nucleophilic attack by strain induction. Metal ion binding as in **LXXXIV** could compress the O-P-O angle α in the ground state and so reduce the activation energy necessary for attainment of the trigonal bipyramidal intermediate (or transition state) in phosphoryl transfer.

$$ROP \underset{O}{\overset{O\diagdown O}{\diagup}} \!\!\!\!\!\! {}_{\alpha} \!\!\!\! M^{n+}$$ LXXXIV

Table 18. Summary of Metal Ion Effects on Phosphorus Compounds

Compound	Reference
Acetyl phosphate	294
Carbamyl phosphate	305
Salicyl phosphate	301, 302, 543
Phenylsalicyl phosphate	306
Phenyllactyl phosphate	306
Phosphoenolpyruvate	307
Methyl phosphate	328
3-Pyridyl phosphate	308, 309
8-Quinolyl phosphate	308, 209
3-Hydroxy-2-pyridylmethyl phosphate	310, 311
2-(4(5)-imidazoyl)phenyl phosphate	299
p-Nitrophenylphosphate	331
AMP	312
Pyridine carbaldoximyl phosphate	313
Phosphorylimidazole	314
Phosphoramidate	315
Phenylglycinyl phosphoramidate	316
Benzyl phosphoroguanidate	317
Cysteamine S-phosphate	318
Diisopropylphosphorofluoridate (a)	322-325
Sarin (b)	324, 326
Tabun (c)	327
ATP and ADP	320, 321

(a) $(C_3H_7O)_2P(O)F$

(b) $(C_3H_7O)(CH_3)P(O)F$

(c) $(C_2H_5O)(Me_2N)P(O)CN$

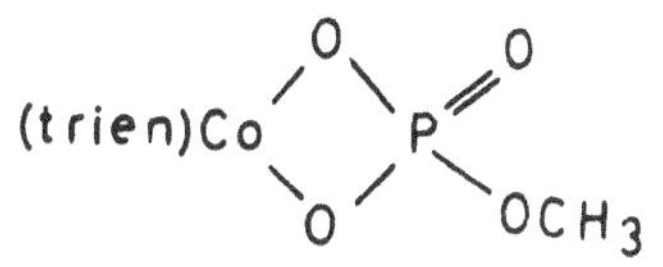

LXXXV

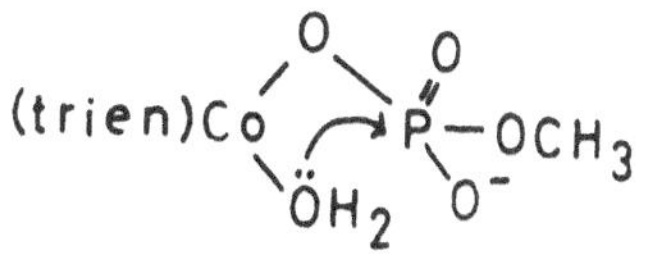

LXXXVI

Similar arguments have been used to account for the extraordinarily high reactivity of five-membered cyclic phosphates such as ethylene phosphate[282,329]. Early work by Lincoln and Stranks[330] had shown that the phosphato complex $[Coen_2PO_4]$ existed in rapid reversible equilibrium with the monodentate complex $[Co(en)_2OH(HPO_4)]$ and the hydrolysis of the phosphatopentammine cobalt(III) in aqueous solution has been studied[343]. Farrell and co-workers[328] studied the rates of hydrolysis of methyl phosphate in the complexes β-$[Co(trien)(O_3POCH_3)]$ and $[Co(NH_3)_5(O_3POCH_3)]$. The ester is bidentate in the former complex and monodentate in the latter. They found that the trien-complex hydrolyzed 130 times faster than $CH_3OPO_3H^-$ (which in turn hydrolyzes faster than $CH_3OPO_3^{2-}$), while the $CO(NH_3)_5^{3+}$ derivative was at least about 10^2 times slower.

The origin of the catalytic effects is unclear since potentiometric titration indicated ring opening of the phosphato complex LXXXV to give the aquo species LXXXVI. The pK_a of the aquo ligand is 6.7. Various reaction schemes involving water attack on the bidentate complex LXXXV and intramolecular attack by water or hydroxide in LXXXVI were considered.

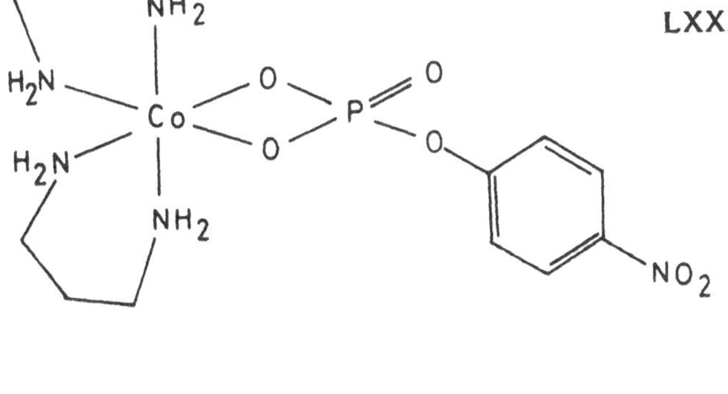

LXXXVII

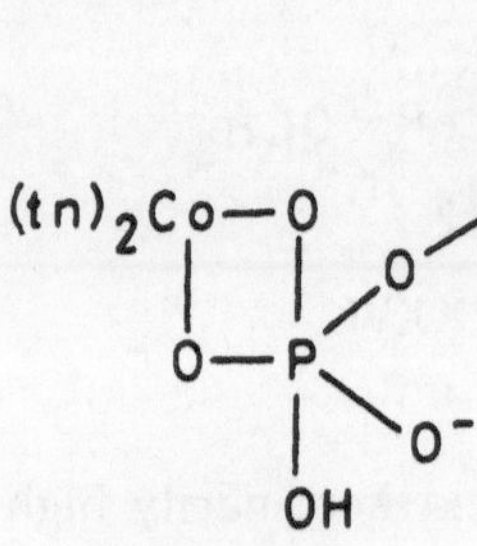

LXXXVIII 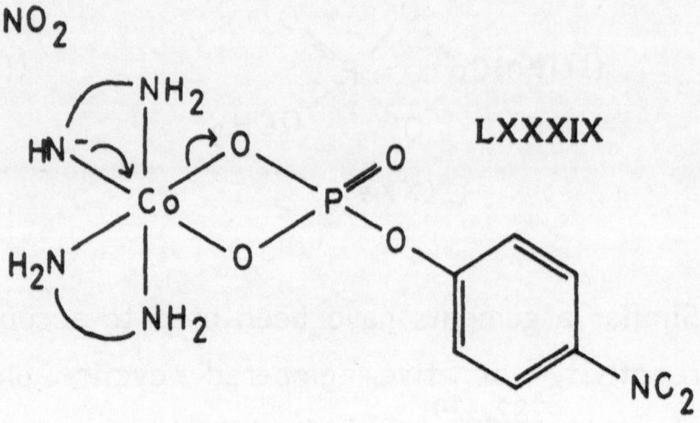   LXXXIX

Sargeson and co-workers[331] have studied the hydrolysis of p-nitrophenyl phosphate in the complex **LXXXVII**. The chelated p-nitrophenylphosphatobis(trimethylenediamine)cobalt(III) hydrolyzes in the pH range 6.5-13.5 to give ca 65% monodentate p-nitrophenylphosphate according to the rate law $k_{obs} = k_1 + k_2[OH^-]$ at 25° and μ = 0.5M, with $k_1 = 7 \times 10^5$ s^{-1} and $k_2 = 5.1$ $M^{-1} s^{-1}$. Tracer studies indicate that ester cleavage and chelate opening on the metal ion occur by different pathways, via the chelated five coordinate phosphorane intermediate **LXXXVIII** and the conjugate base mechanism **LXXXIX** for cobalt(III) complexes respectively. Ester hydrolysis is accelerated 10^9-fold relative to the uncoordinated ester in basic solution. An X-ray study of the crystal structure of $[Co(en)_2PO_4]$ has established the structure **XC** for the phosphatocobalt(III) chelate ring[331]. The in-ring O-P-O angle has been deformed from the unstrained tetrahedral angle of 109.5° to 98.7°, similar to those found in methylethylenephosphate[332] (99.1°) and cytidine 2,3-cyclic phosphate (95.8°)[333].

XC

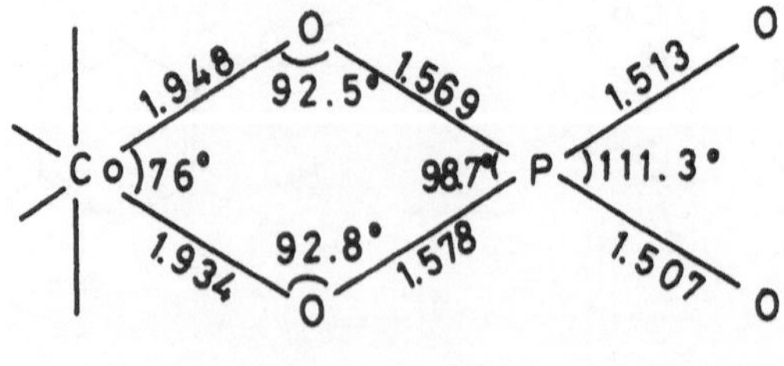

The hydrolysis of adenosine 5'-triphosphate (ATP) in the presence of various cobalt(III) complexes has been studied[334]. Complexes such as $Co(en)_3^{3+}$ which have no available sites for coordination of the substrate display no catalytic activity. Complexes having one site or two sites in a trans-configuration such as tetraethylenepentaminecobalt(III) or bis(dimethylglyoximato)cobalt(III) slightly enhance ATP hydrolysis. However, complexes with two available sites in a cis-configuration such as cis-α- or cis-β-$Co(trien)^{3+}$ exhibit considerable activity. Both the reactions ATP + H_2O ADP + Pi and ATP + H_2O AMP + PPi (Pi = inorganic phosphate; PPi = inorganic pyrophosphate) occur with these systems. The complex $Co(dien)^{3+}$ effectively enhances the hydrolysis of ATP to ADP + Pi. At pH 4.0 the uncatalyzed hydrolysis rate constant for ATP to ADP hydrolysis is 1.18×10^6 s^{-1} at 50°C. For ATP (1×10^{-3} M) and Co(III) $dien^{3+}$ (2×10^{-3} M) at pH 4.0, $k_{obs} = 1.75 \times 10^{-4}$ s^{-1} at 50°C, a rate enhancement of 150 fold.

The substitution-inert complexes between chromium(III) and ATP prepared by DePamphilis and Cleland[335] have been used with considerable success to elucidate the kinetic mechanism of yeast hexokinase[336] and other enzymes[337]. Cornelius and co-workers[338] have recently isolated and characterized the complexes $Co(NH_3)_n HATP$ (n = 2, 3 or 4) and [$Co(NH_3)_n ADP$] (n = 4 or 5) in addition to the complexes $Co(en)_2 HP_2O_7$, $Co(NH_3)_n HP_2O_7$ (n = 4 or 5) and $Co(NH_3)_n H_2P_3O_{10}$ (n = 3 or 4). The ^{31}P nmr spectra of the simpler phosphato complexes provide definitive evidence of pyrophosphate both as a monodentate and as a bidentate ligand and of tripolyphosphate both as a bidentate and as a tridentate ligand. A correlation between O-P-O bond angles and the ^{31}P nmr chemical shift for a number of phosphate esters has been noted[339]. If such a correlation could be extended to cobalt complexes, ^{31}P nmr would be a powerful tool for determining the solution geometry of phosphates bound to cobalt.

$$HO - \overset{\overset{O}{\|}}{\underset{\underset{O}{|}}{P}} - O - \overset{\overset{O}{\|}}{\underset{\underset{O}{|}}{P}} - O - \overset{\overset{O}{\|}}{\underset{\underset{O}{|}}{P}} - OH \qquad XCl$$

$$Co(NH_3)_4$$

The rate of hydrolysis of bidentate triphosphate in $Co(NH_3)_4H_nP_3O_{10}$ has been studied by phosphomolybdate analysis and ^{31}P nmr[340, 77]. Both the X-ray crystal structure[341] and the ^{31}P nmr spectrum of $Co(NH_3)_4H_2P_3O_{10}$ are consistent with the structure XCl in which one terminal phosphate residue is not bonded to the metal center. Kinetic studies establish that hydrolysis of the chelated ligand occurs at some 2/3 of the rate for the free ligand.

Hubner and Milburn[342] have noted a rate increase of 10^5 for cobalt(III) complexes with a 3:1 metal:pyrophosphate stoichiometry. The occurrence of four-membered chelate rings of the type observed with the p-nitrophenylphosphate complex are considered to be important for the catalyzed hydrolysis of pyrophosphate. Clearly important results and developments are to be expected in this experimentally difficult area (see also Section 14.3.1).

6.3   Metal Hydroxide Gels

A number of studies have been reported of the hydrolysis of phosphate esters by metal hydroxide gels[344]. In view of the heterogeneous nature of these reactions it is difficult to come to firm mechanistic conclusions; however, it seems likely that metal-bound hydroxide nucleophiles are involved. Bamann[345] first noted that the hydroxides of lanthanum, cerium and thorium promoted the hydrolysis of α-glycerophosphate in the pH range 7-10 and suggested that the reaction system could be regarded as a model for the metal-containing alkaline phosphatases which cleave phosphate esters around pH 9. When the ester is adsorbed on the hydroxide gel, the cation neutralizes the negative charges on the phosphate dianion $ROPO_3^{2-}$ and allows more facile attack by hydroxide ion. Butcher and Westheimer[347] have investigated similar reactions of this type and have found that the process resembles the enzymatic reaction in that cleavage occurs exclusively at the phosphorus-oxygen bond. The reaction does not occur readily unless the phosphate ester is substitiuted in the β-position, so that the hydrolysis of ethyl phosphate is not greatly promoted by lanthanum hydroxide gel, but the hydrolysis of β-methoxyethyl, β-hydroxyethyl and β-aminoethyl phosphates is strongly catalyzed at pH 8.5 and 78°. The hydrolysis of β-methoxyethyl phosphate is

$$ROPO_3^{2-} + La(OH)^{2+} + H_2O \qquad\qquad \text{Scheme 10}$$

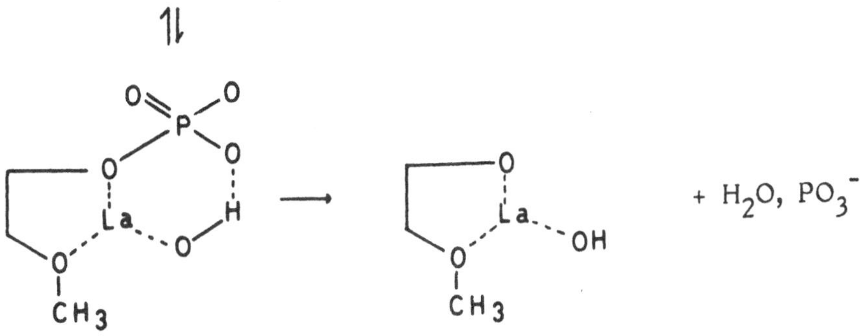

$$\Updownarrow$$

$$+ H_2O, PO_3^-$$

$$PO_3^- + H_2O \longrightarrow H_2PO_4^-$$

regarded as occurring by the steps shown in Scheme 10. 1-Methoxy-2-propyl phosphate is hydrolyzed at pH 4; or by La(OH)$_3$ at pH 8.5 with complete retention of sterochemical configuration and P-O bond cleavage. Catalysis by rare earth ions has been discussed by Trapmann[348].

7.   HYDROLYSIS OF SULFATE ESTERS

The acid-catalyzed hydrolysis of aryl sulfates proceeds by an A-1 mechanism[349]

$$ArOSO_3^- + H^+ \rightleftharpoons ArOSO_3H \rightleftharpoons$$

$$\longrightarrow ArOH + SO_3$$

Nucleophilic catalysis by amines has been observed in the hydrolysis of p-nitrophenyl sulfate[350] and intramolecular carboxyl group catalysis occurs with salicyl sulfate[351] as with salicyl phosphate. Like phosphate ester hydrolysis (Section 6), sulfate ester hydrolysis is also susceptible to metal ion catalysis. The hydrolysis of 8-quinolyl sulfate is subject to pronounced catalysis by copper(II) ions in the pH range 5.4-5.8 at 39.8°C. The 1:1 copper(II) complex hydrolyzes 10^5-10^6 times more rapidly than the free

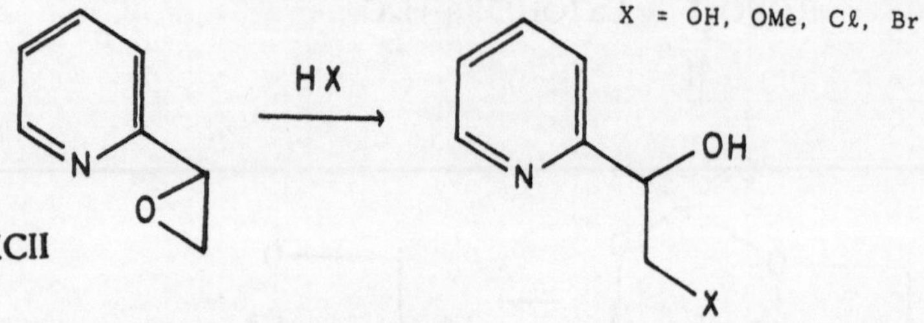

X = OH, OMe, Cℓ, Br

XCII

ligand in the pH range 5-6. Most aryl sulfatases have pH optima in the acidic range (pH 4-6). At the present time there appears to be no evidence for any metal-activated aryl sulfatase.[353] However, the substantial catalytic effects observed in the model reaction, and the mechanistic similarities between phosphate and sulfate ester hydrolysis noted previously, suggest that some such metal activated sulfatases may well be discovered.

8.   HYDROLYSIS OF EPOXIDES

Only a limited amount of work has been published dealing with the hydrolysis of epoxides. Hanzlik and Michaely[355] first observed that in the presence of copper(II) the hydration of 2-pyridyloxiran (XCII) is accelerated by a factor of 1.8×10^4 and its reaction with Cl⁻, Br⁻ and MeO⁻ becomes 100% regiospecific for β-attack. The magnitude of the catalytic effect decreases in the order Cu(II) > Co(II) > Zn(II) >> Mn(II). The pH-rate profile for the copper(II)-catalyzed reaction is a bell-shaped curve with a maximum at ca pH 5.

Both the enzymic and metal ion catalyzed hydration of [^{18}O]-2-pyridyloxirane have been found to involve ⩾ 95% C(2)-O bond cleavage[356]. In the presence of large concentrations of metal binding reagents the enzymic hydration of 1,2-epoxytetradecane is not inhibited and as a result it is considered that a metal ion is not involved at the active site of epoxide

XCIII

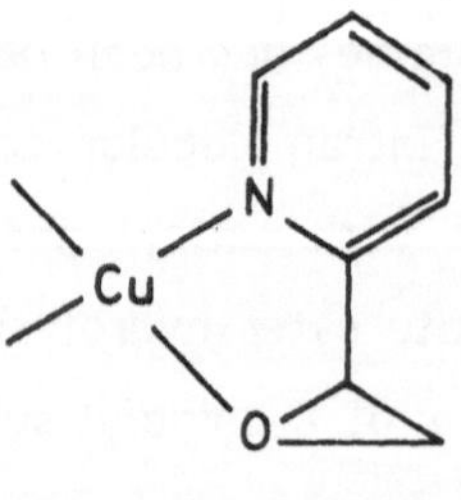

Scheme 11

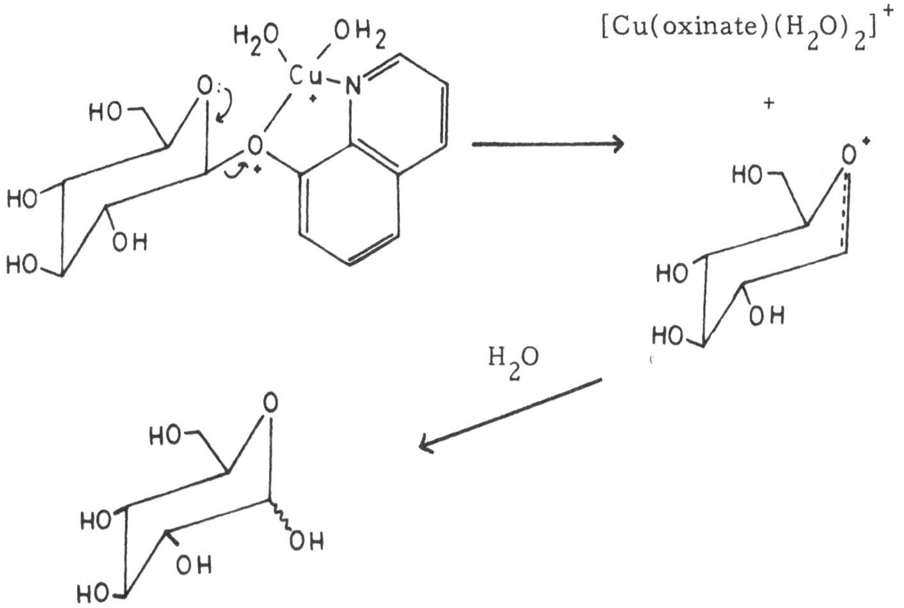

$[Cu(oxinate)(H_2O)_2]^+$

Metal-ion-catalyzed hydrolysis of 8-quinolyl β- D-glucopyranoside.

hydrase[356]. Catalysis by copper(II) is considered to involve a metal complex of the type shown in XCIII. Nucleophilic attack at the least hindered carbon predominates over S_N1-type carbonium ion formation[356].

9. HYDROLYSIS OF GLYCOSIDES AND ACETALS

The hydrolysis of glycosides is susceptible to both general and specific acid catalysis[357-8]. In most theories of glycosidase action it is postulated that the glycoside undergoes hydrolysis in the enzyme-substrate complex with an acidic group of the enzyme providing general acid catalysis and sometimes a basic group also providing nucleophilic catalysis[359-60]. As glycoside hydrolysis is susceptible to acid catalysis it would be expected that cleavage of the glycosidic bond would also be subject to metal ion catalysis. Clark and Hay[361] have found that the hydrolysis of 8-quinolyl β - D-glucopyranoside is subject to pronounced catalysis by copper(II). The copper(II) complex is hydrolyzed 10^5-10^6 times faster than the free glycoside in the pH range 5.5-6.2. The reaction is presumed to occur as

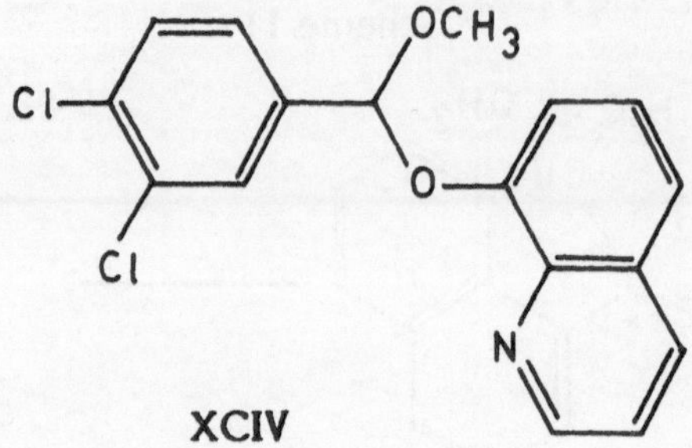

XCIV

shown in Scheme 11. Such effects can also be used for the synthesis of glycosides; thus phenyl 1-thio-D-glucopyranosides in the presence of mercury(II) salts are readily solvolyzed to give alkyl D-glucopyranosides with inverted anomeric configuration[362]. Methanolysis of the β- and α-anomers afforded the methyl α- and β-glycosides which were isolated in yields of 74 and 87% respectively. The approach can be extended to the synthesis of complex glycosides (the α-anomers of which are of special interest) as was illustrated by the preparation of cholestanyl and 1-naphthyl α-D-glucopyranoside and a disaccharide derivative.

The mechanism of the hydrogen ion catalyzed hydrolysis of simple acetals involves the classical A-1 mechanism in which there is rapid pre-equilibrium protonation of the acetal followed by rate-determining breakdown of the conjugate acid to an alcohol and a resonance stabilized carbonium ion[363-4].

Przystas and Fife[365] have recently studied the hydrolysis of substituted benzaldehyde methyl 8-quinolyl acetals such as XCIV in 50% dioxane-water at 30°C. These acetals are subject to both general and specific acid catalysis. A variety of divalent metal ions [Cu(II), Co(II), Ni(II), Mn(II) and Zn(II)] exert a large catalytic effect even though binding to the substrate is very weak. For example, a 0.02M concentration of Ni(II) (1000 fold excess) at pH 7.2 leads to a 2×10^5 enhancement for k_{obs} in the hydrolysis of XCIV, although full complexation of the substrate was not achieved under these conditions. Metal ion catalysis is attributed to a transition state effect in whch the leaving group is stabilized.

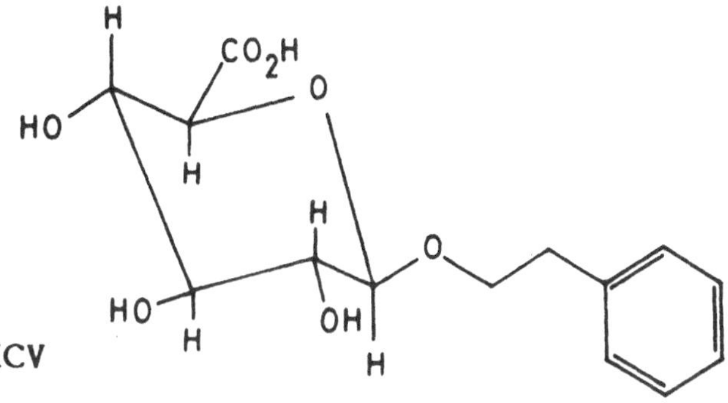

XCV

Some carbohydrate molecules appear to be quite specific in their binding to metal ions. Thus phenylethyl-β-D-glucopyranosiduronic acid (XCV) gives a 1:1 complex with copper(II)[366], but does not interact with Ca(II), Mg(II), Fe(II), Zn(II), Co(II) and Mn(II). Copper(II) is specifically bound to this glucoronide in dilute solution at a pH close to that of blood.

Metal ion catalysis of aldose-ketose isomerizations in acidic solution has also been noted[367]. For example, a variety of transition metal ions catalyze the isomerization of hexose phosphates.

Although only limited information is available on metal ion promoted reactions of carbohydrate derivatives, the available data indicate that interesting developments in this field can be expected. The observed rate enhancements for acetal and glycoside hydrolysis are similar to those seen in the peptide area. The pronounced metal ion catalysis in the hydrolysis of 8-quinolyl β-D-glucoside suggests that similar metal catalyzed processes may be of importance in the action of the glycosidases. Human saliva amylase, for example, requires 1g atom of calcium for full activity and the amylase from B.subtilis at least 4 g atoms. The calcium may be removed by electrodialysis and the resulting apoenzymes have only 5 -10% of the activity of the metalloenzymes[368]. However, present evidence suggests that the calcium is involved in maintaining the tertiary structure of the enzyme rather than taking a direct part in its catalytic action. The hydrolysis of simple aryl glycosides is not catalyzed by metal ions, but this does not exclude the direct participation of a metal ion in the enzymic reaction since the apoenzyme can provide the requisite binding sites for the metal ion.

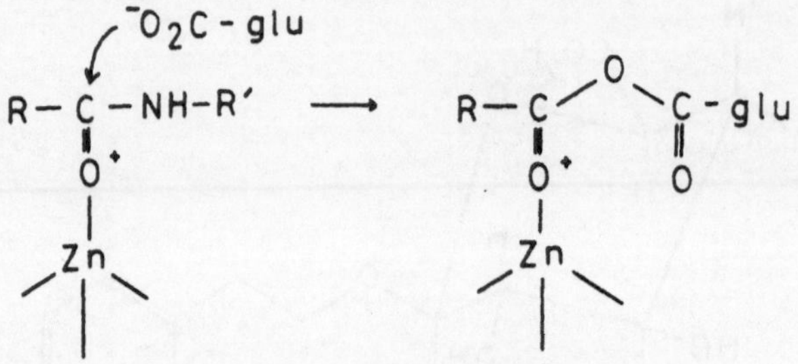

$$(13)$$

10.   HYDROLYSIS OF ANHYDRIDES

One of the most attractive mechanisms for the hydrolysis of peptides or esters by carboxypeptidase A involves two steps with an anhydride (acyl-enzyme) intermediate[369]. In the first step, the zinc(II) activates the substrate carbonyl group toward nucleophilic attack by a glutamate residue, resulting in the production of an anhydride between the enzyme glutamate and the scissile carboxyl group (eq 13). The hydrolysis of this anhydride can only be catalyzed by Zn(II), the sole remaining necessary catalytic group. Of the three identified catalytic groups, the tyrosine hydroxyl appears to be required for peptidase, but not for esterase activity. As the glutamate residue has been incorporated in the anhydride, only Zn(II) is still available for catalysis.

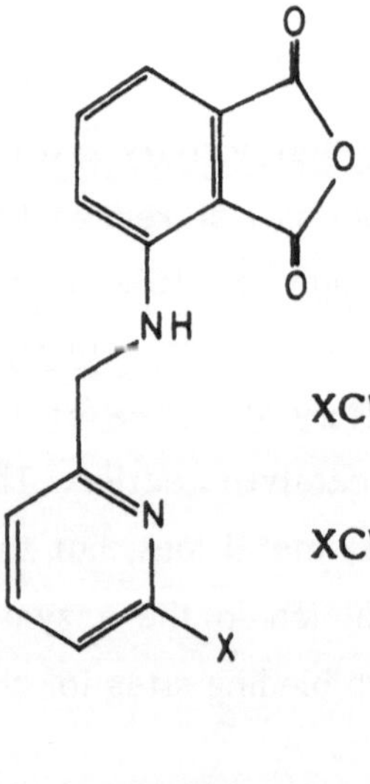

XCVIII

XCVI X = COOH

XCVII X = H

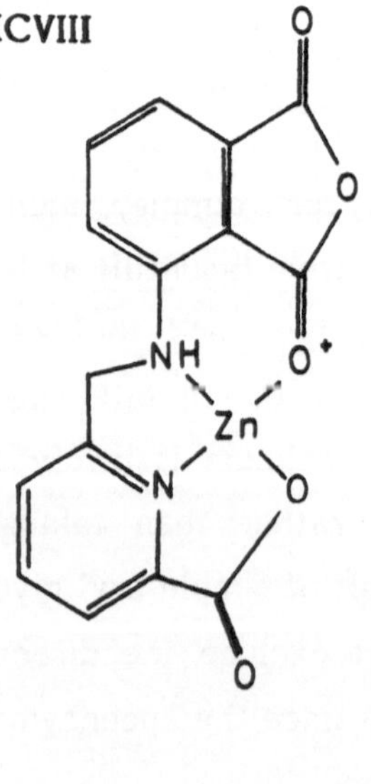

Table 19. Observed First Order Rate Constants for the Hydrolysis of the Anhydrides **XCVI** and **XCVII** and their Metal Complexes at pH 7.50. Data from ref. 370

Substrate	k_{obs} (s^{-1})	k_{rel}
XCVI	2.7×10^{-3}	1.0
CVI + Zn^{2+}	3.0	10^3
XCVII	5.5×10^{-3}	2.0
XCVII + Zn^{2+}	1.5	5×10^2

Breslow and co-workers[370] have studied the hydrolysis of the anhydrides **XCVI** and **XCVII** both in the presence and absence of zinc(II) (and other metal ions). In the absence of metal ion, hydrolyses of the anhydrides is independent of pH in the region 1.0-7.5 as also occurs with phthalic anhydride[371]. However, in the presence of zinc(II) the hydrolysis is first order in hydroxide above pH 5. Table 19 lists values of k_{obs} for the various substrates at pH 7.5. (For the metal complex $k_{obs.} = k_{OH}[OH^-]$). The catalytic effects are the order of 10^3. For **XCVI** + Zn^{2+}, where $k_{obs} = 3$ s^{-1} at pH 7.5, the rate constant is similar to k_{cat} for the enzyme which for several ester substrates lies in the range 0.5-230 s^{-1} [372]. The pH-dependence of the zinc(II) catalysis is consistent with attack by external hydroxide on a complex such as **XCVIII** where the metal acts as a Lewis acid catalyst. A further possibility (**XCIX**) involves attack by coordinated hydroxide on an uncoordinated anhydride carbonyl, and there is some evidence that this is in fact the process which does occur.

XCIX

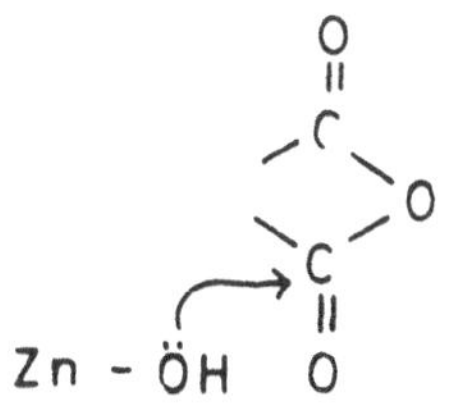

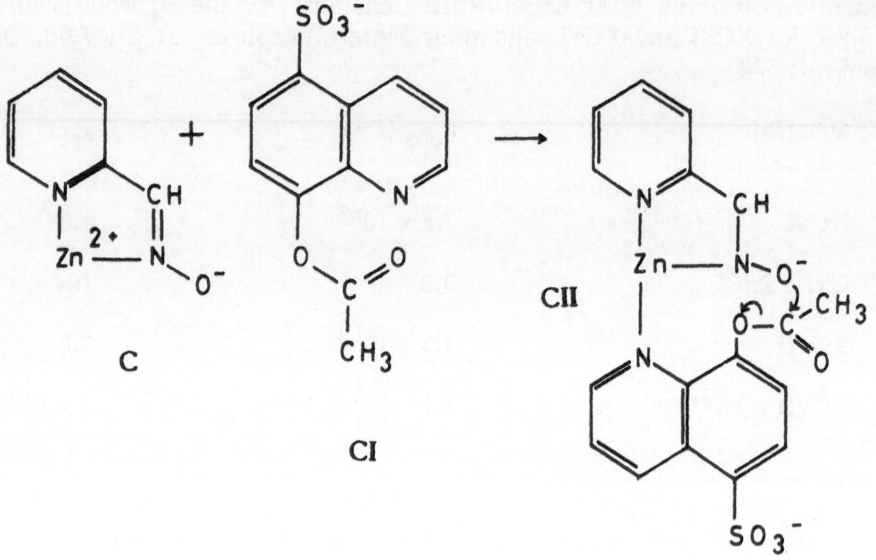

SO_3^-

C + CI → CII

CH_3

C

CI

CII

SO_3^-

Buckingham and Engelhardt[102] have studied the hydrolysis of propionic anhydride by kinetically inert complexes of the type $[M(NH_3)_5OH]^{n+}$; theses reactions are discussed in section 2.1.6.

11.   ACYL TRANSFER REACTIONS

One particularly interesting area which has not been fully exploited is metal-promoted acyl transfer. Studies on esterases and peptidases have shown that acyl transfer occurs to a nucleophilic group of the enzyme within an enzyme-substrate complex and the acyl group is then hydrolyzed in the second step. Many of these enzymes, e.g. carboxypeptidase, contain zinc(II).

Breslow and Chysman[373] have shown that zinc(II)-pyridine-carboxaldoxime anion (C) provides a strong free nucleophile and is a particularly effective nucleophilic catalyst in the hydrolysis of 8-acetoxyquinoline 5-sulfonate (CI). The reaction involves the catalyst-substrate complex CII. Molecular models show that in the mixed ligand complex CII, the N-O⁻ group is in a position to attack the acetyl group of CI. The zinc complex C is also an excellent catalyst for the hydrolysis of p-nitrophenyl acetate; in fact it is comparable in reactivity to hydroxide ion, although its pK_a is only 6.5 (Table 20).

Table 20. Rate Constants for Nucleophilic Attack (k in $M^{-1}s^{-1}$ at 25°C).
Data from ref. 373

Nucleophile	pK_a	AQS	PNA
OH⁻	15.7	2.01	14.8
PCA anion	10.04	6.98	77.2
H_2O	-1.7	7.3×10^{-7}	1.0×10^{-8}
PCA-Zn(II)	6.5	10	10

PCA = pyridine-2-carboxaldoxime

AQS = 8-acetoxyquinoline-5-sulfonate

PNA = p-nitrophenyl acetate

The hydrolysis of 8-acetoxyquinoline (**CIII**) is subject to catalysis by metal ions and detailed kinetic studies of the reaction have been reported[248]. The metal ion could be bound to the carbonyl oxygen (as in **CIV**) or the ether oxygen (as in **CV**) and the actual structure of the catalytically active complex was not unequivocally defined.

Sakan and Mori[374] have shown that 8-acetoxyquinoline reacts with copper(II) chloride in dry tetrahydrofuran to give the green complex **CVI**. The ir spectrum of the complex has νCO at 1790 cm^{-1}, consistent with the ether oxygen acting as a donor. Chelation via the carbonyl oxygen would lead to a substantial decrease in the carbonyl stretching frequency. The

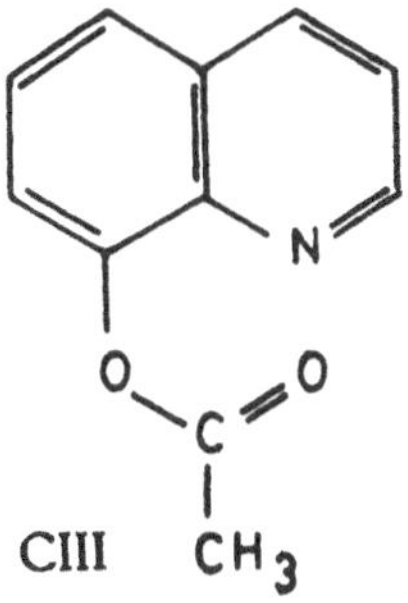

CIII   CH₃   CIV   CH₃   CH₃   CV

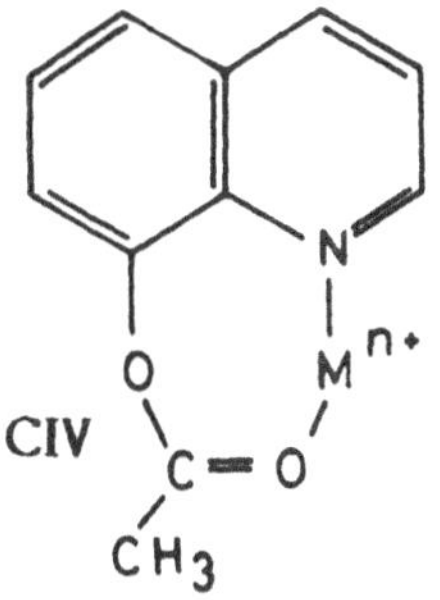

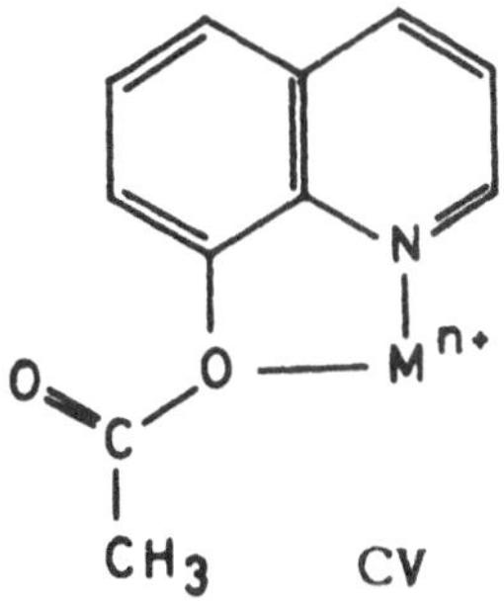

CVI

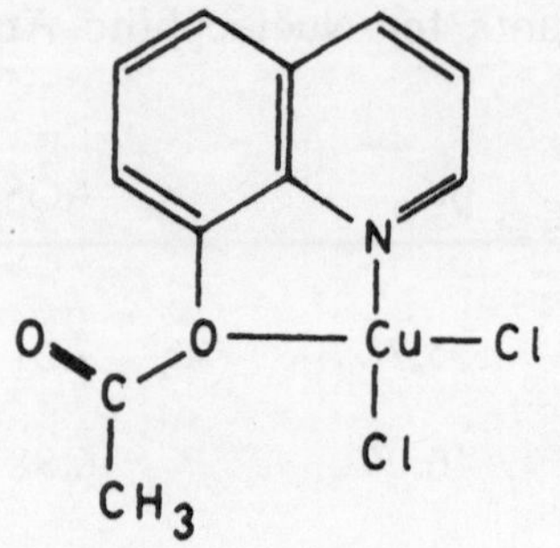

O=C—O—Cu—Cl ... (structure of 8-acetoxyquinoline copper complex)

value of νCO for the free ligand is 1750 cm^{-1} so that an increase is observed. In CVI the carbonyl oxygen is very reactive to nucleophiles owning to polarization of the C-O bond and the substantial thermodynamic stability of the copper(II) complex of 8-hydroxyquinoline (i.e. complexation of the leaving group).

The complex CVI can be used as an acylating agent for alcohols, phenols and amines in THF or benzene solvent[374-5].

$$C_6H_5NH_2 + CVI \xrightarrow[20°C]{THF} CH_3CONHC_6H_5$$

(CVII)

Acetanilide (CVII) can be isolated in 83% yield after 2 days at room temperature. In addition 8-acetoxyquinoline will react with Grignard reagents to give high yields (80%) of ketones. The reaction is believed to proceed via the intermediate magnesium(II) complex CVIII (Scheme 12).

Some kinetic work on the reaction of Bu^nNH_2 with the copper(II) complex of 8-acetoxyquinoline at 25°C in DMF as solvent has been carried out. The second order rate constant is 9.5×10^{-2} M^{-1} s^{-1}[375].

12.   IMINES

12.1   Hydrolysis of Imines

The effect of metal ions on the hydrolysis of imines is of considerable biological interest. A number of papers have discussed the effect of metal ions on the formation and the hydrolysis of such compounds[376-386,393].

Scheme 12

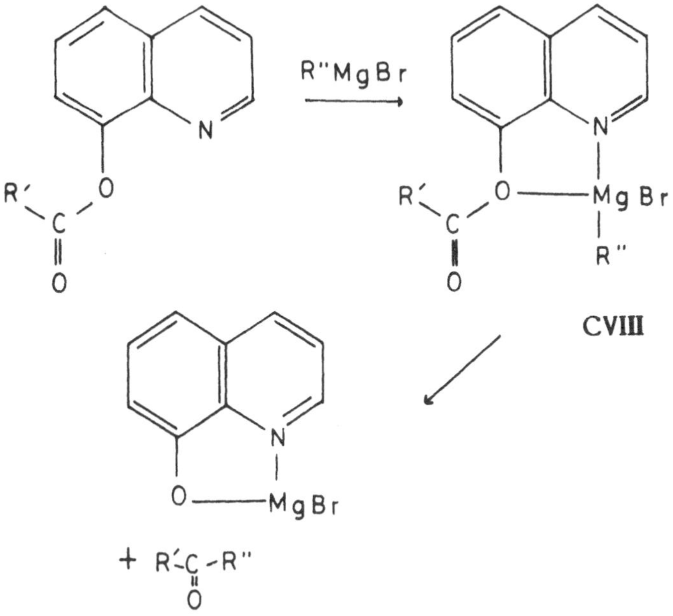

Unfortunately detailed kinetic studies on imine hydrolysis are lacking and there is considerable confusion in the existing literature. In addition, the low solubility of many aromatic imines (which are generally more readily characterized) in aqueous solution has led to the use of mixed alcohol-water solvent systems which can give problems of interpretation.

Dash and Nanda[384] have made a detailed kinetic study of the hydrolysis of N-salicylideneaniline (CIX) in the presence and absence of cobalt(II), nickel(II), copper(II) and zinc(II) using 10% ethanol-water as solvent. The reactivity of the complexes towards hydrolysis was found to follow the reverse sequence of their formation constants, the most thermodynamically stable complexes undergoing the slowest rate of hydrolysis. More recent measurements[387] using high copper(II) to ligand ratios have indicated that the copper(II)-imine is quite stable to hydrolysis at pH 5.

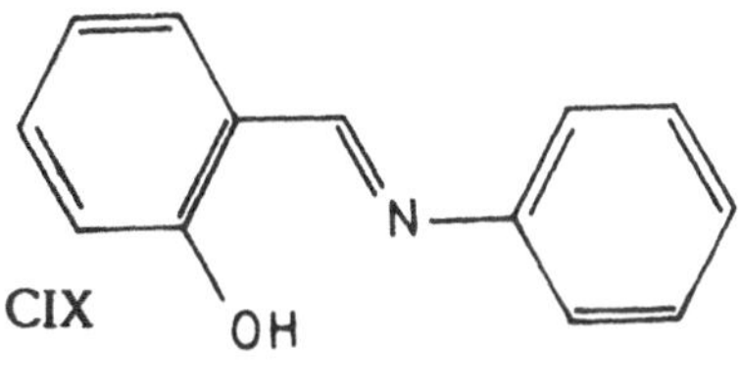

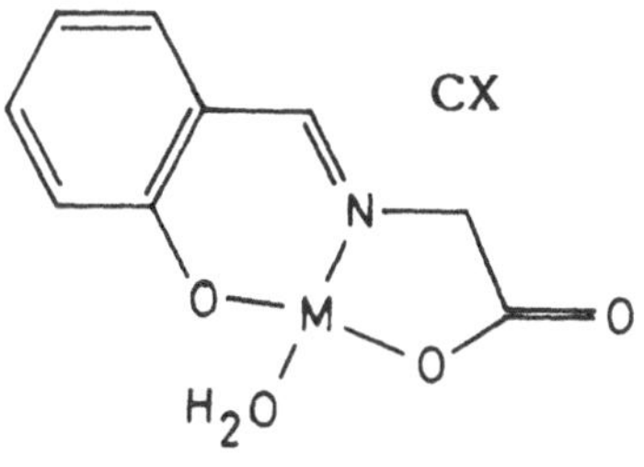

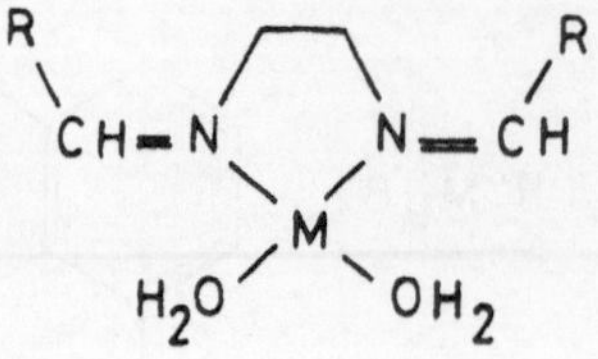

CXI	R = 2-thienyl
CXII	R = 2-furanyl

Imines capable of forming bicyclic chelate rings as in CX are known to be stabilized towards hydrolysis under mildly acidic conditions[379]. However, imines which can form monocyclic chelates such as N,N'-ethylenebis(2-thienylmethyleneimine) (CXI)[377-8,388,390] and N,N'-ethylenebis(2-furanylmethyleneimine) (CXII)[389,391-2] where the S and O atoms respectively do not act as donors are rapidly hydrolyzed in the presence of copper(II) and nickel(II) ions. Hydrolysis of a bicyclic metal complex such as CX involves the rupture of chelate rings, a process which is not normally favored.

Hay and Nolan[394] have carried out a detailed kinetic study on the hydrolysis of N-2-pyridylmethyleneaniline (CXIII) and its copper(II) complex CXIV. Very substantial accelerations were observed in this system. Base hydrolysis of CXIV is some 10^5 times faster than base hydrolysis of CXIII at $25^\circ C$. The rate constants obtained for this system are summarized in Table 21.

Table 21. Rate Constants for the Hydrolysis of N-2-Pyridylmethyleneaniline (L) at $25^\circ C$ and $\mu = 0.1M$. Data from ref. 394

Reaction	k
$HL^+ + H_2O$	$1.26 \times 10^{-1}\ M^{-1}\ s^{-1}$
$L + OH^-$	$2.17 \times 10^{-2}\ M^{-1}\ s^{-1}$
$L + H_2O$ (a) $HL^+ + OH^-$ (a)	$2.8 \times 10^{-4}\ s^{-1}$
$[CuL(OH_2)_2]^{2+} + H_2O$	$3.15 \times 10^{-6}\ M^{-1}\ s^{-1}$
$[CuL(OH_2)_2]^{2+} + OH^-$	$9.72 \times 10^3\ M^{-1}\ s^{-1}$
$[CuL(OH)(OH_2)]^+ + OH^-$	$2.75 \times 10^3\ M^{-1}\ s^{-1}$

(a) These reactions are kinetically indistinguishable

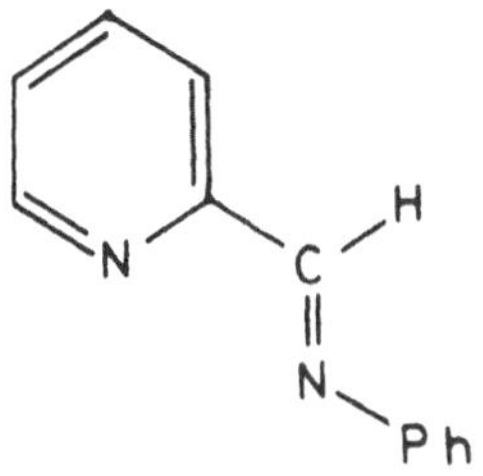

CXIII 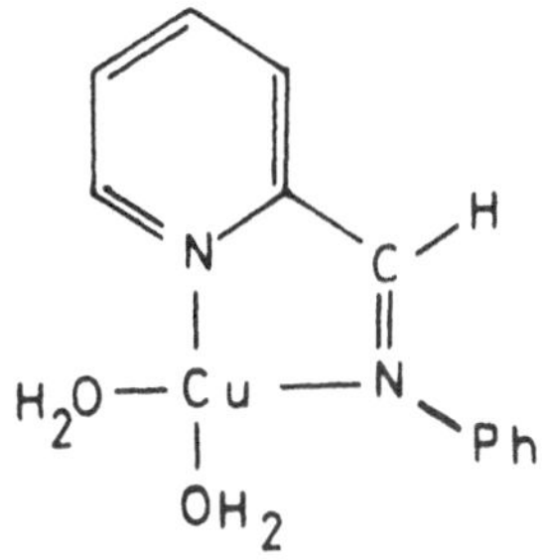   CXIV

12.2 Formation of Imines

It has been shown that under basic conditions coordinated ligands may effect nucleophilic attack at carbonyl centers in organic compounds[395-400]. For example, reaction of cobalt(III) and platinum(IV) ammines with ketones gives the corresponding Co(III) and Pt(IV) imine complexes. A similar reaction between $[Ru(NH_3)_6]^{3+}$ and diones produces the corresponding Ru(II) diimines such as CXV:[401].

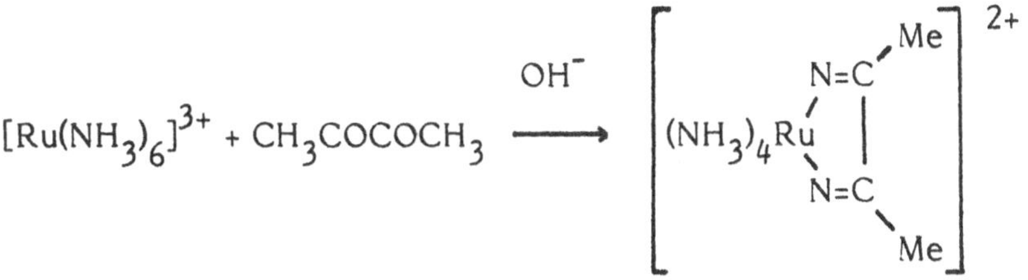

$$[Ru(NH_3)_6]^{3+} + CH_3COCOCH_3 \xrightarrow{\text{OH}^-} \left[(NH_3)_4 Ru \begin{array}{c} N=C \diagup Me \\ | \\ N=C \diagdown Me \end{array} \right]^{2+}$$

CXV

It has also been reported[402] that nitrilepentaammineruthenium(II) complexes $(NH_3)_5 RuNCR$ (R = Me or Ph) can be prepared by the reaction of $[Ru(NH_3)_6]^{3+}$ with the appropriate aldehydes:

$$[Ru(NH_3)_6]^{3+} + RCHO \xrightarrow{\text{OH}^-} [(NH_3)_5 RuN{\equiv}CR]^{2+}$$

This report appears to be the first to describe the formation of a nitrile from the reaction of an aldehyde with a transition metal ammine nucleophile. A pK_a value of approximately 12.4 has been reported[403] for the $[Ru(NH_3)_6]^{3+}$ ion. Solutions of approximately 0.03M $[Ru(NH_3)_6]Br_3$ containing a 100-fold excess of acetaldehyde or benzaldehyde react rapidly (complete in less that 1 min based on color change) to produce the nitrile complexes[402].

CXVI

CXVII

The reaction of $[Ru(NH_3)_5NO]^{3+}$ with several aldehydes in 0.1M NaOH gives trans-$[Ru(NH_3)_4(OH)NO]^{2+}$ as the sole product with the aldehyde serving as a catalyst[404]. The reaction appears to take place by nucleophilic attack of the amido ligand trans to the nitrosyl group at the carbonyl group of the aldehyde to give an unstable imine intermediate, which hydrolyzes at the metal-nitrogen bond to give the hydroxo-complex and regenerate the aldehyde.

Reactions of this type have possibilities for planned organic and chelate synthesis and in some cases regio- and stereospecific condensations have been observed. For example, Gainsford and Sargeson[399] have reported that treatment of CXVI (R = H, Me, CH_2Br, Ph, p-C_6H_4Cl) with base leads to imine complexes where the coordinated aminoacetone has reacted with a bound monoamine RCH_2NH_2 or an ethylenediamine moiety. The product distribution appears to be determined by the relative acidity of the monoamine and ethylenediamine NH centres, condensation being preferred at the most acidic site. Reduction of the imine group in these complexes with $[BH_4]^-$ gives the new complexes $[Co(en)_2(NH_2CH_2CH(Me)NHCH_2R)]^{3+}$ and $[Co(en)(NH_2CH_2CH_2NHCH(Me)CH_2NH_2)(NH_2CH_2R)]^{3+}$ in which the reduced ligand is synthesized sterospecifically.

Coordinated ligand reactions of this type have been used to synthesize a variety of macrocyclic complexes of the clathrochelate variety[233] and somewhat similar reactions have been used quite generally in the macrocycle field[410]. The role of metal ions in the reaction of amino acids with pyridoxal and pyridoxal phosphate is very wide[406]. Hopgood[405] has studied the rates of transamination of 15 amino-acids in the presence of zinc(II) and pyridoxal-5-phosphate (CXVII). On mixing the reagents zinc(II) aldimine complexes are rapidly formed (ca 5 min) and these species subsequently transaminate in a slow second step. A review of some of the work in this area is available[406]. Several papers deal with the characterization of metal complexes with Schiff bases obtained from pyridoxal and amines or amino acids[407-409].

In spite of the wide occurence of Schiff base metal complexes, little mechanistic work on the role of metal ions on the formation of Schiff bases is available. Cordes and Jencks[411] have provided a classic discussion of the mechanism of Schiff base formation and hydrolysis, and Nunez and Eichhorn[412] have studied the effects of nickel(II) and copper(II) on the formation of salicylideneglycine. It was found that formation of the Schiff base is greatly retarded by prior reaction of either nickel(II) or copper(II) with one of the Schiff base components.

13.   POLYMER-METAL COMPLEXES

The advantages inherent in attaching homogeneous catalysts to polymers have stimulated extensive research with respect to techniques of achieving such attachment and studies of the catalytic activity of the resulting polymer-anchored catalysts. Some comprehensive reviews of the field are available[413-4].

Some examples of the use of polymer-metal complexes as catalysts in hydrolytic reactions are discussed below.

The catalytic hydrolysis of oligophosphates by poly-(L-lysine)-copper(II) complexes has been described[415]. Marked catalytic effects were observed with hydrolysis of pentaphosphate occuring exclusively giving orthophosphate as the main product.

Nozawa et al[416] have also found that the poly(L-lysine)-copper(II) complex displays steroselective catalysis of the hydrolysis of phenylalanine esters. The poly(L-lysine)-copper(II)-D-ester complex is apparently more stable than the mixed ligand complex with the L-ester, giving rise to stereoselective effects[417-8].

The hydrolysis of sodium pyrophosphate is catalyzed by some metal complexes of poly(methacrylacetone)[417]. The catalytic activity of the polymer complexes decreased in the order, $Zr(IV)O > U(VI)O_2 > Cr(III) \sim Ce(III) \sim Cu(II)$.

Attempts have been made to develop "artifical" enzymes by combining a metal catalytic group and a hydrophobic binding cavity based on a toroidal cyclohexaamylose[419]. The preparation of such a molecule (CXVIII) and demonstration that an appropriate metal derivative (CXIX) will catalyze the hydrolysis of p-nitrophenyl acetate has been accomplished. The complex CXIX is nearly four times more reactive towards p-nitrophenylacetate than is an equivalent concentration of the nickel(II)-pyridine carboxaldoxime complex, corresponding to an acceleration of greater than 10^3 over the uncatalyzed rate. The p-nitrophenyl group has been shown to bond in the cyclohexaamylose cavity[420]. The increased reactivity is believed to result from binding and reaction with the CXIX-p-nitrophenylacetate complex.

14.   MORE RECENT DEVELOPMENTS

14.1   Reviews

Dixon and Sargeson[421] have discussed the role of the metal ion in the reactions of coordinated substrates and in some metalloenzymes (e.g. urease, carboxypeptidase, aconitase, alkaline and acid phosphatases). The review provides an excellent account of the work of Sargeson's group in this area. A further review deals with models for Zn(II) hydrolases (e.g. carboxypeptidase A, thermolysin, alkaline phosphatase)[422].

14.2   Ester Hydrolysis

α-Amino-acid esters interact with $[Pd(en)(OH_2)_2]^{2+}$,[423] and $[Pd(bipy)(OH_2)_2]^{2+}$,[424] to give mixed ligand complexes of the types CXX and CXXI respectively. These complexes undergo rapid ester hydrolysis, with both water and hydroxide ion acting as nucleophiles. Substantial rate enhancements are observed in base hydrolysis, by factors of 1.6×10^5 for methyl glycinate to 3.3×10^7 for ethyl picolinate in the complex CXXI.

The palladium(II)-promoted hydrolysis of methyl glycylglycinate and isopropyl glycylglycinate has also been studied[425]. The peptide esters act as tridentate ligands, donation occuring via the terminal amino group, the deprotonated amide nitrogen and the carbonyl group of the ester (CXXII). Hydrolysis of the ester function by both water and hydroxide ion occurs. Base hydrolysis of the coordinated peptide esters is ca 10^5 times faster than that of the free ligands.

Amino-acid esters interact with [glycylglycinato(2-)]copper(II) to give ternary complexes. Base hydrolysis of the ester ligands (glyOMe, glyOEt, L-α-alaOEt, or L-PheOEt) is some 50 times faster than for the free e ters[426]. Hydrolysis of α-amino-acid esters in ternary complexes with copper(II)-ethylenediaminemonoacetate has also been studied[427]. Nucleophilic attack by both water and hydroxide ion was observed and base hydrolysis is some 10^3 times faster than for the free esters.

Correlations of λ_{max} values for tridentate copper(II) complexes have been noted both with the log of the rate constant for the copper (II)-complex catalyzed hydrolysis of methyl glycinate, and with the log of the formation constant of the hydroxo complex of the copper chelate[428]. Strong σ donors will decrease the Lewis acidity of the copper (II) ion. The hydrolysis of amino acid esters in ternary complexes with [Cu(imda)]^0 leads to accelerations of ca 10^4 for base hydrolysis[429]. Studies of ester hydrolysis in ternary complexes provide useful biomimetic models for certain metalloenzymes, as the enzyme-substrate complex can be considered as a special type of mixed ligand complex.

Although ester hydrolysis in [Co(NH$_3$)$_5$NH$_2$CH$_2$CO$_2$Et]$^{3+}$ has a half life in 1M CF$_3$SO$_3$H at 25°C in excess of 1 month, the reaction for the corresponding Ru(III) complex under the same conditions is complete in 1 hr.[430] The products in the latter case are [(NH$_3$)$_5$RuOH$_2$]$^{3+}$ + $\overset{+}{N}$H$_3$CH$_2$CO$_2$Et) (30%) and [(NH$_3$)$_5$RuO$_2$CCH$_2$NH$_3$]$^{3+}$ + EtOH) (70%).

$$NH_2CH_2CO_2^-(H)$$

CXXIII

$$OH(H_2O)$$

Oxygen exhange and glycinate ring opening in $[Co(en)_2gly]^{2+}$ have been investigated,[431] and the formation of $[Co(en)_2gly]^{2+}$ from the monodentate complexes derived from **CXXIII** studied in detail[432]. An interesting example of stereospecific co-micelle-promoted hydrolysis of N-acylphenylalanine p-nitrophenyl esters in the presence and absence of cobalt(II) has been reported,[433] and the kinetics of base hydrolysis of the ester ligand in bis(ethyl cysteinato)palladium(II) studied in detail[434].

Carboxypeptidase A is a Zn(II) metalloenzyme which catalyzes the hydrolysis of peptides and O-acyl derivatives of α-hydroxy carboxylic acids. The metal ion interacts with the carbonyl oxygen of peptide substrates, and X-ray crystallographic work at 2Å resolution has also revealed the presence of the carboxyl group of glutamic acid-270 in the active site.[435-6] Both nucleophilic and general base mechanisms have been suggested for the enzyme involving Glu-270. As a result there has been continuing interest in divalent metal ion catalysis by Co(II), Ni(II), Zn(II) and Cu(II) in the hydrolysis of esters where H_2O or OH^- is the nucleophile. A copper(II)-promoted water reaction has been detected at pH < 3 in the hydrolysis of 2-(6-carboxypyridyl)methyl acetate,[437] and the hydrolysis of O-acetyl-2-pyridinecarboxaldoxime is promoted by Cu(II) via both OH^- and H_2O pathways.[438] Zinc(II) catalysis of the latter reaction has also been studied[439] where the rate enhancement is ca 10^3. Divalent metal ions [Ni(II), Co(II) and Zn(II)] exert a large catalytic effect on the hydrolysis of 2-pyridylmethyl hydrogen phthalate.[440]

At saturating concentrations of Ni(II), Co(II) and Zn(II), pH-independent hydrolysis (H_2O attack) is clearly detectable in the hydrolysis of cinnamic 6-carboxypicolinic monoanhydride[441] although metal ion promoted base hydrolysis is very rapid. Rate constants have been

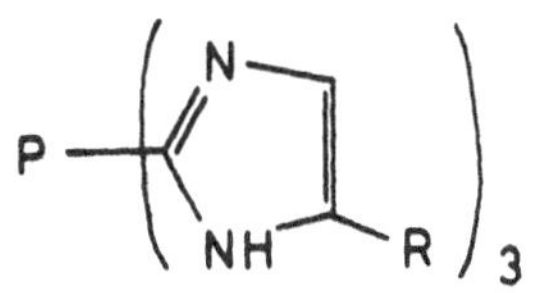

CXXIV   CXXV

R = H   R = CH₂CH₂OH

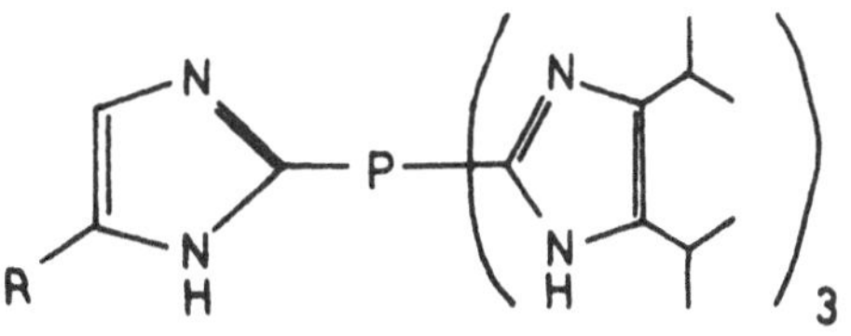

CXXVI   CXXVII

R = H   R = CH₂CH₂OH

CXXVIII

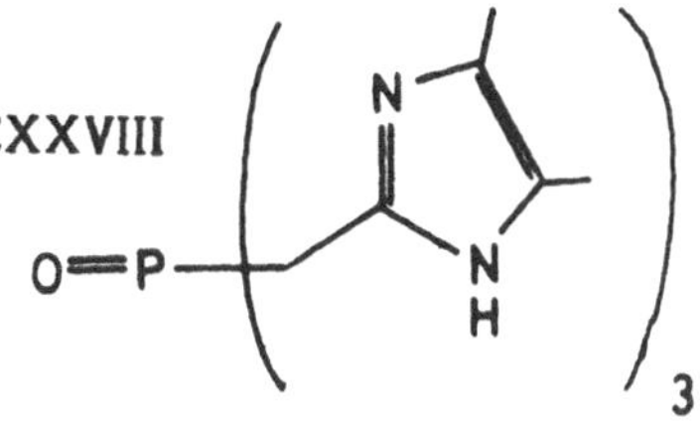

determined[442] for hydrolysis of a series of phenolic and aliphatic esters of picolinic acid. Metal ion promoted attack by both OH^- and H_2O were observed with those esters having leaving groups with pK_a values of 12.4 or less.

A series of tris(imidazole)-containing phosphines have been prepared and their M(II) complexes studied as biomimetic catalysts for the hydrolysis of p-nitrophenyl picolinate.[443] The phosphines CXXIV and CXXV as their Zn(II) complexes promote the hydrolysis of p-nitrophenyl picolinate, their catalytic activities increasing with pH. At a given pH the second order catalytic rate constants are 2-10 fold larger than those for Zn(II) or the ligand alone, indicating a cooperative interaction between the ligand and Zn(II) producing a more active catalyst. Using the phosphine ligands CXXVI and CXXVII it was possible to establish that $[LCo-OH]^+$ is the active species in the cobalt(II)-promoted reaction. The Co(II) complex of tris(2-imidazolylmethyl)phosphine oxide (CXXVIII) also displays good catalytic activity in promoting the hydrolysis of p-nitrophenyl picolinate.[444] Between pH 7.5 and 8.5 saturation kinetics are observed indicating formation of a ternary LCo(II)-ester complex. Rate constants for hydrolysis of p-nitrophenyl picolinate at 25°C in the pH range 6.5-8.5 have been determined in the presence and absence of a variety of divalent ions [(Ni(II), Zn(II),

Co(II), Ca(II) and Mg(II))], and substituted imidazoles or pyridines containing hydroxyl groups in their side chains.[445] In some cases, the presence of both a metal ion [Ni(II) or Zn(II)] and a ligand, leads to high rate enhancements, and values of k_{obs} display saturation kinetics with respect to both the ligand and metal ion concentrations. The ligands complexed with Zn(II) appear to be simple but highly effective models for hydrolytic metalloenzymes.

Steroselective hydrolysis of enantiomeric N-acyl-L- (or D-) phenylalanine p-nitrophenyl esters catalyzed by pentaamine-L-histidineruthenium(III) and anionic surfactants has been studied.[446] The highest selectivity of 4.2 was observed with $CH_3(CH_2)_8CONHCH-(CH_2Ph)CO_2Ar$.

14.3 Phosphate Ester Hydrolysis

14.3.1. Cobalt(III) Complexes

This area has attracted considerable interest. Tracer ^{18}O studies have established[447] that base hydrolysis of coordinated acetyl phosphate in the complex $[(NH_3)_5Co\text{-}OPO_3COCH_3]^+$ [k_{OH} = 0.53 M^{-1} s^{-1} at 25°C and μ= 1.0M ($NaClO_4$)] occurs by exclusive carbon-oxygen bond fission. The hydrolysis of the acetyl phenyl phosphate monoanion is also significantly catalyzed by the exchange-inert hydroxo complex $[(NH_3)_5CoOH]^{2+}$ (k_{MOH} = 2.9 x 10^{-2} M^{-1} s^{-1} at 25°C) which operates by a nucleophilic pathway involving attack at the carbonyl carbon. (Other reactions involving M-OH species are considered in Section 2.1.6)

The hydrolysis of β,γ-$[Co(NH_3)_4H_2P_3O_{10}]$, in which the triphosphate ion is coordinated as a bidentate ligand, has been studied[448] in the presence of $[Co(cyclen)(OH)(OH_2)]^{2+}$ (cyclen = 1,4,7,10-tetraazacyclododecane). In the presence of the macrocyclic complex, the rate of hydrolysis of triphosphate to pyrophosphate is increased by a factor of 5 x 10^5 over the rate for the free ion. The pH rate profile for the reaction indicates that a deprotonation step with a pK of 7.9 was required for the accelerated hydrolysis to occur. The results are consistent with the formation of the binuclear intermediate **CXXIX**, followed by internal nucleophilic attack on phosphorus by coordinated hydroxide. Phosphorus-31 nmr studies[449] have now provided evidence in support of such an intermediate.

CXXIX

$$\text{(cyclen)Co} \begin{array}{c} O \\ \parallel \\ O-P-O-P-O \\ \mid \quad\quad\ \mid \\ O^- \quad\quad O \\ \mid \\ {}^-O-P-O \\ \parallel \\ O \end{array} \text{Co(NH}_3)_4$$

with ÖH on the cyclen Co.

Polyphosphates such as pyrophosphate and triphosphate are interesting ligands. Recently the unidentate and bidentate pyrophosphate complexes $[Co(NH_3)_5HP_2O_7].H_2O$ and $[Co(NH_3)_4HP_2O_7].2H_2O$ have been characterized and their crystal structures determined.[450] The α, β, γ - tridentate $[Co(NH_3)_3(H_2P_3O_{10})]$ complex contains two fused six-membered chelate rings formed by the facial coordination of one O from each of the phosphate residues.[451] The complex $[Co(NH_3)_4(H_2P_3O_{10})].H_2O$ containing the β, γ-bidentate triphosphate ligand has an eight-membered chelate ring in a boat conformation stabilized by two interligand H bonds from the axial ammines above and below the chelate ring to the β- and γ-phosphates.[452]

Sargeson and co-workers[331] initially reported that 4-nitrophenyl phosphate in the complex LXXXVII underwent base hydrolysis some 10^9 fold faster than the uncoordinated ester. Subsequent X-ray work[453] has now shown that the complex LXXXVII and the analogous ethylenediamine derivative do not contain chelated phosphate esters but are dimeric species (CXXX) with a surprisingly stable eight-membered chelate ring. The reactivity of such bridged complexes in basic solution has been discussed in a review[421] and a recent paper.[465] Ester hydrolysis occurs primarily via intramolecular attack by coordinated hydroxide in the ring opened complex CXXXI.

CXXX

$$L_4Co \begin{array}{c} O \ \ OPh \\ \diagdown \diagup \\ O-P-O \\ \diagup \quad\quad \diagdown \\ \quad\quad\quad CoL_4 \\ \diagdown \quad\quad \diagup \\ O-P-O \\ \diagup \diagdown \\ O \ \ OPh \end{array}$$

CXXXI

$$(en)_2Co \begin{array}{c} \quad\quad {}^-O-P-OAr \ (O) \\ OH \\ \diagdown \quad\quad\quad\quad Co(en)_2 \\ O-P-O \\ \parallel \ \diagdown \\ O \quad OAr \end{array}$$

The preparation of cis-[Co(cyclen)PO$_4$] has been described and its reactivity in acidic and basic solution studied.[454] This complex undergoes rapid ring opening in acidic or basic solution to give the monodentate phosphato species. Loss of monodentate phosphate in acidic solution follows at rate expression of the form

$$k_{obs} = k_0 + k_H [H^+]$$

where $k_0 = 5 \times 10^{-4}$ s^{-1} and $k_H = 4.05 \times 10^{-2}$ M^{-1} s^{-1} at 25° (I = 0.49M). The acid-catalyzed reaction displays a solvent deuterium isotope effect k_{D_2O}/k_{H_2O} of ca 1.4, consistent with a mechanism involving a rapid pre-equilibrium protonation followed by rate-determining loss of monodentate phosphate. Loss of monodentate phosphate from the complex [Co(cyclen)OH(OPO$_3$)]$^-$ also takes place in basic solution, and the reaction shows a first order dependence on the hydroxide ion concentration with k_{OH} = 2.7 $\times 10^{-2}$ M^{-1} s^{-1} at 25°C and μ = 0.49M. This work provides supporting evidence for the participation of the intermediate **CXXIX** in the [Co(cyclen)(OH$_2$)$_2$]$^{3+}$-promoted hydrolysis of β,γ-[Co(NH$_3$)$_4$(H$_2$P$_3$O$_{10}$)]. The monodentate intermediate will be sufficiently long-lived in solutions of pH 8-11 to participate in the reaction.

Complex formation between [Co(pn)$_2$(OH$_2$)$_2$]$^{3+}$ and pyrophosphate and its influence on the hydrolysis of pyrophosphate to orthophosphate has been studied by ^{31}P nmr over the pH range 0-12.[342] Rate enhancements of ca 10^5 were observed (pn = 1,3-diaminopropane). There was evidence that the reactive species involved was a 3:1 (pn)$_2$Co(III)-pyrophosphate complex. Similar experiments have been reported on the effects of cobalt(III) complexes on the rate of hydrolysis of the triphosphate ion.[455] The simultaneous coordination of a phosphate moiety and an attacking

CXXXII

nucleophile by a metal ion is shown to be an important factor in determining the rate of phosphate hydrolysis. The hydrolysis of bidentate triphosphate in $[Co(NH_3)_4H_nP_3O_{10}]$ (**CXXXII**) studied by ^{31}P nmr and phosphomolybdate analysis occurs at some two-thirds of the rate for the free ligand.[340] Coordination of the polyphosphate ion by the metal ion is necessarily the first step in any metal-promoted phosphate hydrolysis reaction and coordination has been established by ^{31}P nmr in several cases.[342,456] Cornelius has argued[457] that given the existence in solution of a cobalt(III)-polyphosphato complex, two subsequent reactions are possible, depending on the exact nature of the species on solution. One of these reactions leads to hydrolysis and the other does not (Scheme 13). In Reaction (a), a phosphate group acts as a nucleophile, displacing the hydroxide ion from the coordination sphere of the cobalt(III) ion. This reaction results in the formation of a polyphosphate chelate ring and little or no hydrolysis will occur. In Reaction (b), the nucleophile is the coordinated hydroxide ion which attacks a phosphorus atom in the phosphate chain to give a five-

Scheme 13

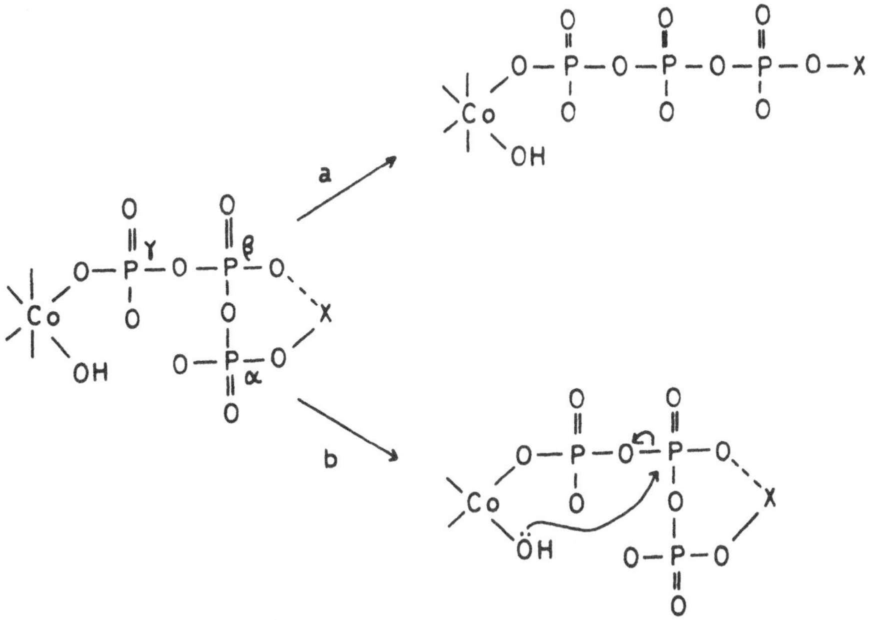

General scheme for the reaction of polyphosphate with a cobalt (III) amine complex. Pathway (a) leads to chelation and pathway (b) to hydrolysis. Protons and charges are omitted for clarity as they are pH-dependent.

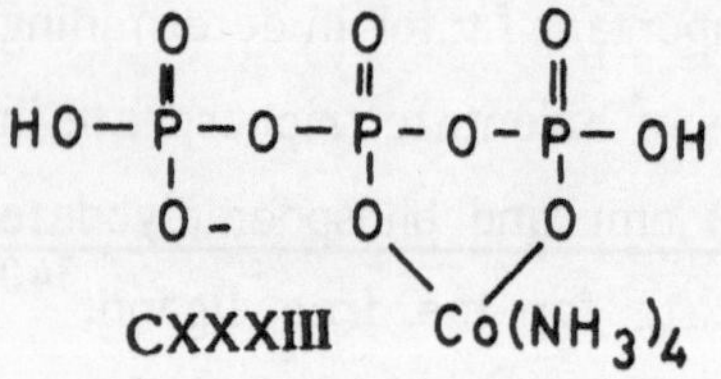

CXXXIII   CXXXIV

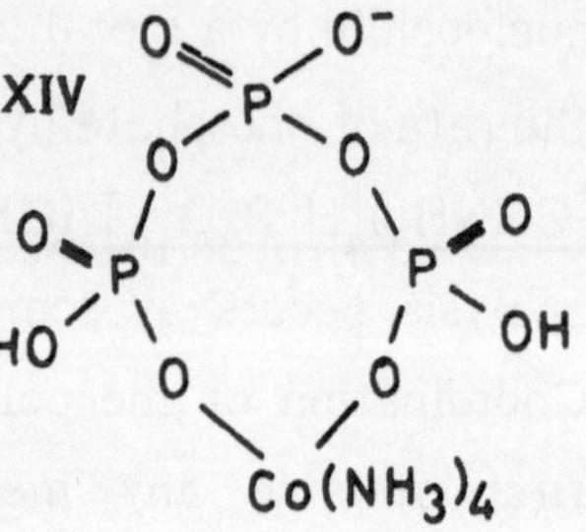

CXXXV   HO-Co(pn)$_2$ (en)$_2$Co   Co(pn)$_2$

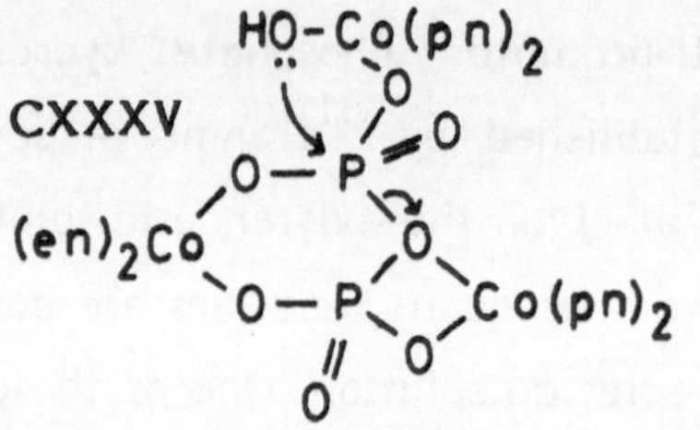

coordinated phosphorane intermediate, a mechanism well established in phosphorus chemistry.[282] The tetrahedral phosphate is then re-formed by cleavage of a P-O-P bond so that this pathway leads predominantly to hydrolysis. The nature of "X" in Scheme 13 determines if chelation or hydrolysis will occur. If "X" is electron withdrawing then the nucleophilic character of the phosphate will be reduced and hydrolysis becomes the dominant pathway. These ideas may be used to rationalize many of the experimental observations in this area.

The isomerization and phosphate hydrolysis reactions of the two linkage isomers of (triphosphato)tetraaminecobalt(II), β,γ-[Co(NH$_3$)$_4$H$_2$P$_3$O$_{10}$] (CXXXIII) and α,γ- [Co(NH$_3$)$_4$H$_2$P$_3$O$_{10}$] (CXXXIV), containing six- and eight-membered chelate rings respectively, have been investigated by ^{31}P nmr and visible spectroscopy.[458] In aqueous solution the isomerization reaction between the β,γ- and α,γ-complexes has an equilibrium constant of 0.07 favoring the β,γ-complex at 40°C and pH 6.5. Interestingly the rates of phosphate hydrolysis for the two isomers are nearly identical despite the difference in stability of the two chelate rings of different size.

Rapid cleavage of P$_2$O$_7^{4-}$ in [Co(en)$_2$P$_2$O$_7$]$^-$ by [Co(pn)$_2$(OH)(OH$_2$)]$^{2+}$ has been observed[459], and the reaction monitored by ^{31}P nmr. The metal ions act in concert to provide a coordinated nucleophile (Co-OH) to assist the leaving group in severing a central P-O bond (CXXXV). The

Scheme 14

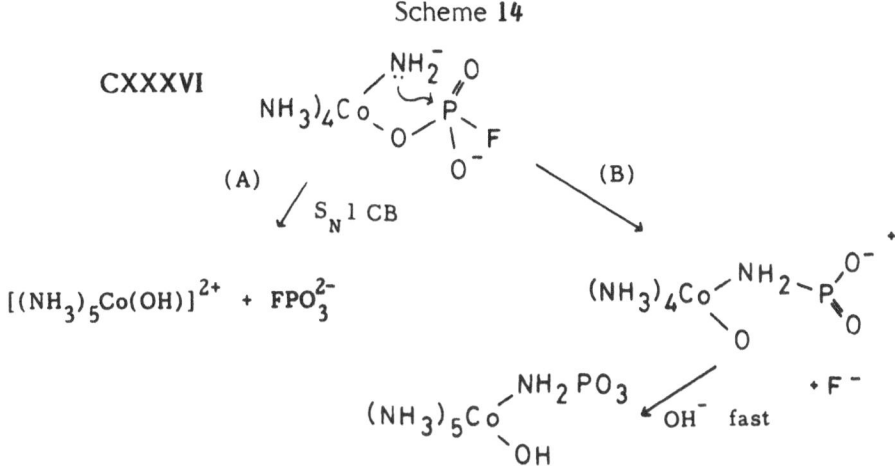

CXXXVI

(A) S_N1 CB   (B)

$[(NH_3)_5Co(OH)]^{2+}$ + FPO_3^{2-}

$[(NH_3)_5CoO_3PF]^+$ ion, containing coordinated fluorophosphate, undergoes base hydrolysis[460] in 0.1-0.3M NaOH at 25°C (I = 1.0M) to generate free FPO_3^{2-}, F^- and $[(NH_3)_4Co(OH)(NH_2PO_3)]$ with k = 3.3 x 10^{-3} M^{-1} s^{-1}. This is a composite rate constant for FPO_3^{2-} release and hydrolysis of F^- with two products being formed in approximately equal amounts. Formation of the conjugate base CXXXVI can lead to S_N1CB displacement [pathway (A)] or intramolecular nucleophilic attack [pathway (B)] (Scheme 14). The measured rate constitutes a ca 10^{10} fold rate enhancement over the base hydrolysis of the free FPO_3^{2-} ion when the concentration of the coordinated nucleophile is taken into account. Facile nucleophilic demethylation of trimethylphosphate in the ion $[NH_3)_5CoOP(OCH_3)_3]^{3+}$ occurs[461] with SCN^-, I^- or $S_2O_3^{2-}$ to give $[NH_3)_5CoO_2P(OCH_3)_2]^{2+}$ and CH_3SCN, CH_3I or $CH_3S_2O_3^-$ respectively. The experimental data support the view that S_N2 substitution at carbon occurs, and establish a novel mode of reaction for a coordinated phosphate(V) ester. The rate enhancement on coordination is ca 150 fold. The base hydrolysis of trimethyl phosphate to give dimethyl phosphate and methanol is accelerated 400 fold in the complex $[(H_3N)_5IrOP(OCH_3)_3]^{3+}$; the reaction occurs via intermolecular attack of hydroxide at the phosphorus center in both the coordinated and free ligand.[462]

Complex formation between ATP (adenosine 5'-triphosphate) and $[Co(pn)_2(OH_2)_2]^{3+}$ and resulting hydrolysis of the ATP to ADP, AMP pyrophosphate and orthophosphate has been studied by ^{31}P nmr.[463] The results support the view that effective metal ion catalysis of ATP hydrolysis requires formation of reactive species involving more than one metal ion per ATP. The Cr(III) and Co(III) complexes of ATP have been shown to be very good analogues for MgATP.[464] Decomposition of $Cr(H_2O)_4ATP$ proceeds predominantly with release of ATP producing lesser amounts of ADP, while the tridentate complex $Cr(H_2O)_3ATP$ produces exclusively ADP. The complex $Co(NH_3)_4ATP$ decomposes more slowly to give amounts of ATP and ADP which are lower than those produced with the analogous $Cr(H_2O)_4ATP$. However, the breakdown of the tridentate $Co(NH_3)_3ATP$ is rapid producing high levels of free ATP and lesser amounts of ADP. Rate constants for hydrolysis are 100-5000 times greater than those for uncomplexed ATP.

The linkage isomers β- and γ-(dihydrogen triphosphato)-pentaamminecobalt(III), $[Co(NH_3)_5H_2P_3O_{10}\cdot H_2O]$, have been synthesized and characterized by ^{31}P nmr spectroscopy and X-ray crystallography, and the phosphate hydrolysis rates studied in the presence of cis-$[Co(cyclen)(OH)_2]^+$.[466] At pH 9.0 and 40°C, the β-isomer hydrolyses some 50 times faster than the corresponding γ-isomer. A mechanism involving competition between chelation and hydrolysis is suggested to account for the observed differences in hydrolysis rates and in hydrogen ion consumption.

The hydrolysis of coordinated 4-nitrophenyl phosphate in cis-$[Co(en)_2(OH)(O_3POC_6H_4NO_2)]$ has been studied[467] over the pH range 7-14 by ^{31}P nmr and by monitoring nitrophenol release at 400nm. Intramolecular attack by ^{18}O-labeled coordinated hydroxide initially gives a five-coordinate phosphorane, which decays to $[Co(en)_2PO_4]$ and nitrophenol with ^{18}O bonded between Co and P. The hydrolysis is 10^5 fold faster than that of the uncoordinated ester in the pH range 9-11.8. This work parallels a previous study[468] in which the base hydrolysis of $[Co(NH_3)_5O_3POC_6H_4NO_2]^+$ was investigated, and where a coordinated aminato ion was the intramolecular nucleophile.

14.3.2. Labile Metal Ions

The accelerations observed in the hydrolysis of adenosine triphosphate (ATP) or triphosphate ($P_3O_{10}{}^{5-}$) by metal ions such as Mg^{2+}, VO^{2+}, Cu^{2+}, Zn^{2+}, and $La(OH)_3$ are not large (ca 10-100 fold).[469-470] Oxidizing VO^{2+} and Mn^{2+} complexes to VO^{3+} and Mn^{3+} does, however, accelerate $P_3O_{10}{}^{5-}$ hydrolysis in acidic conditions by up to 10^5 fold.

Highly selective metal-ion promoted P-O or S-O bond cleavage has been noted in the aminolysis and methanolysis of phenyl phosphosulfate[471]. Magnesium(II) promotes P-O bond cleavage, while Fe(III) promotes S-O fission. The selectivity was almost 100% in each case. Further studies of the methanolysis reaction have appeared.[472] Metal ions with small ionic radii promote the selective O-S bond cleavage, whereas those with larger ionic radii catalyze the selective P-O bond cleavage of phenyl phosphosulfate (CXXXVII) (Scheme 15). Metal ions such as Mg(II) and Zn(II) catalyze the attack of a nucleophile at the phosphorus center of phenyl phosphosulfate (CXXXVII) or 2-pyridyl phosphonosulfate and result in selective P-O bond cleavage at neutral pH.[473]

A novel example of alkali metal ion catalysis in nucleophilic displacement by alkoxide ion in p-nitrophenyl diphenylphosphinate ($Ph_2P(O)OC_6H_4$-p-NO_2) has been reported.[476]

The mechanism of the metal ion facilitated dephosphorylation of nucleoside 5'-triphosphates, including promotion of ATP dephosphorylation by addition of adenosine 5'-monophosphate, has been considered in detail.[477] The effectiveness of the metal ions in promoting ATP dephosphorylation decreases in the order Cu(II) > Cd(II) > Zn(II) > Ni(II) > Mn(II) > Mg(II).

Scheme 15

$$[PhOP(O)_2OSO_3]^{2-}$$

CXXXVII

$$[PhOPO_3]^{2-} + MeOSO_3{}^{-}$$

Be(II), Al(III), Fe(III)

$$[PhOP(O)_2OMe]^{-} + SO_4^{2-}$$

Mg(II), Cr(III), Cu(II), Mn(II), Ca(II)

Nucleophilic attack is believed to occur in an intramolecular manner via an M-OH group.

Recent work has established that in the copper(II) promoted hydrolysis of salicylphosphate (LXXX)[478,543] and 8-quinolylphosphate[479], the metal ion promotes the hydrolysis of the normally unreactive phosphate ester dianion. When allowances are made for intramolecular general acid catalysis in the spontaneous hydrolysis of the phosphate ester, the rate enhancement in the presence of copper (II) is in the range 10^8-10^{10} fold.

14.4   Amide Hydrolysis

Rapid amide hydrolysis has been observed in copper(II) and zinc(II) complexes of CXXXVIII.[257] However, the limited solubility of these complexes above neutral pH hindered detailed analysis. The synthesis and metal promoted hydrolysis of CXXXIX and CXL have now been described.[480] Both ligands form stable complexes with Cu(II), Ni(II), Zn(II) and Co(II). At 50°C the copper(II) promoted hydrolysis of CXXXIX displays a sigmoidal pH rate profile, and the rates increase commensurate with the ionization of a metal-bound water molecule. Similar behavior was observed with the zinc(II) complex of CXL at 70°C. Both Cu(II) and Zn(II) greatly accelerate amide hydrolysis at pH 7, with rate enhancements of 9 x 10^5 and 1.0 x 10^3 for (CXXXIX)-Cu-OH_2 and (CXL)-Zn-OH_2 complexes respectively. Activation parameters for the metal promoted hydrolyses indicated that catalysis results from a substantial increase in $\Delta S^{\ddagger}$. These observations are interpreted in terms of nucleophilic catalysis by a metal-hydroxo species CXLI in basic solution.

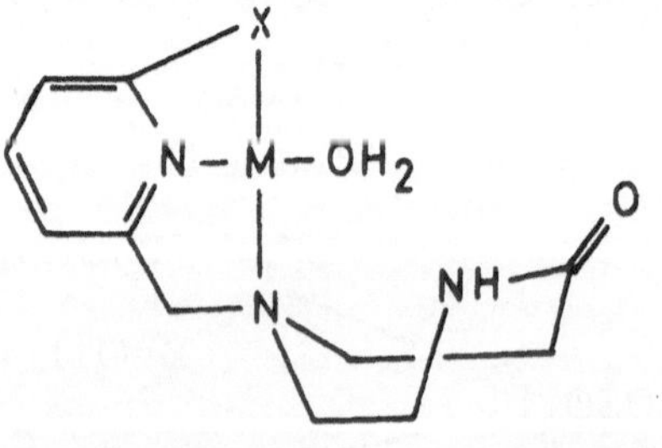

(CXXXVIII)   X   =   -CO_2^-

(CXXXIX)   X   =   -CH_2NMe_2

(CXL)   X   =   -$CH_2N(CH_2CO_2^-)_2$

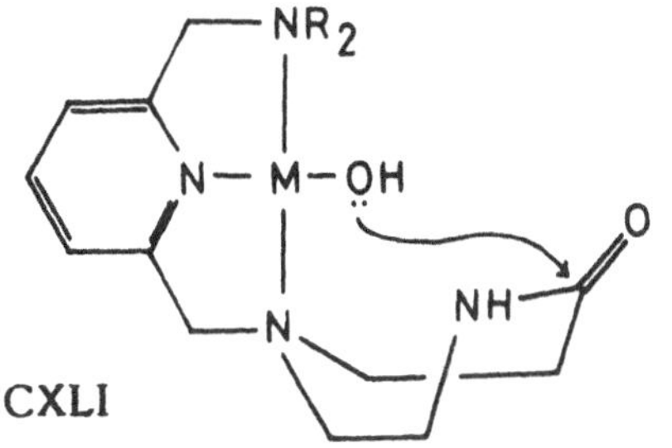

CXLI

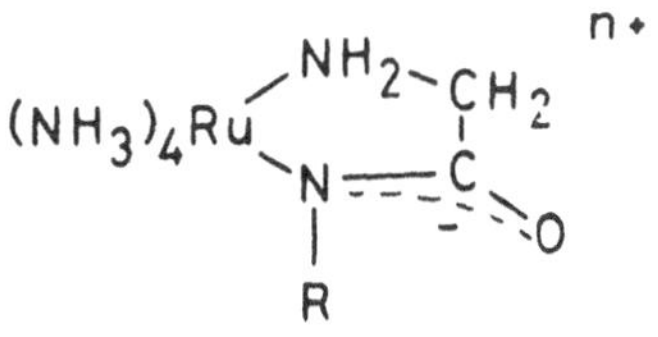

CXLII

In weakly acidic solutions (pH 3), pentaammineruthenium(III) complexes of glycinamide, N'-ethylglycinamide, glycylglycine, glycylglycinamide and ethyl glycylglycinate undergo reaction leading to (N,O)-bound tetraammineruthenium(III) chelates, ammonia being released into solution.[481] In a sense Ru(III) combines the advantages of a substitution-labile centre with those of a substitution inert one and hydrolytic reactions on Ru(III) may present many advantages over Co(III). Reaction of cis-$[Ru(NH_3)_4(OH)(OH_2)]^{2+}$ with glycylglycine, glycinamide and N'-ethylglycinamide gives N,N'-bonded chelates CXLII.[482] The uncatalyzed transformation of these chelates to the (N,O) forms is extremely slow, but the latter can be produced by reducing ruthenium to the 2+ state in acidic solution and reoxidising. The (N,N') form in acidic solution is short lived, $t_{\frac{1}{2}}$ for transformation to the (N,O) form being around 0.2s.

The synthesis, properties and base hydrolysis of $\Delta\Lambda$- and Λ-$[Co(en)_2(S-AlaenH)]$ $(NO_3)_2(ClO_4)_2$ (CXLIII) containing protonated ethylenediamine attached to the acyl function of chelated (S)-alanine has been described.[483] Hydrolysis occurs via the protonated (k_{OH} = 112 M^{-1} s^{-1}) and neutral (k_{OH} = 4.1 M^{-1} s^{-1}) ethylenediamine species. The pK_a of the chelated amide is 11.8. The synthesis of the dimer CXLIV is also described; in this case k_{OH} = 18 M^{-1} s^{-1} and pK_a = 11.3. For hydrolysis of the dipeptide complex $[Co(en)_2(S-ala-glyOC_3H_7)]^{3+}$, k_{OH} = 7 M^{-1} s^{-1} and pK_a = 12.

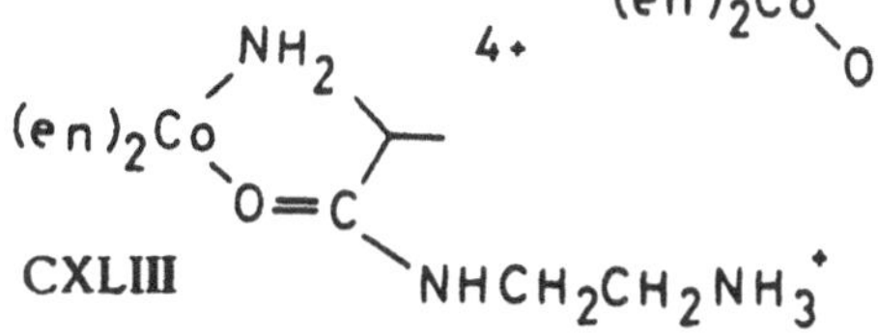

CXLIII

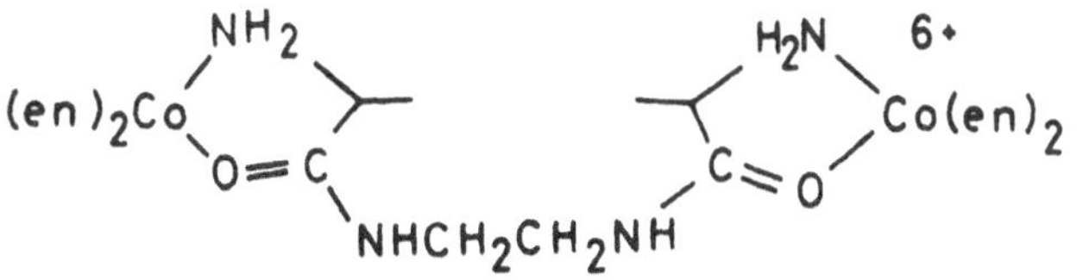

CXLIV

14.5   Cobalt-Hydroxide-Promoted Hydrolysis and Lactonization

The possibility that coordinated hydroxide is directly involved in the catalytic action of some metallo-enzymes such as carbonic anhydrase has prompted a number of investigations of metal hydroxide reactivity towards organic substrates.

Reactions of the $[(NH_3)_5CoOH]^{2+}$ or $[(NH_3)_5CoOH_2]^{3+}$ ions with electron-demanding species including NO^+, NCO^-, CO_2, SO_2, carboxylic esters and anhydrides, acetylacetone, $HSeO_3^-$, WO_4^{2-}, MoO_4^{2-}, $H_2AsO_4^-$ /$HAsO_4^{2-}$ and HIO_3 take place at rates orders of magnitude faster than the rates of normal substitution or water exchange reactions of the Co(III) complexes. In those cases where ^{18}O-tracer experiments have been carried out, it has been shown that the donor oxygen of the product is derived from the initially coordinated water acting as a nucleophile (Scheme 16).

The nucleophilicity of a related series of bases (i.e. oxygen bases or nitrogen bases) is generally related to the pK_a of their conjugate acids (Bronsted relationship). The $M-OH^{(n-1)+}$ complex is thus predicted to be intermediate between H_2O (pK_a, H_3O^+ = -1.74) and OH^- (pK_a, H_2O = 15.74) in its strength as a nucleophile. Some aspects of these reactions have been reviewed.[421]

The $[Co(NH_3)_5OH]^{2+}$ (and other exchange-labile and non-labile metal hydroxide species) promoted hydrolysis of 2,4-dinitrophenyl acetate and 4-nitrophenyl acetate has been studied in some detail.[484] Kinetic studies (25°C, μ = 1.0M) show that these reactions follow shallow Brønsted slopes (β0.33 and 0.40 for the two substrates) which extend over a range of 10^{10} in nucleophile basicity. A correlation is reported which allows the prediction of

Scheme 16

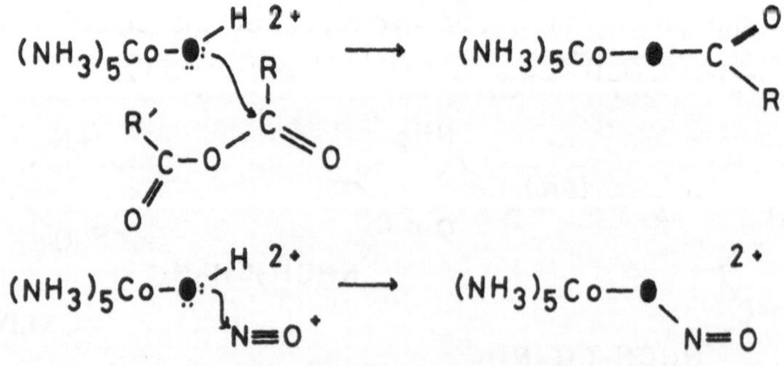

CXLV

CXLVI

CXLVII

reaction rates between $[Co(NH_3)_5OH]^{2+}$ and other activated charge neutral carbonyl substrates such as CO_2 and propionic anhydride.

The $[Co([15]aneN_5)OH]^{2+}$ (CXLV) promoted hydrolysis of 4-nitrophenyl acetate has also been investigated in some detail,[485] ([15]aneN$_5$ = 1,4,7,10,13-pentaazacyclopentadecane, CXVI). The pK_a for the aquo $\rightleftharpoons$ hydroxo equilibrium is 6.3 at $25^\circ C$, not markedly different from that of $[Co(NH_3)_5OH]^{2+}$ where pK_a = 6.4. For the macrocyclic complex, k_{MOH} = 9.3×10^{-3} M^{-1} s^{-1} at $25^\circ C$ and I = 0.1M. In absolute terms the macrocyclic complex is some six times more reactive towards 4-nitrophenyl acetate than $[(NH_3)_5CoOH]^{2+}$, where k_{MOH} = 1.52×10^{-3} M^{-1} s^{-1} at $25^\circ C$. A variety of other studies dealing with the involvement of M-OH species are discussed in the section dealing with phosphate ester hydrolysis.

Tracer studies using ^{18}O have shown that cyclization of cis-$[Co(en)_2(OH_2)(glyOH)]^{3+}$ and cis-$[Co(en)_2(OH_2)(glyO)]^{2+}$ (glyOH = N-bound $NH_2CH_2CO_2H$; glyO = N-bound $NH_2CH_2CO_2^-$) to give $[Co(en)_2(glyO]^{2+}$ (CXLVII) containing chelated glycinate occur intramolecularly without displacement of coordinated water.[486] The rates of these reactions are relatively fast with $t_\frac{1}{2} \sim$ 40s at pH 0-1 and $t_\frac{1}{2} \sim$ 400s at pH 4. Ring closure of cis-$[Co(en)_2(OH_2)(glyO)]^{2+}$ is catalyzed by general acids (including H_3O^+) and is interpreted in terms of rate-determining protonation of an intermediate cyclic species. Interestingly, cyclization of cis-$[Co(en)_2(OH)(glyO)]^+$ is considerably slower ($t_\frac{1}{2} \sim$ 10h) and is not catalyzed by monofunctional buffers. Comparisons with O exchange in glycine indicate large rate enhancements for the metal based system (10^7-$10^{12}M$), and the rates compare favorably with those found for intramolecular lactone formation in purely organic molecules.

Formation of $[Co(en)_2(glyO)]^{2+}$ from monodentate trans-$[Co(en)_2(H_2O/OH)(glyO/H)]^{3+/2+/+}$ species has also been studied.[432] The results support a process involving the synergic displacement of coordinated water or hydroxide by the <u>trans</u>-carboxylic acid or carboxylate anion.

14.6   Nitrile Hydrolysis

The base hydrolysis of coordinated acrylonitrile in $[NH_3)_5CoN\equiv CCH=CH_2]^{3+}$ to give the acrylamide complex has been studied using carbonate buffers.[487] The reaction obeys the rate law

$$k_{obs} = k_{OH}[OH^-] + k_C[CO_3^{2-}]$$

with $k_{OH} = 35 \ M^{-1} \ s^{-1}$, $k_C = 1 \ M^{-1} \ s^{-1}$ at 25°C and $\mu = 1.0M$. Tracer studies using ^{18}O indicate that the mechanism of hydrolysis by CO_3^{2-} involves direct nucleophilic attack at the nitrile group by CO_3^{2-} with subsequent elimination of CO_2. Coordination of malononitrile, cyanoacetic acid and ethyl cyanoacetate to the $[(NH_3)_5Co]^{3+}$ group is also shown to lead to greatly increased acidity of the methylene protons. Thus pK_a of coordinated malononitrile in $[(NH_3)_5CoNCCH_2CN]^{3+}$ is 5.7 compared with 11.3 for the free ligand at 25°C. In basic solution these complexes exist in equilibrium with their deprotonated species. The protonated complexes are hydrolyzed to coordinated amides by both water and hydroxide ion acting as nucleophiles. The deprotonated complexes undergo intramolecular electron transfer to give Co(II) and the ligand radical, and also act as nucleophiles toward appropriate substrates (methyl iodide, methyl pyruvate, or bromine) to give the corresponding substituted species.

Neighboring-group participation in the hydrolysis of coordinated nitriles has also been studied.[488] Base hydrolysis of CXLVIII has been investigated for R = H ($k_{OH} = 1050 \ M^{-1} \ s^{-1}$, $\Delta H^{\ddagger} = 56.9 \ kJ \ mol^{-1}$, $\Delta S^{\ddagger} = 4$ $JK^{-1} \ mol^{-1}$ at 25°C and $\mu = 1.0M$ (LiClO$_4$)) and R = CONH$_2$ ($k_{OH} = 5.8 \times 10^6$ $M^{-1} \ s^{-1}$, $\Delta H^{\ddagger} = 69.1 \ kJ \ mol^{-1}$, $\Delta S^{\ddagger} = 184 \ JK^{-1} \ mol^{-1}$ at 25°C). The unusual kinetic parameters for the latter reaction are attributed to nighboring group participation by the amide group (Scheme 17).

$$[(NH_3)_5CoNC-C_6H_4-\underline{o}-R]^{3+} + OH^- \xrightarrow{k_{OH}}$$

CXLVIII

$$[(NH_3)_5Co.NH.CO-C_6H_4-\underline{o}-R]^{2+}$$

Scheme 17

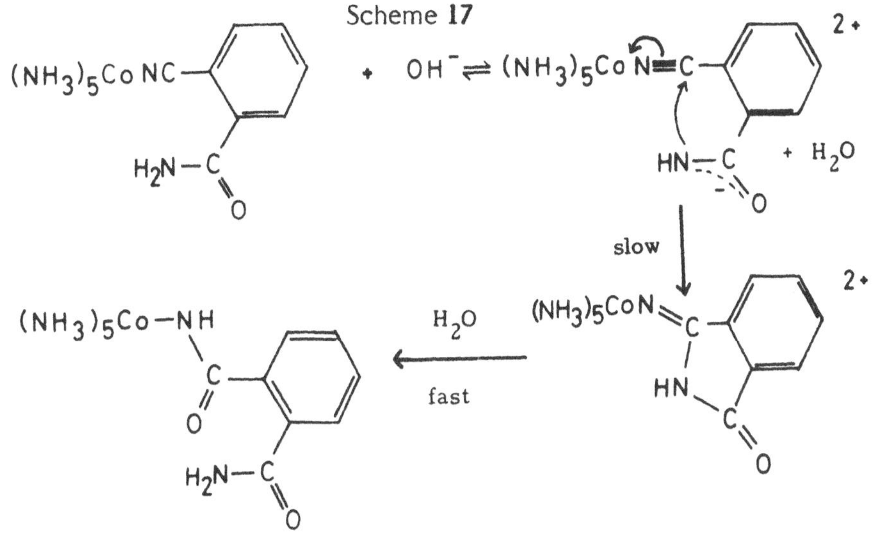

Two new aromatic organonitrile complexes of $Co(NH_3)_5^{3+}$ with 2-and 4-nitrobenzonitrile have been prepared, and their base hydrolysis to coordinated carboxamides studied.[489] The second order rate constants for base hydrolysis are 180 ± 4 M^{-1} s^{-1} and 510 ± 90 M^{-1} s^{-1} at $25°C$ and $\mu = 1.0M$ for the $\underline{o}$- and $\underline{p}$-isomers respectively. Further rate constants for nitrile hydrolysis are incorporated in Table 22 (see also Table 16). Reaction of $[Co(NH_3)_5NCCH_3]^{3+}$ with hydroxide ion gives the deprotonated N-bonded acetamido complex quantitiatively with $k_{OH} = 3.4$ M^{-1} s^{-1} at $25°C$.[265] The benzonitrile derivative gives coordinated benzamide with $k_{OH} = 18.8$ M^{-1} s^{-1}.[264] In each case the rate enhancement compared with base hydrolysis of the uncoordinated nitrile is ca 2×10^6 fold. An inspection of the activation parameters establishes that the acceleration arises almost completely from a more positive value of $\Delta S^{\ddagger}$. The reasons for these effects are yet not fully understood although it is unlikely that they result simply from differential effects of solvation in the transition state.

Table 22. Second Order Rate Constants (k_{OH}, 25°C) and Activation Parameters for the Hydrolysis of Nitriles. Data from Ref. 421

Nitrile	k_{OH} (a)	$\Delta H^{\ddagger}$ (b)	$\Delta S^{\ddagger}$ (c)
CH_3CN	1.60×10^{-6}	18.5	-23
$Cl-C_6H_4-4-CN$	7.2×10^{-6}	17.4	-19
$[NH_3)_5CoNCCH_3]^{3+}$	3.4	18.0	+4
$[(NH_3)_5RhNCCH_3]^{3+}$	1.0		
$[(NH_3)_5RuNCCH_3]^{3+}$	2.2×10^2		
$[NH_3)_5RuNCCH_3]^{2+}$	$<6 \times 10^{-5}$		
PhCN	8.2×10^{-6}	19.3	-15
$[NH_3)_5CoNCPh]^{3+}$	18.8	16.5	+3
$[(NH_3)_5RuNCPh]^{3+}$	2.0×10^3		
$[(NH_3)_5CoNC-C_6H_4-4-O]^{2+}$	0.2	16	-7
$[(NH_3)_5CoNC-8H_4-3-O]^{2+}$	3.6	15	-6
$[(NH_3)_5CoNCCH=CH_2]^{3+}$	35	--	-
$[(NH_3)_5CoNC-N(CH_3)_2]^{3+}$	3.06	-	-

(a) $M^{-1} s^{-1}$
(b) kcal mol^{-1}
(c) cal K^{-1} mol^{-1}

The data in Tables 16 and 22 show that with the exception of the Ru(II) complex a variety of metal ions activate the hydrolysis to a fairly consistent extent (10^6-10^8 fold). The constancy of these values and the anomaly of the Ru(II) complex are consistent with the view that it is the Lewis acidity of the metal ion which is essential to activation. In the Ru(II) complex considerable metal-to-ligand π-bonding is expected, resulting in back-donation of electron density from the metal d-orbitals into the π^* orbitals of the C≡N bond. This view has been confirmed by measurements of the C≡N stretching frequency of free nitriles and their complexes with Ru(II) and various other metal ions.[490] The result is little or no net polarization of the nitrile by the metal ion, and the effect on reactivity is correspondingly small.

Rate constants for the hydrolysis of eleven coordinated aromatic nitriles have been shown[489] to follow a Hammett-type correlation with log $k_{OH} = 1.93\sigma + 1.30$ at 25°C and I = 1.0M. Carbon-13 nmr studies of the free and coordinated nitriles indicate similar chemical shifts so that variation in the slope of the Hammett plots for free and coordinated nitrile hydrolysis is regarded as a transition state effect rather than a ground state phenomenon.

There is growing interest in the use of combinations of metal ions in catalyzing the reactions of coordinated organic substrates. One possibility is that one metal activates the substrate, while the other provides an intramolecularly coordinated nucleophile such as OH⁻ to bring about effective hydrolysis. Intramolecular attack by coordinated hydroxide on one metal ion at acetonitrile bound to the other metal ion in a binuclear μ-amido-octaamminedicobalt(III) species (CXLIX) has been observed under acidic conditions to give a chelated acetamide (CL).[491] The overall rate enhancement in this reaction as a result of Lewis acid catalysis and an intramolecular nucleophile is $\geqslant 10^{15}$ fold. (A large rate enhancement of ca 10^{12} fold at pH 7 is seen for intramolecular attack by coordinated hydroxide on a pendant non-coordinating nitrile in cis-[Co(en)$_2$(OH)(NH$_2$CH$_2$CN)]$^{2+}$).[273] The synthesis and base hydrolysis of

CXLIX   CL

Scheme 18

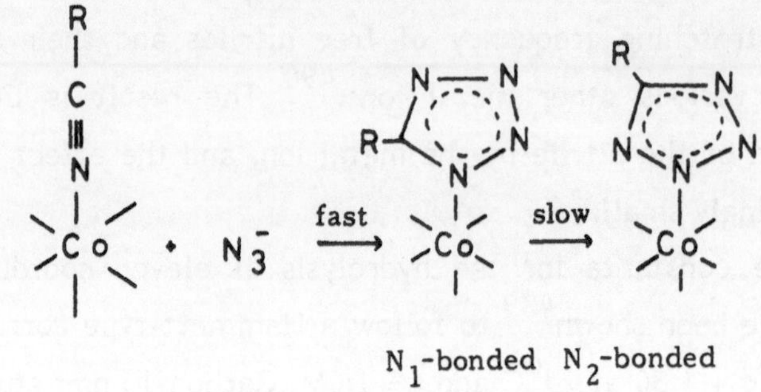

N_1-bonded N_2-bonded

pentaamminerhodium(III) and -iridium(III) complexes of acetonitrile and N,N-dimethylformamide (DMF) have recently been reported[492], and compared with the analogous cobalt(III) and free ligand chemistry. Coordination results in a large increase in the rate of base hydrolysis of the coordinated ligands (ca 10^6 fold).

Coordination of organonitriles to the $[Co(NH_3)_5]^{3+}$ group increases the susceptibility of the nitrile carbon to nucleophilic attack by hydroxide ion and cyanide ion. When aqueous solutions of the complexes $[Co(NH_3)_5(N{\equiv}C\text{-}R)]^{3+}$ (R = Me or Ph) are treated with an excess of NaN_3 at pH 5-6 (to prevent base hydrolysis to the amido complex), significant spectral changes occur, and are attributed to the reactions[493] of Scheme 18. The formation of 5-methyltetrazole from sodium azide and acetonitrile requires a reaction time of 25h at $150^{\circ}C$, compared with only 2h at ambient temperature for coordinated acetonitrile. The subsequent conversion of the N_1-bonded complex to an N_2-bonded complex has been verified as the latter complex has been prepared and its crystal structure determined.[494]

14.7   Amino Acid Synthesis

Intramolecular imine formation between a coordinated aminate ion and a 2-oxo acid has been utilized to synthesize two racemic amino acids, 2-cyclopropylglycine and proline.[495] Thus anation of $[Co(NH_3)_5OH_2]^{3+}$ by $Br(CH_2)_3COCO_2H$ at pH 5 gives two major products, **CLI** and **CLII**. Both are converted to tetraammineiminocarboxylato chelates by attack of an adjacent deprotonated ammonia ligand. The cyclopropylimine complex can,

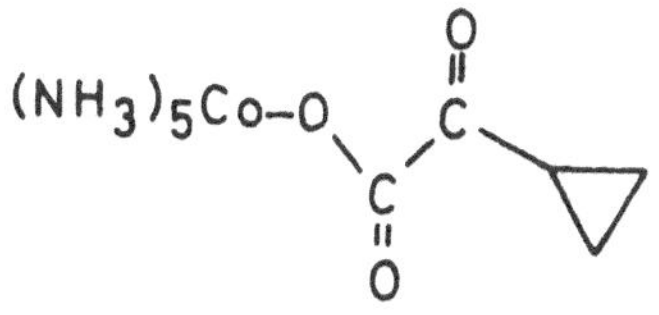

$(NH_3)_5Co-O$ ... C (cyclopropyl structure)

$(NH_3)_5Co-O$ ... C $(CH_2)_3OH$

CLI 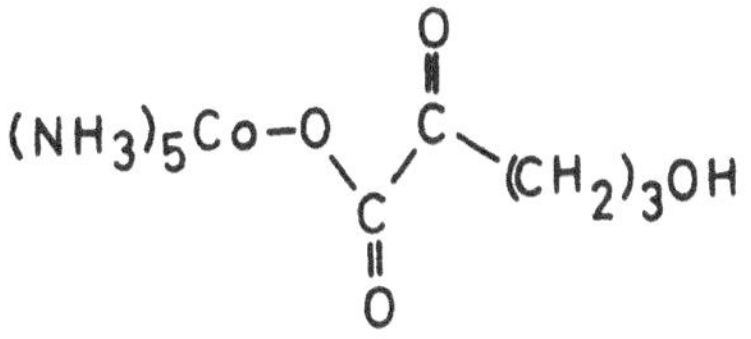   CLII

for example, be reduced by alkaline BH_4^- to give the (RS)-2-cyclopropylglycine complex.

The cobalt(III)-promoted synthesis of β-carboxyaspartic acid has also been reported[496] by the intramolecular addition of coordinated amide ion to the olefinic center in the [3,3-bis(ethoxycarbonyl)-2-propenoato]-pentaamminecobalt(III) ion, $[(H_3N)_5CoOOCCH=C(CO_2Et)_2]^{2+}$. In aqueous solution, an intramolecular addition of the cis-aminate ion at the olefinic center gives the N,O-chelated diester of β-carboxyaspartic acid.

The synthesis of chelated C-formylglycinate ion has been described[497] using the Vilsmeier-Haack adduct derived from (p)-$[Co(trien)(gly)]^{2+}$. The aldehyde is readily reduced with BH_4^- to the serinato complex and adds alcohols to give the hemiacetal.

Base-catalyzed elimination reactions of O-acetyl- and O-sulfonylserine coordinated to cobalt(III) give chelated 2-iminopropanoate.[498] The reaction rates are first order in [OH⁻], and are equal in magnitude to the methine proton exchange rates. Comparison with the rate of elimination from the free ligand O-sulfonylserine in base shows that coordination results in a ca 10^7 fold increase in reactivity. The methyl group in chelated 2-iminopropanoate is deprotonated readily in base and the resulting carbanion reacts rapidly with electrophiles such as aldehydes. The sequence of reactions is represented in Scheme 19.

Scheme 19

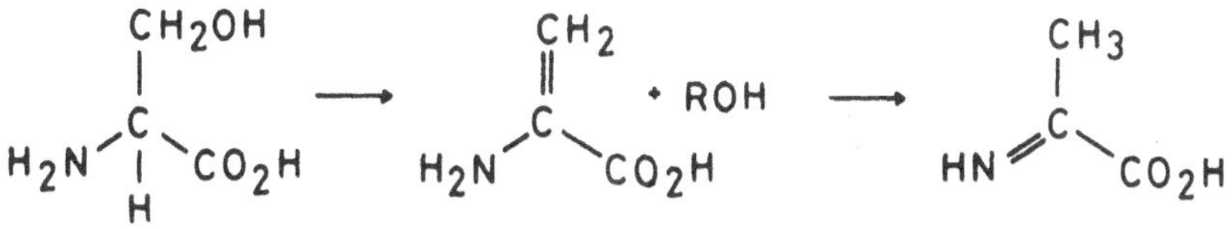

CH_2OH — C — H_2N — CO_2H (with H) → CH_2 = C — H_2N — CO_2H + ROH → CH_3 — C — HN = CO_2H

2-iminopropanoate

14.8 Peptide Synthesis

The rapid aminolysis of cobalt(III)-chelated glycine esters in aprotic solvents has been described in Section 2.2.1. above. The cobalt acts as both an N-protecting and activating group. The synthesis of the chelated amino acid esters has presented some difficulties.[499-500] A recent paper[501] has described the use of methyl trifluoromethane sulfonate for the alkylation of chelated amino acids using dry trimethyl phosphate as solvent (Scheme 20). The cobalt(III) amino acid methyl esters are readily isolated as orange powders following trituration with Et_2O/MeOH. Some have been recrystallised from MeOH as ClO_4^- or I^- salts. Their coupling is rapid in dry DMSO, acetonitrile or methanol. The $[Co(en)_2]^{3+}$ and $[Co(trien)]^{3+}$ moieties are potentially chiral and some specificity is induced when optically active cobalt(III) units are employed.

The synthesis of tetrapeptides and [leu^5]enkephalin by the cobalt(III) technique has been described[502], and the method now provides a viable method for peptide synthesis.

Pentaamminecobalt(III) has now been shown to be a useful C-terminal protecting group for sequential peptide synthesis.[503] The reaction of complexes of the type **CLIII** with BOC-amino acid active esters of BOC symmetric anhydrides gives $[Co(AA)_1(AA)_2BOC]$ (BOC = t-butoxycarbonyl, $Me_3CO.CO.$). The BOC group is removed with 95% CF_3CO_2H to give $[Co(AA)_1(AA)_2]$ which can be used for sequential peptide synthesis. Alternatively the $(NH_3)_5Co$ group is selectively removed by rapid reduction with $NaBH_4$ or NaHS to give the N-protected peptide fragment

Scheme 20 $N_4 = (en)_2$ or trien

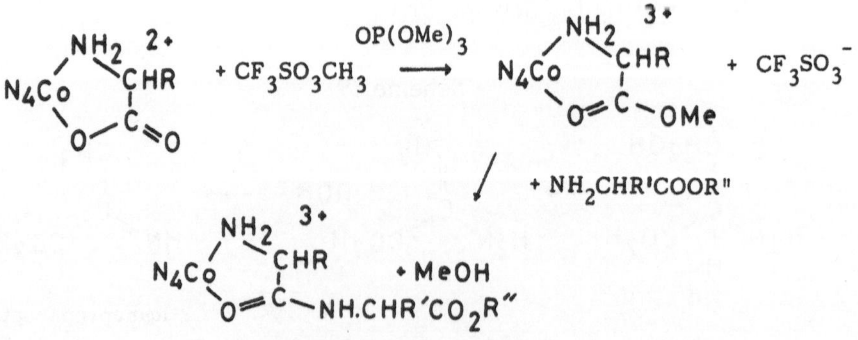

$$[(NH_3)_5CoO_2C.CHR.NH_3](BF_4)_3 \qquad (H_2N)_2CO \rightarrow H_2N\text{-}COOH + NH_3$$

$$\mathbf{CLIII} = [Co(AA)_1] \qquad\qquad \mathbf{CLIV}$$

$[(AA)_1(AA)_2BOC]$ under very mild conditions. The general procedures have been discussed and the preparations of a number of penta- and hexapeptides by this route described.

The pentaammine (L-threonine-O)cobalt(III) ion, $[Co(NH_3)_5(L\text{-}thr)]Br_3$, has been prepared and its structure confirmed by X-ray crystallography.[504] Its applicability to peptide synthesis has been demonstrated by coupling to N-protected L-alanine by standard procedures.

14.9   Ureas

Jack Bean urease is a nickel-metalloenzyme containing two Ni(II) ions[505] which catalyzes the hydrolysis of urea (CLIV) to produce (initially) ammonia and carbamic acid. As a result there has been considerable interest in metal complexes with cyanate, carbamate and urea which may provide some insight into the chemistry of the enzyme. One interesting model system has been described in which nickel(II) promotes the hydrolysis of N-(2-pyridylmethyl)urea[506]. Interconversions of complexes of carbamate, cyanate and urea have been reviewed.[421]

14.10 Schiff Base Complexes, Vitamin B$_6$ and Reactions of Coordinated Amino-acids and Peptides

There has been a marked upsurge of interest in this area. The glycine residue in N-salicylideneglycyl-L-valinato copper(II) reacts with formaldehyde in aqueous solution at pH 8.5. Decomposition of the complex with H_2S at pH 2 gives seryl-L-valine containing optically active serine.[507]

X-ray structural analyses of two metal complexes of O-phospho-DL-threonine-pyridoxal Schiff base, $[Ni(C_{12}H_{15}N_2O_8P)(H_2O)_3].2H_2O$ and $[Cu(C_{12}H_{15}N_2O_8P)(H_2O)].H_2O$, establish that the nickel(II) complex is a monomer with the octahedral nickel ion bonded to the tridentate Schiff base and three water molecules.[508] The copper(II) complex is a dimer with square pyramidal copper bonded to the tetradentate Schiff base involving a phosphate oxygen atom from a neighboring molecule and one water molecule. Structures have also been reported for (pyridoxylidene-DL-

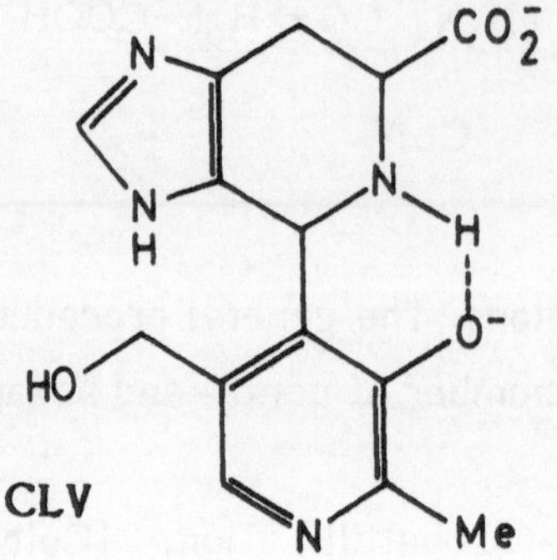

CLV

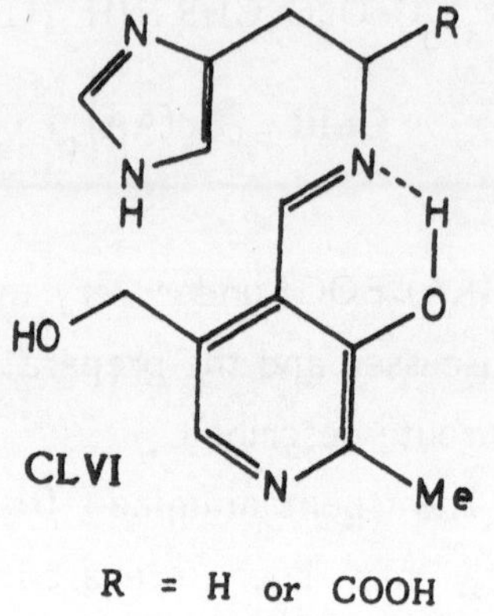

CLVI

R = H or COOH

valinato)copper(II) and (3-methoxy-salicylidene-DL-valinato)-copper(II).[509]

Differential reactivity of the α-methylene protons of bis(pyridoxylideneglycinato)cobalt(III) towards deuterium exchange has been observed.[510] Dunathan[511] had previously suggested that selective catalysis of cleavage of a bond of an amino-acid α-carbon atom can be accomplished by correctly orientating that bond with respect to the π-system of the azomethine group, and the above results are consistent with this view. Pyridoxal reacts with L-histidine, L-histidine methyl ester, histamine, and N'-methylhistamine to give 4,5,6,7-tetrahydropyrido[3,4-d]imidazole compounds CLV via the formation of Schiff-base intermediates. Zn(II) ions can function as a trap for Schiff-base intermediates, and a number of zinc(II) complexes of N-pyridoxylidene-L-amino acids and L-salicylidene-L-amino acids have been obtained.[512] Copper(II) complexes of Schiff bases derived from pyridoxal, salicylaldehyde, or pyruvic acid and histidine, histidine methyl ester, and representative amino acids with non-polar side chains have been prepared by metal ion template synthesis, and their sterochemistry has been studied.[513] Japanese work[514] has also shown that cyclization (intramolecular Mannich reaction) of the pyridoxal Schiff base of histidine or histamine (CLVI) is inhibited by metal ions such as Cu(II) and Zn(II).

Threonine is converted to glycine and acetaldehyde, and serine is converted to glycine and formaldehyde respectively, in pyridoxal-activated enzyme systems. The reaction has also been shown to take place with pyridoxal and metal ions at lower rates.[515] Detailed kinetic studies of the pyridoxal and the pyridoxal-metal-ion-catalyzed dealdolization of threonine, β-hydroxyleucine, and β-hydroxyvaline have now been published[516], and a possible mechanism has been suggested. Under comparable conditions the rates of Al(III) and Ga(III) chelate catalysis were found to be 2-6 times higher than those of the metal-free systems. Pyridoxal-catalyzed β-elimination and dealdolization reactions of β-hydroxyglutamic acid have been found to occur simultaneously.[517] Metal-ion catalysis was found to be less effective than proton catalysis in the promotion of γ-decarboxylation, and the catalytic effects of metal ions were found to decrease with increasing charge on the metal ion.

The kinetics of β-elimination of O-phosphoserine, β-chloroalanine, S-ethylcysteine and β-chloro-α-amino-butyric acid have been investigated by ^{1}H nmr in D_2O in the presence of pyridoxal, and in the presence and absence of Ga(III), Al(III) and Zn(II) ions.[518] Rate constants for pyridoxal catalysis in the presence of metal ions are some ten times greater than those observed in the metal-free systems.

The kinetics and stereochemistry of deuterium exchange of the α-hydrogen in metal complexes of amino-acid Schiff bases with o-hyroxyacetophenone have been studied.[519] In such complexes steric interactions may provide a sufficiently long lifetime for the non-planar chiral carbanion, such that retention of configuration is observed.

It has been shown[520] that in the absence of base, the reaction of formaldehyde with bis(glycinato)zinc(II) monohydrate results in the formation of the complex bis[N-(1,3-dioxa-5-azacyclohexyl)acetato]zinc(II) dihydrate (CLVII). This work has now been extended to the reactions of

CLVII   CLVIII   CLIX

formaldehyde with bis(glycinato) complexes of Ni(II), Co(II) and Cu(II) at pH 4.5.[521] The Ni(II) and Co(II) complexes give the analogues of **CLVII**, but the Cu(II) derivative lacks the water ligands. The copper(II) complex formed by the reaction of bis(glycinato)copper(II) with formaldehyde in the absence of base, when treated with $NaBH_4$, gives the sodium salt of N-(1,3-dioxa-5-azacyclohexyl)acetic acid **(CLVIII)**.[522] The reaction does not cause appreciable cleavage of the heterocyclic ring of the acid. Treatment of the copper complex with H_2S gives glycine and formaldehyde. Condensation of bis(glycinato)nickel(II) with formaldehyde and ammonia gives bis[3N,7N-(1,3,5,7-tetraazabicyclo[3.3.1]nonyl)diaceto]nickel(II) **(CLIX)**,[523] the structure being confirmed by X-ray crystallography. Glycine coordinated to copper(II) reacts with a stoichiometric quantity of aldehydes in basic solution to give β-hydroxyamino acids[524]; thus the use of acetaldehyde and benzaldehyde gives a 66% yield of <u>threo</u> β-phenyl-serine. Polarimetric data have shown that the base-catalyzed reaction of bis(L-serinato)copper(II) with excess formaldehyde proceeds via the initial dissociation of the proton on the nitrogen atom of the amino acid complex.[525] A bis(oxazolidine)copper(II) complex occurs as an intermediate, but this species is not detected polarimetrically at $50^{\circ}C$ and above. The effects of nickel(II), cobalt(III) and a variety of other metal ions on the racemization of free and bound L-alanine have been critically discussed.[526]

Asymmetric synthesis of threonine and partial resolution and retro racemization of α-amino acids via copper(II) complexes of their Schiff bases with (S)-2-N-(N'-benzylprolyl)aminobenzaldehyde and (S)-2-N-(N'-benzylprolyl)aminoacetophenone have been studied.[527] Optically active alanine, valine and leucine have been obtained by a transamination reaction betwen pyridoxamine and the corresponding α-keto acid in the presence of a copper(II) complex with the tridentate ligand 2,6-bis[(3S)-3-phenyl-2-azabutyl]pyridine.[528] In each case the amino acid with the R-configuration was formed preferentially, and the maximum enantiomeric excesses were 54% (alanine), 48% (leucine) and 29% (valine). The stereoselectivity of the reaction was discussed in terms of the possible structure and stability of the intermediate copper(II)-ketimine-ligand complex. Asymmetric synthesis of

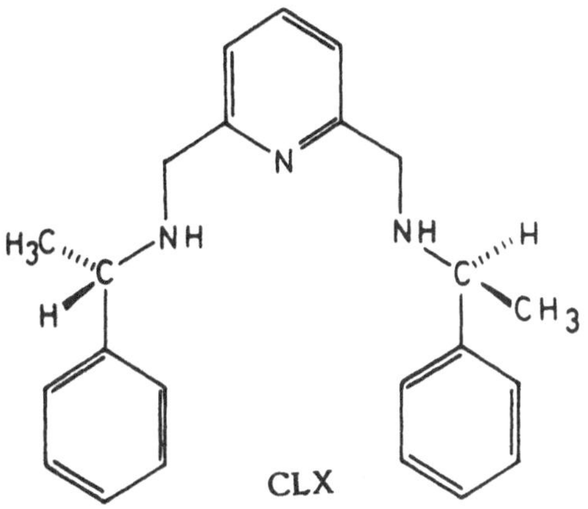

CLX

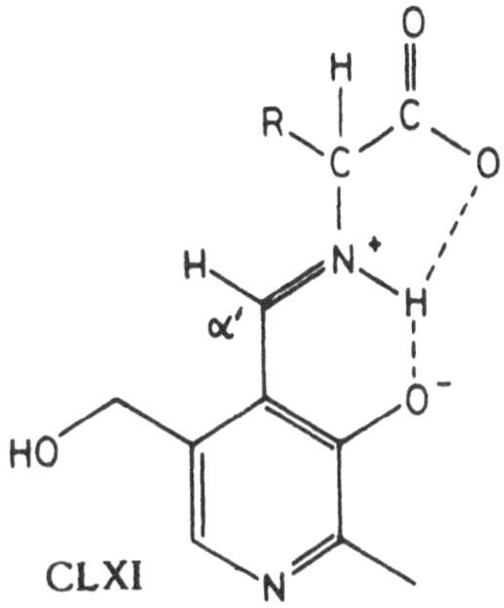

CLXI

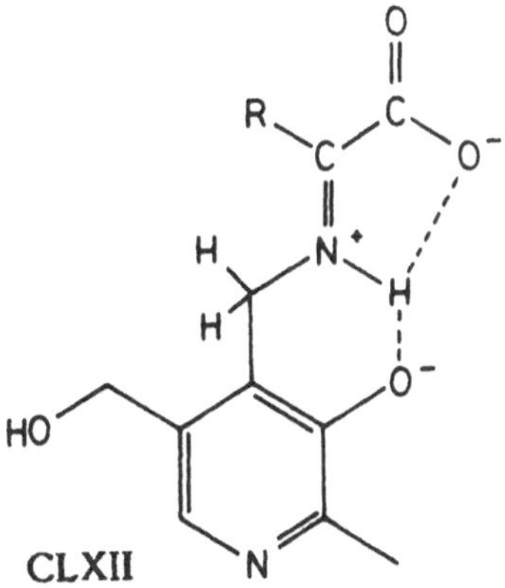

CLXII

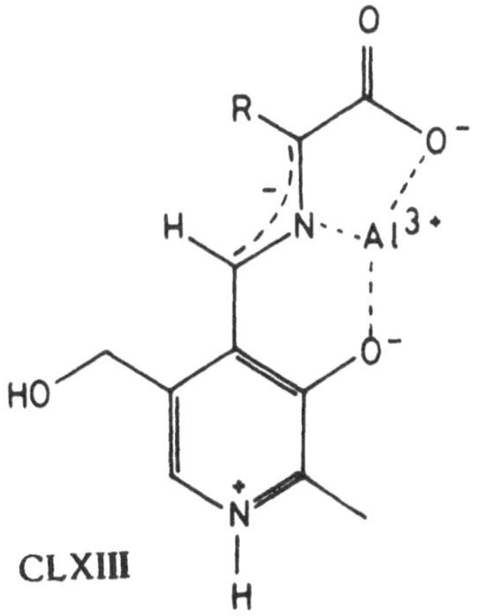

CLXIII

phenylalanine by stereoselective isomerization of the mixed ligand-copper(II) Schiff base formed from pyridoxamine, phenlpyruvic acid and the optically acitve ligand 2,6-bis[(3S)-3-phenyl-2-azabutyl]pyridine (CLX) has also been described.[529] High stereoselectivity with an enantiomeric excess of up to 80% was observed. The zinc(II) complexes derived from the condensation of (1R)-3-hydroxymethylenebornan-2-one and a series of L-amino acids, [ZnL], undergo tautomeric equilibria in solution between enolimine and ketoenamine species.[530] The mechanism of enzymic and non-enzymic vitamin B_6 catalyzed transamination of α-amino and α-keto acids proposed by Metzler et al[240] involves the formation and interconversion of aldimine and ketimine Schiff bases CLXI and CLXII. Nmr evidence for the delocalized α,α'-carbanion of pyridoxal and pyridoxamine Schiff bases [as in

(CLXIII)] as the intermediate in vitamin B_6 catalyzed transamination has been obtained.[531]

Potentiometric titrations have been employed to study the various complex equilibria in systems involving pyridoxamine and histidine with Co(II), Ni(II), Cu(II) and Zn(II).[532] The relevance of these ternary complex equilibria to biological functions was discussed. Thermodynamic equilibria in the system zinc(II)-pyridoxal-5'-phosphate-2-amino-3-phosphonopropionic acid have also been studied[533], as have binary, ternary and quaternary complexes involved in the systems pyridoxamine-glycine-imidazole with the bivalent metal ions Co(II), Ni(II), Cu(II), Zn(II) and Cd(II).[534]

The preparation of mer-[Co(glygly)L(en)]$^{n+}$ complexes (L = H_2O, CN⁻, NCS⁻) has recently been described.[535] Condensation reactions of neutral complexes with acetaldehyde were also examined. On the basis of electronic and ^1H and ^{13}C nmr spectral data, elemental analysis and paper chromatography it was established that the following reaction products were obtained: (a) a mixture of mer-[Co(threogly)L(en)] and mer-[Co(allothreogly)L(en)]. Similar results have also been obtained in the reaction of [Co(glygly)(en)NO$_2$] with acetaldehyde in aqueous solution at pH 11, where four condensation products were obtained.[536]

Copper(II) and nickel(II) complexes of Schiff bases drived from (S)-O-[(N-benzylprolyl)aminoacetophenone] [(S)-bap] and the amino acids, glycine, (R)- and (S)-alanine, (R)- and (S)-valine, (R)- and (S)-adamant-1-ylalanine, and (R)- and (S)-adamant-1-ylglycine have been prepared, and the structures of [Cu{(S)-bap-(S)-val}] and [Ni{(S)-bap-(S)-val}] determined by X-ray crystallography.[537] The kinetic CH-acidity and deuterium exchange of the hydrogen of the amino-acid moiety have also been studied. The deuterium

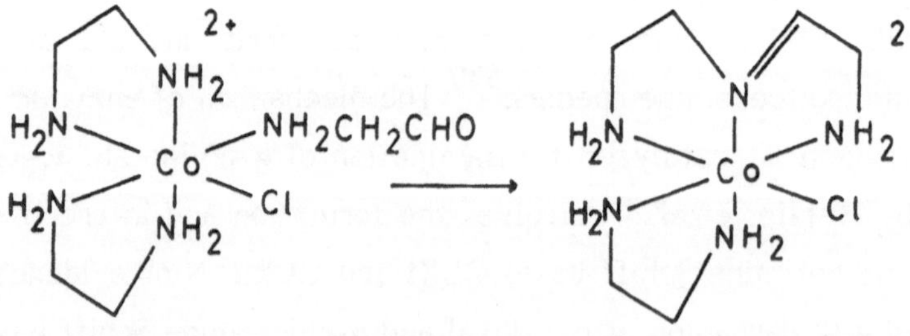

CLXIV   CLXV

exchange is accompanied by epimerization which results in 80% excess of the (S)-2-[^{2}H] amino acid. Stereoselectivity in the nickel complexes was found to be higher than in the copper complexes. Rate constants for epimerization lie in the range 5×10^{-4} to 1.2×10^{-1} M^{-1} s^{-1}.

14.11 Imine Formation and Hydrolysis

Some of the synthetic aspects of intramolecular imine formation are discussed in the section on amino-acid synthesis. The complex cis-[Co(en)$_2$NH$_2$CH$_2$CH(OCH$_3$)$_2$Cl]$^{2+}$ undergoes hydrolysis in dilute HCl solution to give the aminoacetaldehyde complex CLXIV[538] which was isolated in an equilibrium mixture with its hydrated adduct cis-[Co(en)$_2$NH$_2$CH$_2$CH(OH)$_2$Cl]$^{2+}$. In aqueous solution intramolecular imine formation occurs to give CLXV. Reaction takes place with an amino group cis to Cl$^-$ so that the tridentate imine ligand adopts a fac-sterochemistry. The [Co(NH$_3$)$_5$(NH$_2$CH$_2$COCH$_3$)]$^{3+}$ ion undergoes an intramolecular base-catalyzed cyclization reaction to give a coordinated carbinolamine, which undergoes a slower base-catalyzed dehydration to give a chelated imine.[539] The synthesis, characterization and kinetics of formation of various reaction products were described. Several reactions of the imine product have been carried out, including BH$_4^-$ reduction and condensation with methyl vinyl ketone.

The kinetics of hydrolysis of N-salicylidene-2-aminothiazole have been investigated in aqueous solution containing 5% methanol in the presence and absence of cobalt(II), nickel(II), copper(II) and zinc(II).[540] Only copper(II) led to significant retardation of imine hydrolysis.

REFERENCES

1. H. Kroll, J. Am. Chem. Soc. 74:2036 (1952).

2. Q. Fernando, Advan. Inorg. Chem. Radiochem. 7:185 (1965).

3. R. W. Hay, Pure. Appl. Chem. 13:157 (1963).

4. R. W. Hay, J. Chem. Ed. 43:413 (1965).

5. M. M. Jones, "Ligand Reactivity and Catalysis", Academic Press, New York (1968).

6. E. Kimura, Yuki Gosei Kagaku. 29:12 (1971).

7. M. M. Jones and W. A. Connor, Ind. Eng. Chem. 55:15 (1963).

8. M. L. Bender in "Reactions of Coordinated Ligands", Advan. Chem. Ser. 37, American Chemical Society (1963).

9. T. C. Bruice and S. J. Benkovic, "Bio-organic Mechanisms", Vols. 1 and 2, Benjamin, New York (1966).

10. M. L. Bender, "Mechanisms of Homogeneous Catalysis from Protons to Proteins", Wiley Interscience, New York (1971).

11. "Metal Ions in Biological Systems", Vol. 5, H. Sigel, ed., Marcel Dekker, New York (1976).

12. R. P. Hanzlik, "Inorganic Aspects of Biological and Organic Chemistry", Academic Press, New York (1976).

13. R. P. Houghton, "Metal Complexes in Organic Chemistry", Cambridge University Press, Cambridge (1979).

14. A. E. Martell, Xth International Conference on Coordination Chemistry, Butterworths, London (1968).

15. D.P.N. Satchell, Chem. Soc. Rev. 6:345 (1977).

16. J. P. Candlin, K. A. Taylor and D. T. Thompson, "Reactions of Transition Metal Complexes", Elsevier, Amsterdam (1968).

17. A. E. Martell in "Metal Ions in Biological Systems", Vol. 2, H. Sigel, ed., Marcel Dekker, New York, pp. 208-262 (1976).

18. M. M. Taqui Kahn and A. E. Martell, "Homogeneous Catalysis by Metal Complexes", Academic Press, New York (1974).

19. J. E. Coleman in "Progress in Bio-organic Chemistry", Vol. 1, E. T. Kaiser and F. J. Kezdy, eds., Wiley-Interscience, New York, 159 (1971).

20. "Inorganic Biochemistry", Vol. 1 and 2, G. L. Eichhorn, ed., Elsevier, London (1973).

21. "Inorganic Biochemistry", Vol. 1, Chemical Society Specialist Periodical Report (1979).

22. R. W. Hay and P. J. Morris, "Metal Ion-Promoted Hydrolysis of Amino Acid Esters and Peptides" in "Metal Ions in Biological Systems", Vol. 5, H. Sigel, ed., Marcel Dekker, New York (1976).

23. D. A. Buckingham, "Metal-OH and its Ability to Hydrolyze (or Hydrate) Substrates of Biological Interest" in "Biological Aspects of Inorganic Chemistry", A. W. Addison, W. R. Cullen, D. Dolphin and B. R. James, eds., John Wiley, New York (1976).

24. R. W. Hay and D. R. Williams, "Metal Complexes of Amino-acids Peptides and Proteins" in each odd issue of "Amino-acids Peptides and Proteins", R. C. Sheppard, ed., Specialist Periodical Report, The Chemical Society, London. From Volume 17 the topic is reviewed annually.

25. For a discussion see D. D. Perrin and B. Dempsey, "Buffers for pH and Metal Ion Control", Chapman and Hall, London (1974).

26. M. P. Springer and C. Curran, Inorg. Chem. 2:1270 (1963).

27. R. W. Hay and L. J. Porter, Aust. J. Chem. 20:675 (1967).

28. H. Shindo and T. L. Brown, J. Am. Chem. Soc. 87:1904 (1965).

29. Y. K. Sze, A. R. Davies and G. A. Neville, Inorg. Chem. 14:1969 (1975).

30. P. R. Norman and D. A. Phipps, Inorg. Chim. Acta 20:L45 (1976).

31. M. D. Alexander and D. H. Busch, Inorg. Chem. 5:1590 (1966).

32. R. W. Hay, R. Bennett and D. J. Barnes, J.C.S. Dalton Trans. 1524, (1972).

33. K. B. Nolan and A. A. Soudi, J.C.S. Dalton Trans. 1419 (1979).

34. D. A. Buckingham, J. Dekkers and A. M. Sargeson, J. Am. Chem. Soc. 95:4173 (1973).

35. M. L. Bender and B. W. Turnquest, J. Am. Chem. Soc. 79:1889 (1957).

36. R. W. Hay and L. J. Porter, J. Chem. Soc. (B), 1261 (1967).

37. R. W. Hay and P. J. Morris, J. Chem. Soc. (B), 1577 (1970).

38. R. W. Hay and P. J. Morris, J. Chem. Soc. Perkin Trans. II, 1021 (1972).

39. R. W. Hay, L. J. Porter and P. J. Morris, Aust. J. Chem. 19:1197 (1966).

40. J. E. Purdie and N. L. Benoiton, Can. J. Chem. 49:3468 (1971).

41. C. Gustaffson, Ann. Acad. Sci. Fennicae AII, No. 15 (1945).

42. R. P. Bell and B. A. W. Coller, Trans. Faraday Soc. 60:1087 (1964); 61:1445 (1965).

43. F. H. Westheimer and M. W. Shookoff, J. Am. Chem. Soc. 62:269 (1940).

44. R. P. Bell and M. Robson, Trans. Faraday Soc. 60:893 (1964).

45. J. M. White, R. A. Manning and N. C. Li, J. Am. Chem. Soc. 78:2367 (1956).

46. H. L. Conley and R. B. Martin, J. Phys. Chem. 69:2914 (1965).

47. J. E. Hix and M. M. Jones, Inorg. Chem. 5:1863 (1966).

48. C. Regardh, Acta Pharm. Suec. 3:101 (1966).

49. W. A. Connor, M. M. Jones and D. L. Tuleen, Inorg. Chem. 4:1129 (1965).

50. R. W. Hay and P. K. Banerjee, J. C. S. Dalton Trans. 362 (1981); see also Inorg. Chim. Acta. 44:L205 (1980).

51. R. W. Hay and K. B. Nolan, J. C. S. Dalton Trans. 1348 (1975).

52. B. E. Leach and R. J. Angelici, Inorg. Chem. 8:907 (1969).

53. R. J. Angelici and D. Hopgood, J. Am. Chem. Soc. 70:2514 (1968).

54. R. D. Wood, R. Nakon and R. J. Angelici, Inorg. Chem. 17:1088 (1978).

55. J. W. Allison and R. J. Angelici, Inorg. Chem. 10:2338 (1971).

56. R. Nakon, P. R. Rechani and R. J. Angelici, J. Am. Chem. Soc. 96:2117 (1974).

57. R. W. Hay and P. K. Banerjee, unpublished results.

58. R. W. Hay and P. K. Banerjee, J. C. S. Dalton Trans. 2452 (1980).

59. R. W. Hay and P. K. Banerjee, J. C. S. Dalton Trans. 2385 (1980).

60. R. W. Hay and P. K. Banerjee, J. Inorg. Biochem. 14:147 (1981).

61. J. E. Coleman and B. L. Valee, J. Biol. Chem. 236:2244 (1961).

62. S. Lindskog and P. O. Nyman, Biochem. Biophys. Acta. 85:462 (1964).

63. D. E. Newlin, M. A. Pellack and R. Nakon, J. Am. Chem. Soc. 99:1078 (1977).

64. R. W. Hay and P. J. Morris, J. Chem. Soc. (A), 3562 (1971).

65. R. W. Hay and P. J. Morris, J. Chem. Soc. (A), 1518 (1971).

66. N. C. Li, E. Doody and J. M. White, J. Am. Chem. Soc. 79:5859 (1957).

67. H. L. Conley and R. B. Martin, J. Phys. Chem. 69:2923 (1965).

68. L. J. Porter, D. D. Perrin and R. W. Hay, J. Chem. Soc. (A), 118 (1969).

69. J. M. White, R. A. Manning and N. C. Li, J. Am. Chem. Soc. 77:5225 (1955).

70. R. Mathur and N. C. Li, J. Am. Chem. Soc. 86:1289 (1964).

71. R. W. Hay and L. J. Porter, J. Chem. Soc. (A), 127 (1969).

72. T. R. Kelly, Ph.D. thesis, University of Glasgow, (1962).

73. R. W. Hay and P. J. Morris, J. Chem. Soc. (A), 1524 (1971).

74. R. W. Hay and P. J. Morris, J. C. S. Dalton. Trans. 56 (1973); for a preliminary report see idem, Chem. Comm. 732 (1968).

75. R. J. Angelici and B. E. Leach, J. Am. Chem. Soc. 89:4605 (1967).

76. R. J. Angelici and B. E. Leach, J. Am. Chem. Soc. 90:2499 (1968).

77. R. D. Cornelius, Inorg. Chim. Acta 46:L109 (1980).

78. B. E. Leach and R. J. Angelici, J. Am. Chem. Soc. 90:2504 (1968).

79. J. Rodgers and R. A. Jacobson, Inorg. Chim. Acta 13:163 (1975).

80. R. D. Gillard, Inorg. Chim. Acta. Rev. 1:69 (1967).

81. J. E. Hix, Jr. and M. M. Jones, J. Am. Chem. Soc. 90:1723 (1968).

82. R. W. Hay and P. J. Morris, J. Chem. Soc. (A), 1524 (1971).

83. R. W. Hay and P. J. Morris, J. C. S. Chem. Commun. 18 (1969).

84. P. J. Morris and R. B. Martin, J. Inorg. Nucl. Chem. 32:2891 (1970).

85. R. W. Kretsinger, F. A. Cotton and R. F. Bryan, Acta Cryst. 16:651 (1963); M. M. Harding and S. J. Cole, Acta Cryst. 16:643 (1963); K. A. Fraser and M. M. Harding, J. Chem. Soc. (A), 415 (1967); R. Candlin and M. M. Harding, J. Chem. Soc. (A), 421 (1967); M. M. Harding and H. A. Long, J. Chem. Soc. (A), 2554 (1968).

86. R. J. Sundberg and R. B. Martin, Chem. Rev. 74:471 (1974).

87. D. J. Barnes and L. D. Pettit, J. Inorg. Nucl. Chem. 33:2177 (1971).

88. G. Brookes and L. D. Pettit, J. C. S. Chem. Commun. 813 (1974).

89. J. R. Blackburn and M. M. Jones, J. Inorg. Nucl. Chem. 35:1597 (1973).

90. J. R. Blackburn and M. M. Jones, J. Inorg. Nucl. Chem. 35:1605 (1973).

91. J. E. Baldwin, J. C. S. Chem. Commun. 734 (1976).

92. J. E. Baldwin, J. C. S. Chem. Commun. 233 (1977).

93. P. Woolley, Nature 258:677 (1975).

94. R. H. Prince, D. A. Stotter and P. R. Wolley, Inorg. Chim. Acta 9:51 (1974).

95. M. R. Caira, L. R. Nassimbeni and P. R. Woolley, Acta Cryst. B31:1334 (1975).

96. E. Breslow in "The Biochemistry of Copper", J. Peisach, P. Aisen and W. E. Blumberg, eds., Academic Press, New York 149-156 (1966).

97. M. A. Wells, G. A. Rogers and T. C. Bruice, J. Am. Chem. Soc. 98:4336 (1976).

98. E. Chaffee, T. P. Dasgupta and G. M. Harris, J. Am. Chem. Soc. 95:4169 (1973).

99. D. A. Palmer and G. M. Harris, Inorg. Chem. 13:965 (1974).

100. D. A. Buckingham, J. MacB. Harrowfield and A. M. Sargeson, J. Am. Chem. Soc. 95:7281 (1973).

101. J. MacB. Harrowfield, V. Norris and A. M. Sargeson, J. Am. Chem. Soc. 98:7282 (1976).

102. D. A. Buckingham and L. M. Englehardt, J. Am. Chem. Soc. 97:5915 (1975).

103. R. W. Hay, Aust.J.Chem. 17:759 (1964).

104. R. W. Hay and C. R. Clark, Transition Met. Chem. 4:28 (1979).

105. J. A. Bertrand and D. Caine, J. Am. Chem. Soc. 86:2298 (1964).

106. R. P. Houghton and C. S. Williams, Tetrahedron Lett. 5091 (1967).

107. D. S. Sigman and C. T. Jorgenson, J. Am. Chem. Soc. 94:1724 (1972).

108. R. G. Lee, D. A. Long and T. G. Truscott, Trans.Faraday Soc. 65:503 (1969); D. A. Long and T. G. Truscott, Trans. Faraday Soc. 64:1866 (1968); 64:1624 (1968), 59:2316 (1963); 59:1833 (1963).

109. R. D. Gillard, Inorg. Chim. Acta Rev. 1:69 (1967).

110. H. C. Freeman, Advan. Protein Chem. 22:257 (1967).

111. R. D. Gillard, E. D. McKenzie, R. Mason and G. B. Robertson, Coord. Chem. Rev. 1:263 (1966).

112. A. S. Brill, R. B. Martin and R.J.P. Williams in "Electronic Aspects of Biochemistry", B. Pullman, ed., Academic Press, New York (1964).

113. H. C. Freeman, "The Biochemistry of Copper", J. Peisach, P. Aisen and W. E. Blumberg, eds., Academic Press, New York pp. 77 ff (1966).

114. C. A. McAuliffe in Inorganic Biochemistry, Vol. 1, Chemical Society Specialist Periodical Report (1979).

115. R. B. Martin, M. Chamberlin and J. T. Edsall, J. Am. Chem. Soc. 82:495 (1960).

116. H. C. Freeman, J. M. Guss and R. L. Sinclair, Chem. Commun. 485 (1960).

117. D. W. Margerum, L. F. Wong, F. P. Bossu, K. L. Chellappa, J. J. Czarnecki, S. T. Kirksey, Jr. and T. A. Neubecker in "Bioinorganic Chemistry II", K. N. Raymond, ed., Adv. Chem. Ser. 162, A.C.S. Washington (1977).

118. E. W. Wilson and R. B. Martin, Inorg. Chem. 9:528 (1970).

119. H. A. O. Hill and K. A. Raspin, J. Chem. Soc. (A), 619 (1969).

120. E. D. McKenzie, J. Chem. Soc. (A), 1655 (1969).

121. R. D. Gillard, E. D. McKenzie, R. Mason and G. B. Robertson, Nature 209:1347 (1966).

122. M. T. Barnet, H. C. Freeman, D. A. Buckingham, I-Nan Hsu and D. van der Helm, Chem. Commun. 367 (1970).

123. D. L. Rabenstein, Can. J. Chem. 49:3767 (1971).

124. E. Bamann, J. G. Haas and H. Trapmann, Arch. Pharm. 294:569 (1961).

125. E. Bamann and H. Trapmann, Advan. Enzymol. 21:169 (1959).

126. E. Bamann, H. Trapmann and A. Rother, Chem. Ber. 91:1744 (1958); Naturwissenschaften 43:326 (1956).

127. L. Meriwether and F. H. Westheimer, J. Am. Chem. Soc. 78:5119 (1956).

128. I. J. Grant and R. W. Hay, Aust. J. Chem. 18:1189 (1965).

129. T. Nakata, M. Tasumi and T. Miyazawa, Bull. Chem. Soc. Jpn. 48:1599 (1975).

130. M. Tasumi, S. Takahashi, T. Nakata and T. Miyazawa, Bull. Chem. Soc. Jpn. 48:1595 (1975).

131. L. Lawrence and W. J. Moore, J. Am. Chem. Soc. 73:151 (1951).

132. D. A. Long, T. G. Truscott, J. R. Cronin and R. G. Lee, Trans. Faraday Soc. 67:1094 (1971).

133. J. R. Cronin, D. A. Long and T. G. Truscott, Trans. Faraday Soc. 67:2096 (1971).

134. K. Ohkubo and H. Sakamoto, Bull. Chem. Soc. Jpn. 46:2579 (1973).

135. M. D. Alexander and D. H. Busch, Inorg. Chem. 5:602 (1966).

136. M. D. Alexander and D. H. Busch, Inorg. Chem. 5:1590 (1966).

137. M. D. Alexander and D. H. Busch, J. Am. Chem. Soc. 88:1130 (1966).

138. Y. Wu and D. H. Busch, J. Am. Chem. Soc. 92:3326 (1970).

139. D. A. Buckingham, L. G. Marzilli and A. M. Sargeson, J. Am. Chem. Soc. 89:4539 (1967).

140. D. A. Buckingham, D. M. Foster and A. M. Sargeson, J. Am. Chem. Soc. 90:6032 (1968).

141. See for example M. L. Tobe, Acc. Chem. Res. 3:377 (1970).

142. D. A. Buckingham, D. M. Foster and A. M. Sargeson, J. Am. Chem. Soc. 91:4102 (1969).

143. D. A. Buckingham, D. M. Foster, L. G. Marzilli and A. M. Sargeson, Inorg. Chem. 9:11 (1970).

144. D. A. Buckingham, J. Dekkers, A. M. Sargeson and M. Wein, J. Am. Chem. Soc. 94:4032 (1972).

145. D. A. Buckingham, J. Dekkers and A. M. Sargeson, J. Am. Chem. Soc. 95:4173 (1973).

146. K. Nomiya and H. Kobayashi, Inorg. Chem. 13:409 (1974).

147. D. A. Buckingham and M. Wein, Inorg. Chem. 13:3027 (1974).

148. R. W. Hay and K. B. Nolan, J. C. S. Dalton Trans. 1621 (1975).

149. J. R. Fluckiger, C. W. Schlapfer and C. Couldwell, Inorg. Chem. 19:2493 (1980).

150. H. M. Comley and W. C. E. Higginson, J. C. S. Dalton Trans. 2522 (1972).

151. A. C. Dash, R. K. Nanda and S. K. Mohapatra, J. C. S. Dalton Trans. 897 (1975).

152. R. W. Hay, R. Bennett and D. J. Barnes, J. C. S. Dalton Trans. 1524 (1972).

153. R. W. Hay, R. Bennett and D. P. Piplani, J. C. S. Dalton Trans. 1046 (1973).

154. K. B. Nolan, B. R. Coles and R. W. Hay, J. C. S. Dalton Trans. 2503 (1973).

155. D. A. Buckingham, D. M. Foster and A. M. Sargeson, J. Am. Chem. Soc. 91:3451 (1969).

156. D. A. Buckingham, C. E. Davis and A. M. Sargeson, J. Am. Chem. Soc. 92:6159 (1970).

157. D. A. Buckingham, J. MacB. Harrowfield and A. M. Sargeson, J. Am. Chem. Soc. 96:1726 (1973).

158. D. A. Buckingham, C. E. Davis, D. M. Foster and A. M. Sargeson, J. Am. Chem. Soc. 92:5571 (1970).

159. D. A. Buckingham, J. MacB. Harrowfield and A. M. Sargeson, J. Am. Chem. Soc. 96:1726 (1974).

160. D. A. Buckingham, D. M. Foster and A. M. Sargeson, J. Am. Chem. Soc. 92:6151 (1970).

161. D. A. Buckingham, P. J. Morris, A. M. Sargeson and A. Zanella, Inorg. Chem. 16:1910 (1977).

162. C. J. Boreham, D. A. Buckingham, and F. R. Keene, Inorg. Chem. 18:28 (1979).

163. C. J. Boreham, D. A. Buckingham and F. R. Keene, J. Am. Chem. Soc. 101:1409 (1979).

164. J. P. Collman and D. A. Buckingham, J. Am. Chem. Soc. 85:3039 (1963).

165. D. A. Buckingham and J. P. Collman, Inorg. Chem. 6:1803 (1967).

166. D. A. Buckingham, J. P. Collman, D. A. R. Hopper and L. G. Marzilli, J. Am. Chem. Soc. 89:1082 (1967).

167. L. G. Marzilli and D. A. Buckingham, Inorg. Chem. 7:1042 (1967).

168. E. Kimura, S. Young and J. P. Collman, Inorg. Chem. 9:1183 (1970).

169. M. D. Fenn and J. H. Bradbury, Analyt. Biochem. 49:498 (1972).

170. K. W. Bentley and E. H. Creaser, Inorg. Chem. 13:1115 (1974).

171. K. W. Bentley and E. H. Creaser, Biochem. J. 135:507 (1973).

172. E. Kimura, Inorg. Chem. 13:951 (1974).

173. S. K. Oh and C. B. Storm, Bioinorg. Chem. 3:89 (1973).

174. M-J. Rhee and C. B. Storm, J. Inorg. Biochem. 11:17 (1979).

175. J. P. Collman and P. W. Schneider, Inorg. Chem. 5:1380 (1966).

176. P. W. Schneider and J. P. Collman, Inorg. Chem. 7:2010 (1968).

177. R. D. Gillard and D. A. Phipps, Chem. Commun. 800 (1970).

178. Y. Wu and D. H. Busch, J. Am. Chem. Soc. 94:4115 (1972).

179. A. Y. Girgis and J. I. Legg, J. Am. Chem. Soc. 94:8420 (1972).

180. K. Ohkawa, J. Fujita and Y. Shimura, Bull. Chem. Soc. Jpn. 45:161 (1972).

181. L. F. Vilas-Boas, Ph.D. thesis, University of Kent, UK, (1974).

182. R. W. Hay and D. P. Piplani, Kemiai Közlémenyek. 48:47 (1977); see also ref. 542.

183. W. N. Lipscombe, Accounts Chem. Res. 3:81 (1970).

184. L. G. Marzilli and D. A. Buckingham, Inorg. Chem. 7:1042 (1967).

185. R. J. Dellaca, V. Janson, W. T. Robinson, D. A. Buckingham, L. G. Marzilli, I. E. Maxwell, K. R. Turnbull and A. M. Sargeson, J. C. S. Chem. Commun. 57 (1972), see also J. Am. Chem. Soc. 96:1713 (1974).

186. J. P. Collman and E. Kimura, J. Am. Chem. Soc. 89:6096 (1967).

187. R. W. Hay and P. J. Morris, Chem. Commun. 1208 (1969).

188. R. W. Hay, M. L. Jansen and P. L. Cropp, Chem. Commun. 621 (1967).

189. Y. Mitsui, J. Watanabe, Y. Iitaka and E. Kimura, J.C. S. Chem. Commun. 280 (1975).

190. S. C. Chan and F. K. Chan, Aust. J. Chem. 23:1175 (1970).

191. D. A. Buckingham, L. G. Marzilli and A. M. Sargeson, J. Am. Chem. Soc. 89:2772 (1967).

192. D. A. Buckingham, P. A. Marzilli, I. E. Maxwell and A. M. Sargeson, Chem. Commun. 488 (1968).

193. D. A. Phipps, J. Mol. Catal. 5:81 (1979).

194. A. Pasini and L. Casella, J. Inorg. Nucl. Chem. 2133 (1974).

195. A. E. Martell, in "Metal Ions in Biological Systems, "H. Sigel, ed., Marcel Dekker, New York, Vol. 2, p. 208 (1973): E. N. Safonova and V. M. Belikov, Russ. Chem. Rev. 43:745 (1974).

196. (a) R. H. Holm in "Inorganic Biochemistry", G. L. Eichorn, ed., Elsevier, New York, Vol. 2, p. 1137 (1973); (b) D. L. Leussing in "Metal Ions in Biological Systems", H. Sigel, ed., Marcel Dekker, New York, Vol. 5 p. 2 (1976); (c) J. T. Wrobleski and G. J. Long, Inorg. Chem. 16:2752 (1977).

197. (a) M. Sato, K. Okawa and S. Akabori, Bull. Chem. Soc. Jpn. 30:937 (1957); (b) Y. Ikutani, T. Okuda, M. Sato and S. Akabori, ibid. 32:203 (1959).

198. S. Akabori, T. T. Otani, R. Marshall, M. Winitz and J. P. Greenstein, Arch. Biochem. Biophys. 83:1 (1959).

199. Y. Ikutani, T. Okuda and S. Akabori, Bull. Chem. Soc. Jpn. 33:582 (1960).

200. R. D. Gillard and P. M. Harrison, J. Chem. Soc. (A), 1957 (1967).

201. J. P. Aune, P. Maldonado, G. Larcheres and M. Pierrot, Chem. Commun. 1351 (1970).

202. J. R. Brush, R. J. Magee, M. J. O'Connor, S. B. Teo, R. J. Geue and M. R. Snow, J. Am. Chem. Soc. 95:2034 (1973).

203. R. D. Gillard, S. H. Laurie, D. C. Price, D. A. Phipps and C. F. Weick, J.C.S. Dalton Trans. 1385 (1974).

204. R. J. Geue, M. R. Snow, J. Springborg, A. J. Herlt, A. M. Sargeson and D. Taylor, J. C. S. Chem. Commun. 285 (1976).

205. M. J. O'Connor, J. F. Smith and S. B. Teo, Aust. J. Chem. 29:375 (1976).

206. K. Noda, M. Bessho, T. Kato and N. Izumiya, Bull. Chem. Soc. Jpn. 43:1834 (1970).

207. M. Fujioka, Y. Nakao and A. Nakahara, J. Inorg. Nucl. Chem. 39:1805 (1977).

208. M. Murakami and K. Takahashi, Bull. Chem. Soc. Jpn. 32:308 (1959).

209. J. C. Dabrowiak and D. W. Cooke, Inorg. Chem. 14:1305 (1975).

210. D. A. Phipps, Inorg. Chim. Acta 27:L103 (1978).

211. K. Harada and J. Oh-hashi, J. Org. Chem. 32:1103 (1967).

212. T. Ichimawa, S. Maeda, Y. Araki and Y. Ishido, J. Am. Chem. Soc. 92:5514 (1970).

213. T. Ichikawa, S. Maeda, T. Okamoto, Y. Araki and Y. Ishido, Bull. Chem. Soc. Jpn. 44:2779 (1971).

214. T. Ichikawa, T. Okamoto, S. Maeda, S. Ohdan, Y. Araki and Y. Ishido, Tetrahedron Lett. 79 (1971).

215. S. Ohdan, T. Akamoto, S. Maeda, T. Ichikawa, Y. Araki and Y. Ishido, Bull. Chem. Soc. Jpn. 46:981 (1973).

216. S. Ohdan, T. Ichikawa, Y. Araki and Y. Ishido, Bull. Chem. Soc. Jpn. 47:1295 (1974).

217. Yu. N. Belokon, V. M. Belikov, S. V. Vitt, M. M. Dolgaya and T. F. Savel'eva, J. C. S. Chem. Commun. 86 (1975).

218. Yu. N. Belokon, V. M. Belikov, S. V. Vitt, T. F. Savel'eva, V. M. Burbelo, V. I. Bakhmutov, C. G. Aleksandrov and Yu. T. Struchkov, Tetrahedron 33:2551 (1977).

219. L. Benoiton, M. Winitz, R. F. Coleman, S. M. Birnbaum and J. P. Greenstein, J. Am. Chem. Soc. 81:1726 (1959).

220. R. D. Gillard and D. A. Phipps, J. C. S. Chem. Commun. 800 (1970).

221. L. G. Stadherr and R. J. Angelici, Inorg. Chem. 14:925 (1975).

222. R. D. Gillard, P. O'Brien, P. R. Norman and D. A. Phipps, J. C. S. Dalton Trans. 1988 (1977).

223. L. Casella, A. Pasini, R. Ugo and M. Visca, J. C. S. Dalton Trans. 1655 (1980).

224. D. A. Buckingham, L. G. Marizilli and A. M. Sargeson, J. Am. Chem. Soc. 89:5433 (1967).

225. J. I. Legg and D. W. Cooke, J. Am. Chem. Soc. 89:6854 (1967).

226. D. H. Williams and D. H. Busch, J. Am. Chem. Soc. 87:4644 (1967).

227. L. E. Erickson, A. J. Dappen and J. C. Uhlenhopp, J. Am. Chem. Soc. 91:2510 (1967).

228. E. S. Gore and M.L.H. Green, J. Chem. Soc. (A), 2315 (1970).

229. P. R. Norman and D. A. Phipps, Inorg. Chim. Acta 24:L35 (1977).

230. W. E. Keyes and J. I. Legg, J. Am. Chem. Soc. 98:4970 (1976).

231. (a) B. T. Golding, G. J. Gainsford, A. J. Herlt and A. M. Sargeson, Angew. Chem. Int. Ed. Engl. 14:495 (1975); (b) idem., Tetrahedron 32:389 (1976).

232. I. I. Creaser, J. MacB. Harrowfield, A. J. Herlt, A. M. Sargeson, J. Springborg, R. J. Geue and M. R. Snow, J. Am. Chem. Soc. 99:3181 (1977).

233. A. M. Sargeson, Chem. Brit. 15:23 (1979).

234. D. E. Metzler and E. E. Snell, J. Am. Chem. Soc. 74:769 (1952).

235. D. E. Metzler, J. Olivard and E. E. Snell, J. Am. Chem. Soc. 76:644 (1954).

236. K. Ikawa and E. E. Snell, J. Am. Chem. Soc. 76:4900 (1954).

237. J. Olivard, D. E. Metzler and E. E. Snell, J. Biol. Chem. 199:669 (1952).

238. D. Metzler and E. E. Snell, J. Biol. Chem. 198:353 (1952).

239. J. B. Longenecker and E. E. Snell, J. Am. Chem. Soc. 79:142 (1957).

240. D. E. Metzler, M. Ikawa and E. E. Snell, J. Am. Chem. Soc. 76:648 (1954).

241. "Pyrodoxal Catalysis - Enzymes and Model Systems", E. E. Snell, A. E. Braunstein, E. S. Severin and Y. M. Torchinsky, eds., Interscience, New York (1968).

242. E. E. Snell, P. M. Fasella, A. Braunstein and A. Rossi-Fanelli, "Chemical and Biological Aspects of Pyridoxal Catalysis", MacMillan, New York (1963).

243. See ref. 9, Vol. 2.

244. L. F. Lindoy, Quart. Rev. Chem. Soc. 25:379 (1971).

245. H. S. Maslen and T. N. Waters, Coord. Chem. Rev. 17:137 (1975).

246. E. J. Corey and R. L. Dawson, J. Am. Chem. Soc. 84:4899 (1962).

247. C. R. Wasmuth and H. Freiser, Talanta 9:1059 (1962).

248. R. H. Barca and H. Freiser, J. Am. Chem. Soc. 88:3744 (1966).

249. R. W. Hay and C. R. Clark, J.C.S. Dalton Trans. 1993 (1977).

250. R. P. Houghton and R. R. Puttner, Chem. Commun. 1270 (1970).

251. R. W. Hay and C. R. Clark, J.C.S. Dalton Trans. 1866 (1977).

252. K. H. Gerber and R. G. Wilkins, A.C.S. Meeting, Dallas, Texas, Abstract 173 (1973).

253. T. H. Fife and V. L. Squillacote, J. Am. Chem. Soc. 100:4787 (1978).

254. T. H. Fife, T. J. Przystas and V. L. Squillacote, J. Am. Chem. Soc. 101:3017 (1979).

255. For a general discussion of intramolecular catalysis see A. J. Kirby and A. R. Fersht in "Progress in Bioinorganic Chemistry", E. T. Kaiser and F. J. Kezdy, eds., Wiley (1971).

256. T. H. Fife and V. L. Squillacote, J. Am. Chem. Soc. 99:3762 (1977).

257. J. T. Groves and R. M. Dias, J. Am. Chem. Soc. 101:1033 (1979).

258. R. Nakon and R. J. Angelici, Inorg. Chem. 12:1269 (1973).

259. R. W. Hay and K. B. Nolan, J. C. S. Dalton Trans. 2542 (1974).

260. R. W. Hay, K. B. Nolan and M. Shuaib, Transition Met. Chem. 5:230 (1980).

261. R. Breslow, R. Fairweather and J. Keana, J. Am. Chem. Soc. 89:2135 (1967).

262. R. Breslow, "Bioinorganic Chemistry", Adv. Chem. Ser. No. 100, R. F. Gould, ed., p. 23; R. Breslow and M. Schmir, J. Am. Chem. Soc. 93:4960 (1971).

263. K. B. Nolan and R. W. Hay, J. C. S. Dalton Trans. 914 (1974).

264. D. Pinnell, G. B. Wright and R. B. Jordan, J. Am. Chem. Soc. 94:6104 (1972).

265. D. A. Buckingham, F. R. Keene and A. M. Sargeson, J. Am. Chem. Soc. 95:5649 (1973).

266. D. Λ. Buckingham, B. M. Foxman, Λ. M. Sargeson and Λ. Zanella, J. Am. Chem. Soc. 94:1007 (1972).

267. K. Sakai, T. Ito and K. Watanabe, Bull. Chem. Soc. Jpn. 40:1660 (1967).

268. S. Komiya, S. Suzuki and K. Watanabe, Bull. Chem. Soc. Jpn. 44:1440 (1971); K. Watanabe, S. Komiya and A. Suzuki, ibid. 46:2792 (1973).

269. P.F.D. Barnard, J. Chem. Soc. (A), 2140 (1969).

270. (a) A. W. Zanella and P. C. Ford, J.C.S. Chem. Commun. 795 (1974); (b) A. W. Zanella and P. C. Ford, Inorg. Chem. 14:42 (1975).

271. C. R. Clark and R. W. Hay, J. C. S. Dalton Trans. 2148 (1974).

272. D. A. Buckingham, A. M. Sargeson and A. Zannella, J. Am. Chem. Soc. 94:8246 (1972).

273. D. A. Buckingham, P. J. Morris, A. M. Sargeson and A. Zanella, Inorg. Chem. 16:1910 (1977).

274. K. V. Thimann and S. Mahadevan, Arch. Biochem. Biophys. 105:133 (1964); S. Mahadevan and K. V. Thimann, ibid. 107:62 (1964).

275. R. H. Hook and W. C. Robinson, J. Biol. Chem. 239:4257, 4263 (1964).

276. C. Zervos and E. H. Cordes, J. Am. Chem. Soc. 90:6892 (1968).

277. E. Bamann and H. Trapmann, Advan. Enzymol. 21:169 (1959).

278. T. C. Bruice and S. J. Benkovic, "Bioorganic Mechanisms", Vol.2, W. A. Benjamin, New York, Chapters 5, 6 and 7 (1966).

279. J. R. Cox and O. B. Ramsay, Chem. Rev. 64:317 (1964).

280. C. A. Bunton, J. Chem. Ed. 45:21 (1968).

281. J. Emsley and D. Hall, "The Chemistry of Phosphorus", Harper Row, London (1975).

282. F. H. Westheimer, Accounts Chem. Res. 1:70 (1968).

283. P. Gillespie, F. Ramirez, I. Ugi and D. Marquading, Angew. Chem. Int. Ed. Engl. 12:91 (1973).

284. R. F. Hudson, "Structure and Mechanism in Organophosphorus Chemistry", Academic Press, New York (1965).

285. B. J. Walker, "Organophosphorus Chemistry", Penguin, Harmondsworth, Mdx. (1972).

286. R. D. Cook, C. E. Diebert, W. Schwarz, P. C. Turley and P. Haake, J. Am. Chem. Soc. 95:8688 (1973).

287. M. Gallagher, A. Munoz, G. Gence and M. Koenig, J. C. S. Chem. Commun. 321 (1976).

288. J. F. Morrison and E. Heyde, Ann. Rev. Biochem. 41:29 (1972).

289. A. S. Mildvan, Enzymes 2:446 (1970).

290. J. Imsande and P. Handler, Enzymes 5:281 (1961).

291. B. S. Cooperman in "Metal Ions in Biological Systems", Vol. 5, H. Sigel, ed., Marcel Dekker, New York (1976).

292. A. S. Mildvan and G. M. Grisham, Structure and Bonding 1:20 (1974).

293. T. G. Spiro in "Inorganic Biochemistry", Vol. 1, Chapter 17, G. L. Eichhorn, ed., Scientific Publishing Co., New York (1973).

294. P. J. Briggs, D. P. N. Satchell and G. F. White, J. Chem. Soc. (B), 1008 (1970).

295. D. E. Koshland, J. Am. Chem. Soc. 73:4103 (1951); idem, ibid. 74:2286 (1952).

296. J. L. Kurtz and C. D. Gutsche, J. Am. Chem. Soc. 82:2175 (1960).

297. G. D. Sabato and W. P. Jencks, J. Am. Chem. Soc. 83:4393, 4400 (1961).

298. C. H. Oestreich and M. M. Jones, Biochemistry 5:2926 (1966).

299. S. J. Benkovic and L. K. Dunikoski, Jr., J. Am. Chem. Soc. 93:1526 (1971).

300. R. H. Bromilow and A. J. Kirby, J. C. S. Perkin Trans. II 149 (1972).

301. R. Hofstetter, Y. Murakami, G. Mont and A. E. Martell, J. Am. Chem. Soc. 84:3041 (1962).

302. Y. Murakami and A. E. Martell, J. Am. Chem. Soc. 67:582 (1963).

303. J. D. Chanley, E. M. Gindler and H. Sobotka, J. Am. Chem. Soc. 74:4347 (1952); J. D. Chanley and E. M. Gindler, ibid. 75:4035 (1953); J. D. Chanley and E. Feageson, ibid. 77:4002 (1955).

304. M. L. Bender and J. M. Lawlor, J. Am. Chem. Soc. 85:3010 (1963).

305. C. H. Oestreich and M. M. Jones, Biochemistry 5:3151 (1966).

306. J. J. Stettens, E. J. Sampson, I. J. Siewers and S. J. Benkovic, J. Am. Chem. Soc. 95:936 (1973).

307. S. J. Benkovic and K. J. Schray, Biochemistry 7:4096 (1968).

308. Y. Murakami, J. Sunamoto and H. Sadamori, J. C. S. Chem. Commun. 983 (1969).

309. Y. Murakami and J. Sunamoto, Bull. Chem. Soc. Jpn. 44:1827 (1971).

310. Y. Murakami and M. Takagi, J. Am. Chem. Soc. 91:5130 (1969).

311. Y. Murakami, J. Sunamoto and H. Ishizu, Bull. Chem. Soc. Jpn. 45:590 (1972).

312. L. B. Nanninga, J. Phys. Chem. 61:1144 (1957).

313. C. M. Hsu and B. S. Cooperman, J. Am. Chem. Soc. 98:5652 (1976).

314. G. J. Lloyd, C. M. Hsu and B. S. Cooperman, J. Am. Chem. Soc. 93:4889 (1971).

315. W. P. Jencks and M. Gilchrist, J. Am. Chem. Soc. 87:3199 (1965).

316. J. J. Stettens, E. J. Sampson, I. J. Siewers and S. J. Benkovic, J. Am. Chem. Soc. 95:936 (1973).

317. V. M. Clark, A. R. Todd and S. G. Warren, Biochem. Z. 338:591 (1963).

318. S. J. Benkovic and E. M. Miller, Bioinorg. Chem. 1:107 (1972).

319. M. M. Taqui Khan and A. E. Martell, J. Am. Chem. Soc. 84:3037 (1962).

320. M. Tetas and J. M. Lowenstein, Biochemistry 2:350 (1963).

321. G. M. Woltermann, R. A. Scott and G. P. Haight, Jr., J. Am. Chem. Soc. 96:7569 (1974).

322. T. Wagner-Jauregg, B. E. Hackley, T. A. Lies, O. O. Owens and R. Proper, J. Am. Chem. Soc. 77:922 (1955).

323. F. M. Fowkes, G. S. Ronay and L. B. Ryland, J.Phys. Chem. 62:867 (1958).

324. R. C. Courtney, R. L. Gustafson, S. Westerback, H. Hyytiainen, S. Chaberek and A. E. Martell, J. Am. Chem. Soc. 79:3030 (1957).

325. R. L. Gustafson, S. Chaberek and A. E. Martell, J. Am. Chem. Soc. 85:598 (1963).

326. R. L. Gustafson and A. E. Martell, J. Am. Chem. Soc. 84:2309 (1962).

327. K. B. Augustinsson and G. Heimburger, Acta. Chem. Scand. 9:383 (1955).

328. F. J. Farrell, W. A. Kjellstrom and T. G. Spiro, Science 164:320 (1969).

329. E. A. Dennis and F. H. Westheimer, J. Am. Chem. Soc. 88:3422 (1966).

330. S. F. Lincoln and D. R. Stranks, Aust. J. Chem. 21:57 (1968).

331. B. Anderson, R. M. Milburn, J. MacB. Harrowfield, G. B. Robertson and A. M. Sargeson, J. Am. Chem. Soc. 99:2652 (1977).

332. A. Steitz and W. M. Lipscomb, J. Am. Chem. Soc. 87:2488 (1965).

333. C. L. Coulter, J. Am. Chem. Soc. 95:570 (1963).

334. S. Suzuki, S. Kimura, T. Higashiyama and A. Nakahara, Bioinorg. Chem. 3:183 (1974).

335. M. L. De Pamphilis and W. W. Cleland, Biochemistry 12:3714 (1973).

336. K. D. Danenberg and W. W. Cleland, Biochemistry 14:28 (1975).

337. C. A. Janson and W. W. Cleland, J. Biol. Chem. 249:2562, 2567, 2572 (1974); M. I. Schimerlik and W. W. Cleland, ibid. 248:8418 (1973); D. A. Armbruster and F. B. Rudolph, ibid. 251:320 (1976); J. Bar-Tana and W. W. Cleland, ibid. 249:1271 (1974).

338. R. D. Cornelius, P. A. Hart and W. W. Cleland, Inorg. Chem. 16:2799 (1977).

339. D. G. Gorenstein, J. Am. Chem. Soc. 97:898 (1975).

340. R. D. Cornelius, Inorg. Chem. 19:1286 (1980).

341. E. A. Merritt, M. Sundaralingam, R. D. Cornelius and W. W. Cleland, Biochemistry 17:3274 (1978).

342. P. W. A. Hubner and R. M. Milburn, Inorg. Chem. 19:1267 (1980).

343. S. F. Lincoln, J. Jayne and J. P. Hunt, Inorg. Chem. 8:2267 (1969).

344. E. Bamann, Angew.Chem. 52:186 (1939).

345. E. Bamann and M. Meisenheimer, Chem.Ber. 71:1711, 1980, 2086, 2233 (1938).

346. E. Bamann, F. Fischer and H. Trapmann, Biochem. Z. 325:413 (1951).

347. W. W. Butcher and F. H. Westheimer, J. Am. Chem. Soc. 77:2420 (1955).

348. H. Trapmann, Arzneimittel-Forsch. 9:341, 403 (1959).

349. J. L. Kice and J. M. Anderson, J. Am. Chem. Soc. 88:5242 (1966).

350. S. J. Benkovic and P. A. Benkovic, J. Am. Chem. Soc. 88:5504 (1966).

351. S. J. Benkovic J. Am. Chem. Soc. 88:5511 (1966).

352. R. W. Hay, C. R. Clark and J. A. G. Edmonds, J. C. S. Dalton Trans. 9 (1974).

353. K. S. Dodgson, B. Spencer and K. Williams, Biochem. J. 64:216 (1956).

354. K. S. Dodgson and B. Spencer, "Methods of Biochemical Analysis", D. Gluck, ed., Vol. 4, p. 211, Interscience, New York (1956).

355. R. P. Hanzlik and W. J. Michaely, J.C.S.Chem.Commun. 113 (1975).

356. R. P. Hanzlik, M. Edelman, W. J. Michaely and G. Scott, J. Am. Chem. Soc. 98:1952 (1976).

357. J. N. DeMiller, Adv. Carbohydrate Chem. 22:25 (1967).

358. B. Capon, Chem. Rev. 69:407 (1969).

359. I. Wallenfels and P. O. Malhotra, Adv. Carbohydrate Chem. 16:239 (1961).

360. G. Low, Proc. Roy. Soc. (B) 167:431 (1967).

361. C. R. Clark and R. W. Hay, J.C.S.Perkin Trans. II 1943 (1973).

362. R. J. Ferrier, R. W. Hay and N. Vethaviyasar, Carbohydrate Res. 27:55 (1973).

363. T. H. Fife and L. K. Jao, J. Org. Chem. 30:1492 (1965).

364. E. H. Cordes, Prog. Phys. Org. Chem. 4:1 (1967).

365. T. J. Przystas and T. H. Fife, J. Am. Chem. Soc. 102:4391 (1980).

366. J. G. Murphy, J. Pharm. Sci. 61:810 (1972).

367. W. W. Tilley, D. W. Porter and R. W. Gracy, Carbohydrate Res. 27:289 (1973).

368. E. A. Stein, J. Hsiu and E. H. Fischer, Biochemistry 3:56, 61 (1964).

369. For reviews see E. T. Kaiser and B. L. Kaiser, Accounts Chem. Res. 5:219 (1972); W. N. Lipscomb, Tetrahedron 30:1725 (1974).

370. R. Breslow, D. E. McClure, R. S. Brown and J. Eisenach, J. Am. Chem. Soc. 97:194 (1975).

371. J. W. Thanassi and T. C. Bruice, J. Am. Chem. Soc. 88:747 (1966); C. A. Bunton, N. A. Fuller, S. G. Perry and V. J. Shiner, J. Chem. Soc. 2918 (1963).

372. G. Tomalin, B. L. Kaiser and E. T. Kaiser, J. Am. Chem. Soc. 92:6046 (1970); P. L. Hall, B. L. Kaiser and E. T. Kaiser, ibid. 91:485 (1969); E. T. Kaiser and F. W. Carson, ibid. 86:2922 (1966).

373. R. Breslow and D. Chipman, J. Am. Chem. Soc. 87:4195 (1965).

374. T. Sakan and Y. Mori, Chem. Lett. 793 (1972).

375. R. W. Hay and J. F. Ridlington, unpublished.

376. D. L. Leussing, "Metal Ions in Biological Systems", H. Sigel, ed., Vol. 5, Marcel Dekker Inc., New York, p. 1 (1976).

377. G. L. Eichhorn and J. C. Bailar, J. Am. Chem. Soc. 75:2905 (1953).

378. G. L. Eichhorn and I. M. Trachtenberg, J. Am. Chem. Soc. 76:5183 (1954).

379. G. L. Eichhorn and N. D. Marchand, J. Am. Chem. Soc. 78:2688 (1956).

380. D. H. Busch and J. C. Bailar, J. Am. Chem. Soc. 78:1137 (1956).

381. D. F. Martin and F. F. Cantwell, J. Inorg. Nucl. Chem. 26:2219 (1964).

382. L. J. Nunez and G. L. Eichhorn, J. Am. Chem. Soc. 84:901 (1962).

383. D. L. Leussing and C. K. Stanfield, J. Am. Chem. Soc. 88:5726 (1966).

384. A. C. Dash and R. K. Nanda, J. Am. Chem. Soc. 91:6944 (1969).

385. L. F. Lindoy, Quart. Rev. Chem. Soc. 25:379 (1971).

386. C. M. Harris, S. L. Lenzer and R. L. Martin, Aust. J. Chem. 14:420 (1961).

387. R. W. Hay and P. R. Norman, unpublished observations.

388. A. C. Braithwaite, C.E.F. Rickard and T. N. Waters, J. C. S. Dalton Trans. 2149 (1975).

389. E. Hoyer and B. Lorenz, Z. Anorg. Allg. Chem. 336:192 (1965).

390. V. W. Skopenko and E. Hoyer, Z. Anorg. Allg. Chem. 339:214 (1965).

391. E. Hoyer, Naturwissenschaften 1 (1959).

392. E. Hoyer, Z. Chem. 231 (1965).

393. C. V. McDonnell, Diss. Abs. 28B:3242 (1968).

394. R. W. Hay and K. B. Nolan, J. C. S. Dalton Trans. 548 (1976).

395. J. M. Harrowfield and A. M. Sargeson, J. Am. Chem. Soc. 96:2634 (1974).

396. B. T. Golding, J. M. Harrowfield and A. M. Sargeson, J. Am. Chem. Soc. 96:3003 (1974).

397. B. T. Golding, J. M. Harrowfield, G. B. Robertson, A. M. Sargeson and P. O. Whimp, J. Am. Chem.Soc. 96:3691 (1974).

398. J. D. Bell, A. R. Gainsford, B. T. Golding, A. J. Herlt and A. M. Sargeson J. C. S. Chem. Commun. 980 (1974).

399. A. R. Gainsford and A. M. Sargeson, Aust. J. Chem. 31:1679 (1978).

400. D. A. Buckingham, J. M. Harrowfield and A. M. Sargeson, J. Am. Chem. Soc. 95:7281 (1973).

401. I. P. Evans, G. W. Everett, Jr. and A. M. Sargeson, J. C. S. Chem. Commun. 139 (1975).

402. K. Schug and C. P. Guengerich, J. Am.Chem.Soc. 97:4135 (1975).

403. D. Wayshort and G. Navon, J.C.S. Chem.Commun. 1410 (1971). This value may be erroneous; see J. N. Armor, J. Inorg. Nucl. Chem. 35:2067 (1973).

404. C. P. Guengerich and K. Schug, Inorg. Chem. 17:2819 (1978).

405. D. Hopgood, J. C. S. Dalton Trans. 482 (1972) and references therein.

406. T. C. Bruice and S J. Benkovic, "Bio-organic Mechanisms", W. A. Benjamin, New York, Vol. 2, pp. 226-300 (1966).

407. S. Yamada, Y. Kuge and T. Yamayoshi, Inorg. Chim. Acta 8:29 (1974).

408. S. Yamada, Y. Kuge, T. Yamayoshi and H. Kuma, Inorg. Chim. Acta 11:253 (1974).

409. D. Heinert and A. E. Martell, J. Am. Chem. Soc. 85:1334 (1963).

410. For a discussion see "Coordination Chemistry of Macrocyclic Compounds", G. A. Melson, ed., Plenum Press, New York (1979).

411. E. H. Cordes and W. P. Jencks, J. Am. Chem. Soc. 84:832 (1962).

412. L. J. Nunez and G. L. Eichhorn, J. Am. Chem. Soc. 84:901 (1962).

413. Y. Chauvin, D. Comereuc and F. Dawans, Progr. Polym. Sci. 5:95 (1977).

4.14E. Tsuchi and H. Nishide, Advan. Polym. Sci. 24:1 (1977).

415. Y. Moriguchi, Bull. Chem. Soc. Jpn. 37:2656 (1966).

416. T. Nozawa, Y. Akimoto, and M. Hatano, Makromol.Chem. 158:21 (1972).

417. T. Nozawa, M. Hatano and S. Kanbara, Kogyo Kagaku Zasshi 72:373 (1969).

418. For a review of the catalytic activity of poly-L-lysine copper(II) complexes in the steroselective hydrolysis of amino-acid esters see M. Hatano and T. Nozawa in "Metal Ions in Biological Systems", H. Sigel, ed., Vol 5. p. 245 et seq.

419. R. Breslow and L. E. Overman, J. Am. Chem. Soc. 92:1075 (1970).

420. R. L. Van Etten, J. F. Sebastian, G. A. Clowes and M. L. Bender, J. Am. Chem. Soc. 89:3242 (1967); idem. ibid. 89:3253 (1967).

421. N. E. Dixon and A. M. Sargeson in "Zinc Enzymes" T. G. Spiro, ed., John Wiley and Sons, New York, Chapter 7 (1983).

422. R. S. Brown, J. Huguet and N. J. Curtis, Met. Ions. Biol. Syst. 15:55 (1983).

423. R. W. Hay and P. Banerjee, J.C.S. Dalton Trans. 362 (1981).

424. R. W. Hay and A. K. Basak, J.C.S. Dalton Trans. 1819 (1982).

425. R. W. Hay and M. P. Pujari, J.C.S. Dalton Trans. 1083 (1984).

426. R. W. Hay and P. Banerjee, J.C.S. Dalton Trans. 2385 (1980).

427. R. W. Hay and P. Banerjee, J.C.S. Dalton Trans. 2452 (1980).

428. J. K. Walker and R. Nakon, Inorg. Chim. Acta 55:135 (1981).

429. R. W. Hay and P. Banerjee, J. Inorg. Biochem. 14:147 (1981).

430. A. Yeh and H. Taube, J. Am. Chem. Soc. 102:4725 (1980).

431. C. J. Boreham and D. A. Buckingham, Aust. J. Chem. 33:27 (1980).

432. C. J. Boreham and D. A. Buckingham, Inorg. Chem. 20:3112 (1981).

433. K. Ohkubo, K. Kawazoe and M. Toyoda, J. Mol. Catal. 9:219 (1980).

434. R. W. Hay and P. Banerjee, Inorg. Chim. Acta 44:L205 (1980).

435. W. N. Lipscombe, Acc. Chem. Res. 3:81 (1970).

436. R. W. Hay, "Bioinorganic Chemistry", Ellis Horwood, Chichester (1984).

437. T. H. Fife and T. J. Przystas, J. Am. Chem. Soc. 104:2251 (1982).

438. J. Suh, M. Cheong and M. P. Suh, J. Am. Chem. Soc. 104:1654 (1982).

439. J. Suh, E. Lee and E. S. Jang, Inorg. Chem. 20:1932 (1982).

440. T. H. Fife and T. J. Przystas, J. Am. Chem. Soc. 102:7297 (1980).

441. T. H. Fife and T. J. Przystas, J. Am. Chem. Soc. 105:1638 (1983).

442. T. H. Fife and T. J. Przystas, J. Am. Chem. Soc. 107:1041 (1985).

443. R. S. Brown, M. Zamkanei and J. L. Cocho, J. Am. Chem. Soc. 106:5222 (1984).

444. R. S. Brown, D. Salmon, N. J. Curtis and S. Kusuma, J. Am. Chem. Soc. 104:3188 (1982).

445. K. Ogino, K. Shindo, T. Minami, W. Tagaki and T. Eiki, Bull. Chem. Soc. Jpn. 56:1101 (1983).

446. S. Sakaki, Y. Nakano and K. Ohkubo, Chem. Lett. 413 (1983).

447. D. A. Buckingham and C. R. Clark, Aust. J. Chem. 34:1769 (1981).

448. P. R. Norman and R. D. Cornelius, J. Am. Chem. Soc. 104:2356 (1982).

449. R. D. Cornelius and P. R. Norman, Inorg. Chim. Acta 65:L193 (1982).

450. E. A. Merritt and M. Sundaralingam, Acta. Cryst. B36:2576 (1980).

451. E. A. Merritt and M. Sundaralingam, Acta. Cryst. B37:1505 (1981).

452. E. A. Merritt, M. Sundaralingam and R. D. Cornelius, Acta. Cryst. B37:657 (1981).

453. D. R. Jones, L. F. Lindoy, A. M. Sargeson and M. R. Snow, Inorg. Chem. 21:4155 (1982).

454. R. W. Hay and R. Bembi, Inorg. Chim. Acta 78:143 (1983).

455. R. D. Cornelius, Inorg. Chim. Acta 46:L109 (1980).

456. P. R. Norman and R. D. Cornelius, Inorg. Chim. Acta 65:L193 (1982).

457. P. R. Norman, P. F. Gilletti and R. D. Cornelius, Inorg. Chim. Acta 82:L5 (1984).

458. J. Reibenspies and R. D. Cornelius, Inorg. Chem. 23:1563 (1984).

459. I. I. Creaser, G. P. Haight, R. Peachey, W. T. Robinson and A. M. Sargeson, J. C. S. Chem. Commun. 1568 (1984).

460. I. I. Creaser, R. V. Dubs and A. M. Sargeson, Aust. J. Chem. 37:1999 (1984).

461. W. G. Jackson and B. C. McGregor, Inorg. Chim. Acta 83:115 (1984).

462. P. Hendry and A. M. Sargeson, J. C. S. Chem. Commun. 164 (1984).

463. M. Hediger and R. M. Milburn, J. Inorg. Biochem. 16:165 (1982).

464. S. H. McClaugherty and C. M. Grisham, Inorg. Chem. 21:4133 (1982).

465. D. R. Jones, L. F. Lindoy and A. M. Sargeson, J. Am. Chem. Soc. 106:7807 (1984).

466. T. P. Haromy, P. F. Gilletti, R. D. Cornelius and M. Sundaralingam, J. Am. Chem. Soc. 106:2812 (1984).

467. D. R. Jones, L. F. Lindoy and A. M. Sargeson, J. Am. Chem. Soc. 105:7327 (1983).

468. J. MacB. Harrowfield, D. R. Jones, L. F. Lindoy and A. M. Sargeson, J. Am. Chem. Soc. 102:7733 (1980).

469. H. Sigel and F. Hofstetter, Eur. J. Biochem. 132:569 (1983).

470. T. Imamura, D. M. Hinton, R. L. Belford, R. I. Gumport and G. P. Haight, J. Inorg. Biochem. 11:241 (1979).

471. T. Eiki and W. Tagaki, Chem. Lett. 1465 (1981).

472. T. Eiki, T. Ogihara, A. Kato, F. Arai and W. Tagaki, J. C. S. Chem. Commun. 49 (1985).

473. W. Tagaki and T. Eiki, Adv. Chem. Ser. 191:407 (1980).

474. T. Eiki and W. Tagaki, Bull. Chem. Soc. Jpn. 55:1102 (1982).

475. T. Eiki, T. Horiguchi, M. Ono, S. Kawada and W. Tagaki, J. Am. Chem. Soc. 104:1986 (1982).

476. E. Buncel, E. J. Dunn, R.A.B. Bannard and J. G. Purdom, J.C.S. Chem. Commun. 162 (1984).

477. H. Sigel, F. Hofstetter, R. B. Martin, R. M. Milburn, V. Scheller-Krattiger and K. H. Scheller, J. Am. Chem. Soc. 106:7935 (1984).

478. R. W. Hay and A. K. Basak, Inorg. Chim. Acta 79:255 (1983); see also ref. 543.

479. R. W. Hay, A. K. Basak, M. P. Pujari and A. Perotti, in preparation.

480. J. T. Groves and R. R. Chambers Jr., J. Am. Chem. Soc. 106:630 (1984).

481. Y. Ilan and H. Taube, Inorg. Chem. 22:3144 (1983).

482. Y. Ilan and H. taube, Inorg. Chem. 22:1655 (1983).

483. D. A. Buckingham, G. S. Binney, C. R. Clark, B. Garnham and J. Simpson, Inorg. Chem. 24:135 (1985).

484. D. A. Buckingham and C. R. Clark, Aust. J. Chem. 35:431 (1982).

485. R. W. Hay and R. Bembi, Inorg. Chim. Acta 64:L179 (1982).

486. C. J. Boreham, D. A. Buckingham, D. J. Francis, A. M. Sargeson and L. G. Warner, J. Am. Chem. Soc. 103:1975 (1981).

487. I. I. Creaser, J. MacB. Harrowfield, F. R. Keene and A. M. Sargeson, J. Am. Chem. Soc. 103:3559 (1981).

488. R. J. Balahura and W. L. Purcell, Inorg. Chem. 20:4159 (1981).

489. R. L. de la Vega, W. R. Ellis Jr. and W. L. Purcell, Inorg. Chim. Acta 68:97 (1983).

490. R. E. Clarke and P. C. Ford, Inorg. Chem. 9:227 (1970); P. C. Ford, J. C. S. Chem. Commun. 7 (1971).

491. N. J. Curtis, K. S. Hagen and A. M. Sargeson, J.C.S. Chem. Commun. 1571 (1984).

492. N. J. Curtis and A. M. Sargeson, J. Am. Chem. Soc. 106:625 (1984).

493. W. R. Ellis and W. L. Purcell, Inorg. Chem. 21:834 (1982).

494. W. Fleming, J. W. Fronabarger, M. L. Lieberman and V. M. Loyola, Second Chemical Conference of the North American Continent, Las Vegas, Nevada, August 1980, A.C.S., Washington D.C., Abstract INOR 13.

495. P. J. Lawson, M. G. McCarthy and A. M. Sargeson, J. Am.Chem. Soc. 104:6710 (1982).

496. N. E. Dixon and A. M. Sargeson, J. Am. Chem. Soc. 104:6716 (1982).

497. W. G. Jackson, G. M. McLaughlin, A. M. Sargeson and A. D. Watson, J. Am. Chem. Soc. 105:2426 (1983).

498. E. K. Chong, J. MacB. Harrowfield, W. G. Jackson, A. M. Sargeson and J. Springborg, J. Am. Chem. Soc. 107:2015 (1985).

499. H. Wautier, V. Daffe, M-N. Smets and J. Fastrez, J. C. S. Dalton Trans. 2479 (1981).

500. H. Wautier, D. Marchal and J. Fastrez, J.C.S. Dalton Trans. 2484 (1981).

501. C. R. Clark, R. F. Tasker, D. A. Buckingham, D. R. Knighton, D. R. K. Harding and W. S. Hancock, J. Am. Chem. Soc. 103:7023 (1981).

502. D. R. Knighton, D. R. K. Harding, M. J. Friar, W. S. Hancock, G. D. Reynolds, C. R. Clark, R. F. Tasker and D. A. Buckingham, J. Am. Chem. Soc. 103:7025 (1981).

503. S. S. Isied, A. Vassilian and J. M. Lyon, J. Am. Chem. Soc. 104:3910 (1982).

504. S. Bagger, I. Kristjansson, I. Sotofte and A. Thorlacius, Acta. Chem. Scand. A39:125 (1985).

505. N. E. Dixon, P. W. Riddles, C. Gazzola, R. L. Blakeley and B. Zerner, Can. J. Biochem. 58:1335 (1980).

506. R. L. Blakeley, A. Treston, R. K. Andrews and B. Zerner, J. Am. Chem. Soc. 104:612 (1982).

507. S. Suzuki, H. Narita and K. Harada, J.C.S. Chem. Commun. 29 (1979).

508. K. Aoka and H. Yamazaki, J.C.S. Chem. Commun. 363 (1980).

509. H. M. Dawes, J. M. Waters and T. N. Waters, Inorg. Chim. Acta 66:29 (1982).

510. J. R. Fischer and E. H. Abbott, J. Am. Chem. Soc. 101:2781 (1979).

511. H. C. Dunathan, Proc. Natl. Acad. Sci. U.S.A. 55:712 (1966).

512. L. Casella and M. Gullotti, J. Am. Chem. Soc. 103:6338 (1981).

513. L. Casella, M. Gullotti and G. Pacchoni, J. Am. Chem. Soc. 104:2386 (1982).

514. H. Kondo, H. Yoshinaga, K. Morita and J. Sunamoto, Chem. Lett. 31 (1982).

515. See for example J. A. Marcello, A. E. Martell and E. H. Abbott, J.C.S. Chem. Commun. 16 (1975).

516. J. A. Marcello and A. E. Martell, J. Am. Chem. Soc. 104:1087 (1982).

517. K. Tatsumoto, and A. E. Martell, J. Am. Chem. Soc. 103:6203 (1981).

518. K. Tatsumoto, A. E. Martell and R. J. Motekartis, J. Am. Chem. Soc. 103:6197 (1981).

519. Y. N. Belokon, A. S. Melikyan, V. I. Bakhmutov, S. V. Vitt and V. M. Belikov, Inorg. Chim. Acta 55:117 (1981).

520. S.-B. Teo, S.-G. Teoh, J. R. Rodgers and M. R. Snow, J.C.S. Chem. Commun. 141 (1982).

521. S.-B. Teo and S.-G. Teoh, Inorg. Chim. Acta 91:L17 (1983).

522. S.-B. Teo and S.-G. Teoh, Inorg. Chim. Acta 68:107 (1983).

523. S.-B. Teo, S.-G. Teoh and M. R. Snow, Inorg. Chim. Acta 85:L1 (1984).

524. P. Sharrock, Polyhedron 2:111 (1983).

525. S.-B. Teo and M. J. O'Connor, Inorg. Chim. Acta 92:57 (1984).

526. G. G. Smith, A. Khatib and G. S. Reddy, J. Am. Chem. Soc. 105:293 (1983).

527. Y. N. Belokon, I. E. Sel'tzer, V. I. Bakmutov, M. B. Saporovskaya, M. G. Ryzhov, A. I. Yanovsky, Yu. T. Struchkov and V. M. Belikov, J. Am. Chem. Soc. 105:2010 (1983).

528. R. Deschenaux and K. Bernauer, Helv. Chim. Acta 67:373 (1984).

529. K. Bernauer, R. Deschenaux and T. Taura, Helv. Chim. Acta 66:2049 (1983).

530. L. Casella, J. Gullotti and A. Rockenbauer, J.C.S. Dalton Trans. 1033 (1984).

531. A. E. Martell and P. Taylor, Inorg. Chem. 23:2734 (1984).

532. M. S. El-Ezaby and F. M. Al-Sogair, Polyhedron. 1:791 (1982).

533. B. Szpoganicz and A. E. Martell, Inorg. Chem. 23:4442 (1984).

534. M. S. El-Ezaby, M. Rashad and N. M. Moussa, Polyhedron 2:245 (1983).

535. L. R. Solujic, R. Herak, B. Prelesnik and M. B. Celap, Inorg. Chem. 24:32 (1985).

536. L. R. Solujic and M. B. Celap, Inorg. Chim. Acta 67:103 (1982).

537. Y. N. Belokon, V. I. Maleyev, S. V. Vitt, M. G. Ryzhov, Y. D. Kondrashov, S. N. Golubev, Y. P. Vauchskii, A. I. Kazika, M. I. Novikova, P. A. Kratsutski, A. G. Yurchenko, I. L. Dubchak, V. E. Shklover, Yu. T. Struchkov, V. I. Bakmutov and V. M. Belikov, J.C.S. Dalton Trans. 17 (1985).

538. P. D. Ford, K. B. Nolan and D. C. Povey, Inorg. Chim. Acta 61:189 (1982).

539. A. R. Gainsford, R. D. Pizer, A. M. Sargeson and P. O. Whimp, J. Am. Chem. Soc. 103:792 (1981).

540. A. C. Dash, B. Dash and S. Praharaj, J.C.S. Dalton Trans. 2063 (1981).

541. U. Bips, H. Elias, M. Hauroder, G. Kleinhaus, S. Pfeifer and K. J. Wannowius, <u>Inorg. Chem.</u> 22:3862 (1983).

542. R. W. Hay, V.M.C. Reid and D.P. Piplani, <u>Transition Met. Chem.</u> 11:302 (1986).

543. R. W. Hay, A. K. Basak, M. P. Pujari and A. Perotti, <u>J.C.S. Dalton Trans.</u> 2029 (1986).

REACTIONS OF COORDINATED PHOSPHORUS AND SULFUR LIGANDS

D.M.A. Minahan,[a] W.E. Hill[a] and C.A. McAuliffe[b]

[a]Department of Chemistry, Auburn University, Alabama and
[b]Department of Chemistry, University of Manchester Institute of Science and Technology

1. REACTIONS OF PHOSPHORUS AND RELATED LIGANDS

1.1 Introduction

At the start of this Section we would remind readers of a number of excellent review articles and books which give general background to the area covered by this Chapter, viz: a comprehensive survey of P, As, Sb donor chemistry to mid-1972[1], diphosphine complexes[2], an early survey of transition metal complexes,[6] fluoralicyclic-P, As complexes,[7] phosphite, phosponite and amine-phosphine compounds,[8] Lewis base carbonyl complexes,[9] reactions of and catalysis by hydrido complexes,[10] zerovalent nickel, palladium and platinum complexes,[11] cyclophosphines and arsines,[12-13] cyclic phosphine complexes,[14] olefin - group 5B ligand complexes,[15] transition metal cyanide complexes,[16] structure of hydride complexes,[17] oxidative additions to d^8 metal complexes,[18] carbene complexes,[19] ^{31}P nmr spectra of complexes,[20] general structure[21] and structure of nitrosyl complexes,[22,23] reactions of coordinated pnictogens,[24] dioxygen complexes,[25] open-chain tetradentate[26] and tripod tetradentate chelates,[27] homogeneous hydrogenations[28] and general catalysis,[29] and a book which acts as a guide to the whole area.[30]

1.2   Cyclometalation

Since the first reports of cyclometalation in the early 1960's a large number of reactions of this type have been observed. The early work has been summarised in a review by Bruce,[31] in a brief account by Dehand and Pfeffer,[32] and by McAuliffe and Levason.[30] We here cover the literature since these reviews.

1.2.1. Ni, Pd, Pt Complexes

Work continues in Shaw's group on ditertiary phosphines containing bulky t-butyl groups. A volatile cyclometalated species $\overline{PdCl}$ {$Bu_2{}^tPCH_2\overline{CH_2}\overline{CH}CH_2CH_2PBu_2{}^t$} has been formed by reaction of $Bu_2{}^tP(CH_2)_5PBu_2{}^t$ with $PdCl_4{}^{2-}$.[33] The reaction of this ligand with $Pt(PhCN)_2Cl_2$ or $Pt(Bu^tNC)_2Cl_2$ produces $\overline{PtCl}$[$Bu_2{}^tPCH_2CH_2CHCH_2CH_2$ $PBu_2{}^t$} contaminated by another species, possibly $\overline{PtCl}$[$Bu_2{}^tPCH_2\overline{CH}\overline{CH}\overline{CH}CH_2PBu_2{}^t$}, from which it could not be separated. Treatment of {$\overline{Pt[Bu_2{}^tP(CH_2)_5PBu_2{}^t]Cl_2}$}$_x$ with CF_3COOH gives pure $\overline{Pt(CF_3COO)[Bu_2{}^tPCH_2CH_2CHCH_2CH_2PBu_2{}^t]}$, and metathesis of this complex displaces CF_3COO by Cl, Br, I or H. Treatment of the chloro complex with CF_3COOD does not give a deuterio complex, but when the bromo derivative is treated in like fashion deuteriation is observed in both the Bu^t groups and in the methylene chain. Cyclometalated derivatives of platinum have been obtained with $Bu_2{}^tP(CH_2)_6PBu_2{}^t$.

The new tertiary phosphines $RPBu_2{}^t$ and $RPPh_2$ (R = 2,3- or 2,6- $C_6H_3(OMe)_2$) have been complexed with palladium(II) and platinum(II) salts in attempts to effect O- or C-metalation[34]. When R = 2,6-$C_6H_3(OMe)_2$ a methyl group is readily lost, but the 2,3-derivative is less active. Stable complexes of type cis-[Pt{$OC_6H_3(OMe)PPh_2$}$_2$] and trans-[Pt [$OC_6H_3(OMe)PBu_2{}^t$]$_2$] have been prepared in which chelation is achieved through the phenoxy-oxygen and the phosphine. Palladium complexes do not give O-metalated derivatives, all attempts producing metallic palladium. O-metalation appears to be favoured by a polar solvent, but in xylene $Pt(PhCN)_2Cl_2$ reacts with 2,6-$C_6H_3(OMe)_2PPh_2$ to give the C-metalated complex $[PtCl_2[CH_2OC_6H_3(OMe-3)PPh_2]_2]$. Attempts to induce di-C-metalation were unsuccessful.

Heating trans-$PtCl_2(Bu_2{}^tPCH_2OCH_3)_2$ at 125° in 2-methoxyethanol yields $[PtCl(Bu_2{}^tPCH_2OCH_2)(Bu_2{}^tPCH_2OCH_3)]$.[35] If excess NaI and 1,8-bisdimethylaminonaphthalene are added the reaction is accelerated and the iodo complex is obtained. Under the same conditions trans-$PtCl_2(Bu_2{}^tPOCH_2CH_3)_2$ is also metalated at the methyl carbon, but the rate is slower, perhaps indicating that an α-oxygen atom has an accelerating effect on the metalation process. Neither $Bu_2{}^tP(CH_2)_2OCH_3$ nor $Bu_2{}^tPOCH_3$ gave metalated complexes, suggesting that 4- or 6-membered rings are not favored. Steric factors may be important since $Bu_2{}^tPCH(CH_3)OCH_3$ did not yield a cyclometalated derivative.

Reaction of $PBu_3{}^t$ with $PtCl_2$ produces $[Bu_3{}^tPH]_2PtCl_4$[36] and the internally metalated complex trans-$[PtCl(C_4H_8PBu_2{}^t)(PBu_3{}^t)]$. The metalated complex can also be prepared from $Pt(PhCN)_2Cl_2$ or Na_2PtCl_4 and $PBu_3{}^t$, but not from $Pt(COD)Cl_2$. The corresponding Pd derivatives yield trans-$[Pt(PBu_3{}^t)_2Cl_2]$, which converts to the metalated complex trans-$[PdCl(C_4H_8PBu_2{}^t)PBu_3{}^t]$ at room temperature. If the reaction of $PBu_3{}^t$ and $Pt(PhCN)_2Cl_2$ is carried out in CH_2Cl_2 a solvent-dependent equilibrium is established as shown below:[37]

$$[PtCl\{Bu_2{}^tPC(CH_3)_2CH_2\}]_2 + 2PBu_3 \rightleftharpoons$$
$$2[PtCl\{Bu_2{}^tPC(CH_3)_2CH_2\}PBu_3{}^t]$$

Trans-$Pt(H)X(PBu_3{}^t)_2$ (X = Cl, Br, I, CF_3COO) and trans-$PtH_2(PBu_3{}^t)_2$ undergo facile intramolecular metalation to form the P, C - chelated complexes $[PtX\{Bu_2{}^tPC(CH_3)_2CH_2\}PBu_3{}^t]$ and $[PtH\{Bu_3{}^tPC(CH_3)_2CH_2\}]_2$, respectively.[38] The rate of metalation is markedly increased by the presence of trace amounts of alcohol or acids and for $[PtX\{Bu_2{}^tPC(CH_3)_2CH_2\}PBu_3{}^t]$ the rate follows the order I > Br > Cl > CF_3COO.

Reactions of the stilbene ligands o-$R_2PC_6H_4CH=CHC_6H_4PR_2$-o (R = Ph, o-$CH_3C_6H_4$) with divalent nickel, palladium and platinum salts

give the complexes $[MX(\underline{o}\text{-}R_2PC_6H_4C=CHC_6H_4PR_2\text{-}\underline{o}]$ (X = Cl, Br, I).[39]
The analogous methyl derivatives (X = Me) are obtained from $Pt(COD)Me_2$
and the stilbene. $\underline{o}\text{-}Ph_2PC_6H_4CH_2CH_2C_6H_4\text{-}PPh_2\text{-}\underline{o}$ Also yields chelated -
bonded benzyls of formula $[MX(\underline{o}\text{-}Ph_2PC_6H_4CHCH_2C_6H_4PPh_2\text{-}\underline{o}]$.
Hydrogen abstraction from $\underline{o}\text{-}Ph_2PC_6H_4\text{-}\underline{trans}\text{-}CH=CHC(CH_3)C_6H_4PPh_2\text{-}\underline{o}$
by MCl_2 (M = Ni, Pd, Pt) gives [MLX] species in which a hydrogen atom is
lost from the tertiary carbon.[40] The nickel complex is a 1:1 mixture of η^3
allyl isomers, both isomers containing cis phosphines. The palladium
derivative I contains an η^1 ligand in which the double bond has migrated
from the 1- to the 2-position.

Platinum yields two η^1-allyl complexes with the Pt-C bond at either C_1 or
C_2 in the backbone.

1.2.2 Iron, Ruthenium and Osmium

Further work has been reported on cyclometalated complexes of
these elements. Recystallisation of $Fe(dppe)(CH_3)_2$ from toluene yields the
orthometalated complex $[\overline{Fe[C_6H_4}P(Ph)CH_2CH_2PPh_2}](dppe)(CH_3)].$[41]
Treatment with D_2SO_4 yields deuteriated dppe.   Confirmation of
orthometalated triphenylphosphite using ^{13}C nmr for the complexes
$[Cp\overline{Fe[C_6H_4}OP(OC_6H_5)_2}](CO)]$ and $[Cp\overline{Fe[C_6H_4}OP(OC_6H_5)_2}][P(OC_6H_5)_3]]$
has been reported.[42]

Employing size exclusion chromatography[43] to clarify the structures
of " $[M[P(OPh)_3]_4]$ " (M = Fe, Ru) a proposed structure II has emerged:

I   II

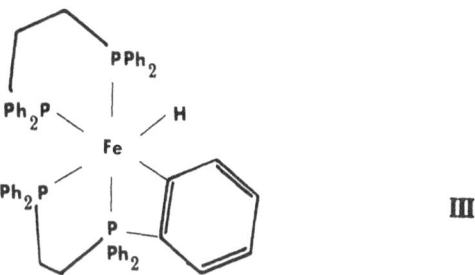

III

The complex $[Fe\{P(OPh)_3\}_2\{C_6H_4OP(OC_6H_5)_2\}_2]$ has been prepared by reaction of iron atoms with $P(OPh)_3$.[44] Reaction of this complex with HCN gives $[NCFe\{C_6H_4OP(OPh)_2\}P(OPh)_3]$, selectively unhooking one ring. Reaction of osmium trichloride trihydrate with $P(OPh)_3$ gives $[Os(C_6H_4OP(OPh)_2)_2(H)[P(OPh)_3]_3]$.

Photolysis of $Fe(dppe)_2(C_2H_4)$ yields **III**.[45] On photolysis in solution, trans-$[Fe(CO)_3\{P(OAr)_3\}_2]$ (Ar = Ph, o-, p-MeC$_6$H$_4$) lost CO to give red products which could not be isolated.[46] Reactions of the red products with CCl$_4$ or N-bromosuccinimide converted the red intermediates to a mixture of $[Fe(CO)_2\{P(OAr)_3\}_2X_2]$ and species of approximate empirical formula $[Fe(CO)_2\{P(OAr)_3\}_2X]$ (X = Cl or Br).

The red products and the monohalides were suggested to be iron(II) compounds, containing o-metalated phosphite and hydride or halide ligands.[46]

1.2.3. Rhodium and Iridium

A large number of cyclometalated complexes containing Rh and Ir have been characterized. Dehydrogenation of an alkane chain has been observed when $Ph_2P(CH_2)_6PPh_2$ is reacted with $[(COD)MCl]_2$ (M = Rh, Ir) in mesitylene.[47] Thus complexes of the type $[MLCl]$ (L = $Ph_2PCH_2CH_2$-trans-CH=CHCH$_2$CH$_2$PPh$_2$) **(IV)** are isolated. In the presence of CO no dehydrogenation occurs. No dehydrogenation was observed for the ligands $Ph_2E(CH_2)_nEPh_2$ (E = P, As; n = 5, 7, 8). The proposed mechanism is shown in Scheme 1. In ethanol $Bu_2^tP(CH_2)_5PBu_2^t$ reacts with $RhCl_3\cdot3H_2O$ to produce a binuclear 16-atom ring complex of formula $[Rh_2H_2Cl_4\{Bu_2^tP(CH_2)_5PBu_2^t\}_2]$, which exists in

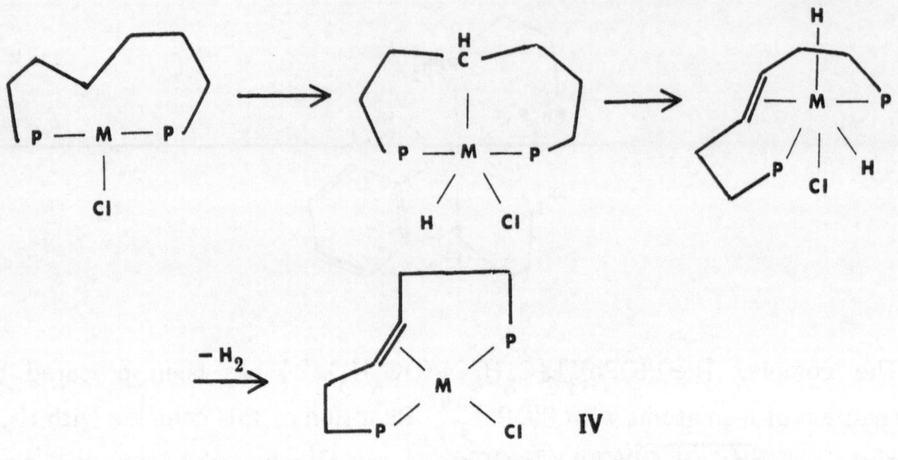

Scheme 1 (Ph groups ommited for clarity)

solution as a mixture of two rotamers, the cyclometalated isomer [RhHCl{Bu$_2$^{t}P(CH$_2$)$_2$CH(CH$_2$)$_2$PBu$_2$^{t}}], and an olefin complex; Bu$_2$^{t}P(CH$_2$)$_2$CH(CH$_3$)(CH$_2$)$_3$PBu$_2$^{t} reacts similarly to produce [RhHCl{Bu$_2$^{t}PCH$_2$CH$_2$C(CH$_3$)CH$_2$CH$_2$PBu$_2$^{t}}] and the related olefin complex. The effect of chain length on cyclometalation is emphasized by the fact that RhCl$_3$·3H$_2$O and Bu$_2$^{t}P(CH$_2$)$_6$PBu$_2$^{t} did not give a cyclometalated product, a mixture of olefin complexes being obtained.[49]

The acetone solvent complexes [M(C$_5$Me$_5$)(Me$_2$CO)$_3$](PF$_6$)$_2$ (M=Rh, Ir) react with P(OPh)$_3$ to give the orthometalated products [M{C$_6$H$_4$OP(OPh)$_2$}{P(OPh)$_3$}(C$_5$Me$_5$)]PF$_6$.[50]

Internal oxidative addition at room temperature of a benzylic C-H bond to Ir(I) has been observed for the complex [IrCl(CO){o-Ph$_2$P-C$_6$H$_4$-(CH$_2$)$_2$C$_6$H$_4$PPh$_2$-o}] forming the chelated metal-carbon σ-bonded complex [Ir(H)(Cl)(CO)(o-Ph$_2$PC$_6$H$_4$CHCH$_2$C$_6$H$_4$PPh$_2$-o)].[51] The arsenic analogue behaves similarly under more forcing conditions. Cationic metalated complexes of the type [Ir(H)(CO)$_2${o-Ph$_2$E-C$_6$H$_4$CH-CH$_2$C$_6$H$_4$EPh$_2$-o}] (E = P, As) have been identified in solution.

The dimeric carbonyl [IrCl(CO)P(o-tolyl)$_3$]$_2$ undergoes bridge splitting reactions with pyridine and γ-picoline to give the bis-cyclometalated complex V[52]

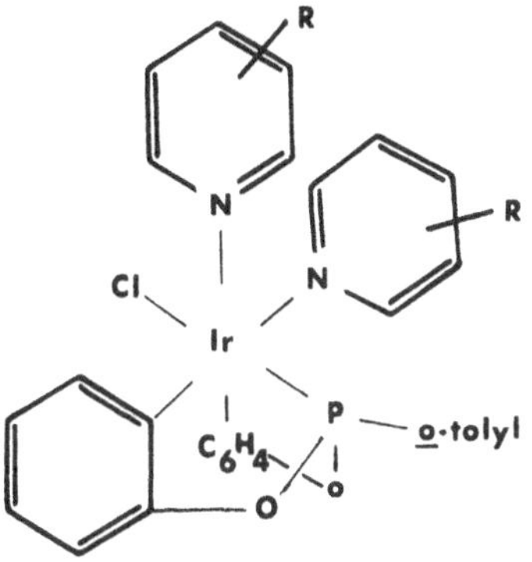

$R = CH_3, H$

V

Reactions of olefinic phosphines of the type $R_2P(allyl)$ ($R = Bu^t, Cy$) with $[Ir(COT)_2Cl]_2$ in the presence of γ-picoline at room temperature yield six-coordinate metalated complexes of type $[(R_2PCH_2CH=CH)IrHCl(R_2PCH_2CH=CH_2)(NC_6H_7)]$.[53] In the case of $Bu_2^tPCH_2CH=CH_2$ with γ-picoline, a five coordinate metallated species is observed. Reaction of $[(COT)_2RhCl]_2$ with the allyl phosphines did not give metalated complexes. Reaction of the allyl phosphines with $(acac)Ir(C_2H_4)_2$ in the presence of γ-picoline or acetonitrile also yields metalated Ir(III) complexes $[(acac)(R_2PCH_2CH=CH)IrHL]$ ($L = \gamma$-picoline or CH_3CN);[54] CO, CNR and $P(OMe)_3$ displace L). With $(acac)Rh(C_2H_4)_2$ no metalated complexes were obtained.

The tertiary aliphatic phosphines $Bu_2^tPPr^n$, $Bu_2^tPBu^n$ and PPr_3^i react with $[(COT)_2IrCl]_2$ in the presence of γ-picoline or acetonitrile to give cyclometalated complexes containing 4- or 5-membered rings but no 6-membered rings.[55] The steric effects of the phosphine and the coordinated ligand determine the composition and stereochemistry of the products.

A series of cationic monometalated and neutral and cationic tridentate dimetalated complexes of iridium(III) with tris-o-tolylphosphite have been prepared, as summarized in Scheme 2.[56]

Scheme 2

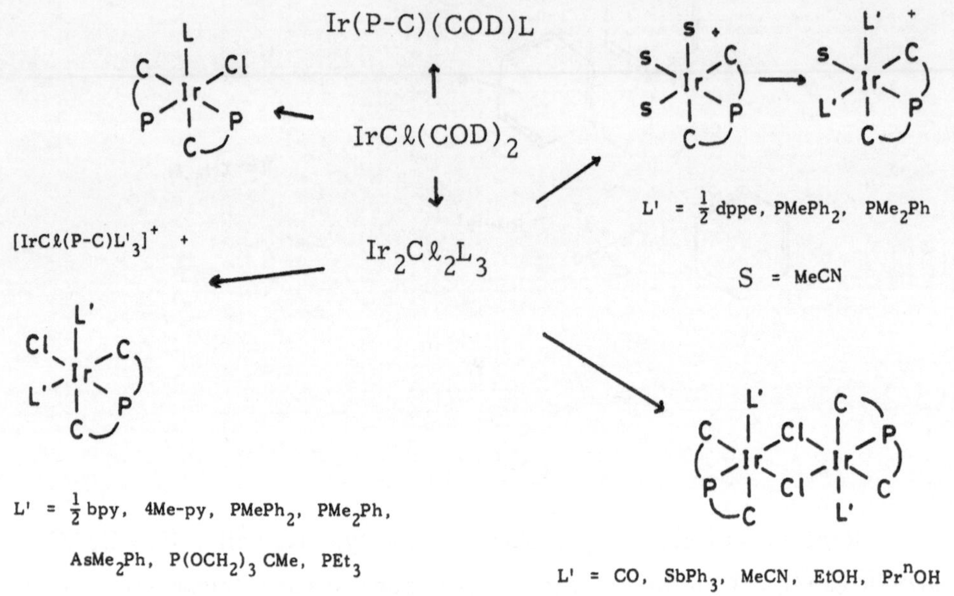

Ir(P-C)(COD)L

IrCℓ(COD)₂

Ir₂Cℓ₂L₃

$[IrC\ell(P-C)L'_3]^+$ +

$L' = \frac{1}{2}$ dppe, PMePh₂, PMe₂Ph

S = MeCN

L' = ½ bpy, 4Me-py, PMePh₂, PMe₂Ph,

AsMe₂Ph, P(OCH₂)₃ CMe, PEt₃

L' = CO, SbPh₃, MeCN, EtOH, Pr^{n}OH

COD = 1,5-cyclooctadiene

L = P(OC₆H₄Me-o)₃

P-C = P(OC₆H₃Me-o)(OC₆H₄Me-o)₂

C-P-C = P(OC₆H₃Me-o)₂ (OC₆H₄Me-o)

The reactions of the monometalated complex [Ir[P(OC₆H₃Me-o) (OC₆H₄Me-o)₂](COD)P(OC₆H₄Me-o)₃] with HX (X = ClO₄, PF₆, BF₄) to give [IrHP(OC₆H₃Me-o) (OC₆H₄Me-o)₂(COD)L]X is proposed to occur by a mechanism involving direct proton attacked at the metalated carbon as shown in Scheme 3.[57]

[Ir(COT)Cl]₂ reacts with Bu₂^{t}PCH₂OCH₃ and Bu₂^{t}PCH(CH₃)OCH₃ to form iridium(III) metalated complexes by oxidative addition to the methyl C-H bond.[35] The 5-coordinate complexes [MCl(CO)L] (L = o-Ph₂PC₆H₄-trans-CH=CH-C₆H₄-PPh₂-o; M = Ir, Rh) are protonated by HCl at the coordinated double bond, the products being chelate -alkyls of Rh(III) and Ir(III).[58] HCl reacts with RhLCl in the presence of Ph₃P to give a dimer, [RhCl₂(o-Ph₂PC₆H₄CHCH₂C₆H₄PPh₂-o)]₂.

Scheme 3

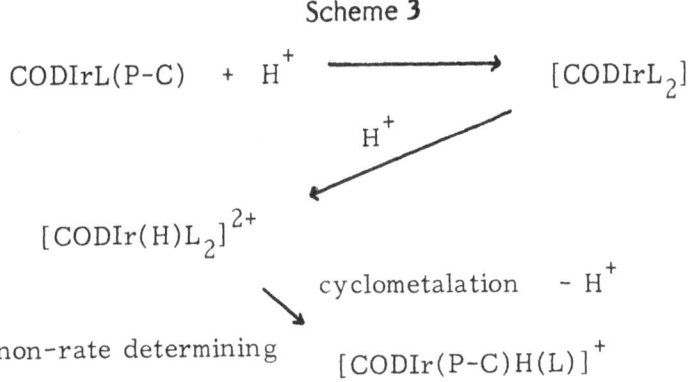

1.3   Reactions involving C=C or C≡C bonds

1.3.1. Alkenyl Reactions.

The complexes $[RhCl(CO)(Ph_2P(CH_2)_nCH=CH_2)_2]$ (n = 1-4) and $[RhCl(CO)(Ph_2P(CH_2)_2CH=CHCH_3)_2]$ add H_2 in methanol. Carbon monoxide acts as an inhibitor, and the reaction rate depends largely on steric factors. Deuteriation experiments show that scrambling does occur.[59] Thermolysis of $Os_3(CO)_{11}(PEt_3)$ gives $[H_2Os_3(Et_2PCCH_3(CO)_9]$ which has been dehydrogenated to $[HOs_3(Et_2P-C=CH_2)(CO)_9]$ by hydride abstraction from the unique CH_3, followed by deprotonation from the metal. The alkene compound isomerizes by a 1,2-hydrogen shift and P-C bond fission to the alkyne compound $HOs_3(PEt_2)(HC≡CH)(CO)_9$. $Os_3(CO)_{11}(PEt_2Ph)$ in hydrocarbon solvents at 150°C gives $HOs_3EtP-(CH=CH_2)(CO)_9$, as well as the ortho-metalated arene derivatives $[HOs_3(Et_2PC_6H_4)(CO)_9]$ and $[HOs_3(C_6H_4)(PEt_2)(CO)_9]$.[60]

o-Vinylphenyldiphenylphosphine is dimerized by $RhCl_3$ in refluxing 2-methoxyethanol to give the rhodium(I) complex $[RhCl\{o-Ph_2PC_6H_4-$ trans-$CH=CH-CH(CH_3)-C_6H_4PPh_2-o\}]$[61], and the iridium analogue has also been prepared. These complexes exist as a mixture of isomers which have different conformations of the chelate ring. The 5-coordinate carbonyl complexes [MCl(CO)L] containing this ligand react with HCl to give chelate 2-butyls, $MCl_2(CO)\{o-Ph_2PC_6H_4CH_2CHCH(CH_3)C_6H_4PPh_2-o\}$, the addition being reversible for rhodium. However [RhClL] gives a chloro-bridged dimer $[RhCl_2\{o-Ph_2PC_6H_4CHCH_2CH(CH_3)C_6H_4PPh_2-o\}]_2$.   [RhClL]   undergoes   oxidative   addition   of chlorine to give [RhCl$_3$L], which loses HCl irreversibly in solution to give the η^3-allyl RhCl$_2${o-Ph$_2$PC$_6$H$_4$CH-CH-C(CH$_3$)C$_6$H$_4$PPh$_2$-o} containing trans phosphorus atoms. The 5-coordinate [Ir(CO)$_2$L]$^+$ can be isolated, but the corresponding ethylene complexes cannot be obtained as the tertiary C-H oxidatively adds to give the η^3-allyl complex [IrH(C$_2$H$_4$){o-Ph$_2$PC$_6$H$_4$CH-CH-C(CH$_3$)C$_6$H$_4$PPh$_2$-o}]BF$_4$.

1.3.2 Alkynyl Reactions

A probable 10-membered RuPC$_8$ heterocycle, formed by the reaction of CpRuH(PPh$_3$)$_2$ with HC$_2$C$_6$H$_5$, is proposed to have the structure   [Ru{C(C$_6$F$_5$)=CHCH=C(C$_6$F$_5$)C$_6$H$_4$PPh$_2$-o}].[62]   Rigid unsymmetrical cis-olefinic diphosphines R$_2$PCH=C(R^1)PR^{2}R^3 (R^1 = CF$_3$, Ph, Bu^t; R^2=R^3=Ph, C$_6$H$_4$CN; R^2=Et, R^3=Ph) can be synthesized in good yields by the stereospecific addition of HPR^{2}R^3 across the triple bond of a phosphorus – coordinated phosphinoacetylene in cis-MCl$_2$(Ph$_2$PC≡CR)$_2$ (M = Pt, Pd).[63-4] The molecular structures of cis-PtCl$_2$(Ph$_2$PC≡CPh)$_2$ 2CH$_3$CN and related species show that the alkynes are forced into steric interaction with one another:[65-6]

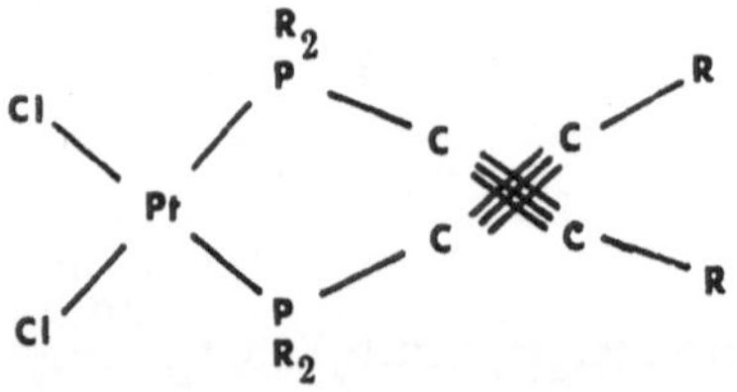

hence it is not suprising that alkyne coupling occurs on refluxing in benzene to give the complex:

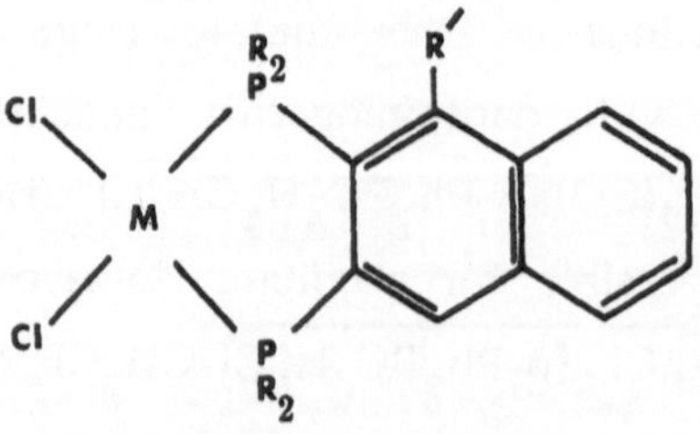

1.4   Oxo Transfer Reactions

Catalytic oxidations of tertiary phosphines are usually considered to go via a process involving oxygen atom transfer from coordinated dioxygen, but oxidation of coordinated $PPh_2(O-i-Bu)$ proceeds mainly via a radical pathway.[67] Thus initial reaction rates for a series of chlorocarbonyl-bis (tertiary phosphine or phosphinite) rhodium(I) complexes with oxygen are independent of complex concentration even though the metal complex is necessary as a catalyst. Addition of hydroquinone reduces the reaction rates.

Reaction of $[Ir(CO)Cl(PPh_3)_2]$ with Bu^tO_2H (2 equivalents) gave CO_2, Ph_3PO and $[Ir(CO)Cl(PPh_3)_2(Bu^tO_2)_2]$. With other carbonyl complexes, CO is oxidized in preference to the phosphine.[68]

Oxygenation reactions of aldehydes, tertiary phosphines and phosphites by the cis-dioxomolybdenum(VI) ethyl-L-cysteinate complex are believed to involve oxo-transfer from molybdenum to trivalent phosphorus.[69]

The reaction of (η^5-cyclopentadienyl)(triphenylphosphine) cobalta-cyclopentadiene complexes, $CpCo(CR^1=CR^2CR^3=CR^4)$ (PPh_3), with phosphites, $P(OR)_3$, gives (η^5-cyclopentadienyl) (1-alkoxyphosphole oxide) cobalt complexes via phosphite-substituted cobaltacyclopentadienes.[70] The proposed mechanism for this reaction is shown in Scheme 4. Complexes of the type $(CH_3O)_3PCo(dmgH)_2X$ (dmgH = monoanion of dimethylglyoxime; X = NO_3, C_2H_5) undergo Arbuzov-type reactions with Br^- to give $(CH_3O)_2P(O)Co$ $(dmgH)_2X$.[71] Similar observations have been made with complexes of the type trans-$[MCl_2L_2]$ (L = $PEt_{3-n}Ph_n$, n = 0 - 3) in their reaction with F_2POR (R = allyl, Pr^n, Bu^n) to give trans-$[MClL_2PF_2O]$ (M = Pt, Pd).[72]

Scheme 4

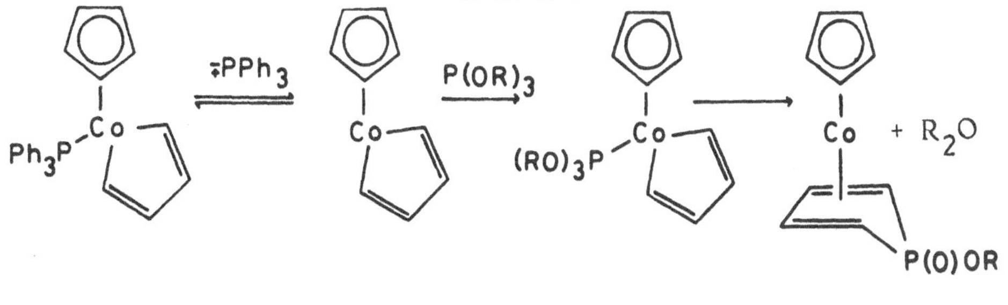

1.5   Miscellaneous Reactions of Coordinated Group V Ligands

Trinuclear complexes $Cp_2Co_3[PO(OR)_2]_6$ react with HBF_4 to produce the salt-like compounds $[CpCo [(P(OR)_2O)_3BF]]BF_4$ where x-ray analysis shows a tris(diethylphosphito)fluoroborate ion to be coordinated to the metal as a tridentate "cage" ligand.[73] $\underline{Cis}$-$Mo(CO)_4(Me_2PLi)_2$, obtained by lithiation of $\underline{cis}$-$Mo(CO)_4(Me_2PH)_2$, reacts with dibromomethylstibine or dichloromethylbismuthine to give the unusual complexes $\underline{cis}$-$(CO)_4Mo(PMe_2$ -SbMe-SbMe-PMe_2) and $\underline{cis}$-$(CO)_4Mo(PMe_2$-BiMe-BiMe-PMe_2);[74] reaction with Bu^tPCl_2 gives $\underline{cis}$-$Mo(CO)_4(PMe_2PBu PMe_2)$.   Metalation of cis-$Mo(CO)_4PhPH_2$ with n-butyl or methyl lithium produces $\underline{cis}$-$Mo(CO)_4(PhPHLi)_2$ which reacts with $ClSiMe_3$, $(CH_3)_2(Cl)SiSi(CH_3)_2(Cl)$, $(C_5H_5)_2TiX_2$, or $(C_5H_5)_2ZrX_2$ to give $\underline{cis}$-$Mo(CO)_4(PhPHSiMe_3)_2$ or $\underline{cis}$-$Mo(CO)_4(PhPH)_2Z$ (Z = $Me_2SiSiMe_2$, $Ti(C_5H_5)_2$, $Zr(C_5H_5)_2$.[75] Further reactions with $\underline{cis}$-$Mo(CO)_4(PhPHSiMe_3)$ or $Mo(CO)_4(PhPH)_2Z$ can be accomplished.

Treatment of $(CO)_5MoPh_2PCl$ with EtMgBr gives the bridged compound $(CO)_5MoPh_2PPPh_2Mo(CO)_5$.[76] $\underline{Cis}$-$[MCl_2(PPh_2Cl)_2]$ (M = Pd, Pt) reacts with one equivalent of H_2NCH_2COOMe in the presence of Et_3N to give $\underline{cis}$-$MCl_2(PPh_2)_2NCH_2COOMe$,[77] but reaction with two equivalents of H_2NCH_2COOMe gives $MCl_2(PPh_2NHCH_2COOMe)_2$. Complete aromatic substitution by chlorine of cyclometalated aromatic rings in a series of iridium triarylphosphite complexes is found to occur rapidly in benzene.[78]

2.   REACTIONS OF SULFUR LIGANDS

2.1   Cyclometalation

Generally sulfur donor ligands tend to form simple coordination complexes with transition metals so that cyclometalation only occurs with the more reactive metal substrates. An example of this is the reaction of $PhCH_2SMe$ which forms $[MCl_2(PhCH_2SMe)_2]$ (M = Pd,Pt) on attempted metalation, while reaction with $MnR(CO)_5$ (R = Me or $PhCH_2$) yields the metalated derivative $Mn(\overline{C_6H_4CH_2SMe})(CO)_4$. Cyclometalation reactions involving sulfur ligands have been reviewed by Dehand and Pfeffer[32], Bruce[79], and Omae.[80]

2.2   S–Dealkylation and S–Alkylation

The S-alkylation of coordinated sulfur donors has been shown to occur when alkyl halides are added to metal-thiolato complexes. A typical example of this type of reaction is the alkylation of the methane- or ethane-thiolato complexes of platinum(II):

$$Pt(SR)_2 + 2RI \longrightarrow Pt(SR_2)_2I_2$$
$$R = CH_3, C_2H_5$$

S-alkylation occurs more easily for nickel chelates than their palladium analogues and under normal conditions will only occur at sulfur atoms which are terminal. Kinetic studies indicate that the sulfur atom remains coordinated during the reaction. The S-alkylation of coordinated sulfur indicates that the sulfur atom retains some of the nucleophilic character present in the free mercaptide ion.

S-dealkylation reactions have been known for over ninety years but it is not until the last fifteen years that many thio-ether chelates of d^8 metal ions were observed to undergo this reaction. It has been postulated that the mechanism of S-dealkylation is similar to that proposed for the Zeisel cleavage of an ether by hydrogen halide, with the metal ion playing the role of the proton, but this mechanism has yet to be confirmed by kinetic studies. Sulfur dealkylation and alkylation reactions reported prior to 1969 have been reviewed by Lindoy.[81]

Meek and co-workers reported the synthesis and characterization of a series of arsenic-sulfur ligands and the complexes of these ligands with Pd(II)[82-83] and Pt(II).[84] These complexes undergo S-demethylation in hot dimethylformamide or hot dimethylformamide-butanol to give dimeric species. Alkylation of these dimeric compounds produces the iodo complexes or the original ligands. Bis-(S-dealkylation) reactions of bis(o-diphenylarsinophenylthio)alkanes are catalyzed by nickel(II) salts,[85] giving the first example of bis(S-dealkylation) at 30°C.

The open chain quadridentate thioether ligands 1,2-bis(o-methyl-thiophenylthio)ethane, 1,3-bis(o-methylthiophenylthio)propane, and 1,4-bis(o-methylthiophenylthio)butane form Pd(II) and Pt(II) complexes which give complicated S-dealkylation reactions.[86] With the first of these ligands three products have been identified and contain the ligands o-methylthiobenzenethiolate and benzene-1,2-dithiolate.

Several mercaptide complexes of nickel(II) and palladium(II) have been prepared[87] by S-demethylation of the corresponding complexes of the perfluoroaryl ligands 2-methylthio-3,4,5,6-tetrafluorophenyldiphenyl-phosphine and bis(2-methylthio-3,4,5,6-tetrafluorophenyl)phenylphosphine. The mono- and bis- chelated palladium(II) complexes of o-methyl-thiophenyldiphenylphosphine, bis(o-methylthiopheny)phenylphosphine and their arsenic analogues undergo S-demethylation reactions in hot dimethylformamide.[88] The bis-demethylated compounds react with a variety of alkylating agents (RCH_2Y) to give a complex of the original or a different thioether ligand.

2.3   Activation of Carbon Disulfide

Carbon disulfide complexes of transition metals react with electrophilic or nucleophilic reagents to produce two different classes of products; attack by alkyl halides gives thioalkylated complexes while reaction with triphenylphosphine leads to sulfur abstraction from one coordinated carbon disulfide and the formation of a thiocarbonyl. On complexation, carbon disulfide can be expected to become more electron rich, especially at the sulfur atoms, suggesting that alkylation can occur at both sulfur sites, although monoalkylation is the most common. Reviews by Butler and Fenster[89] and Yaneff[90] cover the area up to 1976.

2.4   Thiocarbonyl Reactions

Thiocarbonyl complexes can also undergo nucleophilic or electrophilic attack at the thiocarbonyl centre. In general terms the thiocarbonyl carbon has been observed to be more reactive to nucleophilic attack than the carbonyl carbon and, despite the greater π-acceptor capacity of CS relative to CO, if both groups are present in the same molecule the reactive site is always the CS group. This illustrates the basic point that the electrophilic nature of the thiocarbonyl carbon is greater than that of the carbonyl carbon.

The presence of a larger negative charge on sulfur and its lower electronegativity compared to oxygen are reasons to suppose that Lewis acids will form sulfur bound adducts of suitable metal thiocarbonyls under circumstances where the carbonyl ligand does not react. This illustrates the readiness of the coordinated thiocarbonyl group to undergo electrophilic reactions via the coordinated sulfur atom. Work prior to 1977 is reviewed by Yaneff.[90]

Scheme 5

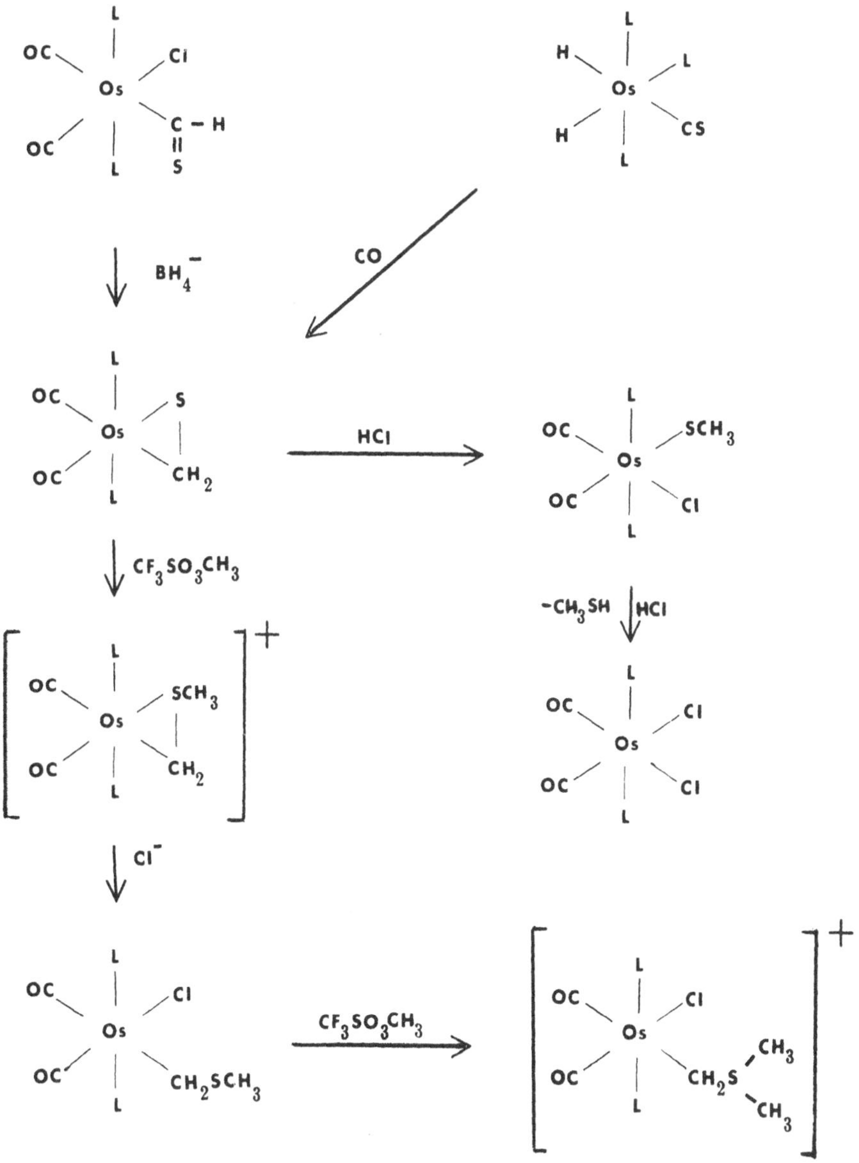

Reaction of $Ir(H)(CS)(PPh_3)_3$ with two equivalents of H_2 produces $IrH_2(SMe)(PPh_3)_3$ and further reaction with HCl releases CH_3SH and forms $[IrH_2(Cl)(PPh_3)_3]$.[91] A thioformyl complex is a likely intermediate and indeed the stable $[Ir(CHS)Cl_2(CO)(PPh_3)_2]$ is obtained by the reaction of $[IrCl_2(CS)(CO)(PPh_3)_2]^+$ with $NaBH_4$. Complexes of the type $OsHX(CS)L(PPh_3)_2$ (X = Cl, Br, L = CO; X = Cl, L = CNp-tolyl) containing mutually cis hydrido and thiocarbonyl groups undergo transfer of the hydrido group to CS and when treated with CO give $OsCl(CHS)(CO)_2(PPh_3)_2$.

A stable complex of the unstable thioformaldehyde results when $[Os(CHS)Cl(CO)_2(PPh_3)_2]$ is reacted with $NaBH_4$.[92-3] The product, $[Os(\eta^2CH_2S)(CO)_2(PPh_3)_2]$, forms methanethiolato derivatives with acids but can be alkylated at sulfur, forming methyl, thiomethyl, and dimethylsulfonium methylide complexes as shown in Scheme 5.

A four-membered osmium metallacycle is formed by the reaction of $[Os(\eta^2-SCN(R)Me)L_2(CS)(CO)]^+$ and BH_4^-, further reaction with CH_3I producing a dicarbene ligand (Scheme 6).[94]

Octahedral osmium complexes of the type $Os(CS)RLX(PPh_3)_2$ (R = p-tolyl; X = Cl, Br, I, O_2CCF_3; L = CO or CNR) with adjacent R and CS ligands undergo rearrangement to dihapto-thioacyl complexes $[Os\{\eta^2-C(S)R\}XL(PPh_3)_2]$, **VI**.[95]

Scheme 6

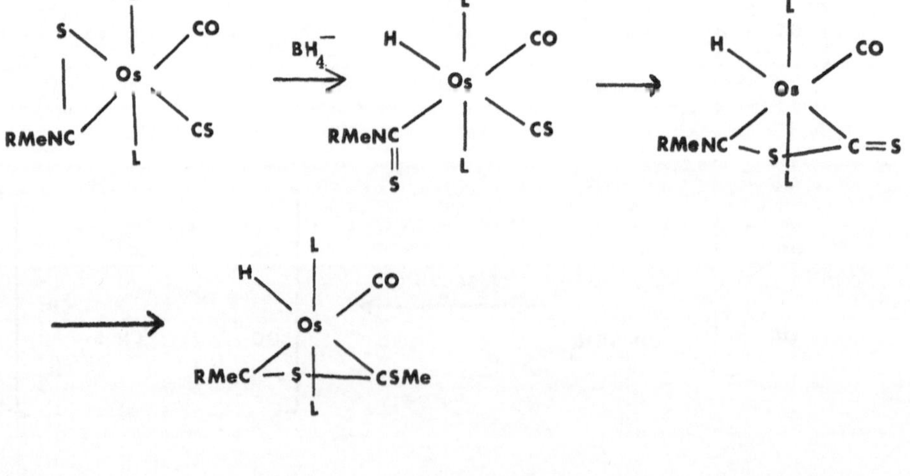

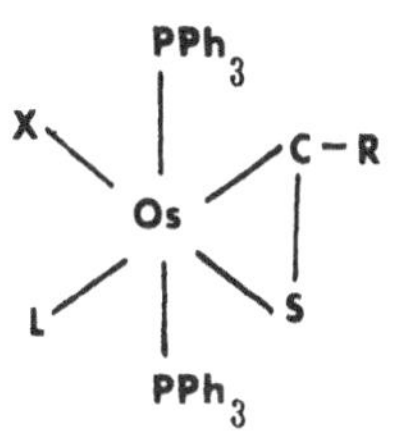

VI   VII

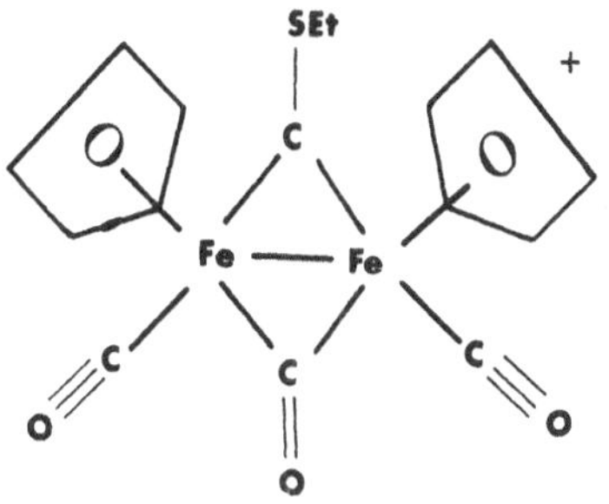

A kinetic study of carbonyl dissociation in $W(CO)_3(CS)(\underline{o}\text{-phen})$ shows labilization by the thiocarbonyl ligand.[96]

Reaction of $[CpRu(CO)_2]^-$ successively with CS_2, CH_3I and CF_3SO_3H yields the cationic thiocarbonyl complex $[CpRu(CO)_2(CS)]^+$.[97] This reacts with N_3^- or N_2H_4 to give $[CpRu(CO)_2NCS]$ and with NCO^- to give $[CpRu(CO)_2CN]$. With NaH, $[CpRu(CO)_2CS]^+$ is reduced to a mixture of $[CpRu(CO)(CS)]_2$ (major product) and $Cp_2Ru_2(CO)_3CS$. In both dimers the CS groups are bridging. The preference of CS for bridging rather than terminal positions in these molecules is rationalized in terms of the weakness of the $C\equiv S$ π-bonds, so that little π-bond stabilization is lost in moving from the terminal $-C\equiv S$ to a bridging $>C=S$ position.

The reaction of $[CpFe(CO)_2]^-$ with $(PhO)_2C=S$ in THF gives $Cp_2Fe_2(CO)_3CS$, also with a thiocarbonyl bridge.[98] The dimer consists of interconverting cisoid and transoid forms in solution. Most reagents which cleave $[CpFe(CO)_2]_2$, such as Br_2, $HgCl_2$ and O_2/HBF_4, do not give simple cleavage reactions with $Cp_2Fe_2(CO)_3CS$. On reaction with alkylating agents, stable S-alkylated cations are formed; the ethyl species has been shown to have the structure VII.

2.5 Sulfur-Assisted Electron Transfer Reactions

Inner sphere electron transfer between two metals via a sulfur bridge is well documented. The thiocyanate group has been the most studied of the sulfur containing ligands and several important points can be made. Reductions of $[Co(NH_3)_5(NCS)]^{3+}$ by Cr^{2+} have been shown to occur by an inner sphere mechanism involving remote attack at the terminal S atom. For $[Co(NH_3)_5SCN]^{3+}$ attack by the Cr^{+2} occurs at both the remote nitrogen and the adjacent sulfur at nearly comparable rates.[99-100] Electron transfer between $[Co(CN)_5]^{2-}$ and $[Co(NH_3)_5(NCS)]^{3+}$ occurs by remote attack at the S atom and ligand transfer, to yield the linkage isomer $[Co(CN)_5(SCN)]^{3-}$.[101]

Inner sphere rate enhancement factors of 35-3400 are observed for thiolate bridging over analogous alkoxide bridging in reactions of $[Cr(III)(en)_2(X-Y)]^{n+}$ [X-Y = $(H)ECH_2CH_2NH_2$, $(H)ECH_2CH_2OH$, $(H)ECH(CH_3)COO$, $(H)SCH_2COO$, (E = O, S), $CH_3SCH_2CH_2NH_2$, CH_3SCH_2COO, and H_2NCH_2COO] with $[Cr(H_2O)_6]^{2+}$ and this enhancement has been attributed to a diminished steric accessibility of the oxygen atom and perhaps a ground state elongation of the Co-N bond <u>trans</u> to an efficient sulfur bridge.[102-3] Substitution control is evident in parallel inner sphere reductions by $[V(H_2O)_6]^{2+}$. For outer sphere reductions with $[Ru(NH_3)_6]^{2+}$ a kinetic advantage for the thiolate class is also observed although at a lower reactivity level.

Rates of reductions of the cations $[Cr(en)_2(SCH_2COO)]^+$, $[Cr(en)_2(SC_6H_4COO)]^+$, $[Cr(en)_2(SCH_2CH_2NH_2)]^{2+}$ and $[Cr(en)_2(C_2O_4)]^+$ with $Cr(H_2O)_6^{2+}$ have been determined.[107] All reactions proceed by inner sphere electron transfer. Reduction of the mercaptoacetato and 2-mercaptobenzoato complexes is catalyzed by acid, whereas the reduction of the 2-mercaptoethylamine and oxalato complexes is essentially independent of acid. The acid-catalyzed redox path is interpreted in terms of a proton-induced <u>cis</u> nonbridging ligand effect. Slower reaction rates for the reduction of thiolatobis(ethylenediamine)chromium(III) complexes when compared to the reduction of thiolatobis(ethylenediamine)cobalt(III) complexes by $[Cr(H_2O)_6]^{2+}$ illustrate the remarkable bridging efficiency of thiolato sulfur in Cr(II)-Cr(III) reactions. This kinetic difference may arise from the fact that thiolato sulfur induces a ground state <u>trans</u> effect in cobalt(III) complexes but not in the analogous chromium(III) complex. Monodentate thiolato Cr(III) products of inner sphere redox reactions undergo hydrolysis or chelate ring closure.[105-6] Spontaneous ring closure of Cr(III) with mercaptoacetate, 2-mercaptoethylamine, and 2-mercapto-propionate occurs rapidly and is subject to acid catalysis. Excess Cr(II) catalyzes the ring closure. Equilibrium constants and rates pertaining to the cleavage and re-formation of the Cr(III)-sulfur bond have been determined:

$$Cr(H_2O)_4(\text{chelate})^{n+} + H_3O^+ \rightleftharpoons Cr(H_2O)_5(\text{chelate} + H)^{(n+1)+}$$

The cleavage pathway is predominantly first order in acid, involving protonation of the coordinated thiolate function according to the following mechanism:

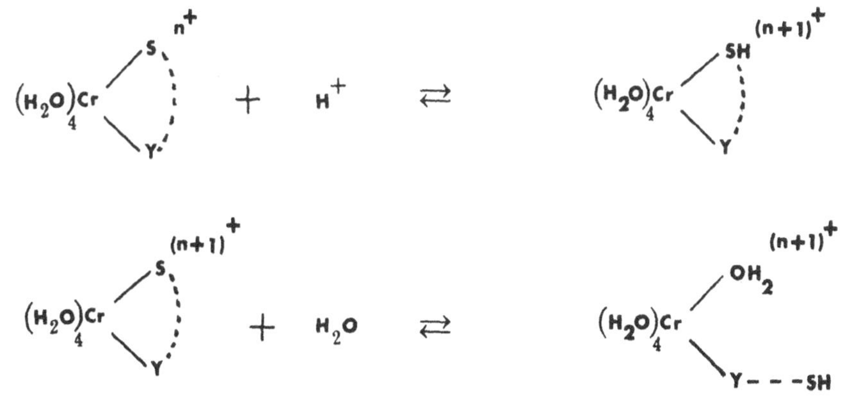

A real acid-independent cleavage is also possible, and has been suggested for 2-mercaptoethylamine and mercaptoacetate. Two mechanistic pathways are proposed, intramolecular protonation of the sulfur followed by chelate ring opening, and chromium-sulfur bond breaking followed by proton abstraction from water. Further clarification is needed to distinguish between the two mechanisms.

The rates of Cr(II) reduction of a series of twelve complexes of type $[(en)_2Co(S(R)CH_2CH_2NH_2)]^{3+}$ have been measured and the reduction is a simple second order process.[107] Detection and characterization of the relatively labile thioether Cr(III) products as well as reactivity patterns show conclusively that the thioether sulfur functions as an electron transfer bridge. The rates of reduction depend primarily on the steric bulk

Scheme 7

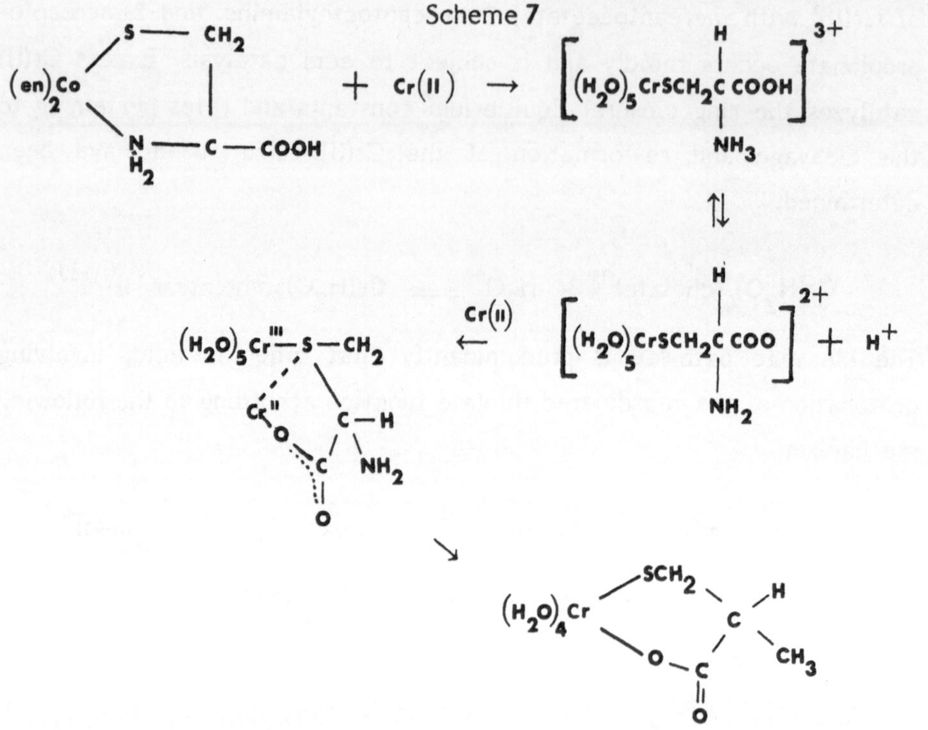

of R. The fact that coordinated sulfur retains a high degree of Lewis basicity is shown by the fact that $[(en)_2Co(SCH_2CH_2NH_2)]^{2+}$ and $[(en)_2Co(SCH_2COO)]^+$ form complexes with the soft acceptors Ag^+, Cu^+ and CH_3Hg^+.[108-9] The affinity of coordinated thiols for soft metal centers is less than that of free thiols but similar to free thioethers. Thus the number of covalent bonding interactions involving sulfur primarily determines its Lewis basicity.

The Cr^{2+} reductions of $[(en)_2Co(methionine)]^{2+}$ and $[(en)_2Co(methylcysteine)]^{2+}$ occur by attack of the Cr^{2+} at the O atom in these O,N-bonded chelates.[110] Attack by Cr^{+2} in $[(en)_2Co(cysteine)]^{2+}$, which is sulfur and nitrogen-bonded, occurs at the sulfur atom according to the mechanism of Scheme 7.

The reduction of the symmetrical cis-dichloro(1,8-diamino-3,6-dithiaoctane)cobalt(III) perchlorate by Fe^{2+} occurs 10^3 times more rapidly than that of cis-$[Co(en)_2Cl_2]^+$ but the inner sphere electron transfer is believed to occur through a chloride bridge.[111] An enhanced rate is also observed for the reduction of cis-oxalato(1,8-diamino-3,6-dithia-octane)cobalt(III) by Fe^{2+} when compared to $[cis-(en)_2Co(C_2O_4)]^+$.[112] Both of these observations suggest that thioether groups positioned trans to a bridging chloride can dramatically enhance the rate. This enhancement is not observed when symmetrical cis-diazido(1,8-diamino-3,6-dithiaoctane)cobalt(III) is reduced by Fe^{+2}.[113]

2.6   Trans Influence and Trans Effect in Sulfur Donors

Substitution reactions involving complexes containing sulfur donor atoms have shown that sulfur exhibits a large trans effect. Thus, substitution of trans-$[Co(NH_3)_5SO_3]^+$ with $^{15}NH_3$ and of cis-$[Co(NH_3)_4(SO_3)_2]^-$ with CN^-, SO_3^{2-}, and NCS^- confirms the specific trans-labilizing influence of the sulfito ligand.[114]

Above pH 12, the substitution of trans-$[Co(en)_2(SO_3)OH]$ by SO_3^{-2} to form trans-$[Co(en)_2(SO_3)_2]^-$ is reversible.[115] The rate law is consistent with a reversible two step limiting S_N1 mechanism. At pH 8.1 the substitution is virtually complete and the rate law is consistent with an S_N1 mechanism. The labilizing effect of the sulfito ligand arises from the unusually low $\Delta H^\ddagger$ values for the dissociative release of the ligand trans to the sulfito group.

The crystal structure of $[Co(NH_3)_5SO_3]Cl.H_2O$ shows the weakening of the Co-N bond trans to the sulfur.[116] Thus the Co-N bond length trans to NH_3 is 1.966Å while the Co-N bond length trans to SO_3^{-2} is 2.055Å. Crystal structures of the salts $[Co(en)_2(SCH_2COO)]Cl.H_2O$, $[Co(en)_2(SCH_2CH_2NH_2)](SCN)_2$[117] and $[Co(en)_2(O_2SCH_2CH_2NH_2)](ClO_4)(NO_3)$[118] also show a lengthening of the Co-N trans to coordinated sulfur. However the lengthening does not appear in $[Cr(en)_2(SCH_2COO)]ClO_4$.[117] The crystal structure of $[Co(3,7-dithianonane-1,9-diamine)(NO_2)Cl]Cl$, readily formed from $[Co(3,7-dithianonane-1,9-diamine (NO_2)_2]ClO_4$ and an aqueous chloride, has shown that the nitro groups trans to the sulfur atom has been displaced, indicating a strong trans influence for the sulfur.[119]

Spontaneous aquation of symmetrical cis-[Co(1,8-diamino-3,6-dithiaoctane)(Cl)H$_2$O]$^{2+}$ under a variety of conditions proceeds with retention of configuration and when compared to [Co(en)$_2$ClH$_2$O]$^{+2}$ shows the unique deactivation role thioether bond atoms play when positioned trans to reactive sites in the complex.[120] This is due to the fact that the secondary nitrogen N-H groups of trien have been replaced by two thioether C-S-C donors. This substitution imparts major constraints on the complexes' ability to isomerize geometrically. Perhaps the larger sulfur donors impose a steric requirement which locks the configuration of adjacent chelate rings. This limited flexing considerably reduces the exchange or movement of solvent molecules contained in the primary solvent shell.

Rates of ligation of trans-(p-tolylsulfinato-S)methanol bis-(dimethylglyoximato)cobalt(III) by a variety of ligands to produce the respective trans ligated complexes have been studied.[121] All reactions are first order in the cobalt complex and are interpreted in terms of a limiting S$_N$1 mechanism. Similar results were obtained for other trans substituted methanol bis(dimethylglyoximato)cobalt(III) complexes of the type RCo(dmgH)$_2$HOCH$_3$ [R = CH$_3$, C$_6$H$_5$, SO$_3$, (CH$_3$O)$_2$P(O)].[122] Labilizing ability of R groups studied is CH$_3$ > C$_6$H$_5$ > SO$_3^-$ > (CH$_3$O)$_2$P(O) > CH$_3$C$_6$H$_4$SO$_2$, where the relative rates of dissociation are 900:300:100:50:1, respectively.

The structure of arylsulfinatopentaamminecobalt(III) complexes shows that the Co-N bond trans to sulfur is significantly longer than the average cis Co-N bond in the complex, and substitution reactions are governed by labilization of the site trans to sulfur.[123]

2.7   Oxidation of Coordinated Sulfur

Oxidation of coordinated sulfur to sulfenato and sulfinato groups can be accomplished without rupturing the metal-sulfur bond. Oxidation of cysteinato-N,S-bis(ethylenediamine)cobalt(III) perchlorate with H$_2$O$_2$ gives the S-bonded isolable complexes [Co(en)$_2${H$_2$NCH(COO)CH$_2$S(O)}]ClO$_4$ and [Co(en)$_2${H$_2$NCH(COO)CH$_2$SO$_2$}]ClO$_4$.[124] Oxidation of other mercapto-amines with H$_2$O$_2$ also produces sulfenato and sulfinato S-bonded

complexes and the mechanism of the reactions has been discussed in terms
of nucleophilic attack by the coordinated sulfur on the O-O bond of
H_2O_2.[125-6]

Reactions of the S-bridged species $[Ni\{Co(SCH_2CH_2NH_2)_3\}_2]Br_2$ or
$[Pb\{Co(SCH_2CH_2NH_2)_3\}_2]$ with H_2O_2 gives $[Co(NH_2CH_2CH_2SO_2)_3]$
containing the chelate S-bonded sulfinato group.[127] The yellow anion
(+)tris-(cysteinesulfinato-(2-)SN)cobaltate(III), $[Co(L-cysn)_3]^{-3}$, is readily
obtained from L-cysteine by the following rapid steroselective reaction[128]

$$[Co(NH_3)_6]Cl_3 + 3 \text{ cysteine} + 3 \text{ KOH} \xrightarrow{H_2O_2} K_3Co(L-cystn)_3$$

Bridging sulfur can also be oxidized to the corresponding oxo
derivative. Thus oxidation of $Pd_2(dppm)_2(\mu\text{-S})Cl_2$ with m-chloroperbenzoic
acid at -60°C gives $Pd_2(dppm)_2(\mu\text{-SO}_2)Cl_2$. No SO-bridged intermediate
was isolated.[129] Oxidation of coordinated mercapto groups by H_2O_2 can
give rise to disulfides.[130] Oxidation of $[Cr(en)_2(SCH_2COO)]^+$ by one
equivalent of Np^{6+} or Ce^{4+} does not give the sulfenato or sulfinato
derivative but gives the monothiooxalato complex:[131]

$$\left[(en)Cr \underset{O-C=O}{\overset{S-C=O}{\diagdown}} \right]^+$$

Oxidations of $[Cr(en)_2(SC_6H_4COO)]^+$ and $[Cr(en)_2(OH_2)(OOCCH_2SH)]^{2+}$
with Np^{+6} consume 1 equivalent of Np^{+6} per mole of complex and can be
considered to give disulfides.[132] The cations $[Cr(en)_2(SCH_2COO)]^+$,
$[Co(en)_2(SCH_2COO)]^+$, $[Cr(en)_2(OH_2)(SCH_2CH_2NH_2)]^{2+}$, $[Co(en)_2(SCH_2-CH_2NH_2)]^{2+}$ and $[Cr(en)_2(SCH_2CH_2COO)]^+$ undergo multiequivalent
multistep oxidations often involving oxidation of the thiol carbon backbone.
Chromium(III) thiolato complexes are oxidized much more readily than the
analogous cobalt(III) complexes. Reactions of mercaptoamine complexes of
cobalt(III) with Np^{+6} (or Co^{+3}aq) in aqueous perchloric acid gives 2-
aminoethyl-N-2-ammonioethyldisulfide-S complexes **VIII**:[133-4]

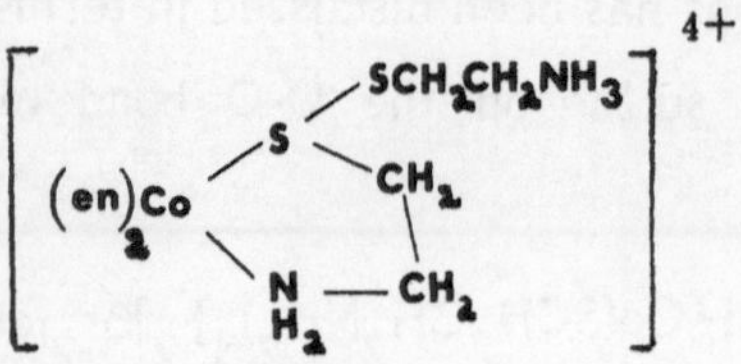

VIII 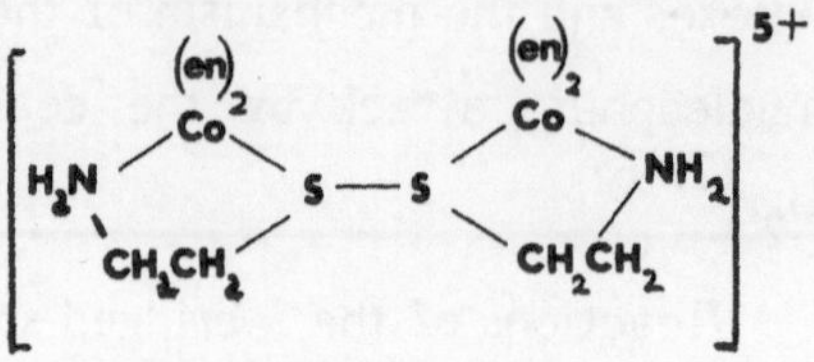   IX

The proposed mechanism involves initial one-equivalent oxidation of the coordinated thiol and reaction of the resultant coordinated thiol radical with an additional equivalent of complex to form a stable radical ion dimer IX, followed by internal electron transfer within IX to give $Co^{+2}aq$ and VIII. Similar mechanisms may operate in the oxidation of S-R to R-S-S-R by the copper in caeruplasmin.[135]

2.8   Reactions Involving Disulfide Cleavage

Cleavage of organic disulfides upon reaction with metal ions has been observed.[81] Thus reactions of disulfide with $Pt(PPh_3)_4$ gives complexes of the type $\underline{cis}$-$(Ph_3P)_2Pt(SR)_2$.[136] These complexes can be converted to $Pt(PPh_3)_2X_2$ and RSH or $RSCH_3$ by the reaction of HCl or CH_3I respectively. Chromium(II) reduction of p-aminophenyl disulfide in aqueous acid gives $[Cr(H_2O)_5SC_6H_4NH_2]^{2+}$ by disulfide cleavage.[137] It is well known that mercaptides react reversibly with Cu(II) to produce Cu(I) and disulfides:

$$2Cu^{2+} + 2RS^- \rightleftharpoons 2Cu^+ + RSSR$$

The position of this equilibrium is dependent on the valence specificity of the coordination environment available to copper. The reactions of bis[2-(2-pyridyl)ethyl]disulfide or bis[2-(N,N-dimethylamino)ethyl]disulfide with Cu^{2+} gives complexes of the disulfides containing Cu^+. Presumably this occurs by cleavage of the disulfide S-S bond by OH^- followed by oxidation of the mercaptide moiety to disulfide, reduction of Cu^{2+} to Cu^+, and complexation with Cu^+.[138-9]

2.9 Reactions involving dithiocarbamates, xanthates and dithiophosphates

Reactions of Pt or Pd complexes of the type $M(S-S)_2$ [S-S = S_2CNR_2, R = Me, Et; S_2COR, R = Et, CH_2Ph; $S_2P(OEt)_2$ and S_2PR_2, R = Me, Et, Ph] with tertiary phosphines occurs by a stepwise cleavage of the metal-sulfur bonds to generate 4-coordinated compounds of formulae $[M(S-S)_2PR'_3]$ and $[M(S-S)(PR'_3)_2](S-S)$ with unidentate and ionic/bidentate coordination respectively.[140] All of the ionic compounds revert to the $[M(S-S)_2PR'_3]$ complex in non-polar solvents. Variable temperature NMR studies on the unidentate-bidentate exchange at ambient temperature for $[Pt(S_2PR_2)_2PR'_3]$ and $[Pt(S_2PR_2)(PR'_3)_2](S_2PR_2)$ suggest a concerted mechanism with both bond-making and bond-breaking steps important.[141] Nucleophilic attack can also occur on a coordinated alkoxy group to give the compounds $[(R'_3P)_2M(S_2CO)]$ and $[(R'_3P)_2MS_2P(O)OEt]$, M = Pt,Pd. For $[M(S-S)(PR'_3)_2](S-S)$ compounds containing S_2CNR_2, the presence of excess PR'_3 catalyzes the reaction between the dithiocarbate ion and dichloromethane giving $CH_2(S_2CNR_2)_2$ and $[M(S_2CNR_2)(PR'_3)_2Cl.H_2O]$.

Reactions of $[Pt(S_2COR)_2]$ with $K[S_2COR]$, followed by the addition of Ph_4AsCl, gives $Ph_4As[Pt(S_2COR)_3]$ which undergoes rapid unidentate-bidentate exchange at ambient temperature.[81] Attempted recrystallization from CH_2Cl_2 or $CDCl_3$ results in an intramolecular rearrangement to give $[Ph_4As][Pt(S_2CO)(S_2COR)]$ and RS_2COR.[142-3] Reaction of $[Pd(S_2COEt)]$ with $K[S_2COEt]$ and Ph_4AsCl gives $[Ph_4AS][Pd(S_2CO)(S_2COEt)]$ as the main product. Reaction of $[Pt(S_2COR)_2]$ with $K[S_2COR]$ and Ph_4AsCl gives $[Ph_4As][Pt(S_2CO)(S_2COMe)]$ or $[Ph_4As]_2[Pt(S_2CO)_2]$, both of which give $[PtL_2S_2CO]$ on treatment with a variety of Lewis bases.

The reaction of $Ni[S_2P(OMe)_2]_2$ with $Ph_2P(CH_2)_2AsPh_2$ in benzene/CH_2Cl_2 gives $[Ni(S_2P(O)OMe)Ph_2P(CH_2)_2AsPh_2]$.[144] The kinetics and mechanism of the reaction of the cobalt analogue $[Co(S_2P(OMe)_2]$ with $Ph_2P(CH_2)_2PPh_2$ in CH_2Cl_2 have been investigated.[145] The second step of the reaction is

$$[Co(S_2P(OMe)_2)_2(dppe)] + [S_2P(OMe)_2]^- \rightarrow [Co(S_2P(OMe)_2)(S_2P(O)OMe)dppe]$$
$$+ \quad P(OMe)_2S(SMe)$$

It is probable that this step occurs by nucleophilic attack of the P-O-C carbon atom on the coordinated $S_2P(OMe)_2^-$.

Reactions of the sodium salt of N,N-di(2-N,N-diethylamino-ethyl)dithiocarbamate (Et_4dien-dtc) with $Na[Au(SCN)_4]$ in CH_3CN resulted in the formation of the linkage isomers:[146] (Et_4dien-dtc)Au(SCN)$_2$, chelated via sulfur atoms, and (dtc-Et_4dien)Au(NCS)(SCN), chelated via nitrogen atoms. The reaction of N,N-diethyldithiocarbamatogold(I) with Br_2, I_2 or $(SCN)_2$ in CS_2 at -78°C permitted isolation of Au(II) complexes of type X.
When warmed to room temperature the complexes rearrange to $[Au(dtc)_2]$ $[AuX(CN)_2]$.

The complex $[(CO)_4Re(\mu\text{-}S_2CS)Re(CO)_5]$ is obtained when $NaRe(CO)_5$ is treated with CS_2 and $ReBr(CO)_5$.[147] Analogous species $[Mn(CO)_4(\mu\text{-}S_2CS)Re(CO)_5]$ and $[(CO)_4Re(\mu\text{-}S_2CS)Mn(CO)_5]$, are prepared similarly.

Addition of sulfur to $[N(PPh_3)_2][Mn(CO)_5]$ gives $[N(PPh_3)_2]$ $[MnS_n(CO)_5]$ which reacts with CS_2 to give $[N(PPh_3)_2][Mn(CO)_4(S_2CS)]$. Methylation of $[N(PPh_3)_2][Mn(CO)_4S_2CS]$ with MeI or $SFO_2(OMe)$ gives $[Mn(CO)_4S_2CSMe]$ in 40% yield.

The reaction of $[CpMn(CO)_2(C_8H_{14})]$ with R_3PCS (R = n-Bu, C_6H_{11}) gives $(Cp)Mn(CO)_2(CS)$ in poor yield.[148]

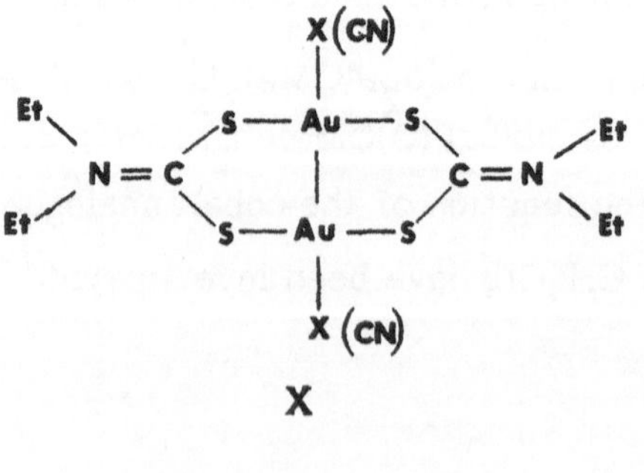

X

2.10   Reactions of Coordinated Disulfur

Only a few reactions of coordinated disulfur have been reported. Alkylation of Cp_2NbS_2Cl with methyl iodide gives CH_3SSCH_3 and Cp_2NbI_2Cl.[149] Reactions of $MoOS_2(S_2CNR_2)_2$ with a variety of nucleophiles are summarized:[150]

$$MoOS_2(S_2CNR_2)_2$$

$P(OC_2H_5)_3$ → $S=P(OC_2H_5)_3 + MoO(S_2CNR_2)_2$

PO_3^- → $S=PO_3^- + MoO(S_2CNR_2)_2$

CH_3NC → $CH_3CNS + MoO(S_2CNR_2)_2$

CN^- → $SCN^- + MoO(S_2CNR_2)_2$

SO_3^{-2} → $S_2O_3^{-2} + MoO(S_2CNR_2)_2$

$C_6H_5S^-$ → $C_6H_5SSC_6H_5 + Mo_2O_2S_2(S_2CNR_2)_2$
$+ C_6H_5SSSC_6H_5$

C_6H_5SH → $C_6H_5SSC_6H_5 + C_6H_5SSSC_6H_5$
$+ Mo_2O_2S_2(S_2CNR_2)_2$

$MoOS_2(S_2CNR_2)_2$ does not react with CH_3I or HCl. Reaction with CH_3OSO_2F gives $Mo[O(SSCH_3)(S_2CNR_2)_2]SO_3F$. $IrS_2(dppe)_2Cl$ does not react with common thiophiles under conditions similar to those used for $MoOS_2(S_2CNR_2)_2$. A very slow reaction of $IrS_2(dppe)_2Cl$ with CH_3SO_3F is observed to give $[Ir(SSCH_3)(dppe)_2]SO_3F$.

REFERENCES

1. "Transition Metal Complexes of Phosphorus, Arsenic and Antimony Donor Ligands", C.A. McAuliffe, ed., Macmillan, London, 1973.

2. W. Levason and C.A. McAuliffe, Adv. Inorg. Chem. Radiochem. 14:173 (1972).

3. G. Booth, Adv. Inorg. Chem. Radiochem. 6:1 (1964).

4. G. Booth, Organic Phosphorus Compounds. 1:434 (1970).

5. W. Levason and C.A. McAuliffe, Coord. Chem. Rev. 19:173 (1976).

6. J.F. Nixon, Adv. Inorg. Chem. Radiochem. 13:363 (1972).

7. W.R. Cullen, Adv. Inorg. Chem. Radiochem. 15:323 (1972).

8. J.R. Verkade and K.J. Koskran, Organic Phosphorus Compounds, 2:1 (1970).

9. T.A. Manuel, Adv. Organomet. Chem. 3:181 (1965).

10. E.L. Muetterties, "Transition Metal Hydrides", Marcel Dekker, New York (1971).

11. L. Malatesta and S. Cenini, "Zerovalent Compounds of Metals", Academic Press, New York (1974).

12. B.O. West, Rec. Chem. Progr. 30, 249 (1969).

13. L.R. Smith and J.L. Mills, J. Organomet. Chem. 81:1 (1975).

14. D.G. Holah, A.N. Hughes and K. Wright, Coord. Chem. Rev. 15: 239 (1975).

15. D.I. Hall, J.H. Ling and R.S. Nyholm, Struct. Bonding. 15:3 (1973).

16. P. Rigo and A. Turco, Coord. Chem. Rev. 13:133 (1974).

17. H.D. Kaesz and R.B. Saillant, Chem. Rev. 72:231 (1972).

18. J.P. Collman and W.R. Roper, Adv. Organomet. Chem. 7:54 (1968).

19. D.J. Cardin, B. Cetinkaya and M.F. Lappert, Chem. Rev. 72:545 (1972).

20. J.F. Nixon and A. Pidcock, Ann. Rev. NMR Spectroscopy 2:345 (1969).

21. D.E.C. Corbridge, "The Structural Chemistry of Phosphorus", Chapter 11, Elsevier, London (1974).

22. J.H. Enemark and R.D. Feltham, Coord. Chem. Rev. 13:339 (1974).

23. N.G. Connelly, Inorg. Chim. Acta Rev. 6:47 (1972).

24. C.S. Kraihanzel, J. Organomet. Chem. 73:173 (1974).

25. J.S. Valentine, Chem. Rev. 73:235 (1973).

26. C.A. McAuliffe, Adv. Inorg. Chem. Radiochem. 17:165 (1975).

27. L.M. Venanzi, Angew. Chem. Int. Ed. Eng. 3:453 (1964).

28. G. Dolcetti and N.W. Hoffman, Inorg. Chim. Acta 9:269 (1974).

29. R.E. Harmon, S.K. Gupta and D.J. Brown, Chem. Rev. 73:21 (1973).

30. C.A. McAuliffe and W. Levason, "Phosphine Arsine and Stibine Complexes of the Transition Elements", Elsevier, Amsterdam (1979).

31. M.I. Bruce, Angew. Chem. Int. Ed. Eng. 16:73 (1977).

32. J. Dehand and M. Pfeffer, Coord. Chem. Rev. 18:327 (1976).

33. N.A. Al-Salem, H.D. Empsall, R. Markam, B.L. Shaw and B. Weeks, J.C.S. Dalton Trans. 1972 (1979).

34. H.D. Empsall, P.N. Heys and B.L. Shaw, J.C.S. Dalton Trans. 257 (1978).

35. B.D. Dombek, J. Organomet. Chem. 169:315 (1979).

36. R.G. Goel and R.G. Monkmayor, Inorg. Chem. 16:2183 (1977).

37. H.C. Clark, A.B. Goel, R.G. Goel, S. Goel and W.O. Ogini, Inorg. Chim. Acta 31:L441 (1978).

38. H.C. Clark, A.B. Goel, R.G. Goel and W.O. Ogini, J. Organomet. Chem. 157:C16 (1978).

39. M.A. Bennett and P.W. Clark, J. Organomet. Chem. 110:367 (1976).

40.   M.A. Bennett, R.N. Johnson, G.B. Robertson, I.B. Tomkins and B.O. Whimp, J. Am. Chem. Soc. 98:3514 (1976).

41.   T. Akariya and A. Yamamoto, J. Organomet. Chem. 118:65 (1976).

42.   R.P. Stewart, L.R. Isbrandt, J.J. Benedict and J.G. Palmer, J. Am. Chem. Soc. 98:3215 (1976).

43.   P.E. Antle and C.A. Tolman, J. Organomet. Chem. 159:C5 (1978).

44.   C.A. Tolman, A.D. English, S.D. Ittel and J.P. Jesson, Inorg. Chem. 17:2374 (1978).

45.   S.D. Ittel, C.A. Tolman and P.J. Krusic, Inorg. Chem. 17:3432 (1978).

46.   S.M. Grant and A.R. Manning, J.C.S. Dalton Trans. 1789 (1979).

47.   P.W. Clark, J. Organomet. Chem. 110:C13 (1976).

48.   P.W. Clark, J. Organomet. Chem. 137:235 (1977).

49.   C. Crocker, R.J. Errington, R. Markam, C.J. Moulton, K.J. Odell and B.L. Shaw, J. Am. Chem. Soc. 102:4373 (1980).

50.   S.J. Thompson, C. White and P.M. Maitlis, J. Organomet. Chem. 136:87 (1977).

51.   M.A. Bennett, R.N. Johnson and I.B. Tomkins, J. Organomet. Chem. 128:73 (1977).

52.   M. Nolte, E. Singleton and E. Van der Stok, J. Organomet. Chem. 142:387 (1977).

53.   S. Heitkamp, D.J. Stufkens and K. Vrieze, J. Organomet. Chem. 122:419 (1976).

54.   S. Heitkamp, D.J. Stufkens and K. Vrieze, J. Organomet. Chem. 134:95 (1977).

55.   S. Keitkamp, D.J. Stufkens and K. Vrieze, J. Organomet. Chem. 139:189 (1977).

56.   E. Singleton and E. Van der Stock, J.C.S. Dalton Trans. 926 (1978).

57.   D.J.A. De Waal, E. Singleton and E. Van der Stock, J.C.S. Chem. Commun. 22:1007 (1978).

58.   M.S. Bennett, R.N. Johnson and I.B. Tomkins, J. Organomet. Chem. 118:205 (1976).

59.   P.W. Clark and G.E. Hartnell, J. Organomet. Chem. 139:385 (1977).

60.   A.J. Deeming, J. Organomet. Chem. 128:63 (1977).

61.   M.A. Bennett, R.N. Johnson and I.B. Tomkins, J. Organomet. Chem. 133:231 (1977).

62.   M.I. Bruce, R.C.F. Gardner, J.A.K. Howard, F.G.A. Stone, M. Welling and P. Woodward, J.C.S. Dalton Trans. 621 (1977).

63.   D.K. Johnson and A.J. Carty, J.C.S. Chem. Commun. 903 (1977).

64.   A.J. Carty, D.K. Johnson and S.E. Jacobson, J. Am. Chem. Soc. 101:5612 (1979).

65.   A.J. Carty, N.J. Taylor and D.K. Johnson, J. Am. Chem. Soc. 101:5422 (1979).

66.   A.J. Carty, T.W. Ng and G.J. Palenik, cited in Ref. 65.

67.   W.R. Cullen, B.R. James and G. Strukul, Inorg. Chem. 17:484 (1978).

68.   I.J. Harrie and F.J. McQuillin, J.C.S. Chem. Commun. 369 (1976).

69.   G. Speier, Inorg. Chim. Acta 32:139 (1979).

70.   K. Yasufuku, A. Hamada, K. Aoki and H. Yamazaki, J. Am. Chem. Soc. 102:4363 (1980).

71.   P.J. Toscano and L.G. Marzilli, Inorg. Chem. 18:421 (1979).

72.   J. Grosse and R. Schmutzler, J.C.S. Dalton Trans. 405 (1976).

73.   W. Klani, H. Neukomm, H. Werner and G. Huttner, Chem. Ber. 110:2283 (1977).

74.   O. Stelzer and E. Unger, Chem. Ber. 110:3430 (1977).

75.   G. Johannsen and O. Stelzer, Chem. Ber. 110:3438 (1977).

76.   C.M. Bartish and C.S. Kraihanzel, J. Organomet. Chem. 112:C31
       (1976).

77.   P.W. Lednor, W. Beck and G. Thiel, Inorg. Chim. Acta 20:L11
       (1976).

78.   M.J. Nolte, E. Singleton and E. van der Stok, J.C.S. Chem.
       Commun. 973 (1978).

79.   M.I. Bruce, reference 31, especially section 3.3.6.

80.   I. Omae, Coord. Chem. Rev. 28:97 (1979).

81.   I.F. Lindoy, Coord. Chem. Rev. 4:41 (1969).

82.   R.L. Dutta, D.W. Meek and D.H. Busch, Inorg. Chem. 9:1215 (1970).

83.   R.L. Dutta, D.W. Meek and D.H. Busch, Inorg. Chem. 9:2098 (1970).

84.   R.L. Dutta, D.W. Meek and D.H. Busch, Inorg. Chem. 10:1820
       (1971).

85.   C.A. McAuliffe, Inorg. Chem. 12:2477 (1973).

86.   W. Levason, C.A. McAuliffe and S.G. Murray, J.C.S. Dalton Trans.
       1566 (1975).

87.   P.G. Eller, J.M. Riker and D.W. Meek, J. Am. Chem. Soc. 95:3540
       (1973).

88.   T.N. Lockyer, Aust. J. Chem. 27:259 (1974).

89.   I.S. Butler and A.E. Fenster, J. Organomet. Chem. 66:161 (1974).

90.   P.V. Yaneff, Coord. Chem. Rev. 23:183 (1977).

91.   W.R. Roper and K.G. Town, J.C.S. Chem. Commun. 781 (1977).

92.   T.J. Collins and W.R. Roper, J. Organomet. Chem. 159:73 (1978).

93.   T.J. Collins and W.R. Roper, J.C.S. Chem. Commun. 901 (1977).

94. G.R. Clark, T.J. Collins, D. Hall, S.M. James and W.R. Roper, J. Organomet. Chem. 141:C5 (1977).

95. G.R. Clark, T.J. Collins, K. Marsden and W.R. Roper, J. Organomet. Chem. 157:C23 (1978).

96. R.A. Pickering and R.J. Angelici, Inorg. Chem. 18:2035 (1978).

97. T.A. Wnuk and R.J. Angelici, Inorg. Chem. 16:1173 (1977).

98. R.E. Wagner, R.A. Jacobson, R.J. Angelici and M.H. Quick, J. Organomet. Chem. 148:C35 (1978).

99. C.J. Shea and A. Haim, J. Am. Chem. Soc. 93:3055 (1971).

100. C.J. Shea and A. Haim, J. Am. Chem. Soc. 96:2635 (1974).

101. C.J. Shea and A. Haim, Inorg. Chem. 12:3013 (1973).

102. R.H. Lane, F.A. Sedor, M.J. Gilroy, P.F. Eisenhardt, J.P. Bennett, R.X. Ewall and L.E. Bennett, Inorg. Chem. 16:93 (1977).

103. R.H. Lane and L.E. Bennett, J. Am. Chem. Soc. 92:1089 (1970).

104. C.J. Weschler and E. Deutsch, Inorg. Chem. 15:139 (1976).

105. R.H. Lane and L.E. Bennett, J.C.S. Chem. Commun. 491 (1971).

106. R.H. Lane, F.A. Sedor, M.J. Gilroy and L.E. Bennett, Inorg. Chem. 16:102 (1977).

107. G.J. Kennard and E. Deutsch, Inorg. Chem. 17:2225 (1978).

108. M.J. Heeg, R.C. Elder and E. Deutsch, Inorg. Chem. 18:2036 (1979).

109. J.K. Farr and R.H. Lane, J.C.S. Chem. Commun. 153 (1977).

110. B.J. Balahura and N.A. Lewis, Inorg. Chem. 16:2213 (1977).

111. J.H. Worrell and T.A. Jackman, J. Am. Chem. Soc. 93:1044 (1971).

112. J.H. Worrell, R.A. Goddard, E.M. Gupton, jr. and T.A. Jackman, Inorg. Chem. 11:2734 (1972).

113. J.H. Worrell, R.A. Goddard and R. Blanco, Inorg. Chem. 17:3308 (1978).

114. L. Richards and J. Halpern, Inorg. Chem. 15:2571 (1976).

115. D.R. Stranks and J.K. Yandell, Inorg. Chem. 9:751 (1970).

116. R.C. Elder and M. Trkula, J. Am. Chem. Soc. 96:2635 (1974).

117. R.C. Elder, L.R. Florian, R.E. Lake and A.M. Yacynych, Inorg. Chem. 12:2690 (1973).

118. B.A. Lange, K. Libson, E. Deutsch and R.C. Elder, Inorg. Chem. 15:2985 (1976).

119. R.W. Hay, P.M. Gidney and G.A. Lawrance, J.C.S. Dalton Trans. 779 (1975).

120. J.H. Worrell, Inorg. Chem. 14:1699 (1975).

121. J.M. Palmer and E. Deutsch, Inorg. Chem. 14:17 (1975).

122. L. Seibler and E. Deutsch, Inorg. Chem. 16:2273 (1977).

123. R.C. Elder, M.J. Heeg, M.D. Paine, M. Trkula and E. Deutsch, Inorg. Chem. 17:431 (1978).

124. C.P. Sloan and J.H. Krueger, Inorg. Chem. 14:1481 (1975).

125. E. Deutsch, I. Kofi Adzamli, J.D. Lydon, D. Nosco and M. Root Abstr. Pap. ACS April Mtg. 324 (1979).

126. I. Kofi Adzamli, K. Libson, J.D. Lydon, R.C. Elder and E. Deutsch, Inorg. Chem. 18:303 (1979).

127. P.R. Butler and E.L. Blinn, Inorg. Chem. 17:2039 (1978).

128. L.S. Dollimore and R.D. Gillard, J.C.S. Dalton Trans. 933 (1973).

129. A.L. Balch. L.S. Benner and M.M. Olmstead, Inorg. Chem. 18:2996 (1979).

130. P.E. Riley and K. Seff, Inorg. Chem. 11:2993 (1972).

131. C.J. Weschler, J.C. Sullivan and E. Deutsch, J. Am. Chem. Soc. 95:2720 (1973).

132. C.J. Weschler, J.C. Sullivan and E. Deutsch, Inorg. Chem. 13:2360 (1974).

133.   M. Woods, J.C. Sullivan and E. Deutsch, J.C.S. Chem. Commun. 749 (1975).

134.   M. Woods, J. Karbwang, J.C. Sullivan and E. Deutsch, Inorg. Chem. 15:1678 (1976).

135.   W. Buyers, G. Curzon, K. Garbett, B.E. Speyer, S.N. Young and R.J.P. Williams, Biochem. Biophys. Acta 310:38 (1973).

136.   R. Zanella, R. Ros and M. Graziani, Inorg. Chem. 12:2736 (1973).

137.   L.E. Asher and E. Deutsch, Inorg. Chem. 11:2927 (1972).

138.   T. Ottersen, L.G. Warner and K. Seff, Inorg. Chem. 13:1904 (1974).

139.   C.L.G. Warner, T. Ottersen and K. Seff, Inorg. Chem. 13:2819 (1974).

140.   J.M.C. Alison and T.A. Stephenson, J.C.S. Dalton Trans. 254 (1973).

141.   D.F. Steele and T.A. Stephenson, J.C.S. Dalton Trans. 2124 (1973).

142.   M.C. Cornock, D.F. Steele and T.A. Stephenson, Inorg. Nucl. Chem. Lett. 10:785 (1974).

143.   M.C. Cornock, R.O. Gould, C.L. Jones, J.D. Owen, D.F. Steele and T.A. Stephenson, J.C.S. Dalton Trans. 496 (1977).

144.   L. Gastaldi, P. Porta and A.A.G. Tomlinson, J.C.S. Dalton Trans. 1424 (1974).

145.   E. Borghi, V. Di Castro, F. Monacelli and A.A.G. Tomlinson, J.C.S. Dalton Trans. 950 (1978).

146.   J.L. Burmeister, D.C. Calabro and R.L. Robert. Abs. Pap. ACS. Mar. Mtg. 175:23 (1978).

147.   I.B. Benson, J. Hurst, S.A.R. Knox and V. Oliphant, J.C.S. Dalton Trans. 1240 (1978).

148.   I.S. Butler and J. Svedman, J. Inorg. Nucl. Chem. 40:1937 (1978).

149.   P.M. Treichel and G.P. Werber, J. Am. Chem. Soc. 90:1753 (1968).

150.   K. Leonard, K. Plute, R.C. Haltiwanger and M.R. Dubois, Inorg. Chem. 18:3246 (1979).

acac	acetylacetonato (2,4-pentanedionato)
ADP	adenosine diphosphate
ala	alanenyl
AMP	adenosine $5'$-monophosphate
ATP	adenosine $5'$-triphosphate
bpdah	$2\text{-py-}CH_2.S.CH_2CH_2.S.CH_2\text{-2-py}$
bpy	$2,2'$-bipyridyl
Cp	$\eta\text{-}C_5H_5$
Cp*	$\eta\text{-}C_5Me_5$
Cy	cyclohexyl
Cyclam	1,4,8,11-tetraazacyclotetradecane
Cyclen	1,4,7,10-tetraazacyclododecane
cysn	cysteinesulfinato
cyst	cysteinyl
depe	$Et_2PCH_2CH_2PEt_2$
diars	see pdma
dien	diethylenetriamine
diphos	see dppe
$dmgH_2$	dimethylglyoxime
DME	1,2-dimethoxyethane
DMF	dimethylformamide
dmpe	$Me_2PCH_2CH_2PMe_2$
dpa	bis(2-pyridylmethyl)amine, $(py\text{-2-}CH_2)_2NH$

dpe	see dppe
dpm, dppm	$Me_2PCH_2PMe_2$
dppe	$Ph_2PCH_2CH_2PPh_2$

EDDA	ethylenediaminediacetate
edma	ethylenediaminemonoacetate
edta	ethylenediaminetetraacetate
eee	$H_2N.CH_2.S.CH_2CH_2.S.CH_2CH_2.NH_2$
EGDA	ethylglycine N,N-diacetate
en	ethylenediamine
EVDA	ethyl valinate-N,N-diacetate

gly	glycinyl

Hb	Hemoglobin
hia	N(O)C(COMe)C(Me)O̲: see Chapter 2, Section 4
His	histidinyl

ile	isoleucinyl
imH	imidazole
imda	iminodiacetate

Mecp	$\eta^5\text{-}MeC_5H_4$
4-Me-py	4-methylpyridine

nta, NTA	nitrilotriacetate

P⌒C	$Ph_2PCH_2\text{-}C_6H_4\text{-}o\text{-}$
pdma	orthophenylenebis(dimethylarsine), DIARS
phe	phenylalanenyl
phen	9,10-phenanthroline
Pi	inorganic phosphate
pic	picolinyl
pn	1,3-diaminopropane
PPi	inorganic pyrophosphate
pr	prolinato
py	pyridine

(8-)quin	quinolin(-8-)olato
Sacsac	pentan-2,4-dithione
sal$_2$en	N,N'-ethylenebis(salicylideneiminato)
Saloph	N,N'-bisalicylidene-$\underline{o}$-phenylenediamino
terpy	2,2',6',2"-terpyridyl
THF, thf	tetrahydrofuran
tmgH$_2$	tetramethylethylenediamine, cyclohexane,1,2-dione dioxime
TMS	tetramethylsilane
tn	trimethylenediamine, 1,3-diaminopropane
tol	tolyl
TPP	tetraphenylporphyrin
tren	2,2 ,2"-triaminotriethylamine, $N(CH_2CH_2NH_2)_3$
trpy	see terpy
ttp	$PhP(CH_2CH_2CH_2PPh_2)_2$
V	Violurato; see Chapter 2, Section 4

INDEX

Esters are generally indexed by parent acid. Peptides and derivatives are indexed under Peptides using conventional abbreviations. Other activated ligand-metal combinations are listed under the relevant metal; thus a reaction at CO of titanocene dicarbonyl is indexed at <u>Titanium</u>, <u>carbonyl</u>.